The Vocabulary of Organic Chemistry

The Vocabulary of Organic Chemistry

Milton Orchin
Fred Kaplan
Roger S. Macomber
R. Marshall Wilson
Hans Zimmer

Organic Division, Department of Chemistry
University of Cincinnati
Cincinnati, Ohio

A Wiley-Interscience Publication

JOHN WILEY & SONS, New York • Chichester • Brisbane • Toronto

Library of Congress Cataloging in Publication Data:

Cincinnati. University. Dept. of Chemistry. Organic Division.
The vocabulary of organic chemistry.

"A Wiley-Interscience publication."
Includes index.
1. Chemistry, Organic–Terminology. I. Orchin, Milton, 1914- II. Title.
QD291.C55 1980 547'.001'4 79-25930
ISBN 0-471-04491-1

Printed in the United States of America

10 9 8 7 6 5 4 3 2 1

Preface

The purpose of this book is to identify the fundamental vocabulary of organic chemistry and then to present concise, accurate definitions, with examples where appropriate, for the words and concepts that make up this vocabulary. This book is not intended to be a dictionary either in form or content. It is not a listing of organic compounds, their structures, and nomenclature except where these illustrate particularly important specific compounds or classes of compounds. The book is organized into chapters, and both the chapters and the material within each chapter are placed in a sequence that makes pedagogical sense to the authors. This approach generated a book in which related terms and concepts appear in close proximity to one another and hence fine distinctions become more understandable.

The idea for this book grew out of discussions by the Organic Division members of the Department of Chemistry of the University of Cincinnati on how to cope with the mass of material that is offered to students in the elementary course in organic chemistry. Recent books intended for this course typically consist of 1000 pages of text and 2500 index entries. It is hopelessly impossible to cover this amount of material during the time allotted for this purpose. Instructors in the course are constantly plagued by the irritating but not unreasonable question asked by many students: "What are we expected to know?" One is always tempted to reply with another question, "What is knowledge without understanding?" It is our hope that this book, by defining terms and concepts, provides both. The earlier book by Saul Patai, *Glossary of Organic Chemistry* (John Wiley & Sons, 1962), attempted to address this problem, but the present book is completely different in organization and purpose.

In the very early stages of the work it became apparent that no general agreement could be reached by the authors either on the specific terms that students at a particular point in their education should be required to know and understand, or on the level and extent of explanation and elaboration of these terms. If the selection and level were aimed primarily at students in the undergraduate chemistry program, the book would be too limiting; if the selections and level were aimed exclusively at the graduate student, then one of the original purposes of the book would be lost; and if the book were oriented exclusively to students and teachers, then the appeal to working professionals would be reduced. Inevitably a compromise was reached. The authors of individual chapters were given the option of including any term or concept that could be found in the index of a modern organic text plus the option of including terms in his particular field of expertise, providing that these latter could be explained within reasonable limits. Thus there are included four chapters of terms containing subject matter not usually treated in such depth in general organic texts: Organometal Compounds (Chapter 12); Natural Products and Biosynthesis (Chapter 13); Polymers (Chapter 14); and Fossil Fuels (Chapter 15). These chapters not only reflect particular interests of some of the authors but represent fields of special current interest and activity. The original version of our manuscript contained chapters on infrared, microwave, and Raman spectra; electronic spectroscopy and photochemistry; nuclear magnetic resonance spectroscopy; and mass spectroscopy. However, the inclusion of these topics, which embrace most of organic spectroscopy, would have expanded the book to an impractical size. If interest warrants, these topics may be covered in a second volume. The present book contains definitions and explanations of about 1350 words and concepts that constitute much if not most of the vocabulary of modern ground state organic chemistry at all levels. We estimate that only between a third and fourth of these should be the required vocabulary of the student in the undergraduate chemistry course.

In addition to students and teachers, the authors hope that this book will also appeal to many others in the chemical profession who, in either the recent or remote past, were familiar with many of these terms, but whose precision of knowledge with respect to their meaning has faded with time. The introduction of new (and sometimes confusing) terminology in recent years, as well as the increased interfacing with other scientific disciplines, have expanded the vocabulary of organic chemistry, and many chemists may not be familiar with such developments. We hope that the format is particularly attractive to all such potential users; not only should they find what they may be looking for without having to wade through a great deal of other material, but they will find it in the context of related items.

Finally, as the authors discovered in their many hours of discussing the best and most accurate definitions of terms, there is frequently disagreement even among authorities. Where pertinent and possible, adherence to recognized definitions such as those prescribed in the many IUPAC (International Union of Pure and Applied Chemistry) publications was observed. The book *Nomenclature of Organic Compounds*, Advances in Chemistry Series 126 (1974), edited by John H. Fletcher, Otis C. Dermer, and Robert B. Fox, was particularly helpful.

We welcome suggestions and criticisms, knowing that despite our best efforts it is likely that we have made errors of fact and interpretation. We would be grateful if these are brought to our attention. Suggestions for inclusions and deletions are also welcome.

MILTON ORCHIN
FRED KAPLAN
ROGER S. MACOMBER
R. MARSHALL WILSON
HANS ZIMMER

Cincinnati, Ohio
August 1980

Acknowledgments

It is a pleasure to acknowledge the generous consultative and critical help of our colleagues, Bruce Ault, Edward Deutsch, James Mark, and Carl Seliskar. Jack Kwiatek and Janet Del Bene graciously reviewed chapters in their specialties. Kurt Loening and Joe Bunnett gave valuable access to IUPAC reports. Arlene Musser helped with some of the early drawings and Don Rehse and David Lankin with some of the final ones; however, in particular we wish to thank Tahsin Chow for his unusual skill in preparing most of the final drawings and R. Duane Satzger for his outstanding photographic work. Early typing of the manuscript was performed by Ann Sparkman, and in the later stages of manuscript preparation Jocelyn McGinnis and Jane Capannari were indispensable. Of course we drew generously on published books and articles, and we are grateful to the many authors who gave unknowingly of their expertise. Finally we wish to thank the Department of Chemistry and the University of Cincinnati Administrations for providing us with an environment that nourished, encouraged, and facilitated the commitment of time and effort required to carry on this enterprise.

Contents

The Vocabulary of Organic Chemistry

1 Symmetry Operations and Symmetry Elements

CONTENTS

In the vocabulary used in this chapter molecules rather than objects are generally used for purposes of definition and examples, although the principles of symmetry are equally applicable to both.

1.010 Equivalent Orientations

Orientations of a molecule which are superimposable on each other, hence indistinguishable.

1.020 Symmetry Operation

Doing something (a manipulative operation) to transform a molecule into any equivalent orientation.

Example. A 180° rotation of the HOH molecule, Fig. 1.020*a* around the axis shown, results in the equivalent orientation, Fig. 1.020*b*, which is indistinguishable from, hence superimposable on, the structure in Fig. 1.020*a*; under the operation the hydrogen atoms transform into each other and the oxygen atom transforms into itself. Accordingly, a symmetry operation has been performed (i.e., it appears that nothing has been done).

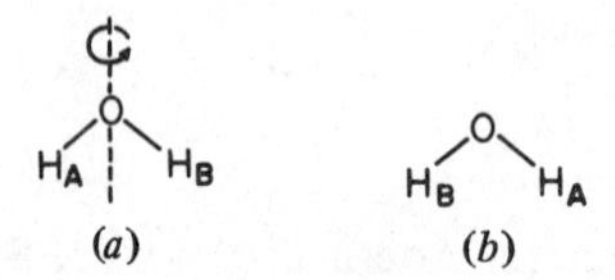

Figure 1.020 The 180° rotation that converts H_A into H_B.

1.030 Identical Orientation

The equivalent orientation that is identical with the original one.

Example. Upon rotating the water molecule by 180° (Fig. 1.020*a*, *b*), an equivalent orientation is obtained. If now the water molecule is rotated again by 180° (in either a clockwise or counterclockwise direction), it is restored to its original (identical) orientation. The overall effect is to do nothing to the molecule. An identical orientation is always obtained when a molecule is rotated by 360° about any imaginary axis through its center of mass. Of course, after the completion of a symmetry operation one cannot distinguish whether the resulting orientation is equivalent or identical. If, however, a symmetry operation is regarded as an instruction to do something to the molecule, then the distinction is real.

1.040 Symmetry Elements

The imaginary line, point, or plane of symmetry, or their combinations, that make symmetry operations possible. In considering individual molecules, the possible symmetry elements are: a rotational axis of symmetry; a center of symmetry; a plane of symmetry; and an alternating axis of symmetry.

1.050 Rotational Axis, C_p

An imaginary axis passing through the center of mass of a molecule, rotation about which by any angle (the operation) results in an orientation that is superimposable on the original; also called a proper axis.

Example. The axis shown by the broken line in Fig. 1.020*a* is a rotational axis; rotation around it by 180° gives an orientation superimposable on the original.

1.060 The "Fold" of a Rotational Axis

The subscript p in C_p, equal to $360°/\Theta°$ where Θ is the smallest angle of rotation leading to an indistinguishable orientation.

Example. On rotation of HOH, Fig. 1.020*a*, around the axis by 180° the molecule repeats itself. This rotational axis is called a 2-fold rotational axis, commonly designated C_2, where C stands for cyclic and subscript 2 is the fold. Benzene has a 6-fold axis, C_6, Fig. 1.060; a 60° rotation around the axis shown transforms the molecule into itself; i.e., all the carbon atoms as well as all the hydrogen atoms are transformed into nearest neighbor carbon and hydrogen atoms. Benzene also possesses three C_2 axes that bisect opposite carbon atoms and three C_2 axes that bisect opposite bonds. One of each of these is shown in Fig. 1.060. These six C_2 axes are in the plane of the molecule whereas the C_6 is perpendicular to the molecular plane. The C_6 axis is coincident with a C_3 and a C_2 axis.

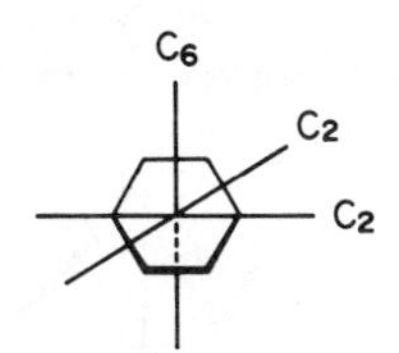

Figure 1.060 Three different rotational axes in benzene.

1.070 Center of Symmetry, *i*

The point at the center of a molecule through which, on reflection, each atom in the molecule encounters an equivalent (indistinguishable) atom. The reflection may be illustrated by drawing a straight line from each atom in a molecule through the center of mass of the molecule and continuing this line an equal distance from this center, whereupon the line should encounter an equivalent atom.

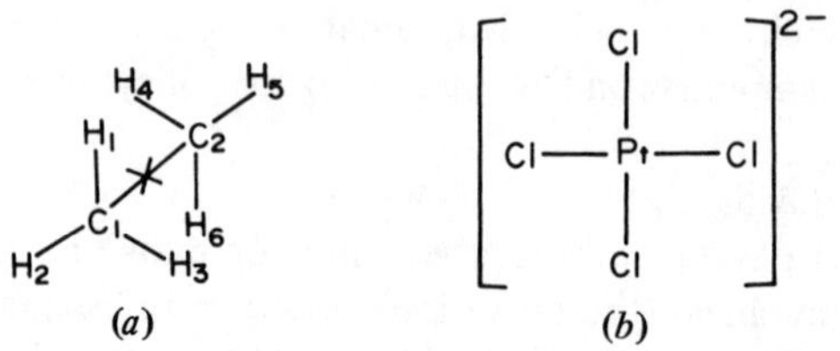

(a) (b)

Figure 1.070 (*a*) A molecule with a center of symmetry; (*b*) a molecule with an atom at the center of symmetry.

Examples. The orientation of ethane shown in Fig. 1.070*a* has a center of symmetry i, the center of the carbon-carbon bond marked x. If one draws a line from H_1 to the center of the molecule and continues an equal distance, one encounters H_6. A similar line from H_2 encounters H_5, H_3 is similarly related to H_4, and the two carbon atoms C_1 and C_2 are similarly related. It is possible to have an atom lying on the center of symmetry, e.g., Pt in the ion $[PtCl_4]^{2-}$, Fig. 1.070*b*.

1.080 Inversion

The symmetry operation of transforming each atom in a molecule into an equivalent atom by reflecting through the center of symmetry i.

Example. In the molecule ethane, shown in Fig. 1.070*a*, inversion transforms H_1 into H_6, H_2 into H_5, H_3 into H_4, C_1 into C_2, and all vice versa. In a Cartesian coordinate system the operation of inversion transforms every point (x, y, z) into $(-x, -y, -z)$.

1.090 Plane of Symmetry, σ

A plane bisecting a molecule such that each atom on one side of the plane, when reflected through the plane, encounters an equivalent (indistinguishable) atom

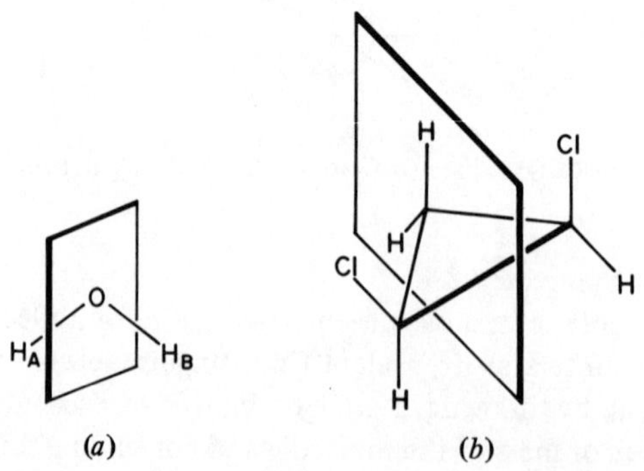

Figure 1.090 Mirror planes in (*a*) H_2O and (*b*) *cis*-1,2-dichlorocyclopropane.

on the other side of the plane. The plane of symmetry is also called a mirror plane. The designation sigma (σ) comes from the German word *Spiegel* (mirror).

Examples. Every planar molecule necessarily has a plane of symmetry, namely the molecular plane. Thus in HOH, Fig. 1.020*a*, the molecular plane is a plane of symmetry. However, the plane that is perpendicular to the molecular plane and which includes the C_2 axis is also a mirror plane, since reflection through it transforms H_A into H_B and 0 into itself; this mirror plane is shown in Fig. 1.090*a*. The mirror plane in *cis*-1,2-dichlorocyclopropane is shown in Fig. 1.090*b*.

1.100 Rotation-Reflection (or Alternating) Axis of Symmetry, S_p

If a molecule is rotated by the angle Θ about some imaginary axis passing through its center of mass and after such rotation the molecule is then reflected through a plane perpendicular to this axis and such a combined operation results in an orientation indistinguishable from the original, the molecule possesses a rotation-reflection axis. Such an axis is also called an alternating axis or an improper axis. The alternating axis is designated as S and subscripted with a number that indicates its fold, i.e., S_p, where $p = 360°/\Theta°$.

Examples. The molecule of ethane in the orientation shown in Fig. 1.070*a* possesses an S_6 axis coincident with the C—C bond axis; clockwise rotation of the molecule by 60° around this axis followed by reflection through a plane perpendicular to the C—C bond axis takes H_1 into the position originally occupied by H_5, H_2 into H_4, and H_3 into H_6 and, of course, C_1 into C_2. Under the S_p operation the equivalent atoms are carried from one side of the reflection plane to the other in an alternating sequence, hence the name alternating axis. Many highly symmetrical molecules contain more than one S_p axis. Thus, for example, methane (CH_4) can be shown to possess multiple S_4 axes. The axis shown in Fig. 1.100*a* passes through the center of the molecules and bisects opposite HCH angles. Clockwise rotation of the molecule by 90° around this axis followed by reflection in a plane perpendicular to it restores the molecule to an orientation superimposable on the original. Inscribing the tetrahedral molecule CH_4 in a

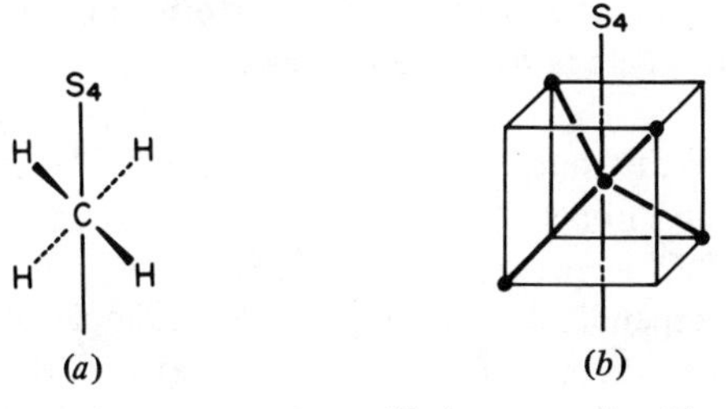

Figure 1.100 (*a*) The S_4 axis of methane; (*b*) the same axis with methane in a cube.

cube, Fig. 1.100*b*, facilitates the recognition of the other S_4 axes. The S_4 shown in the figure bisects opposite sides of the cube, and since there are six equivalent faces of a cube there are three S_4 axes. Note also that each S_4 in CH_4 is also a C_2.

1.110 Newman Projection (M. S. Newman, 1909-)

Originally the representation showing the spatial relationship of atoms in a two-carbon system, obtained by looking down the carbon-carbon bond axis and drawing a projection of the carbons and all the attached groups onto the same plane. In such a drawing the groups attached to the front carbon are represented by solid lines from these atoms meeting at a central point that corresponds to the front carbon atom, while the groups attached to the rear carbon atom (implied but not shown in the drawing) are distinguished by short lines from the atoms to the periphery of a circle around the point. This type of projection drawing is now used more generally to show spatial relationships.

Example. The orientation of staggered ethane (Fig. 1.070*a*) drawn according to the Newman projection is shown in Fig. 1.110. This type of drawing for $CH_3—CH_3$ is particularly useful in depicting the relationships of the six hydrogens under the various symmetry operations. Thus the C_2 shown carries $H_1 \rightarrow H_6; H_3 \rightarrow H_5; H_2 \rightarrow H_4$.

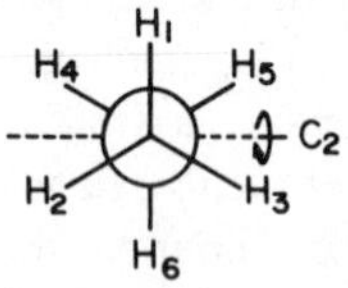

Figure 1.110 The Newman projection of staggered ethane and one of the C_2 axes.

1.120 Classes of Symmetry Operations

Frequently molecules have more than one C_p axis, or more than one σ, or more than one S_p axis. If the different σ or C_p or S_p axes can be taken into each other by another symmetry operation that can be performed on the molecule, these σ or C_p or S_p operations belong to the same class.

Examples. Consider the square planar complex $[PtCl_4]^{2-}$, shown in Fig. 1.120. The dotted lines going through Cl—Pt—Cl are C_2 axes, as are the dashed lines bisecting opposite ClPtCl angles, shown as C_2'. However, the square planar molecule has a C_4 axis perpendicular to the plane of the molecule. Rotation by 90° around this axis takes the two C_2 into each other and the two C_2' into each other. Hence the two C_2 belong to one class of order 2 as do the two C_2', but

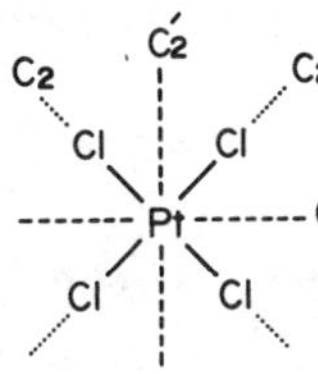

Figure 1.120 The two classes of C_2 in $[PtCl_4]^{2-}$.

since no symmetry operation of the molecule can take C_2 into C_2', these operations belong to different classes.

1.130 Equivalent Symmetry Operations

Different symmetry operations that transform molecules into identical orientations.

Example. The operations i and S_2 are equivalent. The operation i, on the orientation of ethane shown in Fig. 1.110, transforms $H_1 \rightarrow H_6, H_3 \rightarrow H_4$, and $H_2 \rightarrow H_5$. For the S_2 operation consider an axis that is parallel to the C—H_1 bond projection and bisects the C—C. (This is not a C_2 axis!) A 2-fold rotation around this axis followed by reflection in a plane perpendicular to it results in exactly the same transformation as does i, hence $i \equiv S_2$. Another, and very important, relationship is $\sigma \equiv S_1$. These examples illustrate the importance of the alternating axis; indeed there are only two independent point symmetry operations because of the kinds of equivalencies just illustrated, C_p and S_p, or proper and improper axes of rotation, respectively.

1.140 Point Groups

Molecules may be classified on the basis of the symmetry operations that can be performed on them, e.g., C_p, σ, i, S_p. All these operations are called point symmetry operations because under such symmetry operations one point, the center of mass, always remains unchanged. Molecules that possess all the same symmetry elements, and only these elements, belong to the same group, or more exactly, the same point group.

Examples. The molecules HOH, O_3, and phenanthrene, Fig. 1.140*a*, *b*, and *c*, respectively, all belong to the same point group because all three of them possess

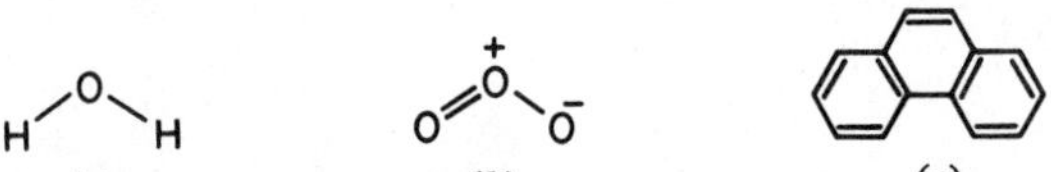

(*a*) (*b*) (*c*)

Figure 1.140 Molecules belonging to the same point group: (*a*) H_2O; (*b*) O_3; (*c*) phenanthrene.

two planes of symmetry and a C_2 axis. The particular point group that is characterized by these symmetry elements is designated C_{2v} because of the C_2 axis and two planes of symmetry that include the axis.

1.150 Local Symmetry of the Methyl Group

The symmetry present in an isolated methyl group of a molecule having the general structure CH_3—A where A can be any group of atoms; generally such symmetry is not characteristic of the entire molecule. Because of free rotation around the C—A axis the three hydrogens are exchangeable by a C_3 axis, hence by local symmetry there is a C_3 axis as well as three planes of symmetry.

Example. When methanol, CH_3OH, is oriented as shown in Fig. 1.150*a*, it has a plane of symmetry, yet in most other orientations it does not even possess that much symmetry. However, the methyl group of CH_3OH, Fig. 1.150*b*, has local C_3 symmetry.

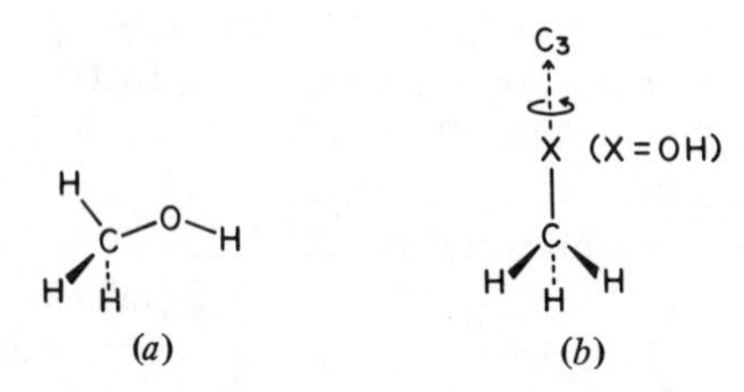

Figure 1.150 (*a*) Methanol (no rotational axis); (*b*) the local C_3 axis.

1.160 Symmetric

If the sign and magnitude of a property of a molecule are unchanged upon application of a symmetry operation appropriate to the molecule, the property is said to be symmetric with respect to that symmetry operation (or the symmetry element associated with it).

Example. Consider the stretching vibration of the water molecule shown in Fig. 1.160. The arrows indicate that both O—H bonds are being stretched. If we now reflect this notion through the plane of symmetry perpendicular to the molecular plane, the arrow on the right goes into the arrow on the left and vice versa. This stretching vibration is thus symmetric with respect to reflection in the mirror plane.

Figure 1.160 The stretching mode of H_2O symmetric with respect to the mirror plane.

1.170 Antisymmetric

If the property of a molecule under investigation (e.g., its oscillating motion) is converted into its negative as a result of a symmetry operation, the property is antisymmetric with respect to that operation.

Examples. Consider the stretching vibration of the water molecule shown in Fig. 1.170. The arrows indicate that while one O—H bond is being stretched, the other is being compressed to the same extent. On reflection through the mirror plane perpendicular to the molecule, the arrows get exchanged but their directions are reversed. This stretching vibration is thus antisymmetric with respect to reflection in the mirror plane.

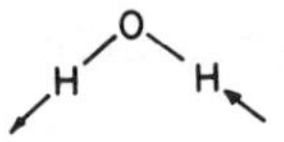

Figure 1.170 The stretching mode of H_2O antisymmetric with respect to the mirror plane.

1.180 Gerade (*g*) and Ungerade (*u*)

(From the German *even* and *uneven*.) In dealing with properties of a molecule with a center of symmetry, symmetric behavior is denoted as *g* (gerade) and antisymmetric behavior as *u* (ungerade).

Examples. The CO_2 molecule is linear and has a center of symmetry. When this molecule undergoes a vibration in which the two oxygens are moving in the same direction, Fig. 1.180*a*, such motion is antisymmetric with respect to reflection through the center of mass. Such an antisymmetric vibration is denoted as *u* when it is classified in symmetry notation. Fig. 1.180*b* shows the symmetric stretch, *g*.

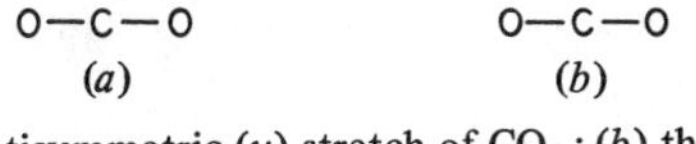

Figure 1.180 (*a*) The antisymmetric (*u*) stretch of CO_2; (*b*) the symmetric stretch (*g*).

1.190 Dissymmetric

A molecule that cannot be superimposed on its mirror image is dissymmetric. Those molecules that possess one or more rotational axes of any fold as the *only* symmetry elements are dissymmetric.

Examples. *trans*-1,2-Dichlorocyclopropane, Fig. 1.190*a*, is dissymmetric; it has a C_2 axis (in the plane of the three carbon atoms) but its mirror image, Fig. 1.190*b*, is not superimposable on the original.

Figure 1.190 (*a*) The dissymmetric molecule *trans*-1,2-dichlorocyclopropane and (*b*) its mirror image.

1.2000 Asymmetric

A molecule that lacks any symmetry element (other than the infinite number of C_1 axes passing through the center of mass which all molecules possess) is asymmetric. All asymmetric molecules are necessarily dissymmetric, but as the example, Fig. 1.190, demonstrates, dissymmetric molecules need not be asymmetric.

Example. A compound that contains a single carbon atom, having four different groups around it, Fig. 1.200*a*, is always asymmetric. This is a sufficient condition for asymmetry. 1,1,2-Trichlorocyclopropane, Fig. 1.200*b*, has two equivalent groups attached to C_1 and C_3 but C_2 is attached to four different groups and hence this molecule is asymmetric.

Figure 1.200 Asymmetric molecules: (*a*) a molecule with four different groups attached to carbon; (*b*) 1,1,2-trichlorocyclopropane.

1.210 Spherical Symmetry

Symmetry characteristic of a sphere; objects possessing it have an infinite number of all symmetry elements.

Example. A ball possesses spherical symmetry, as does an *s* atomic orbital.

1.220 Cylindrical Symmetry

Symmetry characteristic of a cylinder; molecules possessing it have a C_∞ axis passing through the center and a plane of symmetry perpendicular to this axis.

Example. All linear molecules with a center of symmetry have cylindrical symmetry, e.g., O=C=O, HC≡CH. Common objects with cylindrical symmetry are doughnuts, bagels, and footballs.

1.230 Conical Symmetry

Symmetry characteristic of a cone or funnel; molecules possessing it have a C_∞ axis but no plane of symmetry perpendicular to this axis.

Examples. Linear molecules without a center of symmetry have conical symmetry, e.g., H—F, H—C≡N. Common objects other than a funnel with conical symmetry are bottles, saucers, and bowls.

1.240 Symmetry Number, σ

The number of indistinguishable orientations into which a molecule can be turned by rotations around symmetry axes.

Examples. The symmetry numbers for HOH and NH_3 are 2 and 3, respectively, and for ethylene it is 4. The four indistinguishable orientations for ethylene obtained by the indicated 180° rotations (C_2) are shown in Fig. 1.240*a*. The symmetry number is characteristic for each point group. The symmetry number for a molecule that can undergo free rotation is different from that of the same molecule without free rotation. Thus, if ethane were fixed in the orientation shown in Fig. 1.240*b*, its symmetry number would be 6. With free rotation, however, there are three indistinguishable positions of the CH_3 groups with respect to each other (local symmetry) and for each of these there are six positions as for the rigid molecule, hence in all there are 3 × 6 = 18 indistinguishable positions and the symmetry number for ethane is 18. This number is important in certain calculations of entropy.

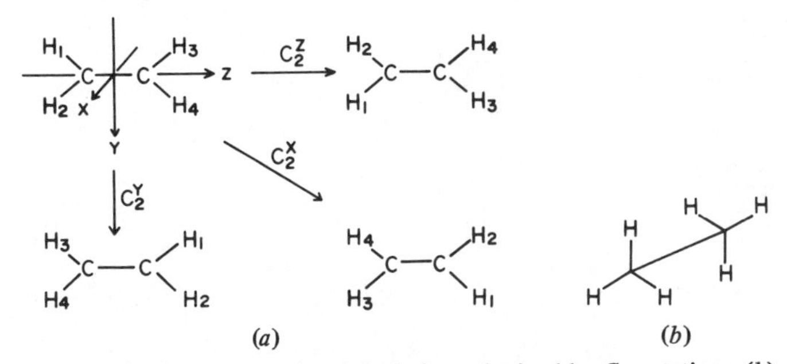

Figure 1.240 (*a*) The four orientations of ethylene obtained by C_2 rotations; (*b*) ethane.

1.250 Symmetry Equivalent Hydrogens

Hydrogen atoms that can be interchanged by *any* symmetry operation. Although there are various types of symmetry equivalent hydrogen atoms depending upon the particular symmetry elements present [(e.g., enantiopic or homotiopic (see Sect. 5.610)], the identification of sets of atoms that can be exchanged by any symmetry operation facilitates determination of the number of compounds that can be derived from a particular compound by substitution of hydrogen by another atom or group of atoms.

Examples. In cyclopropane, Fig. 1.250*a*, the top H's are exchangeable by a C_3 as are the bottom ones, and the top and bottom H's can be exchanged by C_2's in the plane of the C atoms and bisecting C—C bonds, hence all six H's belong to one symmetry set and, accordingly, there can be only one monosubstituted cyclopropane. In monochlorocyclopropane, Fig. 1.250*b*, there are three sets of symmetry equivalent H's, hence three possible dichlorocyclopropanes. When multiple methyl groups are present on the same saturated carbon atom, the methyl groups are usually symmetry equivalent (by local symmetry) and all the hydrogens of all such methyl groups are symmetry equivalent. Thus in 2,5-dimethylhexane, Fig. 1.250*c*, there are three sets of symmetry equivalent H's (one set of 12, one set of 2, and one set of 4), hence three different monochloro-2,5-dimethylhexanes.

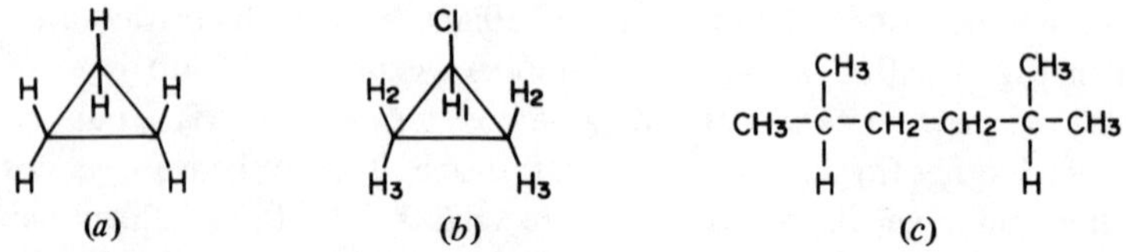

Figure 1.250 Sets of symmetry equivalent hydrogens: (*a*) one set; (*b*) three sets; (*c*) three sets.

2 Wave Mechanics, Quantum Mechanics, Atomic and Molecular Orbitals, Bonding

CONTENTS

2.010 Heisenberg Uncertainty Principle (Werner Heisenberg, 1901–1976) The position and the momentum of an electron cannot be measured simultaneously with precise exactness; mathematically, it is expressed by the equations:

$$\Delta P \cdot \Delta X \propto \frac{h}{2\pi} \quad (2.010a)$$

$$\Delta E \cdot \Delta t \propto \frac{h}{2\pi} \quad (2.010b)$$

where ΔP is the uncertainty in the momentum, ΔX the uncertainty in the position, ΔE the uncertainty in the energy, Δt the uncertainty in time, and h Planck's constant. Equation 2.010*a* informs us that if the position of an electron (or atom) is known with exactness, then its momentum is unknown and the momentum can be known only at the sacrifice of information regarding the position of the electron. Equation 2.010*b* says the same thing with respect to the energy of an electron and the time it spends in a particular energy state. The Uncertainty Principle eliminates the concept of a trajectory for the electron; hence its position is best expressed in terms of the probability of finding it in a particular specified region in space.

2.020 Wave Mechanics

The mathematical description of the behavior of very small particles such as electrons in terms of their wave character. The use of wave mechanics for the description of electrons follows from the experimental observation that electrons have wave as well as particle properties. The wave character results in a probability interpretation of electron behavior (see Sect. 2.160).

2.030 Standing (or Stationary) Waves

The type of wave motion produced, for example, by plucking a string or wire stretched between two fixed points. The amplitude of such a wave is a function of the distance along the wave path; i.e., amplitude is $f(x)$.

Examples. Figure 2.030.

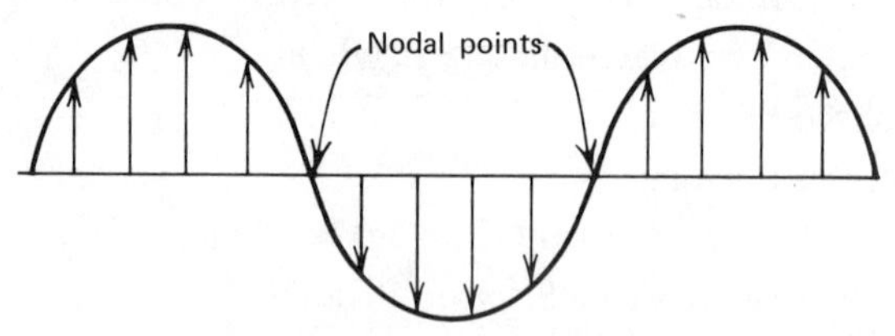

Figure 2.030 A standing wave.

2.040 Nodal Points

The positions in a wave where the direction of the wave changes, hence where the amplitude is zero, Fig. 2.030.

2.050 Wavelength, λ

The minimum distance between two points on the wave having a corresponding sign and amplitude.

Examples. The distance marked λ in Fig. 2.050.

2.060 Frequency, ν

The number of waves (or cycles) that pass a particular point per unit time. Time is usually measured in seconds, hence the unit of frequency is s^{-1}.

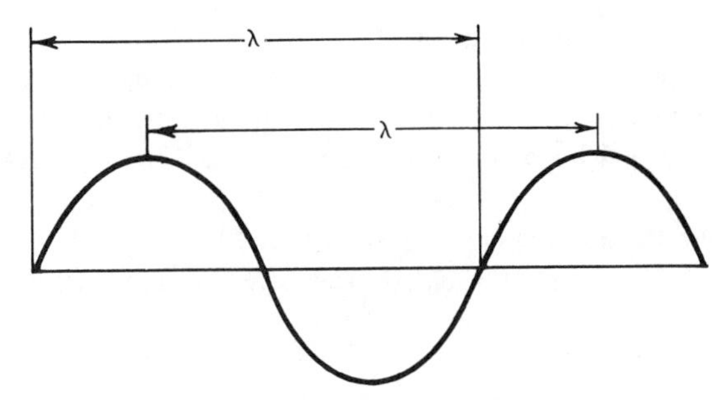

Figure 2.050 The wavelength λ.

Example. Frequency is inversely proportional to wavelength, and with light as with all electromagnetic radiation, the proportionality factor is the speed of light c (3×10^{10} cm s^{-1}), hence $\nu = c/\lambda$.

2.070 Fundamental Wave

The stationary wave with no nodal point; the wave from which the frequency (ν') of all other waves in a set is generated by multiplication of the fundamental frequency ν by an integer n: $\nu' = n\nu$.

Example. In the fundamental wave, Fig. 2.070, the amplitude may be considered to be oriented upward and to continuously increase from either fixed end, reaching a maximum at the midpoint. In this "well-behaved" wave the amplitude is zero at each end and is a maximum at the center.

2.080 First Overtone (Harmonic)

The stationary wave with one nodal point located at the midpoint ($n = 2$ in the equation given in Sect. 2.070).

Example. In the first overtone, Fig. 2.080, the amplitude is zero at the point where the original upward motion changes direction. The two equal segments of the wave form a single wave; they are not independent. The two maxima in the wave come at exactly an equal distance from the ends but are of opposite signs.

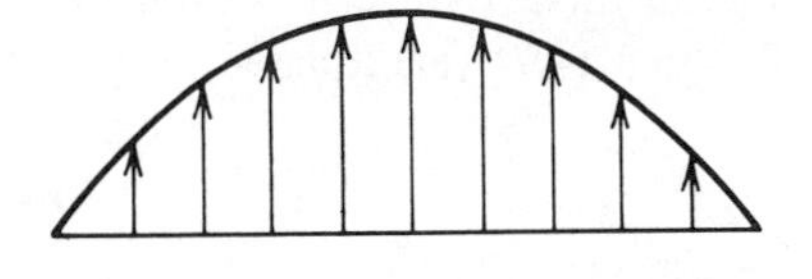

Figure 2.070 A fundamental wave.

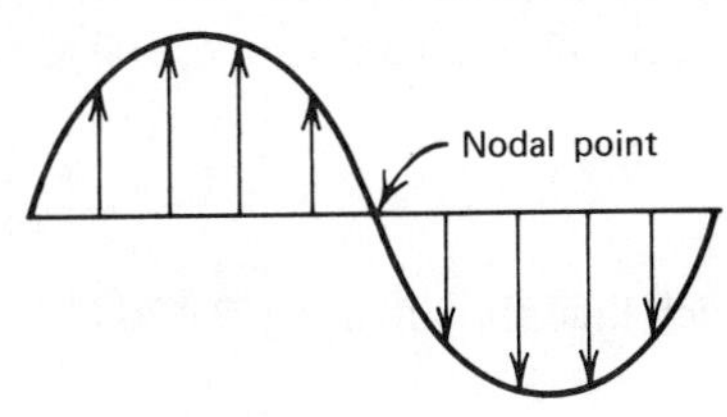

Figure 2.080 The first overtone.

2.090 Duality of Electron Behavior

Electrons may exhibit properties of either particles (they have momentum) or waves (they can be defracted like light). A single experiment may demonstrate either the particle or wave properties of electrons, but not both simultaneously.

2.100 De Broglie Relationship (L. De Broglie, 1892-)

The wavelength of a particle (an electron) is determined by the de Broglie equation:

$$\lambda = \frac{h}{p} = \frac{h}{mv} \tag{2.100}$$

where h is Planck's constant, m is the mass, and v is the velocity of the particle. This relationship makes it possible to relate the momentum of the electron p, a particle property, with the wavelength λ, a wave property.

2.110 Wave Equation

The mathematical equation that describes the amplitude behavior of a wave. In the case of a one-dimensional standing wave, this is a second-order differential equation with respect to the amplitude:

$$\frac{d^2 f(x)}{dx^2} + \frac{4\pi^2}{\lambda^2} f(x) = 0 \tag{2.110}$$

where λ is the wavelength and $f(x)$ is the amplitude function.

2.120 Wave Equation in Three Dimensions

The function $f(x, y, z)$ for the wave equation (Sect. 2.110) in three dimensions, analogous to $f(x)$, which describes the amplitude behavior of the one-dimensional wave. Thus $f(x, y, z)$ satisfies the equation

$$\frac{\partial^2 f}{\partial x^2} + \frac{\partial^2 f}{\partial y^2} + \frac{\partial^2 f}{\partial z^2} + \frac{4\pi^2}{\lambda^2} f = 0 \tag{2.120}$$

2.130 Laplacian Operator (P. S. Laplace, 1749-1827)

In the expression $\partial^2 f/\partial x^2$ the portion $\partial^2/\partial x^2$ is an operator that says "partially differentiate twice with respect to x, that which follows." The sum of the second differential operators with respect to the three Cartesian coordinates in Eq. 2.120 is the Laplacian operator and is denoted as ∇^2 (del squared):

$$\nabla^2 = \frac{\partial^2}{\partial x^2} + \frac{\partial^2}{\partial y^2} + \frac{\partial^2}{\partial z^2} \tag{2.130a}$$

which then simplifies Eq. 2.120 to

$$\nabla^2 f + \frac{4\pi^2}{\lambda^2} f = 0 \tag{2.130b}$$

2.140 Orbital

A wave equation for one electron which is a function of the coordinates of the electron; in terms of Cartesian coordinates this is indicated by $\psi(x, y, z)$, or in spherical coordinates $\psi(r, \theta, \varphi)$. The relationship between the two coordinate systems is shown in Fig. 2.140. An orbital centered on one atom (an atomic orbital) is frequently denoted as ϕ (phi) to distinguish it from an orbital centered on more than one atom, a molecular orbital, designated ψ (psi).

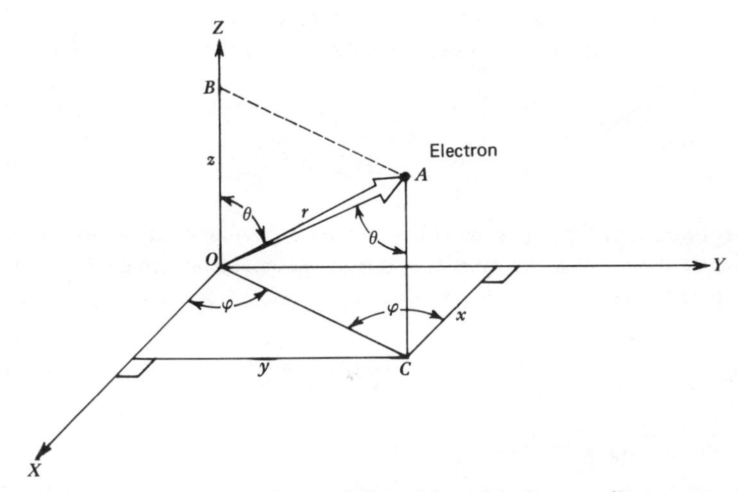

Figure 2.140 The relationship between Cartesian and polar coordinate systems. The projection of r on the z axis is $z = OB$. OBA is a right angle, hence $\cos\theta = z/r$, and $z = r\cos\theta$.

$$\cos\varphi = x/OC, \text{ but } OC = AB = r\sin\theta, \text{ hence } x = r\sin\theta\cos\varphi$$

$$\sin\varphi = y/AB \qquad y = AB\sin\varphi = r\sin\theta\sin\varphi$$

Accordingly, a point x, y, z in Cartesian coordinates is transformed to the spherical coordinate system by the following relationships:

$$z = r\cos\theta$$
$$y = r\sin\theta\sin\varphi$$
$$x = r\sin\theta\cos\varphi$$

2.150 Wavefunction

Used interchangeably with **orbital.**

2.160 Probability Interpretation of the Wavefunction

According to the Bohr interpretation, the wavefunction (or orbital) $\psi(r)$ is the amplitude for the probability distribution of the position of the electron. Thus the probability of finding an electron in a small volume element $d\tau$ surrounding a point r is $\psi(r)^2\, d\tau$. The probability density is therefore proportional to $\psi(r)^2$.

Although ψ has mathematical significance (e.g., it can have negative values), ψ^2, which must always be positive, has physical significance.

2.170 Schrödinger Equation (E. Schrödinger, 1887–1961)
A differential equation whose solution is the wavefunction for the system under consideration. This equation takes the same form as an equation for a standing wave. It is from this form of the equation that the name "wave mechanics" is derived.

The similarity of the Schrödinger equation to a wave equation is demonstrated by first substituting the de Broglie equation (2.100) into Eq. 2.130*b* and replacing *f* by ϕ:

$$\nabla^2\phi + \frac{4\pi^2 m^2 v^2}{h^2}\phi = 0 \tag{2.170a}$$

To incorporate in this equation the total energy E of an electron, use is made of the fact that the total energy is the sum of the potential energy V plus the kinetic energy $\frac{1}{2}mv^2$, or

$$v^2 = \frac{2}{m}(E - V) \tag{2.170b}$$

Substituting Eq. 2.170*b* into Eq. 2.170*a* gives

$$\nabla^2\phi + \frac{8\pi^2 m}{h^2}(E - V)\,\phi = 0 \tag{2.170c}$$

which is the Schrödinger equation.

2.180 Eigenfunction
A hybrid German-English word that in English might be translated as "characteristic function"; an acceptable solution of the wave equation. There are certain conditions that must be fulfilled to obtain "acceptable" solutions of the wave equation, Eq. 2.110 [e.g., $f(x)$ must be zero at each end, as in the case of the vibrating string fixed at both ends; this is the so-called boundary condition]. In general, whenever some mathematical operation is done on a function and the same function is regenerated multiplied by a constant, the function is an eigenfunction, and the constant is an eigenvalue. Thus Eq. 2.110 may be written as

$$\left(\frac{\partial^2}{\partial x^2}\right) f(x) = -\left(\frac{4\pi^2}{\lambda^2}\right) f(x) \tag{2.180}$$

Generally it is implied that wavefunctions, hence orbitals, are eigenfunctions.

2.190 Eigenvalues

The values of λ calculated from the wave equation, Eq. 2.110. If the eigenfunction is an orbital, then the eigenvalue is the orbital energy.

2.200 The Schrödinger Equation for the Hydrogen Atom

An (eigenvalue) equation, the solutions of which in spherical coordinates are

$$\phi(r, \theta, \varphi) = R(r)\,\Theta(\theta)\,\Phi(\varphi). \qquad (2.200)$$

The eigenfunctions ϕ, also called orbitals, are functions of the three variables shown, where in spherical coordinates, r is the distance of a point from the origin, and θ and φ are the two angles required to locate the point (see Fig. 2.140). Associated with each eigenfunction (orbital) is an eigenvalue (orbital energy). An exact solution of the Schrödinger equation is possible only for the hydrogen atom, or any one-electron system. In many-electron systems wavefunctions are generally approximated as products of modified one-electron functions (orbitals). Each solution of the Schrödinger equation may be distinguished by a set of three quantum numbers, n, l, and m, which arise from the boundary conditions.

2.210 Principal Quantum Number, n

The quantum number that is related to the energy of an electron in an orbital (in the hydrogen atom it is solely determined by it). It is represented by an integer n, and can take on the values 1, 2, 3, ... The value of n corresponds to the number of the shell in the Bohr atomic theory.

2.220 Azimuthal Quantum Number, l

The quantum number that determines the angular momentum of the electron resulting from its motion around the nucleus; it is also called the angular momentum number. It can take on values of $l = 0, 1, 2, \ldots, (n - 1)$ and the value determines the shape of the orbital.

Example. Orbitals with azimuthal quantum numbers $l = 0, 1, 2$, and 3, are called s, p, d, and f, respectively. Thus an electron with a principal quantum number of $n = 2$ can take on l values of 0 and 1, corresponding to $2s$ and $2p$ orbitals, respectively. These orbital designations are taken from atomic spectroscopy where the words sharp, principal, diffuse, and fundamental describe lines in atomic spectra.

2.230 Magnetic Quantum Number, m

The quantum number that determines the orientation in space of the orbital angular momentum: it is represented by m (or m_l) and may have values from $+l$ to $-l$ (see Sect. 2.220).

Example. When $n = 2$ and $l = 1$ (the p orbitals), m may thus have values of $+1, 0, -1$, corresponding to three $2p$ orbitals.

2.240 Degenerate Orbitals
Orbitals having equal energies, e.g., the three $2p$ orbitals.

2.250 Electron Spin Quantum Number, *s*
A measure of the intrinsic angular momentum of the electron, usually designated by s. It can have the value of $\frac{1}{2}$ or $-\frac{1}{2}$ only.

2.260 Spin-Orbit Coupling
The interaction of orbital angular momentum of an electron (azimuthal quantum number l, Sect. 2.220) with the spin angular momentum of the electron (quantum number s) to give $s + l$, or spin-orbit coupling.

2.270 *s* Orbital
A spherically symmetrical orbital; i.e., ϕ is a funtion of $R(r)$ only. For s orbitals $l = 0$.

2.280 Atomic Orbitals for Many-Electron Atoms
Modified hydrogenlike orbitals that are used to describe the electron distribution in many-electron atoms. The names of the orbitals, s, p, etc., are taken from the corresponding hydrogen orbitals. The presence of more than one electron in a many-electron atom breaks the degeneracy of orbitals with the same n value. Thus for a given n the orbital energies increase in the order $s < p < d < f < \ldots$.

2.290 1*s* Orbital
The lowest energy orbital of an atom, characterized by $n = 1, l = m = 0$. It corresponds to the fundamental wave and is characterized by spherical symmetry and no nodes. It is drawn as a circle about the nucleus to show that there is a high probability of finding the electron within the sphere represented by the circle.

Example. The numerical probability of finding the hydrogen electron at various distances from the nucleus is shown in Fig. 2.290*a*. The circles represent contours of probability that are on a cut of the sphere (a plane that bisects the sphere). If the contour circle of probability 0.95 is chosen, the electron is 19 times as likely to be inside the corresponding sphere with a radius of 1.7 Å as it is to be outside that sphere. The circle that is usually drawn, Fig. 2.290*b*, to represent the 1*s* orbital is meant to imply that there is a high (unspecified) probability of finding the electron in a sphere of which the circle is a (cross section) cut.

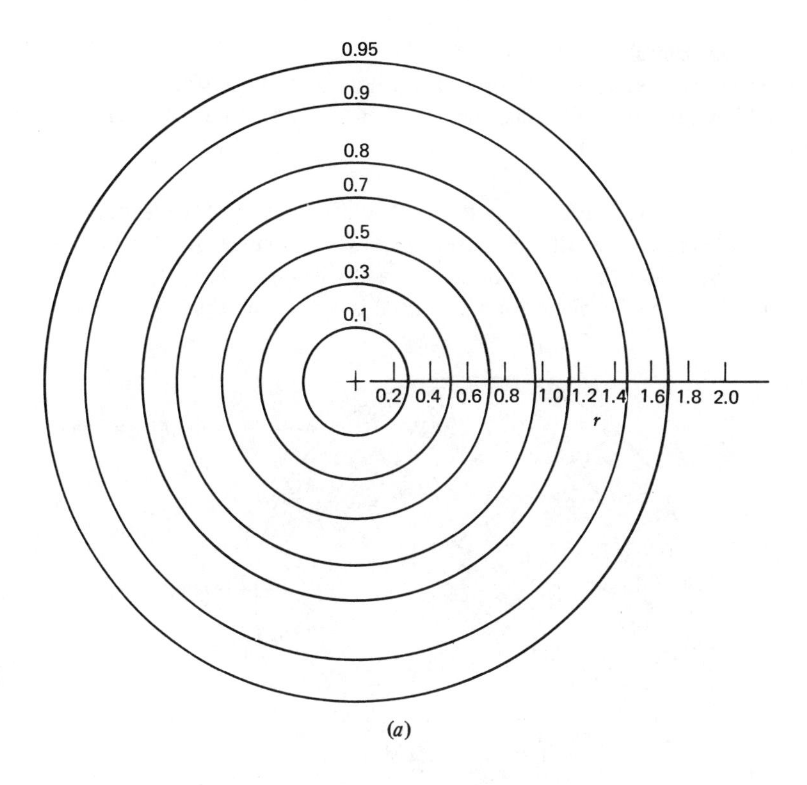

(a)

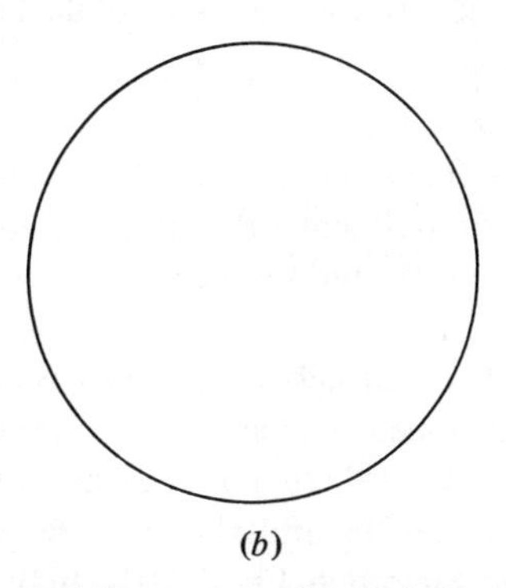

(b)

Figure 2.290 (a) The probability contours for the hydrogen atom; (b) the 1s orbital.

2.300 2*s* Orbital

The spherically symmetrical orbital having one nodal surface; electrons in this orbital have the principal quantum number $n = 2$, but have no angular momentum; i.e., $l = 0, m = 0$.

Example. Figure 2.300 shows the probability density of the 2*s* orbital as a cut of the sphere. The 2*s* orbital is usually drawn as a circle indistinguishable from the 1*s* orbital and the fact that there is a nodal surface is disregarded. However, the 0.95 probability sphere for a 2*s* orbital is larger than that for a 1*s* orbital.

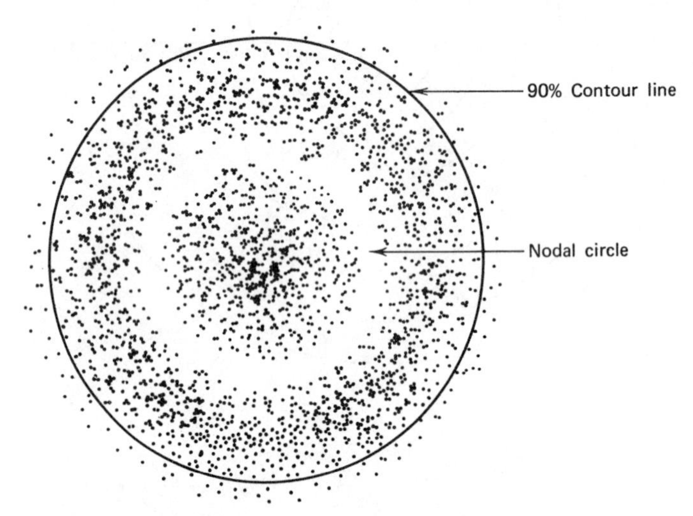

Figure 2.300 Probability density of the 2*s* orbital.

2.310 2*p* Atomic Orbitals

A set of three degenerate (equal energy) atomic orbitals possessing one nodal plane; electrons in these orbitals have the principal quantum number of 2, an azimuthal quantum number of 1, and magnetic numbers of +1, 0, or −1.

Example. The 2*p* orbitals are usually depicted so as to emphasize their angular dependence [i.e., $R(r)$ is assumed constant], hence are drawn for convenience as a planar cross section through a three-dimensional representation of $\Theta(\theta)\phi(\varphi)$. The planar cross section of the $2p_z$ orbital ($\varphi = 0$, see Fig. 2.140) then becomes a pair of circles touching at the origin, Fig. 2.310*a*. In this figure the wavefunction (without proof) is $\phi = \Theta(\theta) = (\sqrt{6}/2) \cos \theta$. Since $\cos \theta$, $90^\circ < \theta < 270^\circ$, is negative, the top circle is positive and the bottom circle negative. However, the physically significant property of an orbital ϕ is its square ϕ^2; the plot of $\phi^2 = \Theta^2(\theta) = \frac{3}{2} \cos^2 \theta$ for the p_z orbital is shown in Fig. 2.310*b*. The shape of

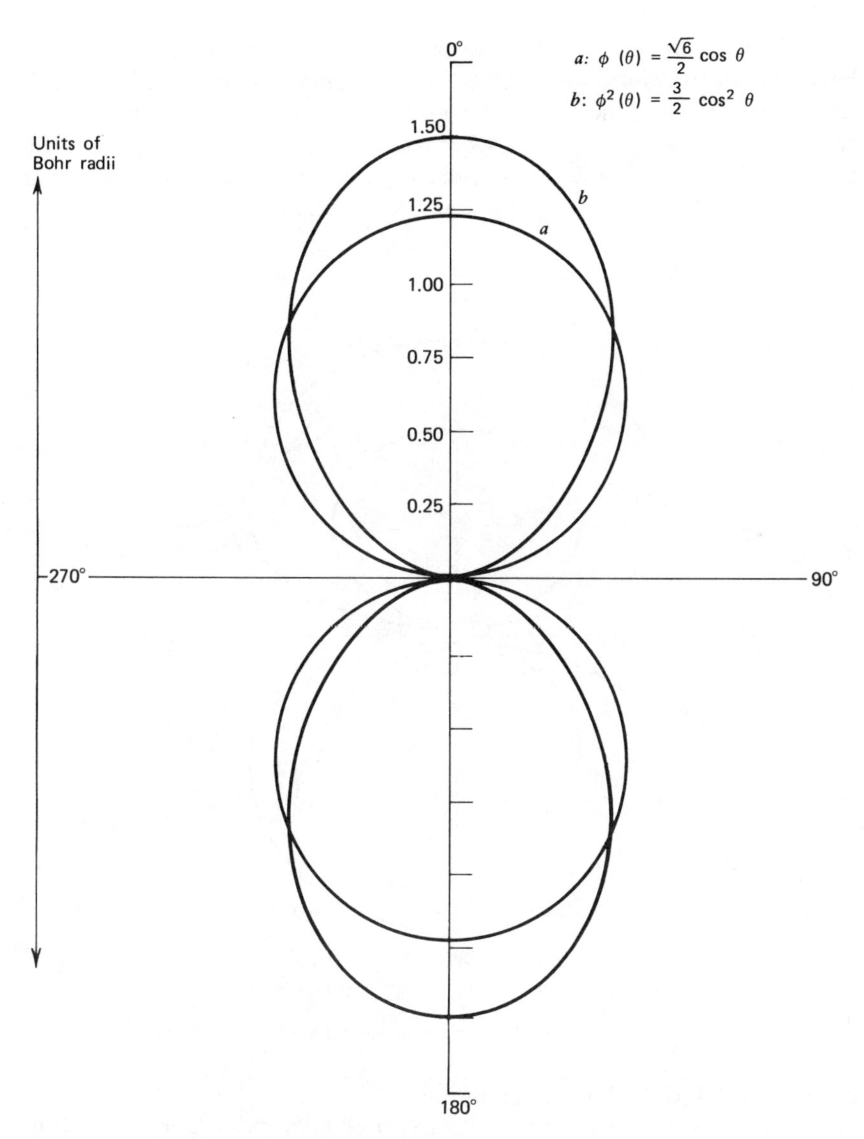

Figure 2.310 (*a*) The p_z orbital showing angular dependence; (*b*) the square of (*a*).

this orbital is the familiar elongated dumbbell-shaped p orbital, but actually both lobes of the orbital are positive. In most common drawings of the p orbitals, the shape of ϕ^2, the physically significant function, is retained but the plus and minus signs are placed in the lobes to emphasize the nodal property. If the function $R(r)$ is included, the oval-shaped contour representation shown in Fig. 2.310*c* results, where ϕ^2 (p_z) is shown as a cut in the *xz* plane.

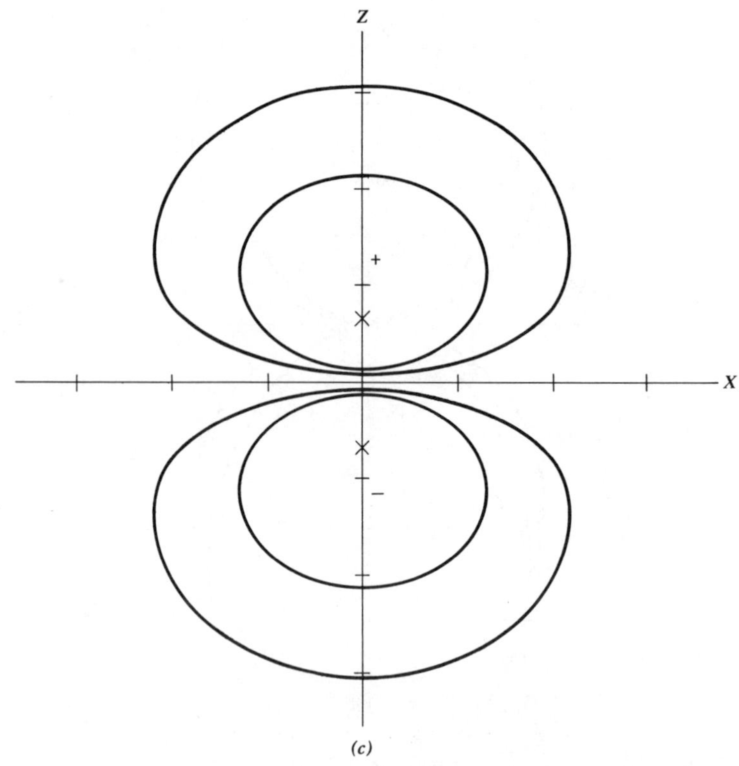

Figure 2.310 (*c*) Contour diagrams including the radial dependence of ϕ.

2.320 Aufbau (*G*. Building Up) Principle

In a many-electron atom electrons must be placed in the orbitals of lower energy, which must be filled before electrons are placed in higher energy orbitals. The sequential placement of electrons must also be consistent with the Pauli exclusion principle and Hund's rule, *vide infra*.

2.330 Pauli Exclusion Principle (W. Pauli, 1900–1958)

A maximum of two electrons can occupy an orbital and then only if the spins of the electrons are paired; i.e., if one electron has $s = +\frac{1}{2}$, the other must have $s = -\frac{1}{2}$.

2.340 Hund's Rule (G. Friederich Hund, 1896–)

A single electron is placed in all orbitals of equal energy (degenerate orbitals) before a second electron is placed in any one of the degenerate set.

Example. The three rules in Sects. 2.320, 2.330, and 2.340 are generally introduced in connection with the construction of the periodic table. The distribution of electrons in the nitrogen atom (atomic number of 7) is shown in Fig. 2.340. Note that the 2*p* orbitals are higher in energy than the 2*s* orbital.

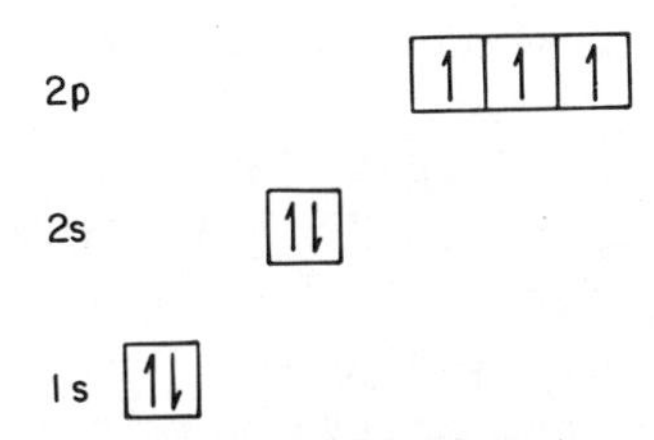

Figure 2.340 The electronic configuration of the nitrogen atom.

2.350 Hybridization of Atomic Orbitals

The (theoretical) mixing of two or more different atomic orbitals to give the same number of new orbitals, each of which has some of the character of the original orbitals. Hybridization requires that the atomic orbitals to be mixed are similar in energy. The resulting hybrid orbitals have directional character, which can lead to more favorable bonding with other atoms.

Example. In much of organic (carbon) chemistry, the 2*s* orbital of carbon is mixed with: (*a*) one *p* orbital to give two hybrid *sp* orbitals; (*b*) two *p* orbitals to give three sp^2 orbitals; or (*c*) three *p* orbitals to give four sp^3 orbitals. The mixing of the C_{2s} with the C_{2p_z} to give two C_{2sp} orbitals is shown pictorially in Fig. 2.350. These two hybrid atomic orbitals have the form $\phi_1 = (s + p_z)$ and $\phi_2 = (s - p_z)$.

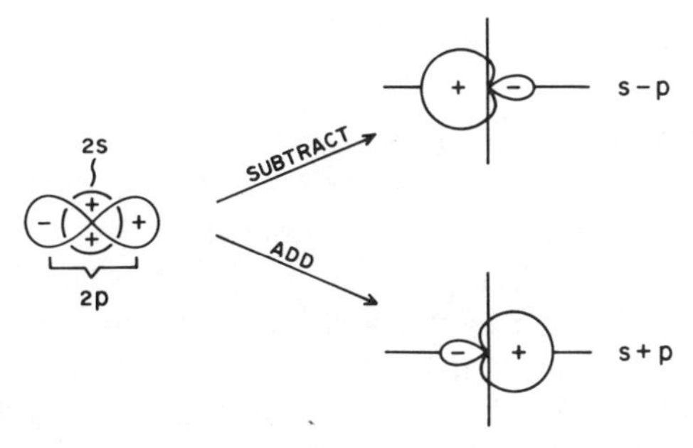

Figure 2.350 The two hybrid *sp* atomic orbitals.

2.360 Hybridization Index
The superscript x on the p in an sp^x hybrid orbital; such an orbital possesses $[x/(1 + x)]$ (100) percent p character.

Example. The hybridization index of an sp^3 orbital is 3 (75% of p); for an $sp^{0.894}$ it is 0.894 (47.2% of p).

2.370 Equivalent Hybrid Atomic Orbitals
A set of hybridized orbitals each member of which possesses precisely the same proportions of the different constituent atomic orbitals.

Example. If the atomic orbitals $2s$ and $2p_z$ are distributed equally in two hybrid orbitals, each resulting orbital will have an equal amount of s and p character; i.e., each orbital will be sp ($s^{1.00}p^{1.00}$). If a $2s$ and two $2p$ orbitals are distributed equally among three hybrid orbitals, three equivalent orbitals result, Fig. 2.370. Mixing of a $2s$ orbital equally with three $2p$ orbitals gives four hybrid orbitals, $s^{1.00}p^{3.00}$ (sp^3); i.e., each of the four sp^3 orbitals has an equal amount of s character, $[1/(1 + 3)] \times 100\% = 25\%$, and an equal amount of p character, $[3/(1 + 3)] \times 100\% = 75\%$.

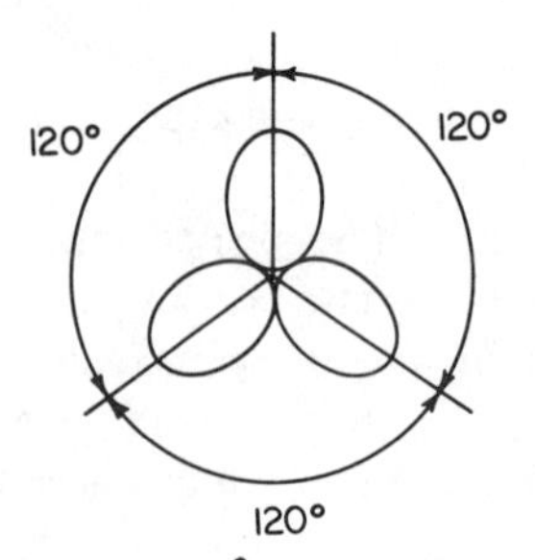

Figure 2.370 The three hybrid sp^2 atomic orbitals (all in the same plane).

2.380 Nonequivalent Hybrid Atomic Orbitals
The hybridized orbitals that result when the constituent atomic orbitals are not equally distributed among a set of hybrid orbitals.

Example. In hybridizing a $2s$ with a $2p$ orbital to form two hybrids it is possible to put more p character and less s character into one hybrid and less p and more s into the other. Thus in hybridizing an s and a p_z orbital it is possible to generate one hybrid that has 52.8% $p(sp^{1.12})$ character. The second hybrid must be 47.2% p and is therefore $sp^{0.89}$ ($x/(1 + x) = 47.2$; $x = 0.89$). Such nonequivalent carbon orbitals are found in CO, where the *sp* carbon hybrid orbital used in bonding to oxygen has more *p* character than the other carbon *sp* hybrid orbital, which contains a lone pair of electrons. If dissimilar atoms are bonded to a carbon atom, the *sp* hybrid orbitals will always be nonequivalent.

2.390 Lone Pair (or Nonbonding) Electrons

A pair of electron in the valence shell of an atom; such electrons do not participate in bonding and usually are written as a pair of dots on the atom.

Example. The oxygen of the water molecule has two lone pairs of electrons, Fig. 2.390.

Figure 2.390 The lone pair electrons on oxygen in H_2O.

2.400 Nonbonding Atomic Orbital

An atomic orbital in which a lone pair of electrons is located.

Example. The two lone pairs on oxygen in H_2O are each in an (approximately) sp^3 nonbonding orbital, Fig. 2.400.

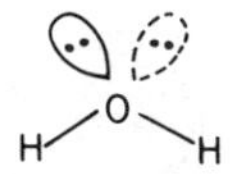

Figure 2.400 The sp^3 lone pair orbitals of H_2O.

2.410 Molecular Orbitals

Orbitals that encompass more than one atom in a molecule.

2.420 Bonding Molecular Orbitals

Molecular orbitals that are lower in energy than the isolated atomic orbitals from which they are made. Electrons in these orbitals help to bond the atoms they encompass.

2.430 Antibonding Molecular Orbitals

Molecular orbitals that are higher in energy than the isolated atomic orbitals from which they are made. Electrons in these orbitals destabilize the bonding between the atoms they encompass.

2.440 Nonbonding Molecular Orbitals

Molecular orbitals that, when occupied by electrons, contribute nothing to the bonding of the atoms they encompass.

2.450 Linear Combination of Atomic Orbitals (LCAO)

A method of generating molecular orbitals by adding and subtracting atomic orbitals; the number of molecular orbitals must be equal to the number of atomic orbitals that were combined.

Example. When two 1*s* hydrogen atomic orbitals, ϕ_A and ϕ_B, are combined, the linear combinations that result are $(\phi_A + \phi_B)$ and $(\phi_A - \phi_B)$. The combinations are shown pictorially in Fig. 2.450*a*. The combination made by addition, ψ_b, is a bonding orbital and that made by substraction, ψ_a, is an antibonding orbital. Figure 2.450*b* shows the end-on combinations of *p* orbitals.

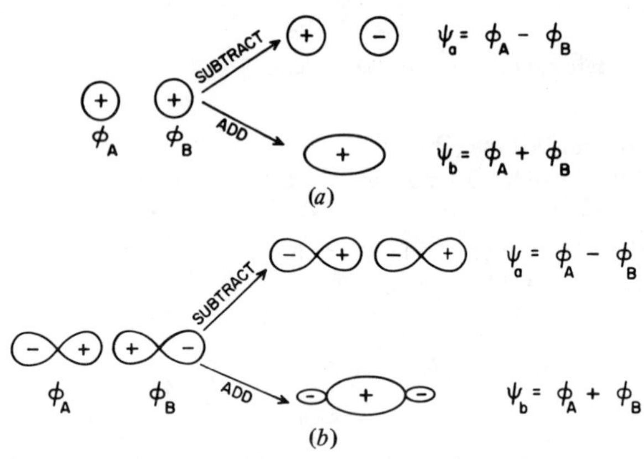

Figure 2.450 (*a*) The linear combination of *s* orbitals; (*b*) the linear combination of end-on *p* orbitals.

2.460 Basis Set of Orbitals

The combination or set of atomic orbitals that is used in generating the molecular orbitals of interest in a molecule.

Examples. The molecular orbitals of the hydrogen molecule (Fig. 2.450*a*) are generated from the 1*s* atomic orbitals of the hydrogen atoms, the basis set. In constructing the four π molecular orbitals of 1,3-butadiene the *p* orbitals on each of the four carbon atoms constitute the basis set (see Sect. 2.540).

2.470 Sigma Bonding Molecular Orbital (σ Orbital); Sigma Bond (σ Bond)

The molecular orbital produced by the in-phase end-on overlap of atomic orbitals on adjacent atoms. Such orbitals have conical or cylindrical symmetry around the internuclear axis, and are lower in energy than the corresponding atomic orbitals. The bond formed as a result of the end-on overlap of atomic orbitals is a sigma bond.

Examples. ψ_b in Fig. 2.450*a*; also the orbital ψ_b produced by end-on overlap of atomic *p* orbitals on adjacent atoms, Fig. 2.450*b*.

2.480 Sigma Antibonding Orbital (σ^* Orbital)

The antibonding molecular orbital generated by the out-of-phase end-on overlap of atomic orbitals on adjacent atoms. Such orbitals have conical or cylindrical symmetry around the internuclear axis, a nodal plane perpendicular to the internuclear axis, and are higher in energy than the corresponding atomic orbitals.

Examples. ψ_a in Figs. 2.450*a* and 2.450*b* are sigma antibonding orbitals. To distinguish these from bonding σ orbitals they are written with an asterisk superscript (σ^*) and called sigma starred orbitals.

2.490 Pi Bonding Molecular Orbital (π Orbital)

The orbital produced by the in-phase overlap of parallel *p* atomic orbitals (*p* orbitals perpendicular to the internuclear axis) on adjacent atoms. Such orbitals are antisymmetric with respect to 180° rotation around the internuclear axis.

Example. In constructing the LCAO of $p\pi$ atomic orbitals of ethylene, the two combinations shown in Fig. 2.490 are generated. The π bonding molecular orbital ψ_b or π orbital is shown in Fig. 2.490*b*, the antibonding ψ_a in Fig. 2.490*a*, and the resultant wave character of ψ_a and ψ_b in Fig. 2.490*c*.

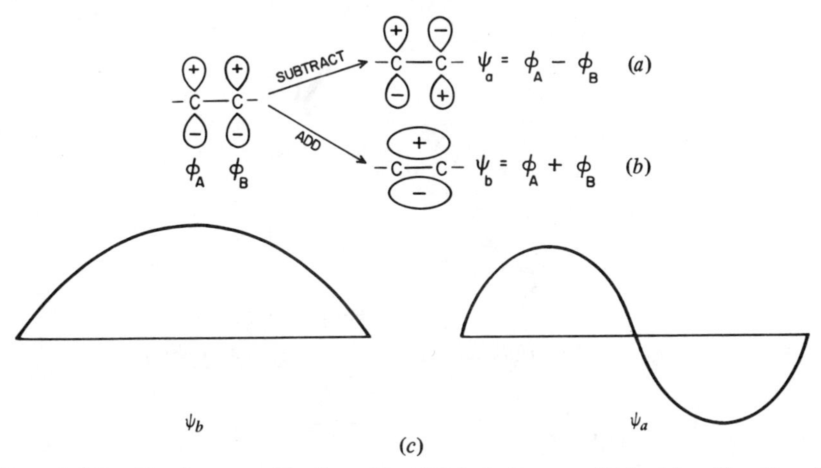

Figure 2.490 The linear combination of *p* orbitals to form π orbitals: (*a*) antibonding; (*b*) bonding; (*c*) the wave character of ψ_a and ψ_b.

2.500 Pi Bond (π Bond)

The type of bond between atoms when electrons occupy a π orbital formed by the overlap of parallel p atomic orbitals on adjacent atoms.

2.510 Pi Antibonding Molecular Orbital (π^* Orbital)

The orbital generated by the out-of-phase overlap of parallel p orbitals on adjacent atoms. Such orbitals are antisymmetric with respect to 180° rotation around the internuclear axis, and have a nodal plane perpendicular to that axis.

Example. Figure 2.490*a*.

2.520 Pi Nonbonding Molecular Orbital

A π orbital that, when occupied by electrons, contributes no bonding energy to the molecule.

Example. When an odd number of atomic $p\pi$ orbitals are combined according to the LCAO procedure, one of the molecular orbitals generated is usually nonbonding. The molecular orbitals for the allyl system are shown in Fig. 2.520; ψ_2 is the nonbonding molecular orbital. This orbital possesses a nodal plane through the central carbon atom, consistent with the requirement that in π nonbonding molecular orbitals there is no interaction between nearest neighbor p atomic orbitals.

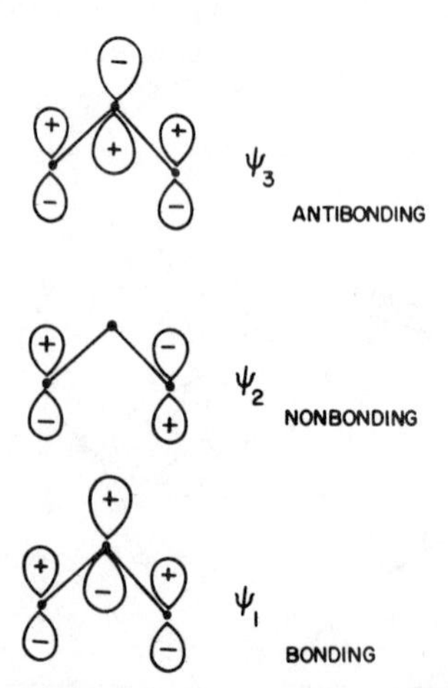

Figure 2.520 The π molecular orbitals of the allyl system.

2.530 $p\pi$ Atomic Orbital

A p orbital that takes part in a π-type bond.

2.540 Delocalized π Molecular Orbital

The molecular orbital that results from the combination of three or more parallel p atomic orbitals on different atoms; usually just called π molecular orbitals. These orbitals encompass all the atoms involved.

Example. The $n\pi$ molecular orbitals generated from the n parallel atomic p orbitals by the LCAO procedure are perpendicular to the molecular plane. When n is even, the number of bonding π molecular orbitals equals the number of antibonding π molecular orbitals, as shown for butadiene in Fig. 2.540.

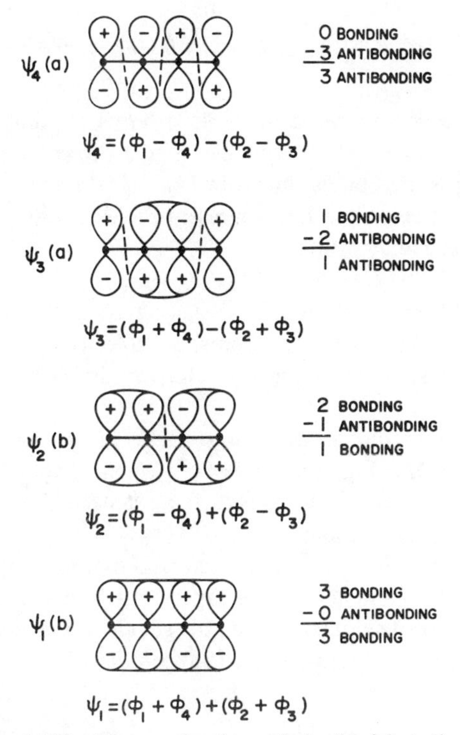

Figure 2.540 The π molecular orbitals of 1,3-butadiene.

2.550 Molecular Orbital Energy Diagram (MOED)

A diagram displaying the relative energies of the atomic orbitals to be combined (the basis set) in the LCAO treatment and the energies of the resulting molecular orbitals.

Example. The MOED for H_2, and for the $p\pi$ combination in ethylene (C_2H_4), are shown in Fig. 2.550*a* and *b*, respectively. In constructing the π MOED for

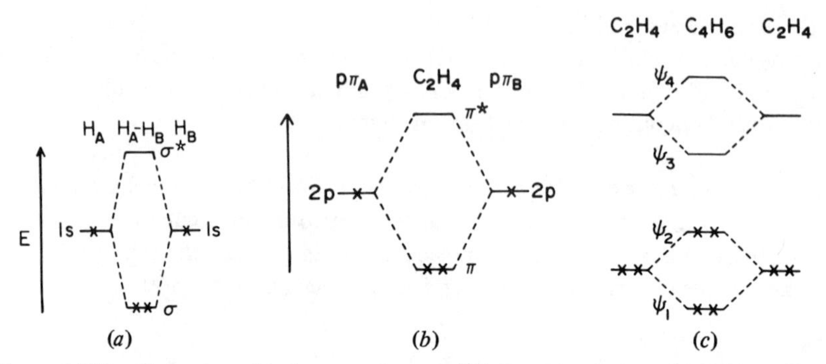

Figure 2.550 Molecular orbital energy diagrams: (*a*) H_2; (*b*) ethylene; (*c*) 1,3-butadiene.

butadiene, C_4H_6, it is convenient to think in terms of combining the π and π^* orbitals of two ethylenes, whereupon the MOED shown in Fig. 2.550*c* results. The relative energies of orbitals shown in Fig. 2.550*b* and *c* are plotted on the same scale. The crosses on the lines represent the electrons in the system, which are placed in lowest energy orbitals.

2.560 Electronic Configuration

The enumeration of the atomic (for atoms) or molecular (for molecules) orbitals occupied by electrons, and the number of electrons in each, as superscripts.

Examples. The electronic configuration of the nitrogen atom, Fig. 2.340, is $1s^2 2s^2 2p^3$; that for H_2, Fig. 2.550*a*, is $1\sigma^2$. When nonbonding atomic orbitals are occupied, the number of electrons in them is usually indicated by adding a superscript to n (for nonbonding).

The electronic configuration for molecules is indicated by writing the occupied molecular orbitals and adding the superscript 1 or 2 as appropriate. If the molecule has a center of symmetry, then the *gerade* or *ungerade* character of the orbital is indicated as a subscript g or u as appropriate. The electronic configuration of H_2 is $1\sigma_g^2$.

2.570 Pi Electron Configuration

The enumeration of the π molecular orbitals occupied by electrons, with the number of electrons in each indicated by a superscript.

Example. In the MOED for butadiene, Fig. 2.550*c*, $\psi_1^2\psi_2^2$.

2.580 Electronic Ground State

That electronic state of an atom or molecule in which the potential energy is a minimum. The electronic configuration for such a state is that in which the electrons are in the orbitals of lowest energy.

Examples. The electronic configurations for the N atom, H_2, and the π electron configurations for C_2H_4 and C_4H_6, shown respectively in Figs. 2.340 and 2.550*a*, *b*, and *c* are the ground state electronic configurations.

2.590 Electronic Excited State

An electronic state of an atom or molecule higher in energy than the ground state.

Example. The MOED for the carbonyl group is shown in Fig. 2.590. Only the 2*p* lone pair orbital on oxygen and the π orbitals of the carbonyl group are shown. The ground state configuration is then $\pi^2 n^2$. If an electron were excited from the n orbital to a π^* orbital, the resulting excited configuration would be $\pi^2 n\pi^*$ and the molecule would be said to be in the $n\pi^*$ excited state.

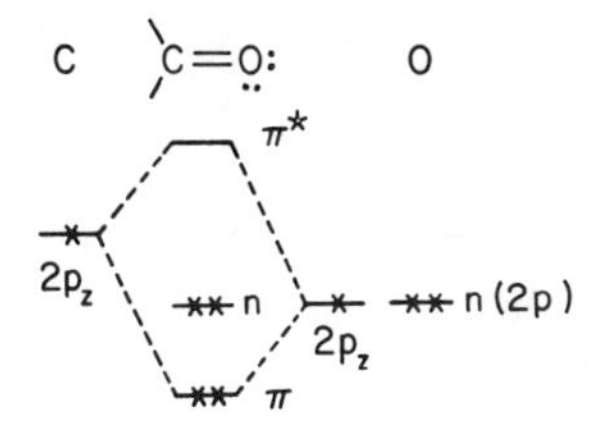

Figure 2.590 The molecular orbital energy diagram for the carbonyl group.

2.600 Virtual Orbital

An orbital that is empty or unoccupied in the ground state.

Example. In the ground state of the carbonyl group, Fig. 2.590, the π^* orbital is a virtual orbital.

2.610 Coefficients of Atomic Orbitals

The number c_{ij} ($-1 \leqq c_{ij} \leqq +1$), which indicates the degree to which each atomic orbital participates in a molecular orbital. This definition and the ones immediately following are those used in simple π electron (Hückel) theory.

Example. In Figs. 2.450*a*, *b* and 2.490 two atomic orbitals have been combined to generate, in each case, one bonding and one antibonding molecular orbital. The correct wavefunctions require that the coefficient on each atomic orbital for all sets of two molecular orbitals be $\pm 1/\sqrt{2}$; thus (without proof) $\psi_b = (1/\sqrt{2})\,\phi_A + (1/\sqrt{2})\,\phi_B$ and $\psi_a = (1/\sqrt{2})\,\phi_A - (1/\sqrt{2})\,\phi_B$. The three molecular orbitals of the allyl system are shown in Fig. 2.610 and are drawn to indicate that ϕ_2 makes a greater contribution to ψ_1 and ψ_3 than do ϕ_1 and ϕ_3. The coefficient of ϕ_2 is zero in ψ_2. When coefficients need to be designated, especially for purposes of calculation, they are indicated by the letter c

$$\psi_3 = \frac{1}{2}\phi_1 - \frac{1}{\sqrt{2}}\phi_2 + \frac{1}{2}\phi_3$$

$$\psi_2 = \frac{1}{\sqrt{2}}\phi_1 - \frac{1}{\sqrt{2}}\phi_3$$

$$\psi_1 = \frac{1}{2}\phi_1 + \frac{1}{\sqrt{2}}\phi_2 + \frac{1}{2}\phi_3$$

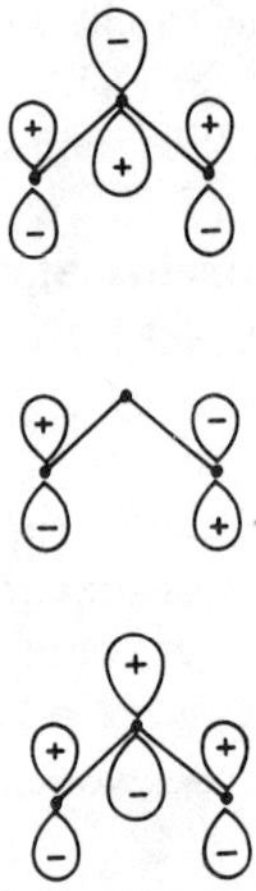

Figure 2.610 The π molecular orbitals of the allyl system, showing relative coefficients of atomic orbitals.

followed by two subscripts ij, where j refers to the jth atomic orbital in the ith molecular orbital. Thus in the molecular orbital ψ_3, Fig. 2.610, $c_{31} = c_{33} = \frac{1}{2}$ and $c_{32} = -1/\sqrt{2}$.

2.620 Sign Inversion

A change in sign of the coefficients on adjacent atomic orbitals making up the molecular wavefunction of a molecule; equivalent to a node.

Examples. The antibonding π molecular orbital of ethylene, Fig. 2.490*a*. The number of sign inversions in a cyclic array of atomic orbitals is of importance in connection with the Hückel and Möbius systems to be discussed later (Sect. 3.460 and 3.660).

2.630 Normalized Orbital

An orbital meeting the requirement that the integration of its square over all space is equal to unity; i.e., $\int \psi^2 \, d\tau = 1$. An alternate statement of this requirement is that the sum of the squares of the coefficients of all atomic orbitals in a molecular orbital is equal to unity; i.e., $\sum_j c_{ij}^2 = 1$. Atomic orbitals ϕ are always assumed to be normalized; i.e., $\int \phi^2 \, d\tau = 1$.

2.640 Normalization

The process of normalizing an orbital, accomplished by multiplying by a number N, so that $\int (N\psi)^2 \, d\tau = 1$.

Example. Neither the orbital ψ_b nor ψ_a, Figs. 2.450*a*, *b* and 2.490, are normalized. To normalize, e.g., $\psi_a = (\phi_A - \phi_B)$: $\int N^2(\phi_A - \phi_B)^2 \, d\tau = N^2 \int (\phi_A^2 - 2\phi_A\phi_B + \phi_B^2)\, d\tau = N^2 [\int \phi_A^2 \, d\tau - 2\int \phi_A\phi_B \, d\tau + \int \phi_B^2 \, d\tau] = 1$. In Sect. 2.630 it was pointed out that $\int \phi^2 \, d\tau = 1$. The integral $\int \phi_A \phi_B \, d\tau$, called the overlap integral, is for most simple approximations assumed to be equal to zero. Hence the above equation simplifies to

$$N^2 [1 + 0 + 1] = 1$$

hence $N = 1/\sqrt{2}$ and

$$\psi_a = \frac{1}{\sqrt{2}} (\phi_A - \phi_B) = \frac{1}{\sqrt{2}} \phi_A - \frac{1}{\sqrt{2}} \phi_B$$

That this orbital is indeed normalized is apparent from

$$\left(\frac{1}{\sqrt{2}}\right)^2 + \left(-\frac{1}{\sqrt{2}}\right)^2 = 1$$

2.650 Orthogonal Orbitals

Any two molecular orbitals ψ_r and ψ_s meeting the requirement that their product integrated over all space is equal to zero; i.e., $\int \psi_r \psi_s \, d\tau = 0$.

Example. The orbitals

$$\psi_b = \frac{1}{\sqrt{2}} \phi_A + \frac{1}{\sqrt{2}} \phi_B \qquad \text{and} \qquad \psi_a = \frac{1}{\sqrt{2}} \phi_A - \frac{1}{\sqrt{2}} \phi_B$$

are seen to be orthogonal, viz.,

$$\int \left(\frac{1}{\sqrt{2}} \phi_A + \frac{1}{\sqrt{2}} \phi_B\right) \left(\frac{1}{\sqrt{2}} \phi_A - \frac{1}{\sqrt{2}} \phi_B\right) d\tau$$

$$= \frac{1}{2}\int \phi_A^2 \, d\tau - \frac{1}{2}\int \phi_B^2 \, d\tau = \frac{1}{2} - \frac{1}{2} = 0$$

2.660 Orthonormal Orbitals

A set of orbitals that are both orthogonal and normalized.

2.670 Hamiltonian Operator, H

A mathematical operator for a system that is the sum of the potential and kinetic energies of that system. The Schrödinger equation may be written as $H\psi = E\psi$, where H is the Hamiltonian, ψ is the wavefunction, and E is the energy (eigenvalue) of the system.

2.680 Coulomb Integral

An integral that corresponds (in the zero-order approximation) to the energy of an electron in orbital ϕ_i in the field of the atom i, i.e., $\int \phi_i H \phi_i \, d\tau$, where H is the Hamiltonian energy operator. This integral is usually denoted by α.

Example. The value of the Coulomb integral α_i is assumed to be independent of any other atom j. It is roughly equivalent to the energy required to remove an electron from the atomic orbital in which it resides (the ionization potential); for a carbon $2p$ orbital this energy is 11.16 eV.

2.690 Overlap Integral

A measure of the extent of overlap of the orbitals on neighboring atoms i and j, i.e., the value of $\int \phi_i \phi_j \, d\tau$. In the Hückel π electron approximation, this integral, frequently denoted by S_{ij}, is taken to be zero.

Example. When i and j are nearest neighbor atoms with $p\pi$ orbitals such as the carbons in ethylene, the actual value of S_{ij} is about 0.25. Even so the consequences of neglecting S_{ij} are not very serious in zero-order approximate calculations.

2.700 Resonance Integral

An integral that determines the energy of an electron in the fields of two atoms i and j involving the orbitals ϕ_i and ϕ_j, i.e., $\int \phi_i H \phi_j \, d\tau$, where H is the Hamiltonian energy operator. This integral is usually denoted by β_{ij}.

Example. When i and j are not adjacent atoms, β_{ij} is neglected and set to zero. The resonance integral is approximately proportional to the actual value of the overlap integral S_{ij}.

2.710 Hückel Molecular Orbital (HMO) Method (E. Hückel, 1896–)

A scheme for calculating the energy of the π electrons in conjugated systems. The scheme includes a number of approximations; e.g., all sigma electrons and their interaction with π electrons are neglected; all overlap integrals (S) are neglected; the Coulomb integrals (α) are all set equal; all resonance integrals (β) are set equal. (All α's and β's are equal only if there are no heteroatoms.)

2.720 Secular Determinant

In HMO theory a mathematical determinant that contains elements consisting of the Coulomb and resonance integrals and the energies of the molecular orbitals of the molecule in question.

Example. The secular determinant for calculating the Hückel molecular orbitals of butadiene, C=C—C=C, has the following form:

$$\begin{vmatrix} \alpha - E & \beta & 0 & 0 \\ \beta & \alpha - E & \beta & 0 \\ 0 & \beta & \alpha - E & \beta \\ 0 & 0 & \beta & \alpha - E \end{vmatrix} = 0$$

If each term is divided by β and then the substitution $x = (\alpha - E)/\beta$ is made, the secular determinant simplifies to

$$\begin{vmatrix} x & 1 & 0 & 0 \\ 1 & x & 1 & 0 \\ 0 & 1 & x & 1 \\ 0 & 0 & 1 & x \end{vmatrix} = 0$$

This reduces to $x^4 - 3x^2 + 1 = 0$, which is readily solved to give the energies of the four π MOs of butadiene. The solution of this determinant permits the calculation of the energies E of molecular orbitals of a molecule in terms of the Coulomb (α) and resonance (β) integrals, and the actual coefficients c_i of the atomic orbitals for each molecular orbital.

2.730 Molecular Orbital Energy

The energy of a molecular orbital. In Hückel π electron theory this is usually expressed in terms of α and β.

Example. The energy of the bonding π molecular orbital in ethylene ψ_b is ($\alpha + \beta$). In the ground state both $p\pi$ electrons are in ψ_b, and therefore the total π molecular orbital energy is $2\alpha + 2\beta$.

2.740 Delocalization Energy

The calculated additional bonding energy that results from delocalization of electrons assumed to be originally constrained to isolated double bonds. In HMO theory this energy is usually expressed in terms of β (the resonance integral) and corresponds to the resonance energy in valence bond theory (Sect. 3.420).

Example. Butadiene has the structure CH_2=CH—CH=CH_2, and the π energies of the two lowest energy molecular orbitals, (both of which encompass all four $p\pi$ atomic orbitals), Fig. 2.540, are: $\psi_1 = \alpha + 1.618\beta$ and $\psi_2 = \alpha + 0.618\beta$ (calculated from the 4 × 4 secular determinant shown in Sect. 2.720). Since two electrons are in each of these orbitals in the ground state, the total π

energy is $4\alpha + 4.472\beta$. The energy of ψ_b for ethylene is $\alpha + \beta$ (Sect. 2.730) and for four electrons in two ψ_b orbitals for two isolated ethylenes it is $4\alpha + 4\beta$. The difference in energy between the more stable butadiene and the two isolated ethylenes is thus 0.472β and this is the value of the delocalization energy.

2.750 *d* Orbitals

The set of five orbitals with azimuthal quantum number $l = 2$, hence $m = \pm 2, \pm 1, 0$. The five *d* orbitals are pictured in Fig. 2.750. Note that all *d* orbitals are *gerade.*

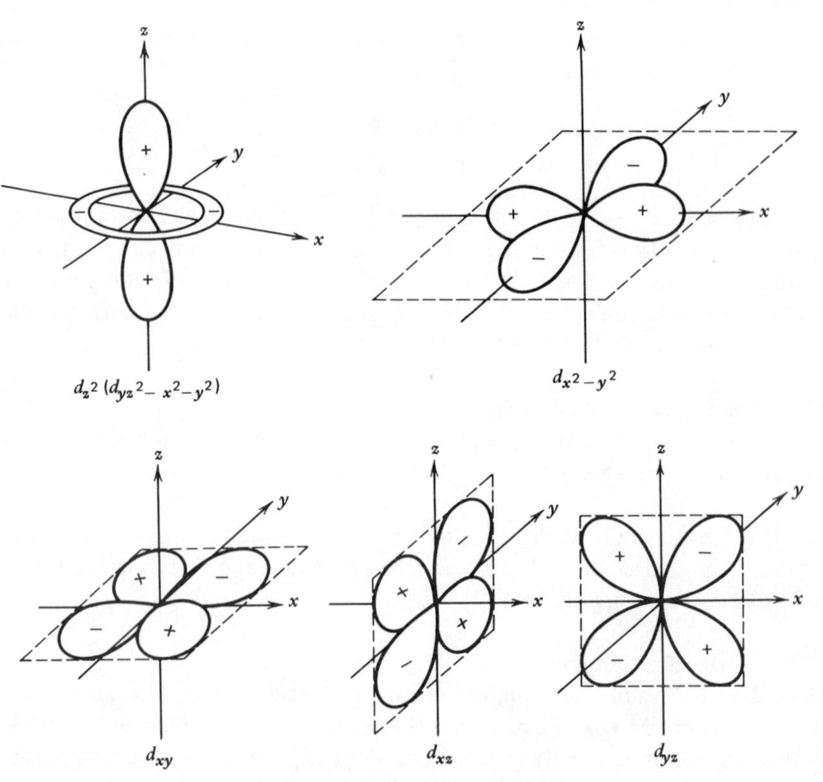

Figure 2.750 The five *d* orbitals.

2.760 *d*-π Bonding

The bonding that results when electrons are in a molecular orbital made from the overlap of a *d* orbital on one atom with a *p* orbital on an adjacent atom.

Example. Some of the bonding between third row elements and second row elements, as in SO_2, can be explained by *d*-π bonding, shown in Fig. 2.760.

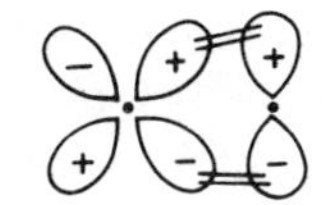

Figure 2.760 d-π bonding.

2.770 Back-Bonding

The bonding between two atoms (or centers) achieved by donation of electrons in the d orbital of one partner into the π^* orbital of the other.

Example. In the metal carbonyls, the one pair of electrons in the sp hybrid orbital on carbon is donated into a metal orbital to form a σ bond, but the metal in turn donates a pair of d electrons to the empty π^* orbital of carbon monoxide. The overlap that makes this back-donation possible is shown in Fig. 2.770.

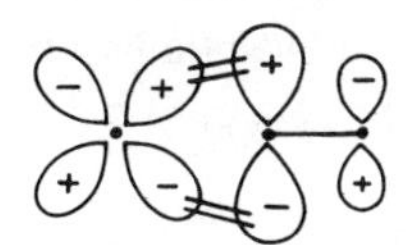

Figure 2.770 d-π^* back-bonding.

2.780 Localized Bonds

The bonds between adjacent atoms only, resulting from a high electron density between them. Such bonds may be treated theoretically by either valence bond theory or molecular orbital theory.

2.790 Valence Bond (VB) Theory for Localized Bonds

The theory that considers the bond between adjacent atoms to be formed by bringing together two distinct valence atomic orbitals on these atoms, each containing one electron of opposite spin.

Example. The wavefunction for the occupied orbital of the hydrogen molecule in this theory may be written: $\psi = \phi_A(1)\,\phi_B(2) + \phi_B(1)\,\phi_A(2)$, where ϕ_A and ϕ_B are the atomic $1s$ orbitals on hydrogens A and B and (1) and (2) refer to electrons of opposite spin. This wavefunction indicates that one electron is associated with one atom and the second electron with the other atom, and that the two electrons are never associated together at one atom. The practice of placing two dots between adjacent atoms for the bond, e.g., H:H, is an expression of this idea. The two dots are generally replaced by a dash, e.g., H—H, to express the pair of electrons responsible for the bonding.

2.800 Molecular Orbital (MO) Theory for Localized Bonds

The theory that considers the bonded atoms to be at equilibrium distance with electrons placed one by one into molecular orbitals that encompass the entire molecule.

Example. In the ground state of the hydrogen molecule both electrons are placed in the low energy molecular orbital, $\psi_b = (1/\sqrt{2})\,\phi_A + (1/\sqrt{2})\,\phi_B$. In the MO approach the two electrons are associated with both hydrogen atomic nuclei equally, whereas in the valence bond approach the two electrons are not simultaneously associated with a single nucleus. Although electrons are automatically delocalized in the MO method, it is possible to transform MOs into a collection of localized orbitals. Both the valence bond and molecular orbital methods can be utilized at different levels of approximations, but in the end they can be shown to be completely equivalent.

2.810 Two Center–Two Electron Bond (2*c*–2*e*)

A localized bond, i.e., a bond between adjacent atoms involving two electrons. This is the bond implied by the single dash between atoms in the structure of the molecule.

2.820 Delocalized Bonding

The bonding that results when three or more atomic nuclei are held together by electrons encompassing these atoms.

Example. The molecular orbital treatment of such bonding is particularly attractive. Thus the π bonding in butadiene is best described in terms of the π electrons in the orbitals ψ_1 and ψ_2 (Fig. 2.540).

2.830 Multicenter Bonding

In its broadest sense synonymous with delocalized bonding. The term is commonly restricted to the bonding is stable species which results when a series of atoms are linked together but fewer valence electrons are available than are required for separate localized bonds.

2.840 Three Center–Two Electron Bonding (3*c*–2*e*)

The multicenter bonding involving the bonding between three atoms by two electrons.

2.850 Open Three Center–Two Electron Bonding

Three center–two electron bonding in which the two outside atoms (or groups) use orbitals pointed toward the middle atom (or group); the two outside atoms are not directly bonded to each other.

Examples. The B—H—B bonding in boron hydrides such as B_2H_6, Fig. 2.850*a*. (In 3*c*–2*e* bonds only two, not four, electrons are present to bond the three atoms. This "deficiency" of electrons is apparent, since in B_2H_6 there are 2(3) + 6(1) = 12 valence electrons; eight electrons are involved in conventional H—B bonds, leaving two electrons for each of the two three-center B—H—B bonds. The atomic orbitals used in this type of bonding are illustrated in Fig. 2.850*b*. The LCAO treatment in such situations gives

$$\psi_1 = \frac{1}{2}\phi_1 + \frac{1}{\sqrt{2}}\phi_2 + \frac{1}{2}\phi_3 \qquad \text{(strongly bonding)}$$

$$\psi_2 = \frac{1}{\sqrt{2}}\phi_1 - \frac{1}{\sqrt{2}}\phi_3 \qquad \text{(nonbonding)}$$

$$\psi_3 = \frac{1}{2}\phi_1 - \frac{1}{\sqrt{2}}\phi_2 + \frac{1}{2}\phi_3 \qquad \text{(antibonding)}$$

The two available electrons occupy ψ_1.

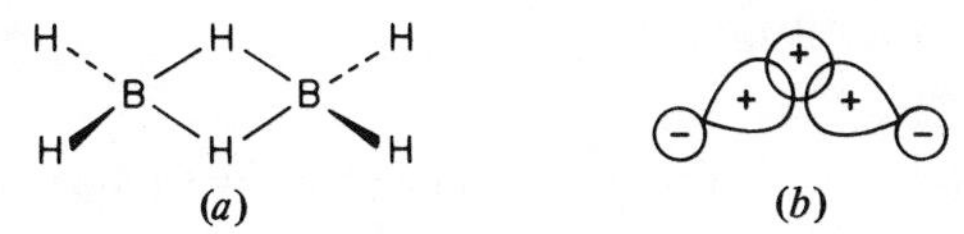

Figure 2.850 (*a*) The structure of B_2H_6; (*b*) the atomic orbitals involved in open three-center bonding.

2.860 Closed Three Center–Two Electron Bonding

Three center–two electron bonding in which the three atoms (or groups) use orbitals directed mutually at each other; all three atoms are directly bonded to each other.

Examples. The boron bonding shown in Fig. 2.860*a*. The atomic orbitals used in this bonding to generate the molecular orbitals are shown in Fig. 2.860*b*; the two electrons occupy the MO $\psi_1 = (1/\sqrt{3})\,\phi_1 + (1/\sqrt{3})\,\phi_2 + (1/\sqrt{3})\,\phi_3$. In some of the complicated boron hydrides both open and closed 3*c*–2*e* bonds may be present.

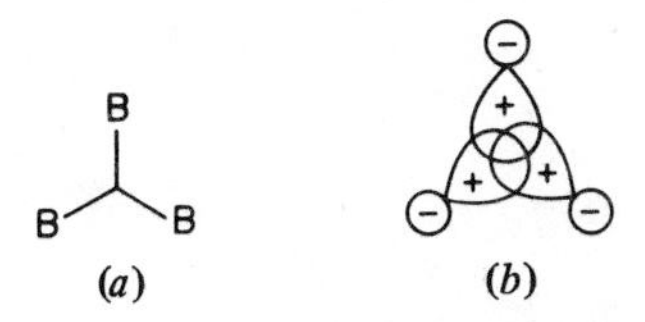

Figure 2.860 (*a*) The arrangement of atoms in closed three-center bonding; (*b*) the atomic orbitals involved in closed three-center bonding.

3 Hydrocarbons

CONTENTS

3.000 Hydrocarbons
Compounds containing only the elements of carbon and hydrogen.

3.010 Saturated Hydrocarbons
Hydrocarbons in which all atoms are bonded to each other by single (sigma) bonds: i.e., no multiple bonds are present.

3.020 Acyclic Saturated Hydrocarbons
Saturated hydrocarbons in which no rings are present.

3.030 Alkanes, C_nH_{2n+2}
Synonomous with acyclic, saturated hydrocarbons.

Example. The simplest alkane, CH_4, is methane. The tetrahedral arrangement of methane and one of its S_4 axes has been shown in Fig. 1.100*a* and *b*; all H—C—H angles are 109°28′. It is instructive to consider the geometry of a tetrahedron, Fig. 3.030*a*, in which the CH_4 molecule may be inscribed. When a tetrahedron is used to represent methane, it is understood that the four hydrogen atoms are located at the corners *a*, *b*, *c*, and *d* and that the carbon atom is exactly in the center of the solid figure. Each of the four faces of the tetrahedron is an equilateral triangle, characterized by 3-fold symmetry. Since there are four faces, methane possesses four C_3 axes, each one of which coincides with a C—H bond and penetrates the center of the opposite face. Thus, for example, one of the C_3 axes coincides with the C—H_d bond and passes through the center of face *acb* in Fig. 3.030*a*; another coincides with the C—H_a bond and passes

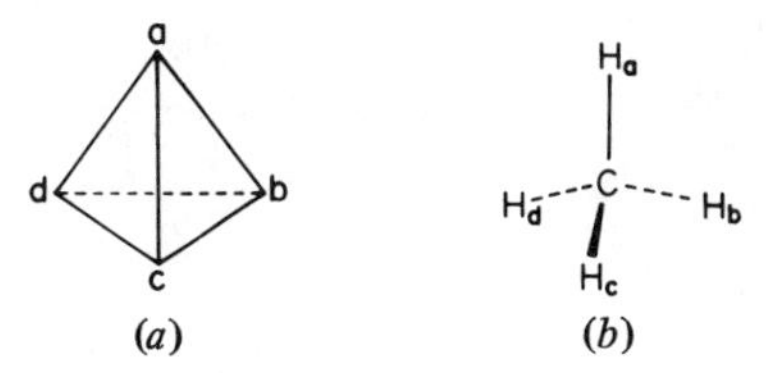

Figure 3.030 (*a*) The tetrahedron; (*b*) perspective drawing of methane.

through the center of face *bcd*. This latter axis may also be seen by drawing the structure of methane as in Fig. 3.030*b*.

3.040 Aliphatic Hydrocarbons

Synonymous with alkanes.

3.050 Paraffins

Synonymous with alkanes, but sometimes limited to those alkanes that are solid at room temperature.

3.060 Straight Chain Alkanes

Alkanes in which all carbon atoms except the terminal ones are bonded to two other carbon atoms only.

Examples. $CH_3CH_2CH_3$; $CH_3CH_2CH_2CH_2CH_2CH_3$. It is a common practice to represent carbon atoms in chains by means of points connected by straight lines. Thus these two compounds may be abbreviated as in Fig. 3.060*a* and *b*, respectively.

Figure 3.060 Carbon skeleton drawings of straight chain alkanes: (*a*) propane; (*b*) hexane.

3.070 Branched Chain Alkanes

Acyclic, saturated hydrocarbons in which at least one of the carbon atoms is attached to either three or four other carbon atoms.

Examples. 2-Methylpropane, Fig. 3.070*a*, and 3,3,4-trimethylhexane, Fig. 3.070*b*; these compounds may be abbreviated as shown in Fig. 3.070*c* and *d*, respectively.

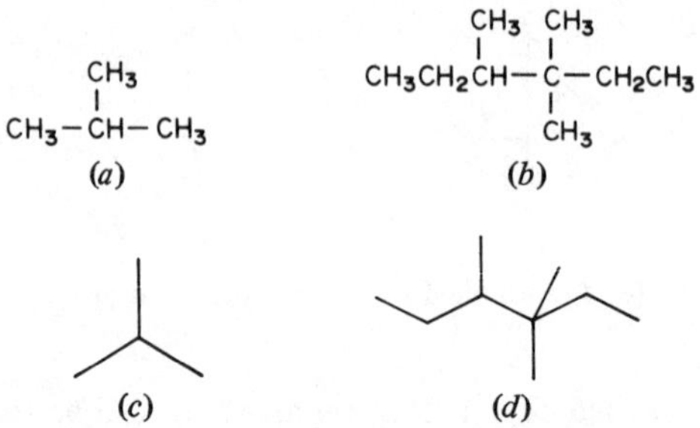

CH_3
$CH_3-CH-CH_3$
(*a*)

CH_3 CH_3
$CH_3CH_2CH-C-CH_2CH_3$
CH_3
(*b*)

(*c*)

(*d*)

Figure 3.070 Carbon skeleton drawings of branched chain alkanes: (*a*) 2-methylpropane; (*b*) 3,3,4-trimethylhexane. Abbreviated forms: (*c*) 2-methylpropane; (*d*) 3,3,4-trimethylhexane.

3.080 Primary, Secondary, Tertiary, and Quaternary Carbon Atoms

Carbon atoms singly bonded to only one other carbon atom; those singly bonded to two other carbons; those singly bonded to three other carbons; and those bonded to four other carbon atoms; respectively.

Example. In propane, $CH_3CH_2CH_3$, also Fig. 3.060*a*, there are two primary carbon atoms and one secondary one. In the compound shown in Fig. 3.070*b* there are five primary carbon atoms, two secondary, one tertiary, and one quaternary carbon atom. In the compound neopentane (2,2-dimethylpropane), Fig. 3.080, there are four primary and one quaternary carbon atoms; the 12 hydrogen atoms are all equivalent.

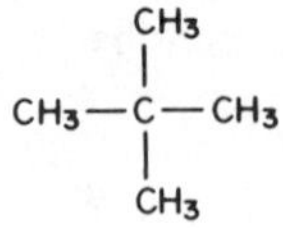

CH_3
CH_3-C-CH_3
CH_3

Figure 3.080 Neopentane.

3.090 Primary, Secondary, and Tertiary Hydrogen Atoms

The hydrogen atoms bonded to primary, secondary, and tertiary carbon atoms, respectively.

Example. 2-Methylbutane (isopentane), Fig. 3.090, has nine primary, two secondary, and one tertiary hydrogen atoms.

CH_3
$CH_3-CH-CH_2CH_3$

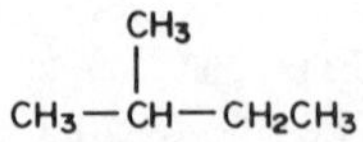

Figure 3.090 2-Methylbutane.

3.100 The IUPAC System of Nomenclature

The nomenclature approved by the organization called the International Union of Pure and Applied Chemistry. Committees of this organization work continuously on systematic schemes for naming all chemical compounds.

3.110 Homologs

Two compounds that differ only by a —CH_2— group.

Example. Propane, $CH_3CH_2CH_3$, and butane, $CH_3CH_2CH_2CH_3$; methanol, CH_3OH, and ethanol, CH_3CH_2OH, are alkane and alcohol homologs, respectively.

3.120 Homologous Series

A series of compounds that are homologs of each other.

3.130 Alkyl Groups

The group of atoms remaining after one of the hydrogen atoms is removed from an acyclic, saturated hydrocarbon. There will be as many different alkyl groups as there are nonequivalent hydrogen atoms in the corresponding acyclic, saturated hydrocarbon. Cycloalkyls are derived from cycloalkanes (Sect. 3.190).

Example. All the hydrogen atoms in methane and ethane are equivalent; therefore only one alkyl group can be derived from them and these are called, respectively, methyl (CH_3—) and ethyl (CH_3CH_2—) groups. In propane there are two different sets of hydrogen atoms, giving rise to two different alkyl groups, namely, $CH_3CH_2CH_2$— (*n*-propyl) and CH_3—CH—CH_3 (isopropyl). With the five carbon, straight chain, saturated hydrocarbon pentane, there are three different possible alkyl groups, Fig. 3.130.

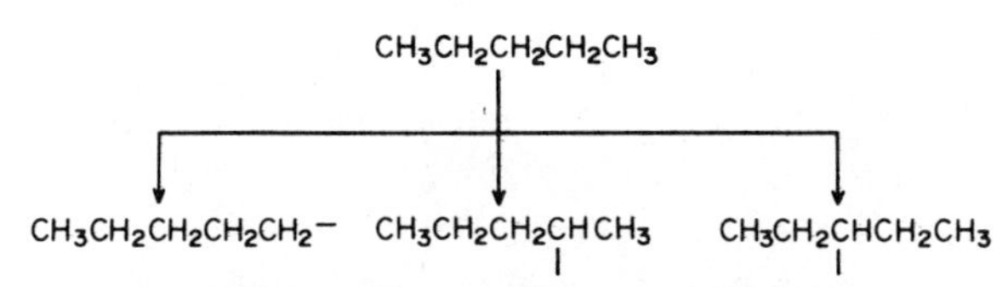

Figure 3.130 Alkyl groups derived from pentane.

3.140 Normal Alkyl Groups

The alkyl group derived by removing a hydrogen atom from the terminal methyl group of a straight chain alkane, abbreviated *n*.

Example. $CH_3CH_2CH_2$— is *n*-propyl.

3.150 Isoalkanes

A branched chain alkane having a single methyl branch on the penultimate carbon atom.

Example. Isopentane, Fig. 3.090.

3.160 Isoalkyl Group

The group derived by removal of a hydrogen atom from the methyl group at the nonbranched end of an isoalkane.

Examples. The general structure shown in Fig. 3.160*a*, where $n = 0, 1, 2, \ldots, n$; isopropyl group, Fig. 3.160*b*; isobutyl, Fig. 3.160*c*.

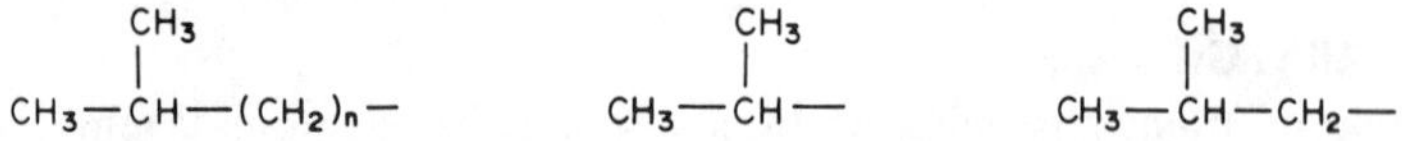

$$CH_3-\overset{\overset{\displaystyle CH_3}{|}}{C}H-(CH_2)_n- \qquad CH_3-\overset{\overset{\displaystyle CH_3}{|}}{C}H- \qquad CH_3-\overset{\overset{\displaystyle CH_3}{|}}{C}H-CH_2-$$

(*a*) (*b*) (*c*)

Figure 3.160 (*a*) The isoalkyl group; (*b*) isopropyl; (*c*) isobutyl.

3.170 Methylene Group

The triatomic group consisting of one carbon atom and two attached hydrogens, i.e., $-CH_2-$ or $=CH_2$. As the name implies, the methylene group may be considered to be derived from the methyl group, CH_3- (which is itself derived from methane, by removal of a hydrogen), by removal of a hydrogen atom to give a doubly hydrogen deficient carbon atom, hence the suffix *ene*.

Examples. CH_2Cl_2 is frequently called methylene chloride, although dichloromethane is preferred; methylenecyclohexane, Fig. 3.170.

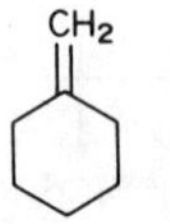

CH_2

Figure 3.170 Methylenecyclohexane.

3.180 Methine Group

The diatomic group consisting of one carbon atom and one attached hydrogen, i.e., $-\underset{|}{C}H-CH-$.

Example. Isopentane, Fig. 3.090, contains one methine group.

3.100 The IUPAC System of Nomenclature
The nomenclature approved by the organization called the International Union of Pure and Applied Chemistry. Committees of this organization work continuously on systematic schemes for naming all chemical compounds.

3.110 Homologs
Two compounds that differ only by a $—CH_2—$ group.

Example. Propane, $CH_3CH_2CH_3$, and butane, $CH_3CH_2CH_2CH_3$; methanol, CH_3OH, and ethanol, CH_3CH_2OH, are alkane and alcohol homologs, respectively.

3.120 Homologous Series
A series of compounds that are homologs of each other.

3.130 Alkyl Groups
The group of atoms remaining after one of the hydrogen atoms is removed from an acyclic, saturated hydrocarbon. There will be as many different alkyl groups as there are nonequivalent hydrogen atoms in the corresponding acyclic, saturated hydrocarbon. Cycloalkyls are derived from cycloalkanes (Sect. 3.190).

Example. All the hydrogen atoms in methane and ethane are equivalent; therefore only one alkyl group can be derived from them and these are called, respectively, methyl ($CH_3—$) and ethyl ($CH_3CH_2—$) groups. In propane there are two different sets of hydrogen atoms, giving rise to two different alkyl groups, namely, $CH_3CH_2CH_2—$ (*n*-propyl) and $CH_3—\underset{|}{C}H—CH_3$ (isopropyl). With the five carbon, straight chain, saturated hydrocarbon pentane, there are three different possible alkyl groups, Fig. 3.130.

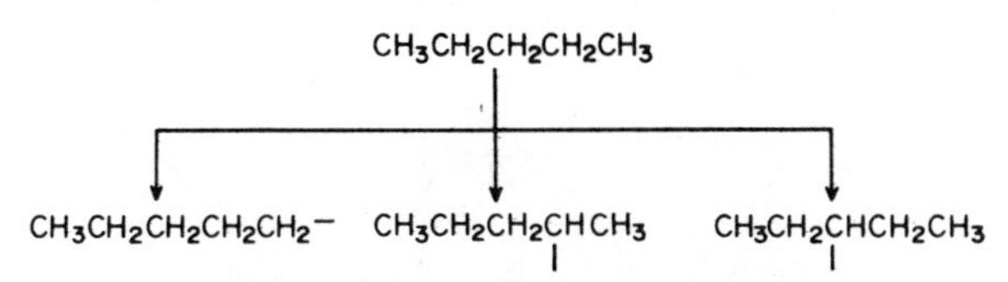

Figure 3.130 Alkyl groups derived from pentane.

3.140 Normal Alkyl Groups
The alkyl group derived by removing a hydrogen atom from the terminal methyl group of a straight chain alkane, abbreviated *n*.

Example. $CH_3CH_2CH_2—$ is *n*-propyl.

3.150 Isoalkanes

A branched chain alkane having a single methyl branch on the penultimate carbon atom.

Example. Isopentane, Fig. 3.090.

3.160 Isoalkyl Group

The group derived by removal of a hydrogen atom from the methyl group at the nonbranched end of an isoalkane.

Examples. The general structure shown in Fig. 3.160*a*, where $n = 0, 1, 2, \ldots, n$; isopropyl group, Fig. 3.160*b*; isobutyl, Fig. 3.160*c*.

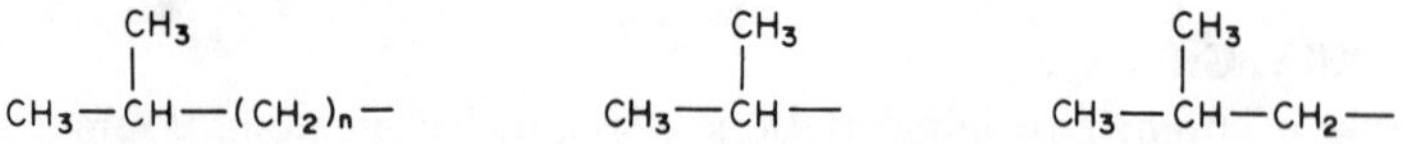

$$CH_3-\underset{}{\overset{CH_3}{\overset{|}{C}}}H-(CH_2)_n- \qquad CH_3-\overset{CH_3}{\overset{|}{C}}H- \qquad CH_3-\overset{CH_3}{\overset{|}{C}}H-CH_2-$$

(*a*) (*b*) (*c*)

Figure 3.160 (*a*) The isoalkyl group; (*b*) isopropyl; (*c*) isobutyl.

3.170 Methylene Group

The triatomic group consisting of one carbon atom and two attached hydrogens, i.e., $-CH_2-$ or $=CH_2$. As the name implies, the methylene group may be considered to be derived from the methyl group, CH_3- (which is itself derived from methane, by removal of a hydrogen), by removal of a hydrogen atom to give a doubly hydrogen deficient carbon atom, hence the suffix *ene*.

Examples. CH_2Cl_2 is frequently called methylene chloride, although dichloromethane is preferred; methylenecyclohexane, Fig. 3.170.

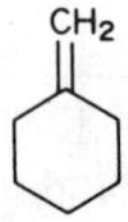

CH_2

Figure 3.170 Methylenecyclohexane.

3.180 Methine Group

The diatomic group consisting of one carbon atom and one attached hydrogen, i.e., $-\underset{|}{C}H-CH-$.

Example. Isopentane, Fig. 3.090, contains one methine group.

3.190 Cycloalkanes, C_nH_{2n}
Saturated hydrocarbons containing one ring.

Examples. Cyclobutane, Fig. 3.190*a*; 1,1,2,2-tetramethylcyclobutane, Fig. 3.190*b*.

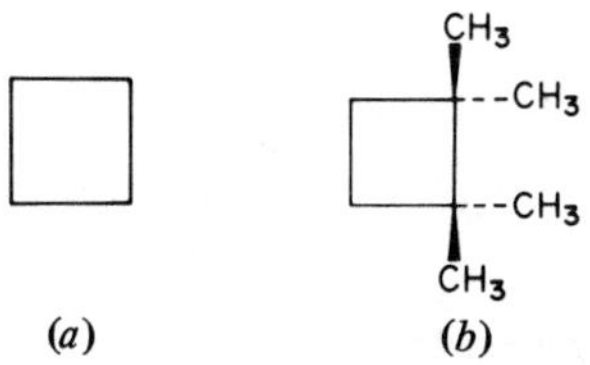

(*a*) (*b*)

Figure 3.190 Cycloalkanes: (*a*) cyclobutane; (*b*) 1,1,2,2-tetramethylcyclobutane.

3.200 Alicyclic Compounds
Synonymous with cycloalkanes.

3.210 Olefins
Hydrocarbons containing a carbon-carbon double bond.

3.220 Alkenes
Synonymous with olefins.

Example. 2,3-Dimethyl-2-butene, commonly called tetramethylethylene, Fig. 3.220. It is sometimes useful to think of a C=C as a two-membered ring.

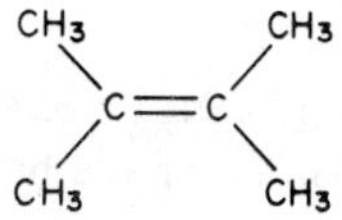

Figure 3.220 2,3-Dimethyl-2-butene.

3.230 Vinyl Group, $CH_2{=}CH{-}$
The group remaining after one hydrogen is removed from ethylene.

Example. 4-Vinylcyclohexene, Fig. 3.230.

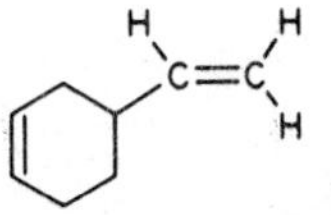

Figure 3.230 4-Vinylcyclohexene.

3.240 Cycloalkenes
Cyclic hydrocarbons containing a double bond.

Example. 3,3-Dimethylcyclohexene, Fig. 3.240.

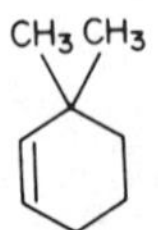

Figure 3.240 3,3-Dimethylcyclohexene.

3.250 Bridged Cycloalkane Hydrocarbons
Cyclic hydrocarbons containing one (or more) pairs of carbon atoms common to two (or more) rings.

Example. Bicyclo[2.2.2] octane, Fig. 3.250.

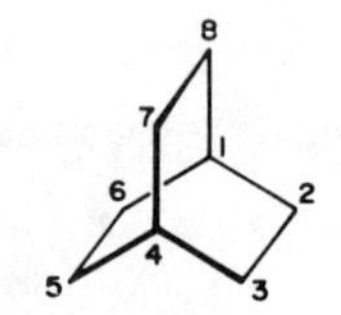

Figure 3.250 Bicyclo[2.2.2] octane.

3.260 Bridgehead Carbon Atoms
The carbon atoms common to two or more rings.

Example. The carbon atoms numbered 1 and 4, Fig. 3.250, are bridgehead carbons; in naming such compounds one of the bridgehead carbon atoms is always numbered 1.

3.270 Bridges
A bond or atom or chain of atoms joining bridgehead carbons.

Examples. In Fig. 3.250 there are three bridges, each consisting of two carbon atoms. In *trans*-bicyclo[4.4.0] decane (also called *trans*-decahydronaphthalene or

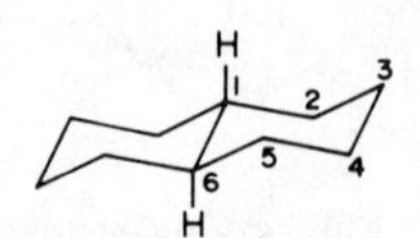

Figure 3.270 *trans*-Bicyclo[4.4.0] decane.

decalin), Fig. 3.270, there are also three bridges but in this case one of them consists of a bond only.

3.280 Angular Groups

Atoms or groups attached to bridgehead carbons.

Example. The hydrogen atoms at positions 1 and 6 in Fig. 3.270. Most commonly this nomenclature is applied to angular methyl groups in many naturally occurring compounds.

3.290 Bicyclic Compounds

Compounds containing one pair of bridgehead carbon atoms; these have the general structure shown in Fig. 3.290*a*.

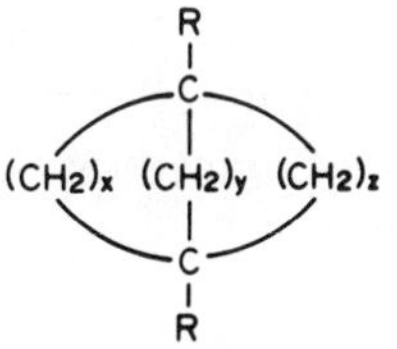

R
C
$(CH_2)_x$ $(CH_2)_y$ $(CH_2)_z$
C
R

(*a*)

Example. Bicyclo[1.1.0]butane, Fig. 3.290*b*. The prefix bicyclo is part of the **von Baeyer system** of naming bridged alicyclic hydrocarbons. In this system the number of rings is equal to the minimum number of bond scissions required to convert the bridged ring system into an acyclic hydrocarbon having the same number of carbon atoms. In our example two cuts or scissions does this. Saturated, bicyclic hydrocarbons have the empirical formula C_nH_{2n-2}.

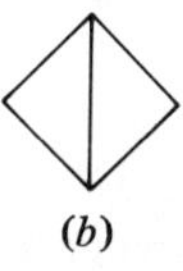

(*b*)

Figure 3.290 (*a*) General structure of bicyclo compounds; (*b*) bicyclo[1.1.0]butane.

3.300 Tricyclic Compounds

Compounds possessing either four bridgehead carbons each common to two rings or a pair of bridgehead carbons common to three rings.

Examples. **Adamantane**, Fig. 3.300*a*, the tetrahedral molecule whose systematic name is tricyclo[$3.3.1^{1,5}.1^{3,7}$]decane. The four bridgehead carbon atoms are 1, 3, 5, and 7. The derivation of the systematic name is apparant if one first

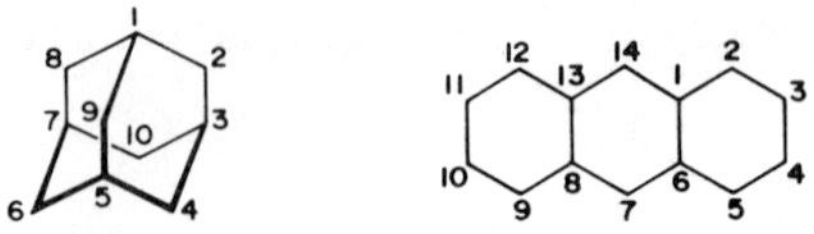

(*a*) (*b*)

Figure 3.300 (*a*) Adamantane; (*b*) perhydroanthracene.

removes (figuratively, of course) carbon atom number 10, whereupon carbons 1 and 5 are the only bridgehead carbons in a bicyclo[3.3.1] system. The addition of carbon atom 10 between atoms 3 and 7 gives the third ring, which contains a one-carbon bridge. The tricyclo nature of a saturated compound can be deduced from the empirical formula $C_{10}H_{16}$. In adamantane, bond scissions between atoms 1 and 9, 4 and 5, and 7 and 8 produce an acyclic compound, hence adamantane is tricyclic. A compound such as Fig. 3.300*b* is also technically a tricyclic compound (tricyclo[8.4.$0^{1,6}$.$0^{8,13}$]tetradecane), but when such compounds have an aromatic counterpart, all compounds having this nucleus are named as derivatives of the parent aromatic; e.g., Fig. 3.300*b* is tetradecahydroanthracene or perhydroanthracene.

3.310 Spirans

Compounds containing one carbon atom common to two rings.

Example. The compound shown in Fig. 3.310 is spiro[3.3]heptane; the two small rings joined by the common tetrahedral carbon atom are at right angles to each other.

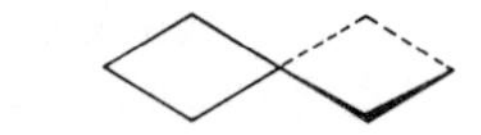

Figure 3.310 Spiro[3.3]heptane.

3.320 Conjugated Dienes and Polyenes

Hydrocarbons possessing two carbon-carbon double bonds separated by a single bond are conjugated dienes. Compounds possessing a system of alternating carbon-carbon double and single bonds are conjugated polyenes. Conjugation requires overlap of the *p* orbitals involved.

Examples. 1,3-Butadiene, Fig. 3.320*a*, is the simplest example of a conjugated diene. 1,3,5-Hexatriene, Fig. 3.320*b*, is a conjugated triene. However, the definition of a conjugated polyene excludes the six-membered cyclic triene, 1,3,5-cyclohexatriene, Fig. 3.320*c*, (and related compounds) because it has special properties uncharacteristic of an acyclic conjugated triene, hence, it is put in a different class of compounds called aromatic hydrocarbons (Sect. 3.470).

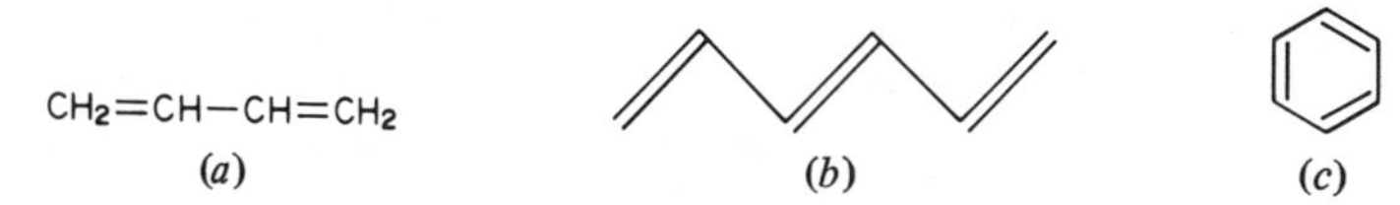

Figure 3.320 (*a*) 1,3-Butadiene; (*b*) *trans* (or *E*)-1,3,5-hexatriene; (*c*) 1,3,5-cyclohexatriene (benzene).

3.330 Exocyclic and Endocyclic Double Bonds

Double bonds connected to, hence external to, a ring are exocyclic, and double bonds that are part of a ring are endocyclic.

Examples Methylenecyclohexane, Fig. 3.170, possesses an exocyclic double bond. In the compound shown in Fig. 3.330 the two double bonds are exocyclic to ring *A* and endocyclic in ring *B*.

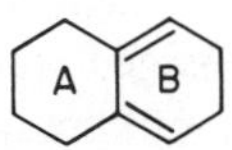

Figure 3.330 Two excocyclic double bonds to ring *A* and two endocyclic bonds in ring *B*.

3.340 Cumulenes

Compounds containing two or more successive double bonds.

Example. Figure 3.340 shows 1,2,3-butatriene. The internal carbon atoms can be thought of as spiran atoms, hence this compound may be thought of as having three two-membered rings, each ring at right angles to the ring(s) to which it is attached.

$$CH_2{=}C{=}C{=}CH_2$$

Figure 3.340 1,2,3-Butatriene.

3.350 Alkynes

Compounds containing a carbon-carbon triple bond.

Example. 5-Methyl-2-hexyne, Fig. 3.350.

$$CH_3C{\equiv}CCH_2\underset{}{\overset{\overset{\displaystyle CH_3}{|}}{C}}HCH_3$$

Figure 3.350 5-Methyl-2-hexyne.

3.360 Resonance Structures

Two or more structures of the same compound having identical geometry, possessing the same number of paired electrons, but differing in the pairing arrangement of these electrons. These structures are conventionally shown as related to each other by a single double-headed arrow (↔) to emphasize that such structures have no physical reality or independent existence and are not different substances in equilibrium.

Examples. Benzene is the classic example used to illustrate resonance structures. Two resonance structures of benzene are shown in Fig. 3.360*a*; these two structures are completely equivalent. The three structures of CO_3^{2-} shown in Fig. 3.360*b* are likewise equivalent, i.e., indistinguishable. The importance of resonance structures becomes apparent in explaining that each C—C bond in benzene is intermediate between a single and double bond, as is each C—O bond in CO_3^{2-}. Resonance structures need not be equivalent; three of the resonance structures of butadiene shown in Fig. 3.360*c* indicate this. The first structure is a lower energy structure than the other two, which are charge-separated equivalent species. The first structure is said to make a larger contribution to the ground state structure than either of the charged structures. Thus the uncharged structure is the single best representation of butadiene, although the fact that, experimentally, butadiene shows some double bond character between atoms 2 and 3 indicates that the charged structures also make a contribution to the ground state.

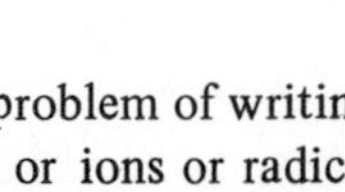

(*a*) 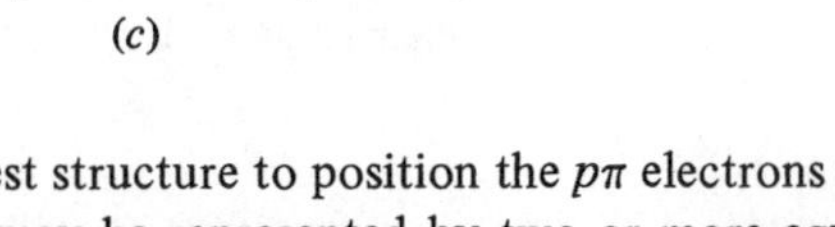 (*b*)

$$CH_2{=}CH{-}CH{=}CH_2 \leftrightarrow \overset{+}{C}H_2{-}CH{=}CH{-}\overset{-}{C}H_2 \leftrightarrow \overset{-}{C}H_2{-}CH{=}CH{-}\overset{+}{C}H_2$$

(*c*)

The problem of writing one best structure to position the $p\pi$ electrons in compounds or ions or radicals that may be represented by two or more equivalent structures is frequently solved by the use of broken lines connecting the atoms over which the $p\pi$ electrons are delocalized. Thus for benzene the single structure, Fig. 3.360*d*, and for CO_3^{2-} the single structure, Fig. 3.360*e*, are representations showing π electron distribution. In the case of a cyclic sextet of delocalized π electrons a solid circle inside the ring, as shown for benzene, Fig. 3.360*f*, is used. Originally the circle notation referred to a sextet of electrons but now it has been extended to indicate any monocyclic aromatic system containing $(4n + 2)$ π electrons as well as to indicate the electrons in the rings of polycyclic aromatic compounds.

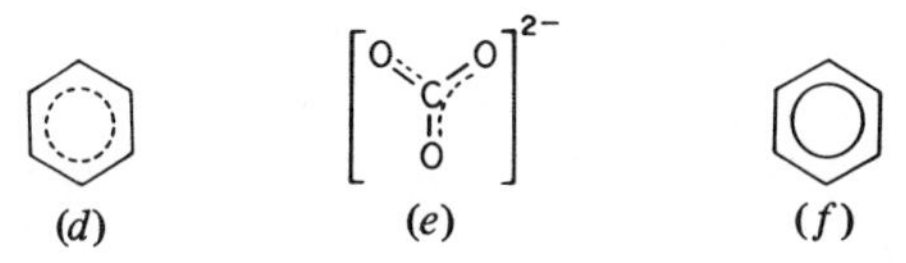

Figure 3.360 Resonance structures: (*a*) benzene; (*b*) $(CO_3)^{2-}$; (*c*) butadiene; (*d*) broken line representation for benzene; (*e*) broken line representation for $(CO_3)^{2-}$; (*f*) circle notation for benzene.

3.370 Kekulé Structures (F. A. Kekulé, 1829–1896)

The structures of cyclic conjugated systems represented by conventional alternating single and double bonds.

Examples. The two Kekulé resonance structures of benzene are shown in Fig. 3.360*a*.

3.380 Dewar Structures (J. Dewar, 1842–1923)

Resonance structures of benzene (or other aromatics) containing a single "long" bond between opposite atoms.

Examples. The three possible Dewar structures of benzene are shown in Fig. 3.380.

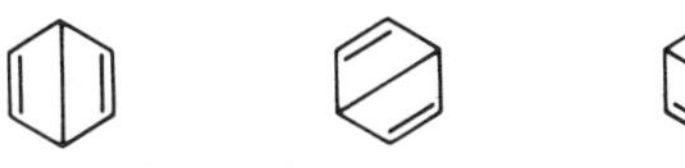

Figure 3.380 Dewar structures of benzene.

3.390 Canonical Forms

Almost, but not quite, synonymous with resonance structures. The mathematical requirements (e.g., orthogonality) for a canonical set of structures of a cyclic compound preclude structures having bonds that intersect.

Example. The rules for drawing proper resonance forms when applied to the π system of benzene permit not only the three Dewar structures, but also the unfamiliar crossed structure, Fig. 3.390. However, such a structure is not part of a canonical set.

Figure 3.390 A resonance structure that is not a canonical structure.

3.400 Essential Single Bonds

A single bond in a conjugated system which can be written only as a single bond in all resonance structures not involving charge-separated species.

Example. The single bond in butadiene can be written as a double bond only by involving charge-separated structures (or long bonds), as shown in Fig. 3.360*c*. On the other hand the single bonds in either of the benzene resonance structures, Fig. 3.360*a*, are not essential single bonds because the single bonds in one structure become a double bond in the other noncharged structure.

3.410 Essential Double Bonds

Double bonds in a conjugated system that can be written only as double bonds in all resonance structures not involving charge-separated species.

Example. The double bonds in the uncharged butadiene structure, Fig. 3.360*c*, are essential double bonds, whereas those in the benzene structures, 3.360*a*, are nonessential double bonds.

3.420 Resonance Energy

Additional stabilization energy over that calculated for a single resonance structure. The resonance stabilization arises as a result of the delocalization of the electrons over a conjugated system and is essentially equivalent to the delocalization energy (Sect. 2.740).

Example. Benzene is considerably more stable than one would expect if each of its three double bonds were individually similar to the one in cyclohexene. Experimentally, the difference can be shown, e.g., by measuring the heat of hydrogenation of benzene to cyclohexane (208.4 kJ mol^{-1} or 49.8 kcal mol^{-1}) and comparing it with the heat liberated by the hydrogenation of three moles of cyclohexene [361 kJ $(3\ mol)^{-1}$ or 86.4 kcal $(3\ mol)^{-1}$]. The difference in these heats (36.6 kcal) is equivalent to the resonance energy of 1 mole of benzene. Usually the greater number of equivalent structures (structures possessing the same number of single and double bonds) that can be written for a compound, the greater is its resonance energy.

3.430 Cross-Conjugated Hydrocarbons

Hydrocarbons possessing three double bonds, two of which are conjugated to the third but are not formally conjugated to each other.

Example. 3-Methylene-1,4-pentadiene, Fig. 3.430. The central double bond is individually conjugated to each of the other terminal bonds but the two end double bonds are not formally conjugated. The classical resonance structures

```
                CH2
                ||
CH2=CH—C—CH=CH2
```

Figure 3.430 A cross-conjugated polyene.

would show only the interaction of the conjugated double bonds, whereas in molecular orbital theory the π electrons are delocalized over the entire six-carbon-atom system.

3.440 Hyperconjugation

Overlap between σ orbitals and π (or p) orbitals.

Example. The overlap of the C—H σ bonds of the methylene group with the π bonds of cyclopentadiene, Fig. 3.440a. This overlap is more readily recognized by combining the two methylene hydrogen atoms into the combinations $[H_1(s) + H_2(s)]$ and $[H_1(s) - H_2(s)]$, called group orbitals; this latter orbital has the same symmetry as the π system, hence can overlap, Fig. 3.440b.

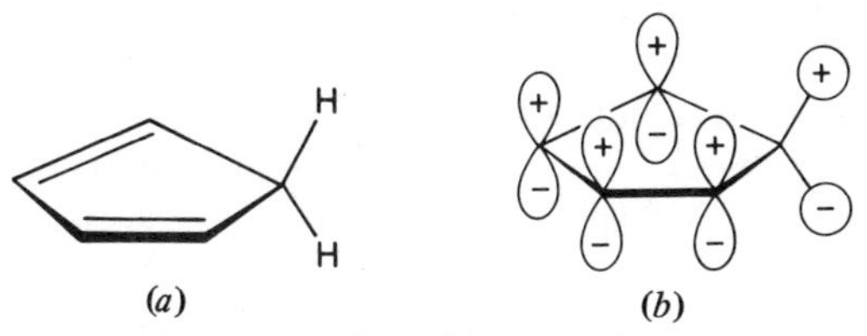

Figure 3.440 (a) Cyclopentadiene; (b) σ-π overlap in hyperconjugation.

3.450 Benzenoid

Having the electronic character of benzene.

Example. Benzene has physical and chemical properties entirely different than those expected of a conjugated cyclic triene. The special properties are reflected in equivalent C—C bond distances, the ultraviolet spectrum, and in its relative chemical inertness. These properties arise because the $6p\pi$ electrons are delocalized over all carbon atoms of the ring.

3.460 Hückel's Rule (E. Hückel, 1896-)

The rule that states that a planar monocyclic system possessing $(4n + 2)\,p\pi$ electrons delocalized over the ring, where n is any integer, 0, 1, 2, 3, . . . , possesses unusual thermodynamic stability; that is, it will be aromatic (*vide infra*).

Examples. Benzene, cyclopentadienyl anion, cyclopropenyl cation, cycloheptatrienyl cation, Fig. 3.460a–d, respectively, are compounds and ions that

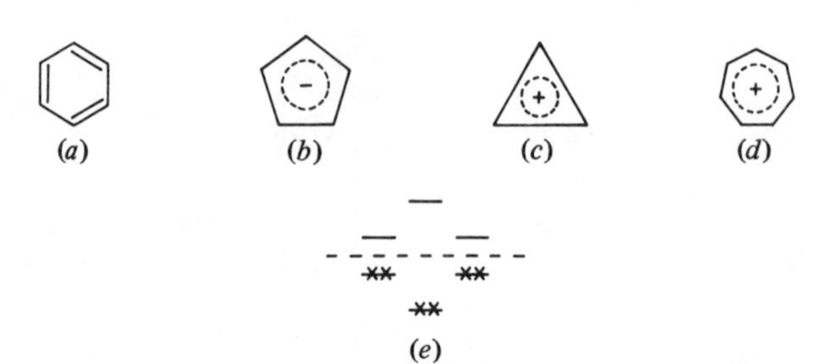

Figure 3.460 Some $4n + 2$ π-electron (Hückel) species: (*a*) benzene; (*b*) cyclopentadienyl anion; (*c*) cyclopropenyl cation; (*d*) cycloheptatrienyl cation; (*e*) the MOED for benzene.

obey this rule (n = 1, 1, 0, 1, respectively), hence are characterized by large resonance energies. Although the rule is strictly applicable only to monocyclic carbocycles, it has also been applied to polycyclic compounds. The rule arises from the molecular orbital treatment of the π bonding in a monocyclic conjugated system which gives rise to one low-lying orbital accommodating two electrons and then pairs of degenerate orbitals accommodating four electrons. The MOED (Sect. 2.550) for benzene is shown in Fig. 3.460*e*. (See also Sect. 3.680.)

3.470 Aromatic Hydrocarbons (Arenes)

The class of hydrocarbons, of which benzene is the first member, consisting of assemblages of cyclic conjugated carbon atoms and characterized by large resonance energies.

Examples. Naphthalene, Fig. 3.470*a*; phenanthrene, Fig. 3.470*b*; anthracene, 3.470*c*; and benz[*a*]anthracene, 3.470*d*; are all aromatic hydrocarbons. When the aromatic hydrocarbons consist of six-membered rings with no more than two carbon atoms common to two rings, the number of $p\pi$ electrons obeys Hückel's rule.

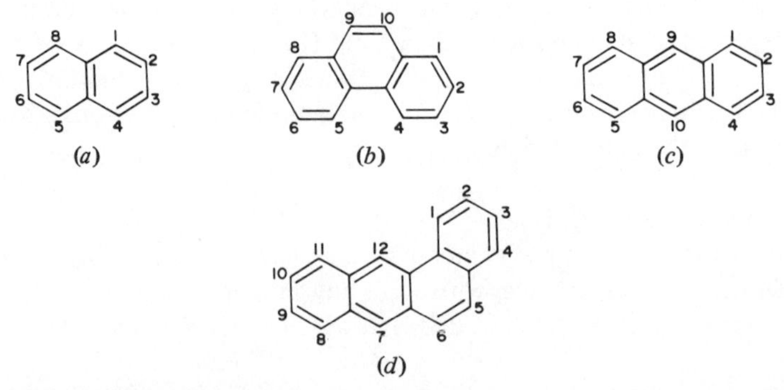

Figure 3.470 Aromatic hydrocarbons: (*a*) naphthalene; (*b*) phenanthrene; (*c*) anthracene; (*d*) benz[*a*]anthracene.

3.480 Aryl Group

Class name for the aromatic grouping that remains after the conceptual removal of a hydrogen from the ring position of an arene nucleus (strictly, from a benzene or substituted benzene nucleus but extended to all arenes by common usage).

3.490 *ortho*, *meta*, and *para*

Prefixes used to distinguish the three possible disubstitution patterns of benzene; *ortho* meaning substituents on adjacent carbons, (1,2-); *meta* with one intervening open position (1,3-); and *para* with substituents opposite each other (1,4-); abbreviated *o*-, *m*-, and *p*-, respectively.

Examples. *o*-Xylene, *m*-xylene, *p*-xylene, Figs. 3.490*a–c*, respectively.

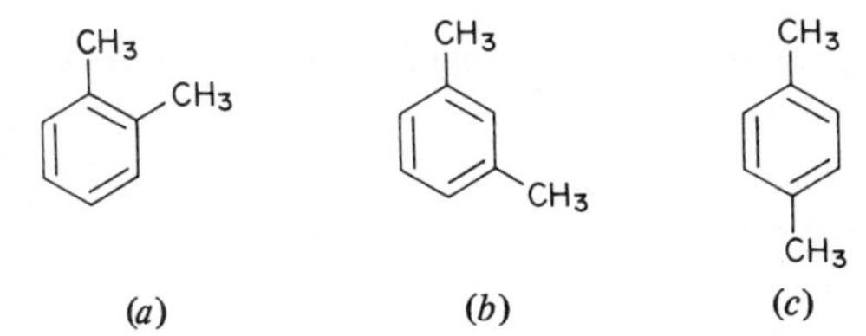

Figure 3.490 The three xylenes: (*a*) *ortho*; (*b*) *meta*; (*c*) *para*.

3.500 Aromaticity

Characterized by aromatic character, i.e., possessing the properties usually associated with compounds that obey Hückel's rule.

3.510 Kekulé Ring (F. A. Kekulé, 1829–1896)

A six-membered ring containing three conjugated double bonds.

Example. The three most important resonance structures of naphthalene are shown in Fig. 3.510*a–c*. It will be noted that in Fig. 3.510*a* rings A and B are both Kekulé rings, but in structure 3.510*b* only ring A, and in structure 3.510*c* only ring B, is Kekulé. In placing double bonds in polycyclic (also called polynuclear) aromatic compounds, that structure which can be written with the maximum number of Kekulé rings is the preferred structure. Thus of the three reso-

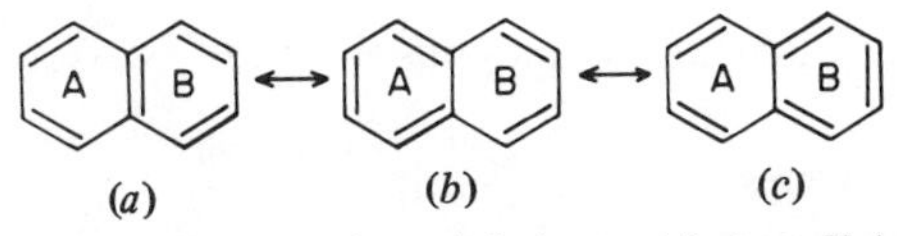

Figure 3.510 Resonance structures of naphthalene: (*a*) two Kekulé rings; (*b, c*) one Kekulé and one quinoidal ring.

nance structures of naphthalene Fig. 3.510*a* is the preferred disposition of double bonds.

3.520 Quinoidal Ring

A fused six-membered ring in a polycyclic aromatic compound that possesses two conjugated double bonds inside the ring (endocyclic) and two conjugated double bonds exocyclic to the ring.

Examples. Ring B in Fig. 3.510*b* and ring A in Fig. 3.510*c*. The name is derived from the formal relationship to a class of compounds called quinones, e.g., *ortho*- or 1,2-benzoquinone, Fig. 3.520.

Figure 3.520 *ortho*-Benzoquinone (or 1,2-benzoquinone).

3.530 Fused Aromatic Rings

Aromatic rings that have two carbon atoms common to two or more rings.

Example. All the compounds shown in Fig. 3.470. Pyrene, Fig. 3.530*a*, has four fused rings and here the central carbon atoms are each common to three rings. Biphenyl, Fig. 3.530*b*, is not a fused ring aromatic compound.

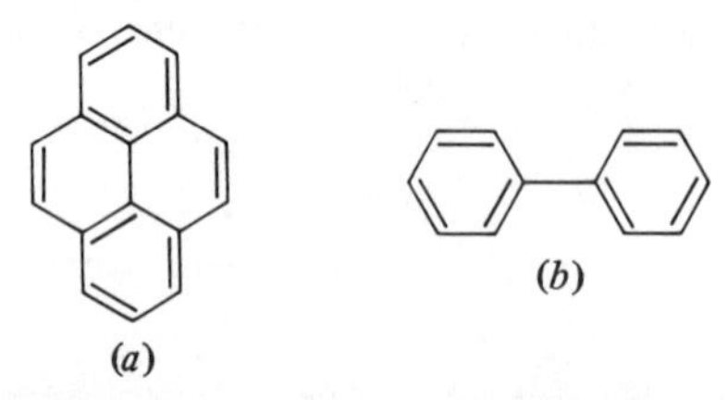

Figure 3.530 (*a*) Pyrene; (*b*) biphenyl.

3.540 Cata-Condensed Polycyclic Aromatic Compounds

These compounds have no carbon atom belonging to more than two rings, thus implying that every carbon atom is on the periphery and that there are no "internal" carbon atoms.

Examples. All the polycyclic compounds shown in Fig. 3.470.

3.550 Polycyclic Acenes

Cata-condensed polycyclic compounds in which all the rings are fused in a linear array.

Examples. Anthracene, Fig. 3.470*c*, and naphthacene, Fig. 3.550, (also called tetracene) are cata-condensed acenes.

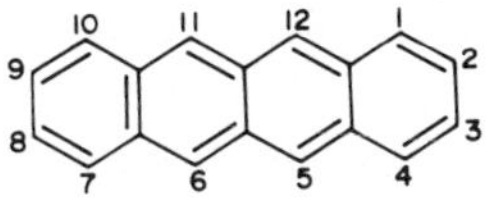

Figure 3.550 Naphthacene.

3.560 Polycyclic Phenes

Cata-condensed polycyclics containing an angular or bent system of rings.

Examples. Phenanthrene, Fig. 3.470*b*, and benz[*a*]anthracene, Fig. 3.470*d*, are cata-condensed phenes.

3.570 Peri-Condensed Polycyclic Aromatic Compounds

Polycyclic aromatic compounds in which some carbon atoms are common to three or more rings.

Examples. Pyrene, Fig. 3.530*a*; perylene, Fig. 3.570*a*; and anthanthrene, Fig. 3.570*b*, respectively. Such aromatic systems need not obey the Hückel Rule.

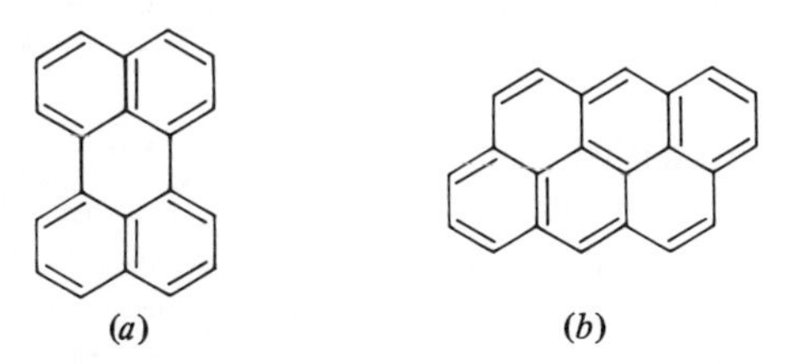

Figure 3.570 (*a*) Perylene; (*b*) anthanthrene.

3.580 The Ring Index

A book published (1960) by the American Chemical Society and written by A. M. Patterson, L. T. Capell, and D. F. Walker which contains a systematic listing of the structures and names of practically all known polycyclic ring systems.

3.590 Meso Carbon Atoms

In polycyclic aromatics, the carbon atoms on the middle ring of a linear three-ring system that are not shared with other rings. (See also Sect. 13.890).

Examples. The 9,10 carbon atoms of anthracene, Fig. 3.470*c*, and the 5,6,11,12 carbon atoms of tetracene, Fig. 3.550.

3.600 K, L and Bay Regions of Polycyclic Hydrocarbons

Positions in polycyclic compounds as shown in Fig. 3.600, which have special significance in correlations of the structures of polycyclic aromatic compounds with their carcinogenic properties. The L region corresponds to the meso positions in anthracene, and the K region to the 9,10 position in phenanthrene, Fig. 3.470*b*. Rather recently, a different molecular area of polycyclic hydrocarbons, the bay region, has been implicated in structure-carcinogenic activity relationships. The **bay region** is defined as a concave exterior region of a polycyclic aromatic hydrocarbon bordered by three phenyl rings, at least one of which is a terminal ring. Such a region in benz[*a*]anthracene is shown in Fig. 3.600 along with the K and L regions in this molecule.

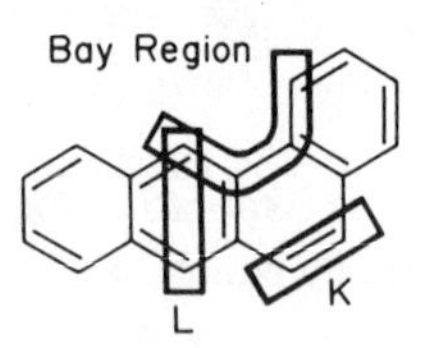

Figure 3.600 The K, L, and bay regions in polynuclear aromatics (PNA).

3.610 Alternant Hydrocarbon

Any conjugated hydrocarbon that does not include an odd-membered carbon ring system.

Examples. 1,3,5-Hexatriene, Fig. 3.320*b*; benzene, Fig. 3.460*a*; naphthalene, Fig. 3.470*a*; phenanthrene, Fig. 3.470*b*. This classification of hydrocarbons is significant in molecular orbital theory because alternant hydrocarbons possess complementary bonding and antibonding orbitals; i.e., the bonding levels and antibonding levels have a mirror image relationship. Recall the MOED for benzene, Fig. 3.460*e*.

3.620 Even Alternant Hydrocarbons

An alternant system with an even number of carbon atoms.

Examples. The examples given in Sect. 3.610 are all even alternants.

3.630 Odd Alternant Hydrocarbons

Alternant hydrocarbons with an odd number of carbon atoms.

Examples. The allyl and benzyl cations, Fig. 3.630*a* and *b*, respectively, are odd alternants.

Figure 3.630 Odd alternant hydrocarbons: (*a*) allyl cation; (*b*) benzyl cation.

3.640 Nonalternant Hydrocarbons

Hydrocarbons containing odd-membered conjugated rings.

Examples. Azulene, fulvene, and fluoranthene, Fig. 3.640*a*, *b*, and *c*, respectively. A more fundamental distinction between alternant and nonalternant hydrocarbons is the property that in the former all atoms in the molecule can be divided into two classes or sets, one starred and the other unstarred, such that no two atoms of the same set are bonded together. This procedure is shown for the even alternant hydrocarbon naphthalene and for the odd alternant benzyl ion in Fig. 3.640*d* and 3.640*e*, respectively. In the starring procedure for nonalternant hydrocarbons, one will always end up with either bonded starred or bonded unstarred atoms, as shown for azulene, Fig. 3.640*a*, where there is a bond between two starred atoms, or Fig. 3.640*f*, where there is a bond between unstarred atoms.

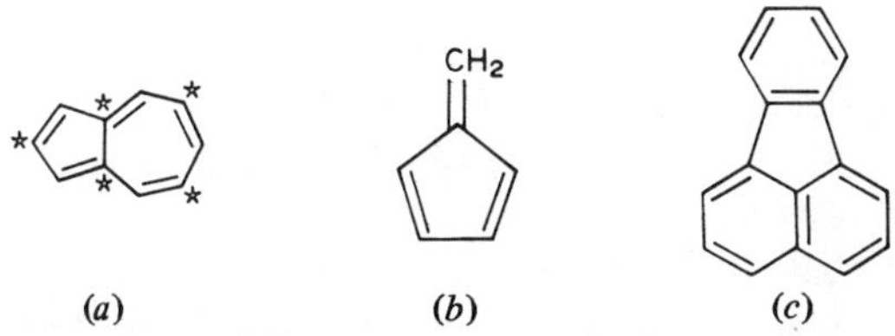

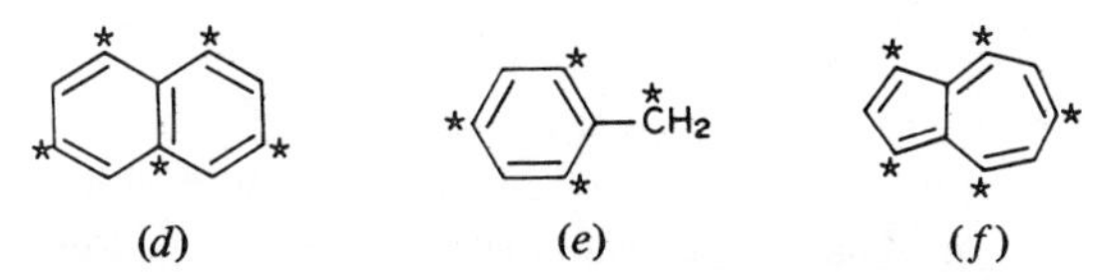

Figure 3.640 Classification of conjugated hydrocarbons: (*a–c*) nonalternant hydrocarbons; (*d*, *e*) starring procedure for alternants; (*f*, *a*) its failure in nonalternants.

3.650 Antiaromatic Compounds

Conjugated cyclic hydrocarbons in which π electron delocalization results in an *increase* in energy (destabilization); monocyclic systems with $4n$ π electrons, hence also designated "Hückel-antiaromatic" in contrast to Hückel aromatic systems, which possess $(4n + 2)$ π electrons.

Examples. Cyclobutadiene and cyclopropenyl anion, Fig. 3.650*a* and *b*, respectively. The delocalization energy of 1,3-butadiene is .472β (Sect. 2.730). The π electronic energy of cyclobutadiene is $4\alpha + 4\beta$, corresponding to zero delocalization energy. Refined calculations show that cyclobutadiene is less stable than two isolated ethylenes.

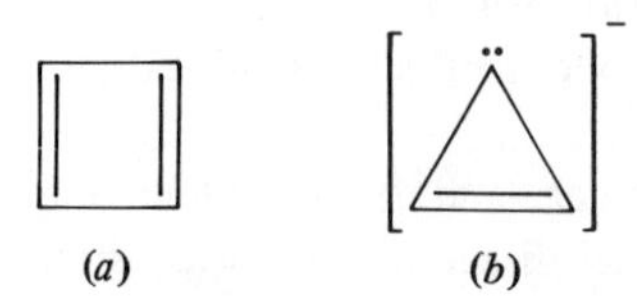

Figure 3.650 Antiaromatic species: (*a*) cyclobutadiene; (*b*) cyclopropenyl anion.

3.660 Möbius System (A. Möbius, 1790–1868)

A cyclic array of atomic orbitals with an odd number of sign inversions (Sect. 2.620). The Möbius designation is derived from the "Möbius strip." Such a strip results, e.g., by taking a strip of paper, giving the strip a half twist (180°), and then joining the two ends, Fig. 3.660*a*. In the Möbius strip the two original surfaces, front and back, become one continuous uninterrupted surface. The one-half twist results in the equivalent of one sign inversion in the cycle.

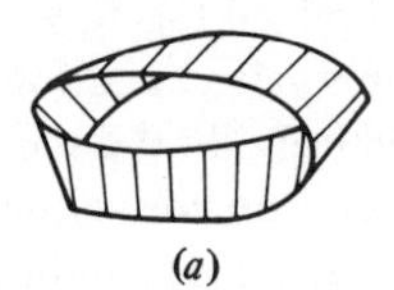

Example. Hückel aromatic stability is associated with $(4n + 2)$ electrons (Sect. 3.460) in a cyclic array of atomic orbitals. Möbius "aromatic" stability on the other hand is associated with $4n$ electrons. These stabilities arise as a consequence of the energies of the molecular orbitals characteristic of the two arrays. Without going into the derivation, the relative energies in the two systems for the four molecular orbitals of cyclobutadiene are shown in Fig. 3.660*b* and *c*. The figure shows that the levels are such that the Möbius system, Fig. 3.660*b* is more stable (hence preferred) for the $4n$ electron case.

(*b*) (*c*)

Figure 3.660 (*a*) A Möbius strip; (*b*) Möbius and (*c*) Hückel MOEDs.

3.670 Nonbenzenoid Aromatic Compounds

Compounds having considerable resonance stabilization or aromatic character but not possessing a benzene nucleus, or in the case of a fused ring system, having one (or more) ring(s) that is not a benzene ring.

Examples. Azulene, Fig. 3.640*a*, and acenaphthylene, Fig. 3.670.

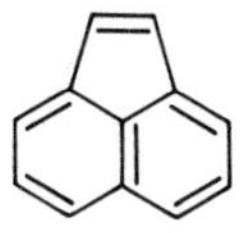

Figure 3.670 A nonbenzenoid aromatic compound, acenaphthylene.

3.680 Annulenes

Monocyclic conjugated hydrocarbons having the general formula $[-CH{=}CH-]_n$: if the annulene has $(4n + 2)$ delocalized π electrons, and n is less than seven, it possesses aromatic character.

Examples. [18] Annulene, Fig. 3.680. The number in brackets denotes the number of carbon atoms in the ring. [18] Annulene has aromatic character and is considered a nonbenzenoid aromatic compound.

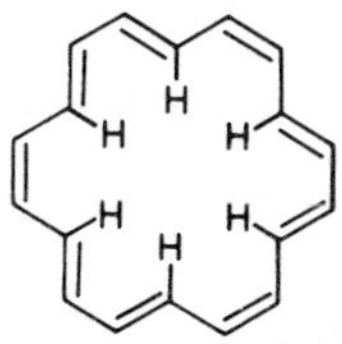

Figure 3.680 [18] Annulene.

3.690 Pseudoaromatic Compounds

Unstable cyclic conjugated hydrocarbons possessing two or more rings. This nomenclature is not widely used in recent literature.

Examples. Pentalene, Fig. 3.690*a*, and heptalene, Fig. 3.690*b*. Although both of these ring systems can be written in two equivalent resonance structures, neither resonance structure can be written with a central double bond and both possess $4n$ π electrons. These compounds then resemble a cyclic conjugated polyene and presumably this accounts for their instability. (Such molecules do not possess a totally symmetric ground state wave function in the valence bond quantum mechanical treatment).

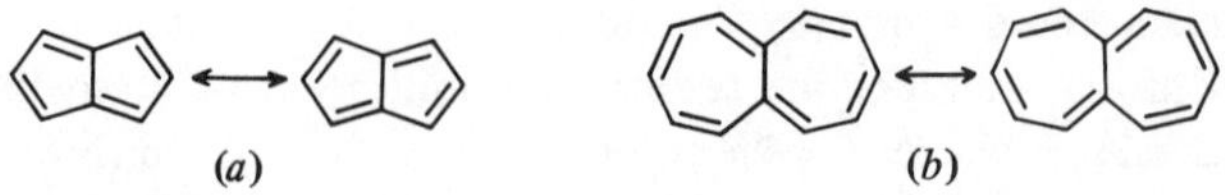

(*a*) (*b*)

Figure 3.690 Pseudoaromatic compounds: (*a*) pentalene; (*b*) heptalene.

3.700 Cyclophanes

Compounds having a benzenoid ring that is disubstituted in the *meta* or *para* positions by a closed chain of carbon atoms, usually methylene groups. The carbon chain may or may not be interrupted by other benzenoid rings similarly disubstituted.

Examples. [3.4]Paracyclophane, Fig. 3.700*a*; the methylene groups in this compound connect *para* positions of benzenes and contain 3 and 4 methylene carbon atoms, hence the numbers in brackets; [2.3]metacyclophane, Fig. 3,700*b*; [10] paracyclophane, Fig. 3.700*c*.

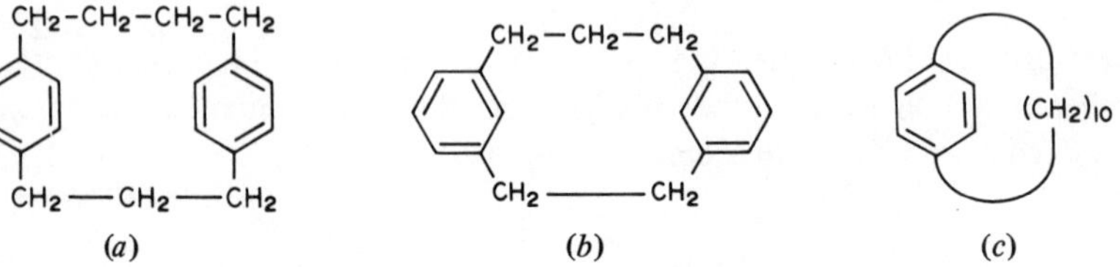

$CH_2-CH_2-CH_2-CH_2$ $CH_2-CH_2-CH_2$ $(CH_2)_{10}$

$CH_2-CH_2-CH_2$ CH_2-CH_2

(*a*) (*b*) (*c*)

Figure 3.700 Cyclophanes: (*a*) [3.4]paracyclophane; (*b*) [2.3]metacyclophane; (*c*) [10] paracyclophane.

3.710 Heteroaromatic Compounds

Aromatic compounds in which one or more —CH= groups of an aromatic hydrocarbon are replaced by a hetero atom, most frequently by nitrogen, oxygen, or sulfur. Oxygen and sulfur are usually part of a five-membered cycle but —N= commonly replaces —CH= in all types of ring systems.

Examples. The five membered ring systems pyrrole, furan, and thiophene, Fig. 3.710*a–c*, respectively. These compounds are all iso-π-electronic with the cyclopentadienide ion, Fig. 3.710*d*, and all possess 6 $p\pi$ electrons, the heteroatoms contributing one pair.

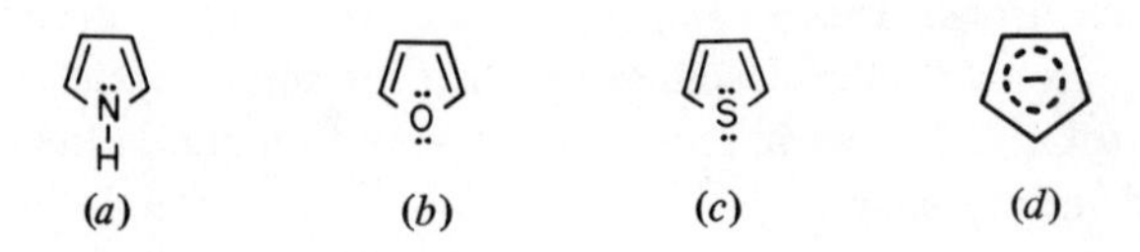

The six-atom ring system containing one nitrogen is pyridine, Fig. 3.710*e*. There are three possible six-membered rings containing two nitrogens, 1,2-, 1,3, and 1,4-diazabenzene, Fig. 3.710*f–h*, respectively. Polynuclear heteroaromatics may contain one or more nitrogen atoms; common examples are isoquinoline, Fig. 7.710*i*, and phenanthroline, Fig. 3.710*j*.

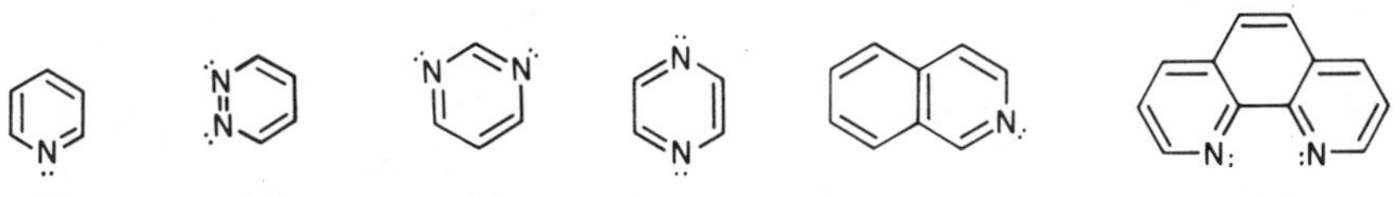

(*e*) (*f*) (*g*) (*h*) (*i*) (*j*)

Figure 3.710 Heterocyclic systems: (*a–c*), (*d*) cyclopentadienide ion; (*e–h*) nitrogen heterocycles with six atoms; (*i*) isoquinoline; (*j*) phenanthroline.

3.720 π-Electron-Excessive Compounds

Compounds with more delocalized π electrons than atoms over which they are delocalized.

Examples. Pyrrole and furan, Fig. 3.710*a* and *b*, respectively, each possess six *p*π electrons delocalized over five atoms and thus the ring carbons are more electron rich than in benzene.

3.730 π-Electron-Deficient Compounds

Heteroaromatics in which the π electron density on the ring carbons is less than that in the analogous aromatic hydrocarbon.

Example. Pyridine, Fig. 3.730; its resonance structures show the relative positive charge of the ring carbons, hence the compound's relative inertness toward electrophilic substitution.

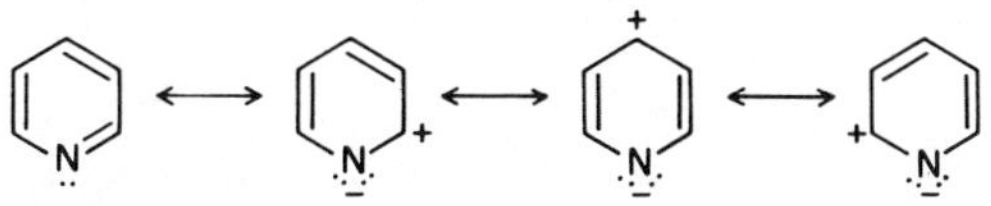

Figure 3.730 Resonance structures of pyridine.

3.740 Unsaturated Hydrocarbons

Hydrocarbons that possess one or more multiple bonds.

3.750 Index of Unsaturation (Hydrogen Deficiency), *i*

A number whose value indicates the number of rings and/or double bonds (a triple bond is counted as two double bonds) present in a molecule of known molecular formula. For a hydrocarbon it may be calculated from the formula

$$i = \frac{(2C + 2) - H}{2}$$

where C and H represent the *number* of carbon and hydrogen atoms, respectively, in the molecule.

Examples. C_7H_{14} must have

$$\frac{(14 + 2) - 14}{2} = 1 \text{ ring or } 1 \text{ double bond}$$

For a generalized molecular formula $A_I B_{II} C_{III} D_{IV}$ where I, II, III, and IV equal the total number of all monovalent atoms A (H, F, Cl, etc.), divalent atoms B (O, S, etc.), trivalent atoms C (N, P, etc.) and tetravalent atoms D (C, Si, etc.) respectively, the index of hydrogen deficiency is

$$i = \mathrm{IV} - \tfrac{1}{2}(\mathrm{I}) + 0(\mathrm{II}) + \tfrac{1}{2}(\mathrm{III}) + 1$$

Thus for morphine, $C_{17}H_{19}O_3N$

$$i = 17 - \tfrac{1}{2}(19) + \tfrac{1}{2}(1) + 1 = 9$$

indicating a sum total of nine rings and double bonds.

4 Classes of Organic Compounds Other Than Hydrocarbons

CONTENTS

4.010 Functional Groups

Organic compounds are frequently classified on the basis of the functional group(s) present in the molecule. A functional group is an atom or group of atoms that is considered to have replaced a hydrogen atom of a given hydrocarbon. Its presence confers characteristic physical and chemical properties on the molecule in which it occurs. It is useful to indicate the alkyl or aryl group that remains after the (theoretical) removal of a hydrogen atom as R— and attach the functional group to it, e.g., R—NO_2, R—Cl, R—CO_2H. When two or more R groups are involved it is assumed that they may be the same or different and may be either alkyl or aryl.

In this chapter we consider classes of compounds that possess one functional group, and further restrict the coverage to those compounds that, in addition to C and H, may possess only the additional elements of B, N, O, S, P, and the halogens, either alone or in combination.

4.020 Alkyl Halides, RX

Derivatives of alkanes in which one of the hydrogen atoms has been replaced by halogen.

Examples. 2-Chloropropane, chlorocyclohexane, 2-chloro-1-pentene, dichloromethane, Fig. 4.020*a–d*, respectively. The halogenated compounds may also be named as alkyl salts of the halogen acids; *tert*-butyl iodide, benzyl fluoride, isopropyl bromide, Fig. 4.020*e–g*, respectively.

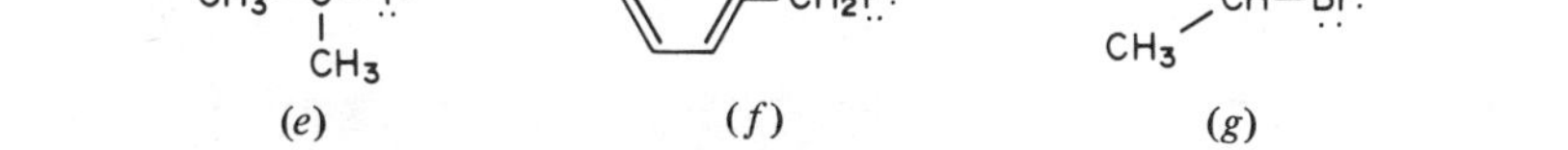

CH_3CHCH_3 with :Cl: on C-2 (*a*); cyclohexane–Cl: (*b*); $CH_2{=}C\,CH_2CH_2CH_3$ with :Cl: on C-2 (*c*); H–C(Cl:)(H)–Cl: (*d*)

CH_3–C(CH_3)(CH_3)–I: (*e*); C_6H_5–CH_2F: (*f*); (CH_3)$_2$CH–Br: (*g*)

Figure 4.020 Alkyl halides: (*a*) 2-chloropropane; (*b*) chlorocyclohexane; (*c*) 2-chloro-1-pentene; (*d*) dichloromethane; (*e*) *tert*-butyl iodide; (*f*) benzyl fluoride; (*g*) isopropyl bromide.

4.030 Radicofunctional Names

A system of naming compounds consisting of two major words, the first word corresponding to the name(s) of the alkyl or aryl group(s) (loosely called radicals) involved and the second corresponding to the functional group present.

Examples. The names of the alkyl halides written as salts of the halogen acids are radicofunctional names, e.g., isopropyl bromide, methyl iodide. Indeed the name "alkyl halides" constitutes a generic radicofunctional name. Other examples include ethyl alcohol, dimethyl ether, and ethyl methyl ketone.

4.040 Aryl Halides, RX

Derivatives of aromatic hydrocarbons in which one of the ring hydrogens has been replaced by halogen. Frequently the aryl group is designated by Ar to distinguish it from alkyl, R. Thus aryl halides may be written ArX but in general we use R to denote either alkyl or aryl.

Examples. 1-Chloronaphthalene and bromobenzene, Fig. 4.040*a* or *b*, respectively.

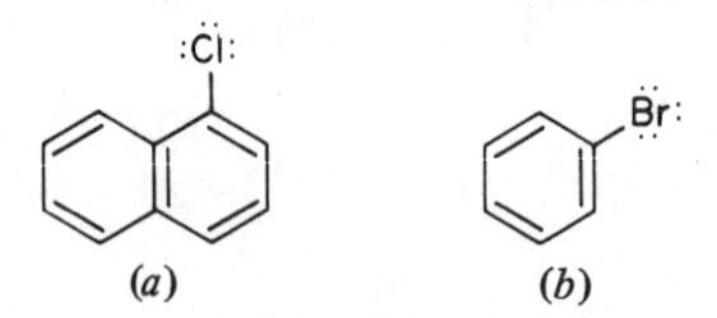

Figure 4.040 (*a*) 1-Chloronaphthalene; (*b*) bromobenzene.

4.050 Alcohols, ROH (R = Alkyl)

Compounds in which a hydrogen atom of a hydrocarbon has been replaced by hydroxy, —OH. Alcohols are classified as *primary*, *secondary*, or *tertiary* depending on whether the —OH group is attached to a primary, secondary, or tertiary carbon atom; these may be indicated as having the structures RCH_2OH, R_2CHOH, and R_3COH, respectively.

Examples. Ethanol, 2-propanol, 2-methylpropanol (IUPAC names), Fig. 4.050*a–c*, are respectively a primary, secondary, and tertiary alcohol. Such com-

$CH_3CH_2—\ddot{O}—H$ (*a*)

$CH_3—CH(CH_3)—\ddot{O}—H$ (*b*)

$CH_3—C(CH_3)_2—\ddot{O}—H$ (*c*)

pounds may also be named according to the radicofunctional name, i.e., ethyl alcohol, isopropyl alcohol, and *tert*-butyl alcohol, respectively. In a third system of nomenclature the alcohols may be named as derivatives of the first member of the series, methanol, CH_3OH. For this purpose methanol is called carbinol and other alcohols are considered as derivatives in which one or more hydrogen atoms of the methyl group are replaced by alkyl groups. According to this system the compounds shown in Fig. 4.050*a–c* are called methylcarbinol, dimethylcarbinol, and trimethylcarbinol, respectively. Figure 4.050*d* shows *sec*-butyl alcohol (radicofunctional), 2-butanol (IUPAC), or ethylmethylcarbinol (derived).

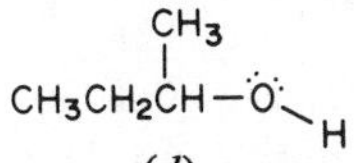

Figure 4.050 Alcohols: (*a*) primary; (*b*) secondary; (*c*) tertiary; (*d*) *sec*-butyl alcohol.

4.060 Alkoxy Group, RO—

The group remaining when the hydrogen attached to oxygen in an alcohol is removed.

Examples. Methoxy, Fig. 4.060*a*, and isopropoxy, Fig. 4.060*b*.

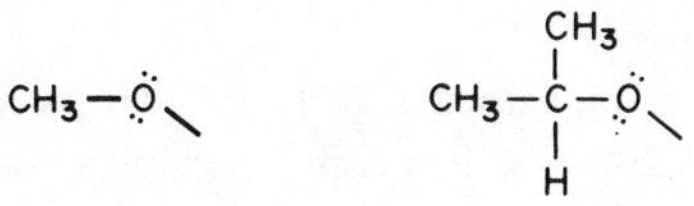

Figure 4.060 (*a*) Methoxy; (*b*) isopropoxy.

4.070 Phenols, ROH(R = Aryl)

Compounds possessing a hydroxy group attached directly to an aromatic ring; named after the first member of the series PhOH, phenol, Fig. 4.070*a*. Ph is frequently used to denote the phenyl group, C_6H_5—.

Examples. 1-Naphthol and 9-phenanthrol, Fig. 4.070*b* and *c*, respectively.

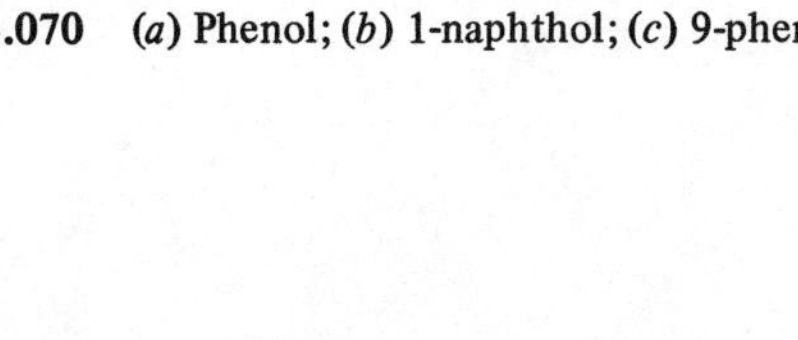

Figure 4.070 (*a*) Phenol; (*b*) 1-naphthol; (*c*) 9-phenanthrol.

4.080 Substitutive Nomenclature

The system of naming compounds in which the functional or replacement group, called the *substituent*, is cited as a prefix or suffix to the name of the compound or nucleus to which it is attached; the latter is called the parent.

Examples. 2-Chloropropane, Sect. 4.020. The compounds shown in Fig. 4.070*b* and *c* may be called 1-hydroxynaphthalene and 9-hydroxyphenanthrene, respectively, by this system; the parents in these examples are naphthalene and phenanthrene. Compounds containing the —OH function can also be named by attaching the suffix *ol* to the parent, e.g., 2-butanol and 1-naphthol. Use of either the appropriate prefix or suffix in substitutive nomenclature has IUPAC approval.

4.090 Ethers, ROR

Compounds that may be considered as derived from hydrocarbons by replacement of a hydrogen atom by an alkoxy (or aroxy) group.

Examples. The common anesthetic $C_2H_5OC_2H_5$, diethyl ether. The compound shown in Fig. 4.090*a* is isopropyl methyl ether, and Fig. 4.090*b* is methyl phenyl ether, also known by its common name, anisole.

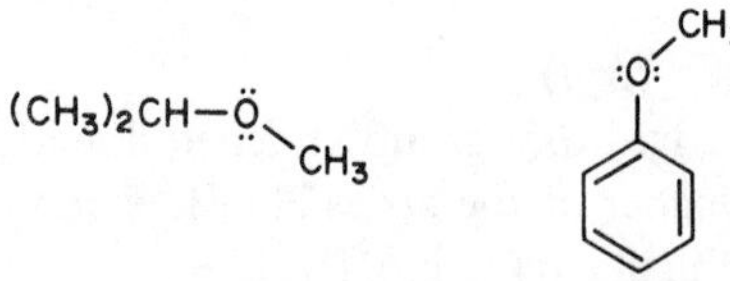

$(CH_3)_2CH{-}\ddot{O}{-}CH_3$ (*a*)

$C_6H_5{-}\ddot{O}{-}CH_3$ (*b*)

Figure 4.090 (*a*) Isopropyl methyl ether; (*b*) anisole.

4.100 Carbonyl Group, —CO—

An oxygen atom doubly bonded to a carbon atom.

Examples. The carbonyl function is one of the most important groupings in organic chemistry and is present in aldehydes, ketones, esters, etc. (*vide infra*). There are two lone pairs of electrons on the oxygen atom of the carbonyl group. It is usually assumed that one electron pair is in an *sp* orbital and is located opposite the carbon-oxygen sigma bond and that one electron pair is in the p_y orbital, which is perpendicular (orthogonal) to the $p\pi$ orbital. The π bond of the carbonyl is formed from overlap of the $p\pi$ orbitals on carbon and oxygen, Fig. 4.100*a*. The MOED for the carbonyl group involving the $p\pi$ electrons and the p_y lone pair on oxygen is shown in Fig. 4.100*b*.

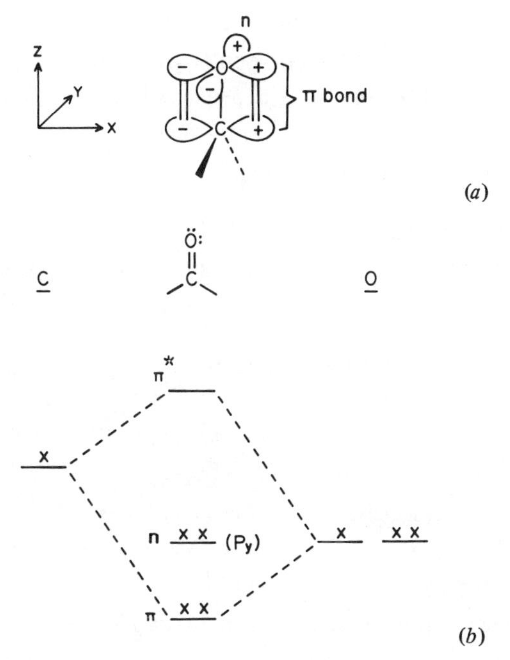

Figure 4.100 (*a*) The π bond; (*b*) the MOED for carbonyl groups.

4.110 Aldehydes, RCHO

Compounds that have a hydrogen atom and an alkyl (or aryl) group bonded to a carbonyl group.

Examples. The simplest aldehyde is $H_2C{=}O$, formaldehyde, and the simplest RCHO is CH_3CHO, acetaldehyde (or ethanal). The simplest aromatic aldehyde is benzaldehyde, PhCHO, Fig. 4.110.

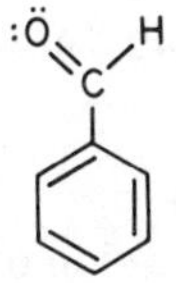

Figure 4.110 Benzaldehyde.

4.120 Hemiacetals, RCH(OH)(OR)

The compounds that result from the addition of an alcohol to the carbonyl group of an aldehyde.

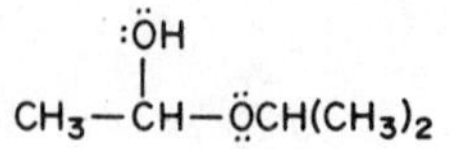

:ÖH
|
CH₃—CH—ÖCH(CH₃)₂

Figure 4.120 1-Isopropoxyethanol.

Example. 1-Isopropoxyethanol, Fig. 4.120.

4.130 Acetals, $RCH(OR)_2$

Compounds in which the carbonyl oxygen of an aldehyde is replaced by two alkoxy groups.

Examples. 1,1-Diethoxypropane, Fig. 4.130.

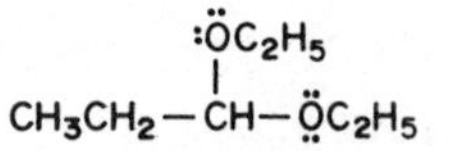

:ÖC₂H₅
|
CH₃CH₂—CH—ÖC₂H₅

Figure 4.130 1,1-Diethoxypropane.

4.140 Ketones, RCOR

Compounds having a carbonyl group flanked by two R groups.

Examples. The simplest ketone is CH_3COCH_3, acetone (common name) or propanone (IUPAC name). The carbonyl carbon may also be in a ring, as in cyclopentanone. The two R groups may be different; e.g., $CH_3COCH(CH_3)_2$ is isopropyl methyl ketone (common name) or 3-methyl-2-butanone (IUPAC), and $PhCOCH_3$ is methyl phenyl ketone (or acetophenone), Fig. 4.140.

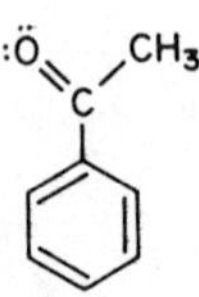

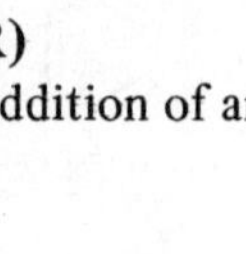

Figure 4.140 Acetophenone.

4.150 Hemiketals, $R_2C(OH)(OR)$

Compounds that result from the addition of an alcohol to the carbonyl group of a ketone.

4.160 Ketals, $R_2C(OR)_2$

Compounds in which the carbonyl oxygen of a ketone is replaced by two alkoxy groups.

Example. 1,1-Dimethoxy-*n*-butylbenzene, Fig. 4.160.

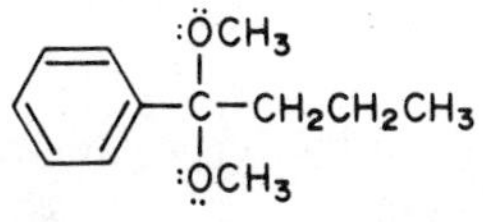

$$\mathrm{C_6H_5{-}C(\ddot{O}CH_3)_2{-}CH_2CH_2CH_3}$$

Figure 4.160 1,1-Dimethoxy-*n*-butylbenzene.

4.170 Cyanohydrins, RCH(OH)(CN)

Compounds that result from the addition of HCN to the carbonyl group of an aldehyde or (less frequently) a ketone.

4.180 Enols, $\mathrm{HO\overset{|}{C}{=}\overset{|}{C}{-}}$

Compounds in which one of the hydrogen atoms attached to an olefinic linkage is replaced by a hydroxy group; usually in equilibrium with the corresponding carbonyl compound.

Examples. 2-Hydroxy-1-propene (the enol of acetone), Fig. 4.180*a*. This enol is present to a very small extent (estimated to be about 0.00025%) in pure acetone. 1-Cyclohexenol, Fig. 4.180*b*, is the enol of cyclohexanone, Fig. 4.180*c*. It should be noted that phenol, Fig. 4.070*a*, has the enolic grouping; it is so stable that it exists essentially completely in the enol form.

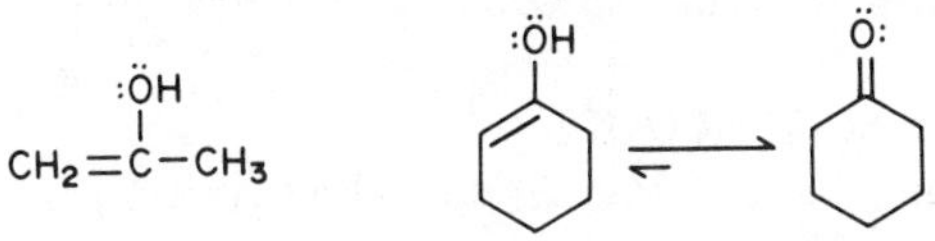

$$\mathrm{CH_2{=}C(\ddot{O}H){-}CH_3}$$

Figure 4.180 (*a*) The enol of acetone; (*b*) 1-cyclohexenol, the enol of cyclohexanone; (*c*) cyclohexanone.

4.190 Keto-Enol Isomers

Isomers that involve reversible hydrogen migration from a carbon atom adjacent to a carbonyl group to the oxygen of the carbonyl group; they have the structural relation:

$$\mathrm{H{-}\overset{|}{\underset{|}{C}}{-}\overset{|}{C}{=}O \rightleftharpoons {-}\overset{|}{C}{=}\overset{|}{C}{-}O{-}H.}$$

Examples. This equilibrium between isomers (also called proton tautomerism, see Sect. 5.070) is characteristic of all carbonyl-containing functional groups possessing an α hydrogen atom; e.g., acetaldehyde and its enol, Fig. 4.190*a* and *b*, respectively; also cyclohexanone and its enol, Fig. 4.180*c* and *b*, respectively.

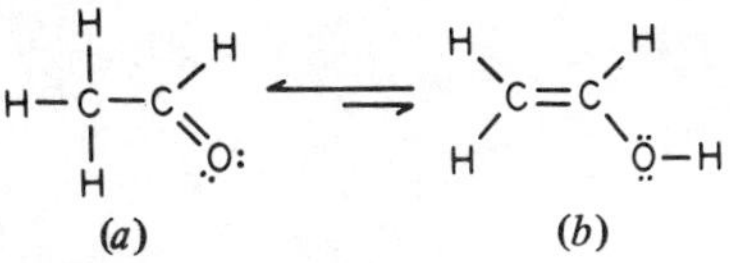

$$\mathrm{CH_3{-}CH{=}\ddot{O}: \rightleftharpoons H_2C{=}CH{-}\ddot{O}{-}H}$$

Figure 4.190 (*a*) Acetaldehyde; (*b*) its enol.

4.200 Enol-Ethers, $RO\overset{|}{C}=\overset{|}{C}-$

Compounds in which one of the hydrogen atoms of an olefinic linkage is replaced by an alkoxy (RO—) group.

Example. Methyl vinyl ether, Fig. 4.200.

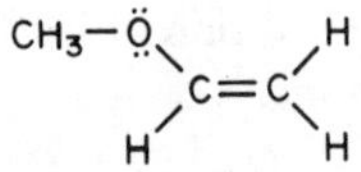

Figure 4.200 Methyl vinyl ether.

4.210 Aldehyde and Ketone Derivatives

Used to characterize these carbonyl compounds, they may be regarded in a formal way as resulting from a splitting out of water by loss of the oxygen of the carbonyl group and two hydrogen atoms from the derivatizing agent. The hydrogens are usually on a nitrogen atom and a typical reaction may be written

$$R_2C{=}O + H_2N{-}Z \longrightarrow R_2C{=}N{-}Z + H_2O$$

Table 4.210 gives the names and structures of various derivatizing agents and the names and structures of the derivatives.

4.220 Carboxylic Acids, RCO_2H

Compounds containing a carbonyl group and a hydroxy group both attached to a single carbon atom. The $-CO_2H$ group is called the carboxy group.

Examples. The first member of the series is HCO_2H, formic acid (R = H in RCO_2H). The second member of the series is CH_3CO_2H, acetic acid (common name) or ethanoic acid (IUPAC), Fig. 4.220*a*. $CH_3CH_2CO_2H$ is propionic acid (common name) or methylacetic acid (derived name) or propanoic acid (IUPAC). When a carboxy group is attached to a ring, the compound is named by adding the suffix carboxylic acid to the name of the ring system to which the carboxy group is attached (substitutive nomenclature, Sect. 4.080). Thus the compounds shown in Fig. 4.220*b* and *c* are, respectively, cyclohexanecarboxylic acid and 1-phenanthrenecarboxylic acid.

(*a*) (*b*) (*c*)

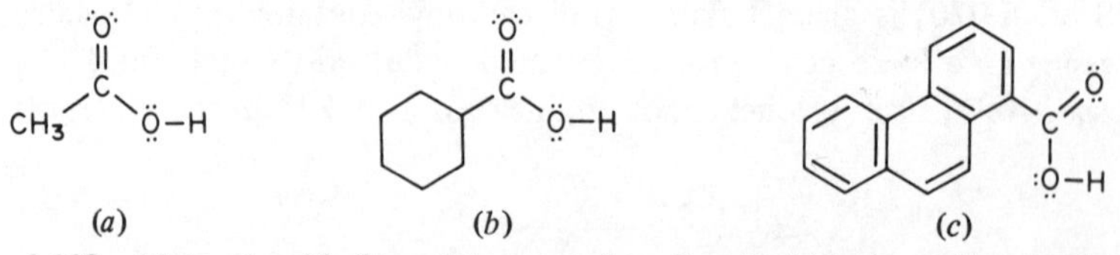

Figure 4.220 (*a*) Acetic acid; (*b*) cyclohexanecarboxylic acid; (*c*) 1-phenanthrenecarboxylic acid.

Table 4.210 Some derivatives of carbonyl compounds RRC=O (R may be H) used for characterization purposes

Reagent		Derivative	
Name	Structure	Generic name	Structure
Hydroxylamine	$H_2N—OH$	Oxime	RRC=N—OH
Hydrazine	$H_2N—NH_2$	Hydrazone	$RRC{=}N—NH_2$
Phenylhydrazine	$H_2N—NHPh$	Phenylhydrazone	RRC=N—NHPh
Tosylhydrazine	$H_2N—NH—SO_2—C_6H_4—CH_3$	Tosylhydrazone	RRC=N—NHTs
2,4-Dinitrophenylhydrazine	$H_2N—NH—C_6H_3(NO_2)_2$ (2-O_2N, 4-NO_2)	2,4-Dinitrophenylhydrazone	$RRC{=}N—NHC_6H_3(NO_2)_2$
Semicarbazide	$H_2N—NHCONH_2$	Semicarbazone	$RRC{=}N—NHCONH_2$
Primary amine	$H_2N—R$	Schiff base	RRC=N—R
Sodium bisulfite[a]	$NaHSO_3$	Sodium bisulfite addition compound	$R—CH(OH)—SO_3Na$

[a]Reacts only with aldehydes, cyclic ketones and methyl ketones to give addition product without loss of water.

4.230 Carboxylate Anion, RCO_2^-

The anion formed by the loss of a proton from a carboxylic acid.

Example. Acetate anion, Fig. 4.230. The ion can be represented by either of the equivalent resonance structures or by the structure in which a dotted line represents delocalization over the three-atom system. The carbon atom to which the carboxylate anion is attached as well as the three atoms of the carboxylate anion all lie in the same plane. In this anion there are four $p\pi$ electrons; the two electrons associated with the carbonyl group (one from carbon, one from oxygen) and a pair of $p\pi$ electrons on the oxygen atom. There are three atoms involved in the π system O—C—O, and thus there are three π-type molecular orbitals. One of these orbitals is very strongly bonding, one approximately nonbonding or slightly antibonding, and the third, strongly antibonding. The four electrons occupy the two lower energy π orbitals.

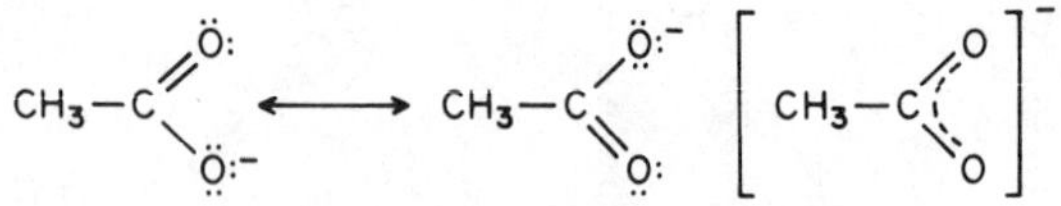

Figure 4.230 Acetate anion.

4.240 Carboxylic Acid Derivatives, RCOZ

Compounds of the structure RCOZ where Z = H, R, and OH are, respectively, aldehydes, ketones, and acids, but when Z is replaced by other atoms or groups of atoms, the resulting classes of compounds are frequently considered as derived from the carboxylic acids.

4.250 Acyl and Aroyl Halides, RCOX

Compounds in which the OH of the corresponding acid has been replaced by halogen. When R is aliphatic these compounds are acyl halides, and when R is aromatic they are aroyl halides.

Examples. Acetyl fluoride and benzoyl bromide, Fig. 4.250*a* and *b*, respectively. These compounds are named as salts of HF and HBr; the organic part of the name is derived from the corresponding acids, acetic and benzoic acids, by replacing the *ic* by *yl.*

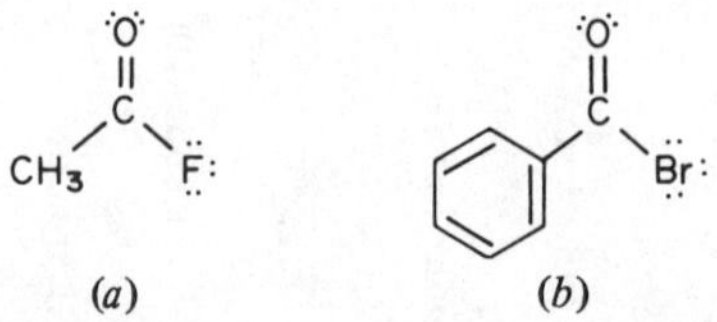

Figure 4.250 (*a*) Acetyl fluoride; (*b*) benzoyl bromide.

4.360 Nitrones, $RCH{=}\overset{O^-}{\overset{|}{N^+}}R$ ($RCH{=}\overset{O}{\overset{\uparrow}{N}}R$)

Compounds in which an oxygen atom is bonded to the nitrogen atom of a Schiff base (RCH=NR, see Table 4.210).

Example. Figure 4.360, ethylidenemethylazane oxide, the preferred name, but also called *N*-methyl-α-methylnitrone. Because of the double bond system, *E-Z* isomers are possible.

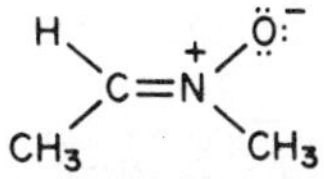

H, CH_3 \C=N$^+$/ $\ddot{O}$:$^-$, CH_3

Figure 4.360 (*E*)-Ethylidenemethylazane oxide (*N*-methyl-α-methylnitrone).

4.370 Quaternary Ammonium Salts, $R_4\overset{+}{N}\overset{-}{X}$

Ammonium salts in which the N^+ atom shares its four valence electrons with carbon atoms, usually formed by the reaction between a tertiary amine and RX. The nitrogen atom carrying the positive charge is called a quaternary nitrogen.

Examples. $(CH_3)_4\overset{+}{N}\overset{-}{Cl}$ is tetramethylammonium chloride. Figure 4.370 is pyridinium hydrochloride.

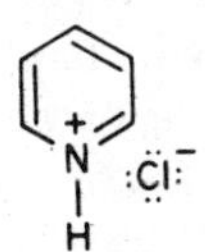

Figure 4.370 Pyridinium hydrochloride.

4.380 Nitriles, RCN

Compounds with a terminal nitrogen atom triply bonded to carbon.

Examples. CH_3CN is most commonly called by the trivial name acetonitrile derived from the corresponding CH_3CO_2H having the common or trivial name of acetic acid; $CH_2{=}CHCN$ is acrylonitrile. Generally the substitutive name is preferred for less common nitriles. Figure 4.380*a* is pentanenitrile. When a ring system is attached to —CN, the compound is named substitutively by adding the suffix carbonitrile to the name of the ring system. Figure 4.380*b* is cyclohexane-

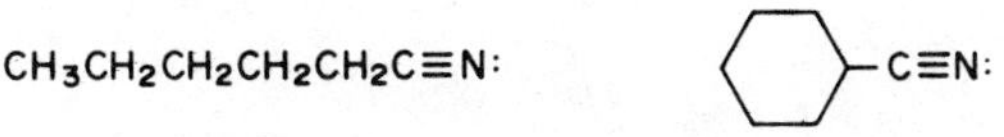

$CH_3CH_2CH_2CH_2CH_2C{\equiv}N{:}$ (*a*)

(cyclohexyl)—$C{\equiv}N{:}$ (*b*)

Figure 4.380 (*a*) Pentanenitrile; (*b*) cyclohexanecarbonitrile.

carbonitrile. In some cases (e.g., multiple functions) where radicofunctional names are used, —CN is called the cyano group.

4.390 Isocyanides, $R\overset{+}{N}\equiv\overset{-}{C}:$

Compounds in which the hydrogen of a hydrocarbon has been replaced by the —NC group; any particular RNC is isomeric with the corresponding RCN, hence isocyanides are sometimes called *isonitriles* (unapproved).

Example. Phenylisocyanide, Fig. 4.390. However, the substitutive name carbylaminobenzene is preferred to the radicofunctional name.

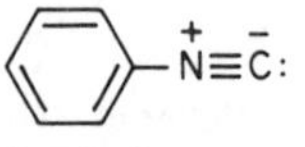

Figure 4.390 Carbylaminobenzene (phenyl isocyanide).

4.400 Nitroalkanes, RNO_2

Compounds in which the hydrogen atom of an alkane is replaced by a nitro group $—NO_2$.

Example. CH_3NO_2, Fig. 4.400*a*, is nitromethane. The resonance structure, 4.400*b*, is equivalent and structure 4.400*c* attempts to display this. The σ bond skeleton of RNO_2 involves the use of three sp^2-like orbitals on nitrogen, leaving a pure p orbital that can overlap with the $p\pi$ orbitals on each of the oxygen atoms. From a molecular orbital point of view, the three MOs are: one strongly bonding, 4.400*d*; one nonbonding, 4.400*e*; and one antibonding, 4.400*f*; these are analogous to the isoelectronic carboxylate anion RCO_2^-. The four $p\pi$ electrons occupy the bonding and nonbonding molecular orbitals.

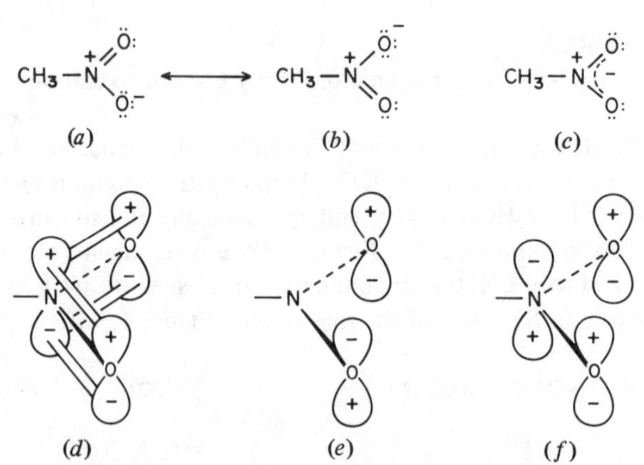

Figure 4.400 Nitromethane: (*a*, *b*) resonance structures; (*c*) a composite representation; (*d-f*) the three π molecular orbitals.

4.410 Nitroso Compounds, RN=O

Compounds in which the hydrogen atom of a hydrocarbon is replaced by the —NO group.

Example. Nitrosobenzene, Fig. 4.410. The only stable alkyl nitroso compounds are the tertiary ones, $R_3C—N=O$. If a hydrogen atom is present on the α-carbon atom, the tautomeric oxime is formed irreversibly: —CH—N=O → —C=N—OH.

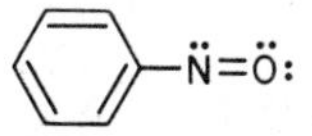

Figure 4.410 Nitrosobenzene.

4.420 Diazoalkanes, $R_2C=N_2$

Alkanes in which two hydrogen atoms attached to the same carbon are replaced by the diazo group, $=N_2$.

Examples. Diazomethane and diphenyldiazomethane, Fig. 4.420*a* and *b*, respectively.

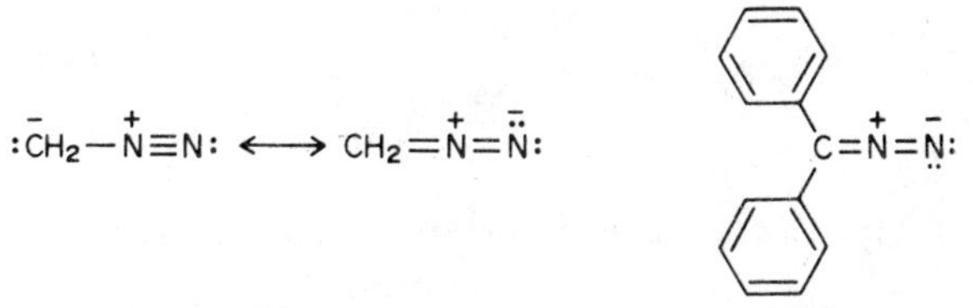

(*a*) (*b*)

Figure 4.420 (*a*) The main resonance structures of diazomethane; (*b*) diphenyldiazomethane.

4.430 α-Diazoketones, $RCOCR'N_2$

Compounds in which two hydrogens on a carbon atom adjacent to a carbonyl group are replaced by a diazo group, $=N_2$; R′ may be H.

Examples. α-Diazoacetophenone, Fig. 4.430*a*, and diazomethyl ethyl ketone, Fig. 4.430*b*.

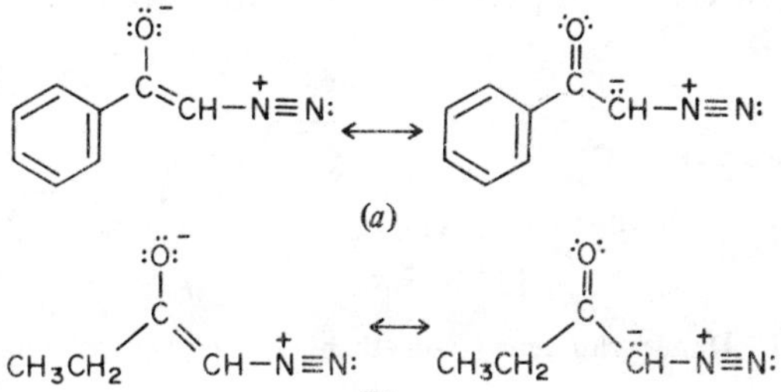

(*a*)

(*b*)

Figure 4.430 (*a*) α-Diazoacetophenone; (*b*) diazomethyl ethyl ketone.

4.440 Azides, RN_3

Compounds in which a hydrogen atom of a hydrocarbon is replaced by the azido, $—N_3$, group; alternatively, alkyl or aryl derivative of hydrazoic acid, HN_3.

Example. Methyl azide, Fig. 4.440.

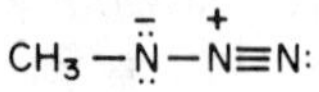

$$CH_3—\overset{-}{N}—\overset{+}{N}{\equiv}N:$$

Figure 4.440 Methyl azide.

4.450 Acid (or Acyl) Azides, $RCON_3$

Compounds in which the hydroxy group of a carboxylic acid is replaced by the azido, $—N_3$, group; alternatively, acyl or aroyl derivatives of hydrazoic acid, HN_3.

Examples. Pentanoyl azide and benzoyl azide, Fig. 4.450*a* and *b*, respectively.

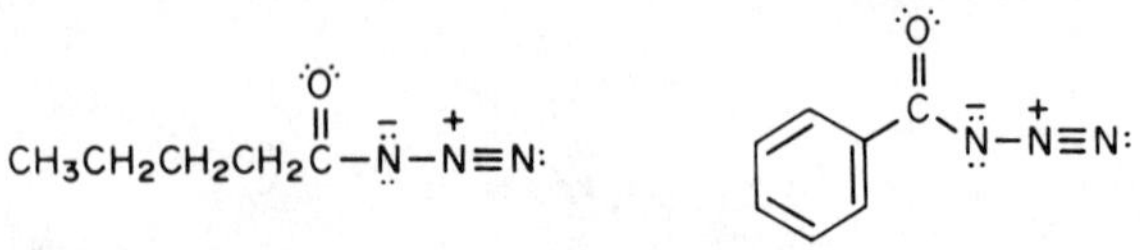

$$CH_3CH_2CH_2CH_2\overset{O}{\overset{\|}{C}}—\overset{-}{N}—\overset{+}{N}{\equiv}N:$$

(*a*) (*b*)

Figure 4.450 (*a*) Pentanoyl azide; (*b*) benzoyl azide.

4.460 Azo Compounds, RN=NR

Compounds in which two R groups are separated by a —N=N— group.

Examples. Azomethane, Fig. 4.460*a*. The preferred systematic name is dimethyldiazene. In this nomenclature system the dinitrogen-containing saturated compound commonly called hydrazine, $H_2N—NH_2$, is called diazane (similar to alkane, the saturated hydrocarbon series) and HN=NH is a diazene (commonly called diimine). Hence azomethane is preferably called dimethyldiazene, and azobenzene, Fig. 4.460*b*, is diphenyldiazene. Both *cis*(*Z*) and *trans*(*E*) isomers are possible.

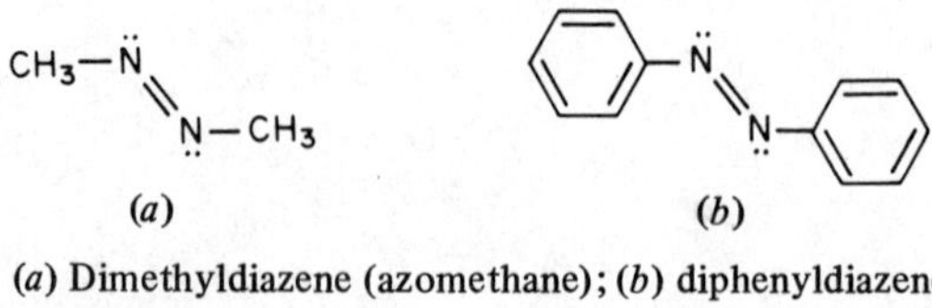

$$CH_3—N{=}N—CH_3$$

(*a*) (*b*)

Figure 4.460 (*a*) Dimethyldiazene (azomethane); (*b*) diphenyldiazene (azobenzene).

4.470 Azoxy Compounds, $RN^{+}(O^{-})=NR$

Compounds in which an oxygen atom is bonded to one of the nitrogen atoms of an azo compound.

Example. Azoxybenzene, Fig. 4.470.

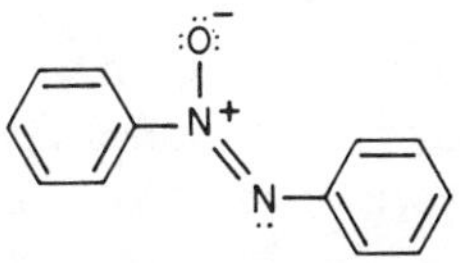

Figure 4.470 Azoxybenzene.

4.480 Diazonium Compounds, $R\overset{+}{N}_2\overset{-}{X}$

Salts in which a $-\overset{+}{N}_2$ group is bonded directly to an alkyl or aryl group.

Example. Benzenediazonium chloride, Fig. 4.480.

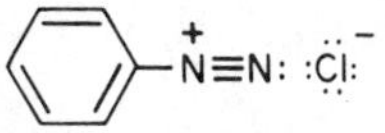

Figure 4.480 Benzenediazonium chloride.

4.490 Cyanates, ROCN

Compounds in which the hydrogen of a hydrocarbon is replaced by —O—C≡N, the cyanato group.

Examples. *n*-Propyl cyanate, Fig. 4.490*a*; this compound is named as an ester of cyanic acid HO—C≡N. Phenyl cyanate, Fig. 4.490*b*.

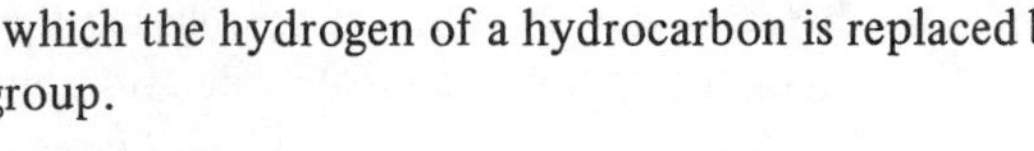

$CH_3CH_2CH_2-\ddot{O}-C\equiv N:$

(*a*) (*b*)

Figure 4.490 (*a*) *n*-Propyl cyanate; (*b*) phenyl cyanate.

4.500 Fulminates, $RO\overset{+}{N}\equiv\overset{-}{C}:$

Compounds in which the hydrogen of a hydrocarbon is replaced by $-O-\overset{+}{N}\equiv\overset{-}{C}:$, the fulminate group.

Example. *n*-Propyl fulminate, Fig. 4.500. This compound is named as an ester of $HO-\overset{+}{N}\equiv\overset{-}{C}:$, fulminic acid.

$CH_3CH_2CH_2-\ddot{O}-\overset{+}{N}\equiv\overset{-}{C}:$

Figure 4.500 *n*-Propyl fulminate.

4.510 Isocyanates, RNCO

Compounds in which the hydrogen atom of a hydrocarbon is replaced by —N=C=O, the isocyanato group.

Example. Phenyl isocyanate, Fig. 4.510. This compound is not the salt of an acid, hence, although it is commonly called phenyl isocyanate, it should be correctly called by the substitutive name carbonylaminobenzene.

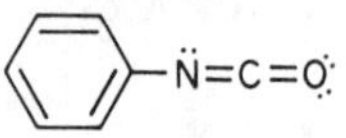

Figure 4.510 Carbonylaminobenzene (phenyl isocyanate).

4.520 Carbodiimides, RN=C=NR

Compounds formally derived from the parent carbodiimide, HN=C=NH (the stable form of which is the tautomer, $H_2N—C\equiv N$, cyanamide), by replacement of hydrogens by R— groups.

Example. Dicyclohexylcarbodiimide, Fig. 4.520. The preferred name for this compound is dicyclohexylmethanediimine and for the parent, HN=C=NH, methanediimine.

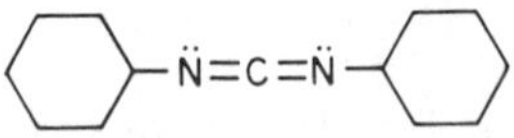

Figure 4.520 Dicyclohexylmethanediimine [dicyclohexylcarbodiimide (DCCD)].

4.530 Heterocyclic Systems

Ring compounds containing atoms of at least two different elements as ring members. Because organic chemistry is concerned principally with carbon atom chemistry, the other element in a heterocyclic system is called the **hetero atom.**

4.540 Oxa-Aza or Replacement Nomenclature

The system of naming heterocyclic compounds by relating them to the corresponding carbocyclic system. The hetero element is denoted by prefixes ending in *a*.

Examples. Oxygen is denoted as oxa, nitrogen as aza, sulfur as thia, phosphorus as phospha, boron as bora, etc. Figure 4.540*a* is 1,4-dioxacyclohexane

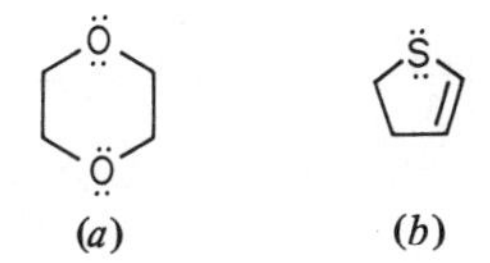

Figure 4.540 (*a*) 1,4-Dioxacyclohexane (dioxane); (*b*) thia-2-cyclopentene.

(common name, dioxane); Fig. 4.540*b* is thia-2-cyclopentene. Heteroaromatics may also be named by this method; e.g., pyridine, Fig. 3.700*e*, is azabenzene and Fig. 3.700*f* is 1,2-diazabenzene.

4.550 Hantzsch-Widman Heterocyclic Nomenclature

A system of naming heterocyclics, proposed by the named chemists, which consists of adding identifying suffixes to the prefix used in replacement nomenclature; see Table 4.550.

Table 4.550 Examples of simple heterocyclic systems

Number of atoms in saturated ring (hetero atom)	Replacement name	H.W.[a] name	Common name
3 (N)	Azacyclopropane	**Aziridine**	Ethylenimine
3 (O)	Oxacyclopropane	**Oxirane**	Ethylene oxide[b]
3 (S)	Thiacyclopropane	**Thiirane**	Episulfide
4 (N)	Azacyclobutane	**Azetine**	Trimethyleneimine
4 (O)	Oxacyclobutane	**Oxetane**	Trimethylene oxide
4 (S)	Thiacyclobutane	**Thietane**	Trimethylene sulfide
5 (N)	Azacyclopentane	**Azolidine**	Pyrrolidine
5 (O)	Oxacyclopentane	**Oxolane**	Tetrahydrofuran
5 (S)	Thiacyclopentane	**Thiolane**	Tetramethylene sulfide

[a]Hantzsch-Widman.

[b]The cyclic ethers having three or four atoms in the ring are frequently called 1,2- and 1,3-**epoxides**.

4.560 Lactones, $R-\overset{}{CH}-(CH_2)_x-C{=}O$ (ring closed through O between CH and C)

Intramolecular (cyclic) esters; compounds formed by (formal) elimination of water from a hydroxy and carboxy group in the same molecule.

Examples. The number of methylene groups x can be any number, but the most common lactones possess a total of six, five, or four atoms, ($x = 3, 2, 1$),

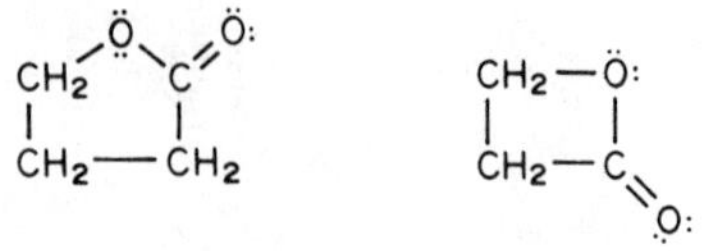

(*a*) (*b*)

Figure 4.560 (*a*) γ-Butyrolactone; (*b*) β-propiolactone.

e.g., γ-butyrolactone, Fig. 4.560*a*, and β-propiolactone, Fig. 4.560*b*. In these examples R = H.

4.570 Lactams, R—CH—(CH$_2$)$_x$—C=O (ring closed through —NH—)

Intramolecular (cyclic) amides; compounds formed by the (formal) elimination of water from an amino and carboxy group in the same molecule.

Example. Caprolactam, Fig. 4.570.

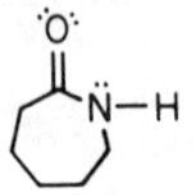

Figure 4.570 Caprolactam.

4.580 Boranes

Derivatives of boron hydrides, B_xH_y. BH_3 and B_2H_6 are borane and diborane, respectively (Sect. 2.850). Hydrides containing more than one boron atom are named by using the familiar multiplying prefixes and adding an Arabic numeral in parenthesis as a suffix to denote the number of hydrogens present in the parent; e.g., B_5H_9 is pentaborane (9).

4.590 Organoboranes

Compounds in which one or more R groups replace the hydrogens in a boron hydride.

Examples. 1,2-Dimethyldiborane(6), Fig. 4.590*a*, and dimethoxyisopropylborane, Fig. 4.590*b*.

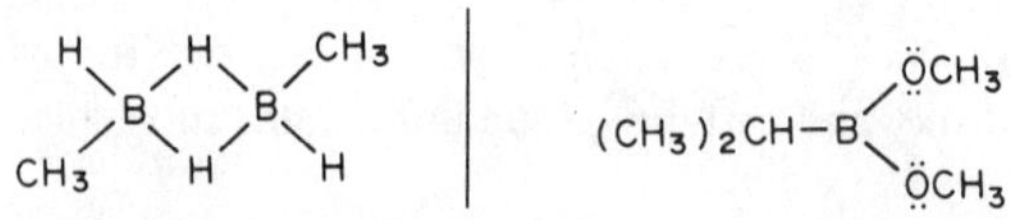

(*a*) (*b*)

Figure 4.590 (*a*) *trans*-1,2-Dimethyldiborane(6); (*b*) dimethoxyisopropylborane.

4.600 Organoborates, $(RO)_3B$, $(RO)_2B(OH)$, and $ROB(OH)_2$

Esters of boric acid, $B(OH)_3$, in which one or more of the hydrogen atoms is replaced by —R.

Examples. Triethyl borate, Fig. 4.600*a*, diethyl hydrogen borate, Fig. 4.600*b*, and ethyl dihydrogen borate, Fig. 4.600*c*.

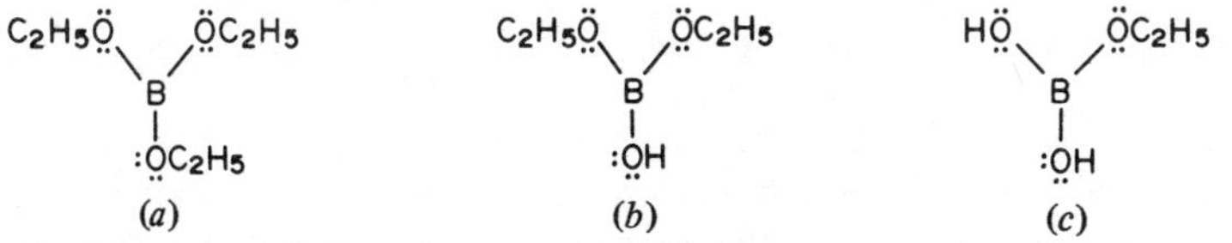

$C_2H_5\ddot{O}$ $\ddot{O}C_2H_5$ B :$\ddot{O}C_2H_5$ (*a*)

$C_2H_5\ddot{O}$ $\ddot{O}C_2H_5$ B :$\ddot{O}H$ (*b*)

$H\ddot{O}$ $\ddot{O}C_2H_5$ B :$\ddot{O}H$ (*c*)

Figure 4.600 (*a*) Triethyl borate; (*b*) diethyl hydrogen borate; (*c*) ethyl dihydrogen borate.

4.610 Heterocyclic Boranes

Organic ring systems containing one or more boron atoms.

Example. The replacement (oxa-aza) nomenclature is commonly used, e.g., 1,4-diboracyclohexane, Fig. 4.610.

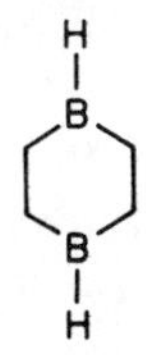

Figure 4.610 1,4-Diboracyclohexane.

4.620 Carboranes

A shortened form of carbaborane, denoting a class of compounds in which one or more carbon atoms replace boron in polyboron hydrides.

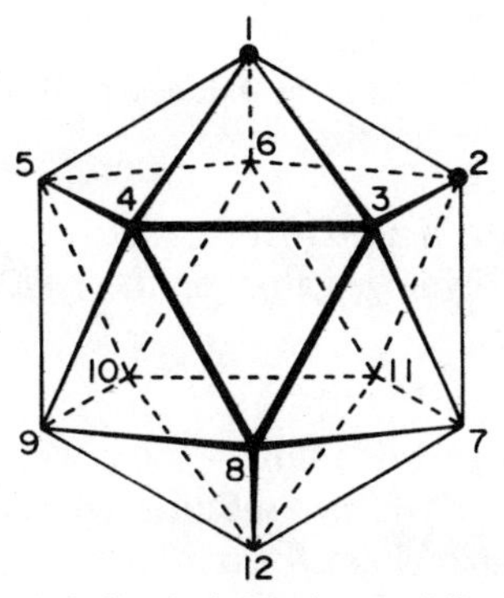

Figure 4.620 The carborane 1,2-dicarbadodecaborane(12); dots represent carbon atoms.

Examples. 1,2-Dicarbadodecarborane(12), $B_{10}C_2H_{12}$, Fig. 4.260; there are three isomers of this compound; the 1,2-isomer (location of the carbon atoms) is shown. The triangulated polyhedral structures have the formula $B_{n-2}C_2H_n$.

4.630 Hydroperoxides, ROOH

Compounds in which the hydrogen of a hydrocarbon is replaced by an —OOH group.

Example. *tert*-Butyl hydroperoxide, Fig. 4.630. The group —OOH is called the hydroperoxy group, hence compounds containing it may be named by the substitutive nomenclature. Thus the compound shown in Fig. 4.630 is also called 2-hydroperoxy-2-methylpropane.

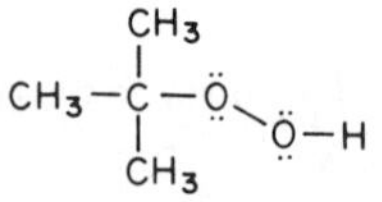

Figure 4.630 *tert*-Butyl hydroperoxide (2-hydroperoxy-2-methylpropane).

4.640 Peracids, RCO_3H

Carboxylic acids in which an oxygen atom is inserted between the carbon of the carbonyl group and the oxygen of the hydroxy group.

Example. Perbenzoic acid, Fig. 4.640; the preferred name for this compound is peroxybenzoic acid.

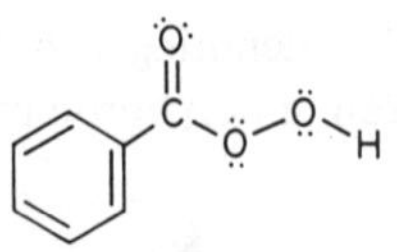

Figure 4.640 Peroxybenzoic acid (perbenzoic acid).

4.650 Disubstituted Peroxides, ROOR

Compounds in which the hydrogen atom on the —OOH of the hydroperoxide is replaced by another R group.

Examples. Di-*tert*-butyl peroxide, Fig. 4.650*a*. When the two R groups are different, peracids are preferably named substitutively as dioxy derivatives of a parent. Thus Fig. 4.650*b* is methyldioxybenzene.

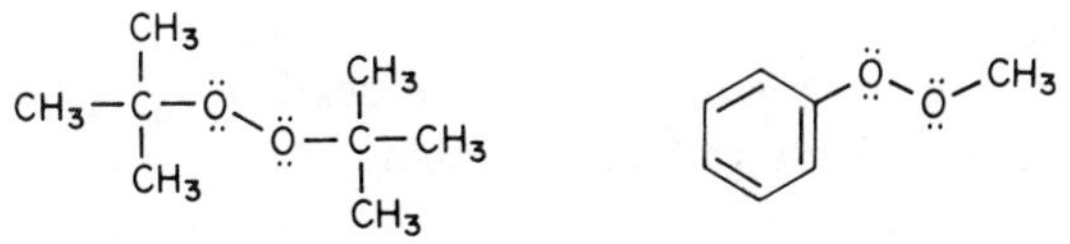

Figure 4.650 (*a*) Di-*tert*-butyl peroxide; (*b*) methyldioxybenzene.

4.660 Thiols, RSH

Compounds in which the hydrogen atom of a hydrocarbon is replaced by a mercapto (—SH) group.

4.670 Thioalcohols, RSH (R = Alkyl)

Thiols in which the R group is alkyl or cycloalkyl.

Examples. Ethanethiol, Fig. 4.670*a*, and phenylmethanethiol, Fig. 4.670*b*. These compounds are also known as ethyl mercaptan and benzyl mercaptan, respectively, but such radicofunctional names are not officially (IUPAC) sanctioned. Cyclohexanethiol, Fig. 4.670*c*.

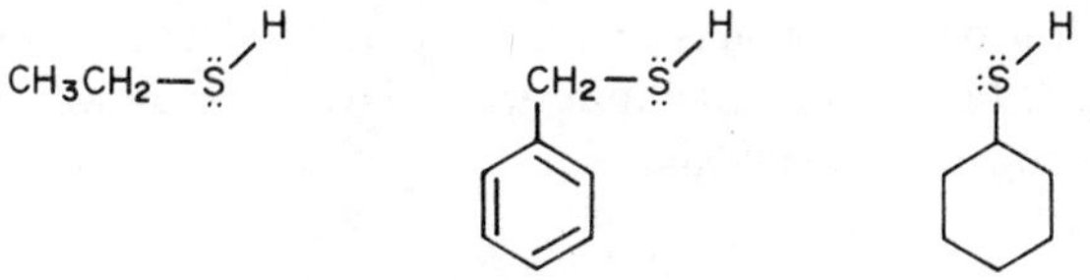

Figure 4.670 (*a*) Ethanethiol (ethyl mercaptan); (*b*) phenylmethanethiol (benzyl mercaptan); (*c*) cyclohexanethiol.

4.680 Thiophenols, RSH (R = Aryl)

Thiols in which the R group is aryl.

Examples. Benzenethiol, Fig. 4.680*a*; also called phenyl mercaptan but this name is no longer sanctioned (IUPAC): 2-naphthalenethiol, Fig. 4.680*b*.

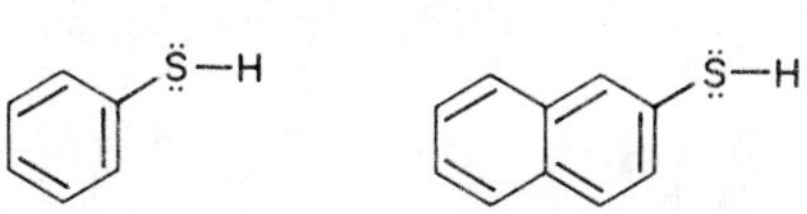

Figure 4.680 (*a*) Benzenethiol (thiophenol); (*b*) 2-naphthalenethiol.

4.690 Sulfides, RSR

Compounds in which the hydrogen of a hydrocarbon is replaced by an —SR group, or alternatively, hydrocarbons in which a sulfur atom has been inserted into a carbon-carbon bond.

Examples. Ethyl methyl sulfide, Fig. 4.690*a*, and cyclopentyl phenyl sulfide, Fig. 4.690*b* [also called (cyclopentylthio)benzene].

$CH_3—\ddot{S}—CH_2CH_3$

(*a*) (*b*)

Figure 4.690 (*a*) Ethyl methyl sulfide; (*b*) cyclopentyl phenyl sulfide [(cyclopentylthio)-benzene].

4.700 Sulfoxides, $R_2S(O)$ $(R_2\overset{+}{S}—\bar{O})$

Compounds in which an oxygen atom is bonded to the sulfur atom of a sulfide.

Example. Ethyl methyl sulfoxide, Fig. 4.700, also called (methylsulfinyl)ethane. (The prefixes alkylthio and alkylsulfinyl represent R—S— and R—$\overset{+}{S}$—$\bar{O}$, respectively.) The sulfur-oxygen bond is a partial double bond, as indicated by the resonance structure. In the older literature the —S(O)— group was called *thionyl;* $SOCl_2$ is still known as thionyl chloride.

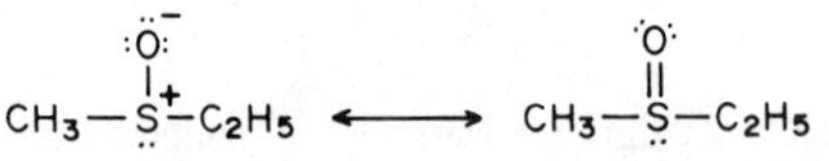

$$CH_3—\overset{:\ddot{O}:^-}{\overset{|}{\overset{+}{S}}}—C_2H_5 \longleftrightarrow CH_3—\overset{\ddot{O}}{\overset{\|}{S}}—C_2H_5$$

Figure 4.700 Ethyl methyl sulfoxide [(methylsulfinyl)ethane].

4.710 Sulfones, R_2SO_2 $(R_2\overset{+}{S}—\bar{O})$ with $=O$ on S

Compounds in which the sulfur of a sulfide is bonded to two oxygen atoms.

Example. Isopropyl methyl sulfone, Fig. 4.710, also called 2-methylsulfonylpropane (IUPAC). Various resonance structures for the $—SO_2—$ group may be written. In the older literature the $—SO_2—$ group was called *sulfuryl.*

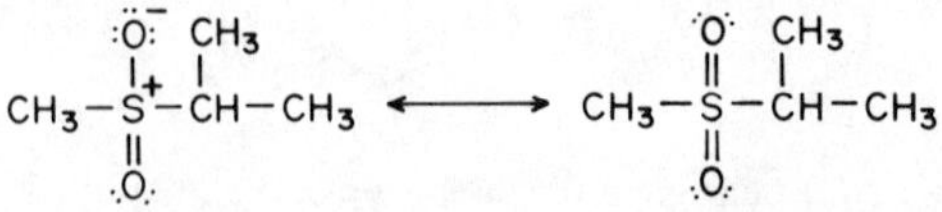

$$CH_3—\underset{\underset{\ddot{O}}{\|}}{\overset{:\ddot{O}:^-}{\overset{|}{\overset{+}{S}}}}—\overset{CH_3}{\overset{|}{C}H}—CH_3 \longleftrightarrow CH_3—\underset{\underset{\ddot{O}}{\|}}{\overset{\ddot{O}}{\overset{\|}{S}}}—\overset{CH_3}{\overset{|}{C}H}—CH_3$$

Figure 4.710 Isopropyl methyl sulfone (2-methylsulfonylpropane).

4.720 Sulfonic Acids, RSO_2OH

Compounds in which the hydrogen atom of a hydrocarbon is replaced by a sulfo ($-SO_2OH$) group.

Example. Benzenesulfonic acid, Fig. 4.720. Replacement of the OH group by halogen gives a **sulfonyl halide**, RSO_2Cl.

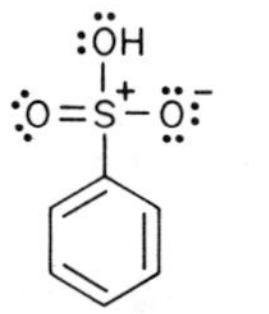

Figure 4.720 Benzenesulfonic acid.

4.730 Sultones, $RCH-(CH_2)_x-SO_2$ (with O bridging RCH and SO_2)

The dehydration product of an hydroxysulfonic acid; analogous to lactones, which are dehydration products of hydroxycarboxylic acids.

Example. γ-Hydroxysulfonic acid sultone, or 1,3-propanesultone, Fig. 4.730.

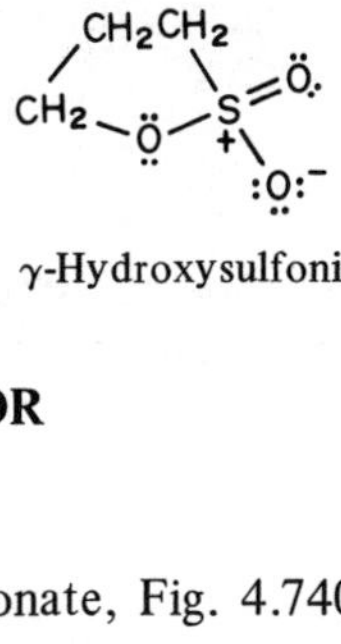

Figure 4.730 γ-Hydroxysulfonic acid sultone.

4.740 Sulfonic Esters, RSO_2OR

Esters of sulfonic acids.

Examples. Methyl ethanesulfonate, Fig. 4.740*a*, and isopropyl benzenesulfonate, Fig. 4.740*b*.

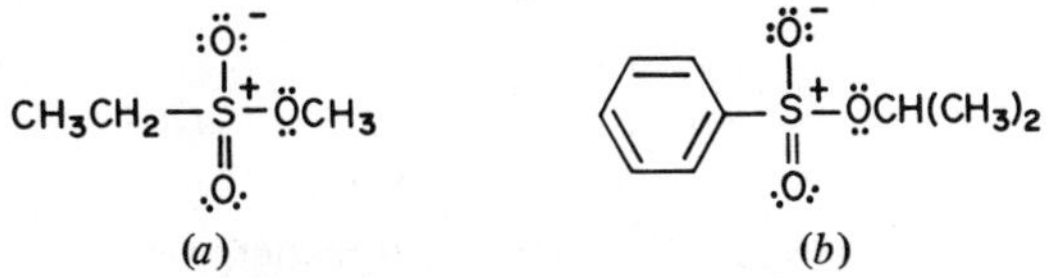

(*a*) (*b*)

Figure 4.740 (*a*) Methyl ethanesulfonate; (*b*) isopropyl benzenesulfonate.

4.750 Sulfate Esters, $(RO)_2SO_2$

Esters of sulfuric acid.

Example. Dimethyl sulfate, Fig. 4.750.

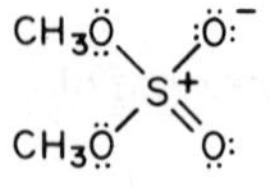

Figure 4.750 Dimethyl sulfate.

4.760 Bisulfate Esters, $ROSO_2OH$

Mono esters of sulfuric acid.

Examples. Methyl hydrogen sulfate, Fig. 4.760.

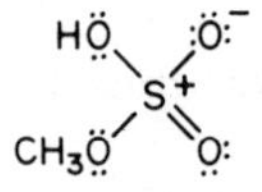

Figure 4.760 Methyl hydrogen sulfate.

4.770 Sulfinic Acids, RS(O)OH

Compounds in which the hydrogen of a hydrocarbon is replaced by a sulfino ($—SO_2H$) group.

Example. Ethanesulfinic acid, Fig. 4.770; replacement of the OH group by halogen gives a **sulfinyl halide,** RS(O)X.

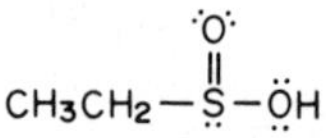

Figure 4.770 Ethanesulfinic acid.

4.780 Sulfinic Esters, RS(O)OR

Esters of sulfinic acid.

Example. Ethyl methanesulfinate, Fig. 4.780.

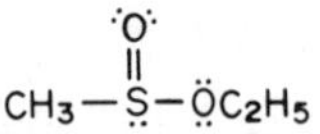

Figure 4.780 Ethyl methanesulfinate.

4.790 Sulfenic Acids, RSOH

Compounds in which the hydrogen of a hydrocarbon is replaced by a sulfeno (—SOH) group.

Example. Butanesulfenic acid, Fig. 4.790; replacement of the OH group by halogen gives a **sulfenyl halide,** RSX.

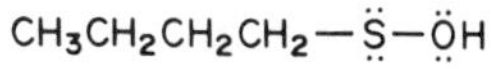

$CH_3CH_2CH_2CH_2{-}\ddot{\underset{..}{S}}{-}\ddot{\underset{..}{O}}H$

Figure 4.790 Butanesulfenic acid.

4.800 Sulfenic Esters, RSOR

Esters of sulfenic acid.

Example. Methyl propanesulfenate, Fig. 4.800.

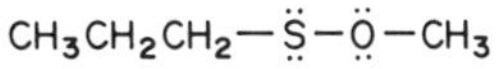

$CH_3CH_2CH_2{-}\ddot{\underset{..}{S}}{-}\ddot{\underset{..}{O}}{-}CH_3$

Figure 4.800 Methyl propanesulfenate.

4.810 Thiocarboxylic Acids, RC(S)OH or RC(O)SH

A generic name for carboxylic acids in which either of the oxygen atoms of the carboxy group has been replaced by sulfur.

4.820 Thioic *S*-Acids, RC(O)SH

Carboxylic acids in which the oxygen of the hydroxy group has been replaced by sulfur.

Example. Propanethioic *S*-acid, Fig. 4.820.

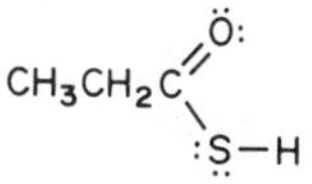

$CH_3CH_2C({=}\ddot{O}{:}){-}\ddot{\underset{..}{S}}{-}H$

Figure 4.820 Propanethioic *S*-acid.

4.830 Thioic *S*-Esters, RC(O)SR

Esters of thioic *S*-acids.

Example. *S*-Methyl ethanethioate, Fig. 4.830.

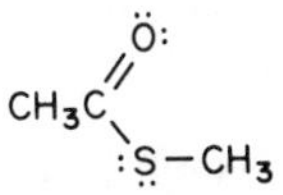

$CH_3C({=}\ddot{O}{:}){-}\ddot{\underset{..}{S}}{-}CH_3$

Figure 4.830 *S*-Methyl ethanethioate.

4.840 Thioic *O*-Acids, RC(S)OH

Compounds in which the carbonyl oxygen of the carboxy group is replaced by sulfur.

Examples. Propanethioic *O*-acid, Fig. 4.840*a*, and benzenecarbothioic *O*-acid, Fig. 4.840*b*.

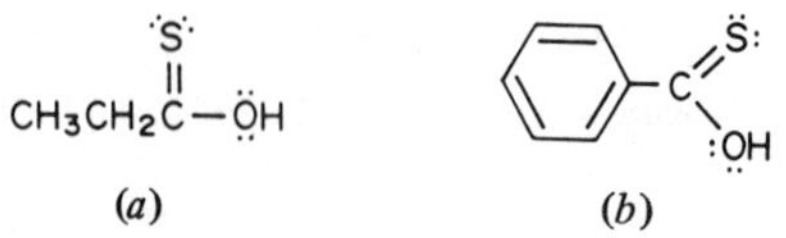

$CH_3CH_2C(=S)-OH$ (a) $C_6H_5C(=S)OH$ (b)

Figure 4.840 (*a*) Propanethioic *O*-acid; (*b*) benzenecarbothioic *O*-acid.

4.850 Thioic *O*-Esters, RC(S)OR

Esters of thioic *O*-acids.

Example. *O*-Ethyl ethanethioate, Fig. 4.850.

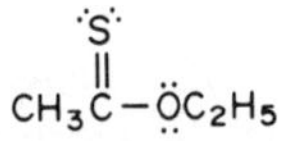

$CH_3C(=S)-OC_2H_5$

Figure 4.850 *O*-Ethyl ethanethioate.

4.860 Dithioic Acids, RCSSH

Compounds in which both of the oxygen atoms of the carboxy group are replaced by sulfur.

Examples. Butanedithioic acid, Fig. 4.860*a*, and 1-naphthalenecarbodithioic acid, Fig. 4.860*b*.

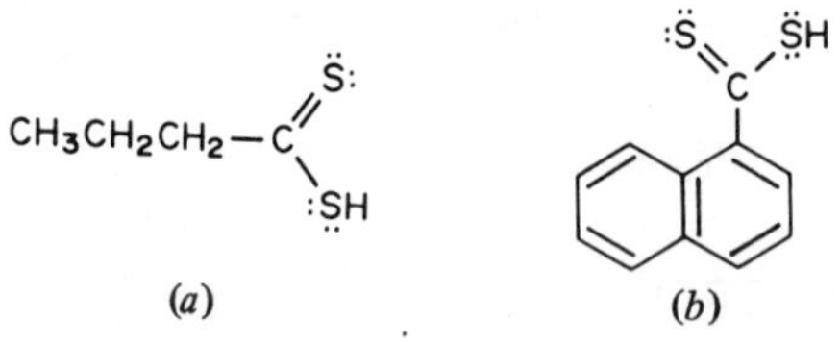

$CH_3CH_2CH_2-C(=S)SH$ (a) (b)

Figure 4.860 (*a*) Butanedithioic acid; (*b*) 1-naphthalenecarbodithioic acid.

4.870 Dithioic Esters, RCSSR

Esters of dithioic acids.

Example. Methyl 1-naphthalenecarbodithioate, Fig. 4.870.

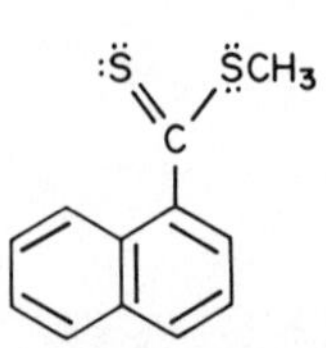

Figure 4.870 Methyl 1-naphthalenecarbodithioate.

4.880 Thioaldehydes, RCHS

Aldehydes in which the carbonyl oxygen is replaced by sulfur; always dimeric or trimeric.

Examples. Ethanethial, Fig. 4.880*a*; the older name of thioacetaldehyde is not sanctioned (IUPAC); benzenecarbothialdehyde, Fig. 4.880*b*.

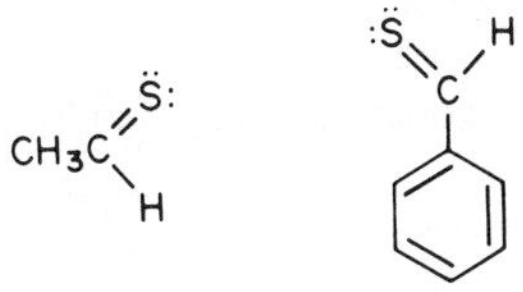

(*a*) (*b*)

Figure 4.880 (*a*) Ethanethial (thioacetaldehyde); (*b*) benzenecarbothialdehyde.

4.890 Thioketones, RCSR

Ketones in which the carbonyl oxygen is replaced by a sulfur atom; often trimeric or dimeric.

Example. 2-Propanethione, Fig. 4.890; the name dimethyl thioketone is also acceptable but the name thioacetone is not approved (IUPAC).

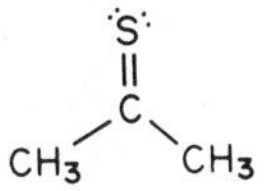

Figure 4.890 2-Propanethione (dimethyl thioketone).

4.900 Thiocyanates, RSCN

Compounds in which the hydrogen atom of a hydrocarbon is replaced by the thiocyanato —S—C≡N group.

Example. 2-Thiocyanatopropane, Fig. 4.900.

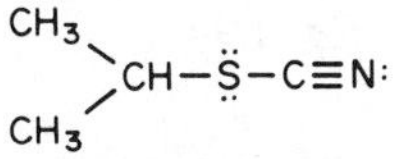

Figure 4.900 2-Thiocyanatopropane.

4.910 Isothiocyanates, RNCS

Compounds in which the hydrogen of a hydrocarbon is replaced by an isothiocyanato group —N=C=S.

Example. Isothiocyanatobenzene, Fig. 4.910.

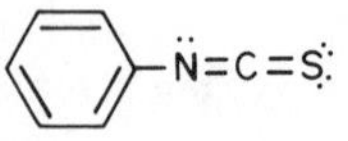

$C_6H_5-\ddot{N}=C=\ddot{S}:$

Figure 4.910 Isothiocyanatobenzene.

4.920 Sulfonamides, RSO_2NH_2

Amides of sulfonic acids.

Examples. 1-Naphthalenesulfonamide, Fig. 4.920*a*, and *N*-ethylmethanesulfonamide, Fig. 4.920*b*.

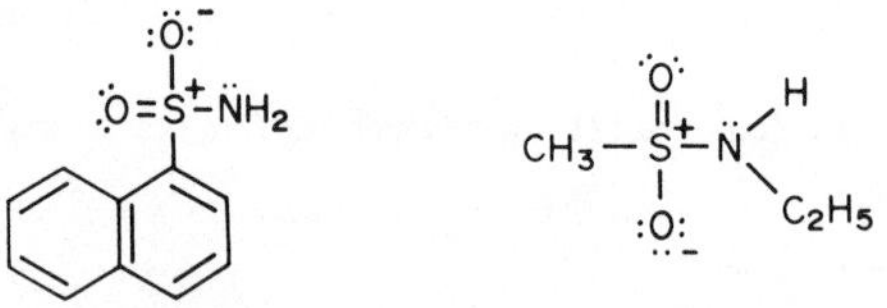

:Ö:⁻ Ö=S⁺–N̈H₂ CH₃–S⁺–N̈ H C₂H₅ :Ö: Ö

(*a*) (*b*)

Figure 4.920 (*a*) 1-Naphthalenesulfonamide; (*b*) *N*-ethylmethanesulfonamide.

4.930 Sulfinamides, $RS(O)NH_2$

Amides of sulfinic acid.

Example. 2-Propanesulfinamide, Fig. 4.930.

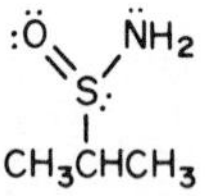

:Ö N̈H₂ S: CH_3CHCH_3

Figure 4.930 2-Propanesulfinamide.

4.940 Sulfenamides, $RSNH_2$

Amides of sulfenic acid.

Example. Ethanesulfenamide, Fig. 4.940.

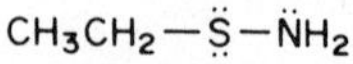

$CH_3CH_2-\ddot{S}-\ddot{N}H_2$

Figure 4.940 Ethanesulfenamide.

4.950 Thioyl Halides, RCSX

Compounds in which the oxygen of the carbonyl group of an acyl halide is replaced by sulfur.

Example. Ethanethioyl chloride, Fig. 4.950.

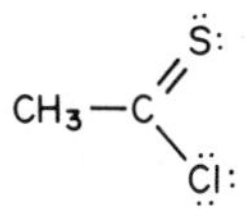

$CH_3-C(=S)-Cl$

Figure 4.950 Ethanethioyl chloride.

4.960 Sulfur Heterocyclics

Compounds that contain one or more sulfur atoms in a ring system.

Examples. There are hundreds of such compounds. They are usually named by the replacement nomenclature system: thiacyclopentane, Fig. 4.960*a*; 1-thiacyclopenta[*b*] naphthalene, Fig. 4.960*b*. In the latter compound the letter in brackets denotes the side of naphthalene to which the thiacyclopentane ring is fused. The sides are lettered starting with side *a* between positions 1 and 2 of the parent hydrocarbon and then lettering each side consecutively clockwise.

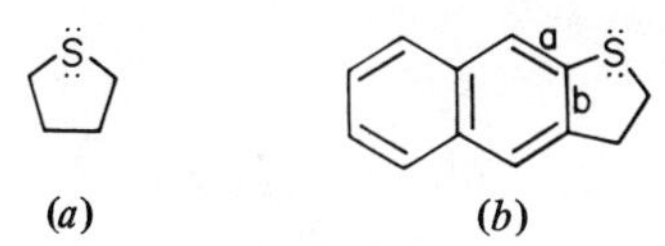

(*a*) (*b*)

Figure 4.960 (*a*) Thiacyclopentane; (*b*), 1-thiacyclopenta[*b*] naphthalene.

4.970 Thiohemiacetals, RCH(OH)(SR)

Compounds arising from the addition of a molecule of a thiol to the carbonyl group of an aldehyde. The generic name is ambiguous, see Sect. 4.980. When the H attached to C is R the compound is a **thiohemiketal.**

Example. 1-(Ethylthio)-1-propanol, Fig. 4.970.

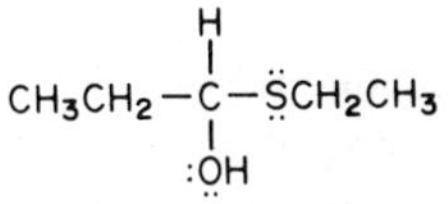

$CH_3CH_2-CH(OH)-SCH_2CH_3$

Figure 4.970 1-(Ethylthio)-1-propanol.

4.980 Dithiohemiacetals, RCH(SH)(SR)

Compounds that may be regarded as arising from the addition of a molecule of thiol to a thioaldehyde; sometimes also called thiohemiacetals. When the H attached to C is R the compound is a **dithiohemiketal.**

Example. 1-(Methylthio)-1-butanethiol, Fig. 4.980.

```
                  H
                  |    ..
CH3CH2CH2 — C — S — CH3
                  |    ..
                 :SH
                  ..
```

Figure 4.980 1-(Methylthio)-1-butanethiol.

4.990 Thioacetals, $RCH(SR)_2$

Compounds in which the oxygen atom of an aldehyde is replaced by two alkylthio (or arylthio) groups. When H is R the compound is a **thioketal.**

Example. 1,1-Bis(ethylthio)propane, Fig. 4.990.

```
           H
           |   ..
CH3CH2 — C — S — CH2CH3
           |   ..
          :S — CH2CH3
           ..
```

Figure 4.990 1,1-Bis(ethylthio)propane.

4.1000 Sulfuranes, R_4S

Neutral, four-coordinated organosulfur compounds.

Example. Tetraphenylsulfurane, Ph_4S, Fig. 4.1000.

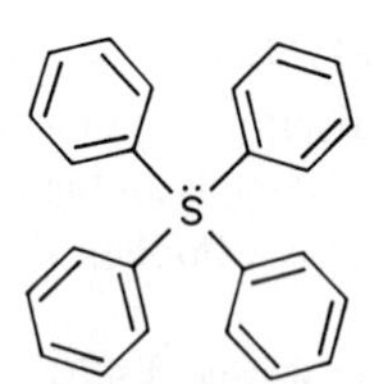

Figure 4.1000 Tetraphenylsulfurane.

4.1010 Organophosphorus Compounds

Compounds containing at least one phosphorus atom linked to at least one carbon atom. More than 100,000 such compounds are known and a seven volume series of books is devoted exclusively to them.

4.1020 Phosphines, RPH_2, R_2PH, R_3P and RPO

Compounds in which one or more of the hydrogen atoms of phosphine, PH_3, are replaced by R groups or compounds in which the two hydrogen atoms of RPH_2 are replaced by an oxygen atom.

Examples. Dimethylphosphine, Fig. 4.1020*a*, and ethyloxophosphine, Fig. 4.1020*b*. When H_2P— is treated as a substituent, the H_2P— group is given the generic prefix name phosphino, e.g., 1-chloro-4-dimethylphosphinobenzene, Fig. 4.1020*c*.

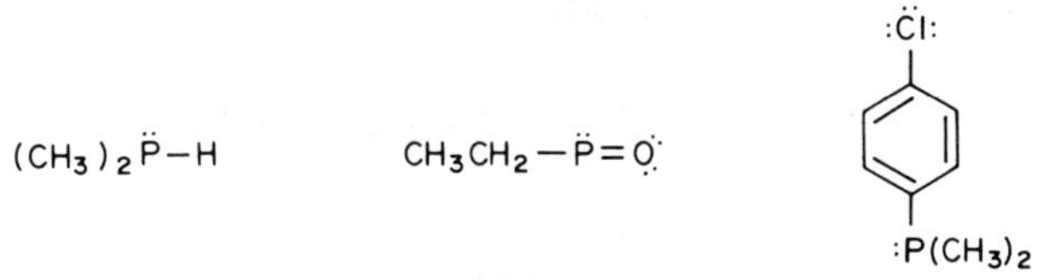

$(CH_3)_2\ddot{P}-H$ $CH_3CH_2-\ddot{P}=\ddot{O}$:Cl: / $:P(CH_3)_2$

(*a*) (*b*) (*c*)

Figure 4.1020 (*a*) Dimethylphosphine; (*b*) ethyloxophosphine; (*c*) 1-chloro-4-dimethylphosphinobenzene.

4.1030 Phosphoranes, R_5P

Compounds in which one or more of the hydrogen atoms of phosphorane, PH_5, are replaced by R groups.

Examples. Triethylphosphorane, Fig. 4.1030*a*. When H_4P— is treated as a substituent, the H_4P— group is given the generic prefix name *phosphoranyl*, e.g., 2,3-dimethyl-1-ethylphosphoranylbenzene, Fig. 4.1030*b*.

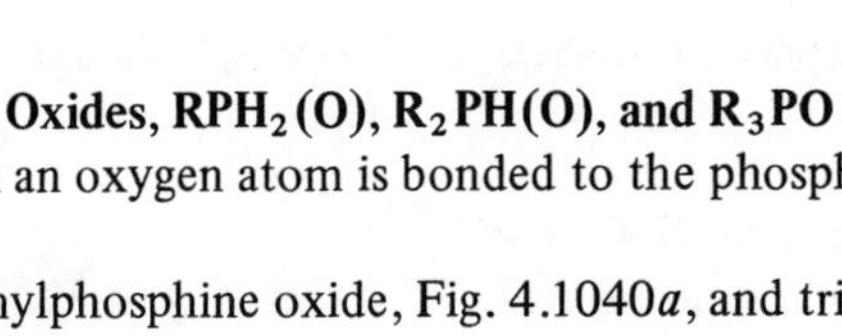

$(CH_3CH_2)_3PH_2$ H H H-P-CH_2CH_3 CH_3 CH_3

(*a*) (*b*)

Figure 4.1030 (*a*) Triethylphosphorane; (*b*) 2,3-dimethyl-1-ethylphosphoranylbenzene.

4.1040 Phosphine Oxides, $RPH_2(O)$, $R_2PH(O)$, and R_3PO

Phosphines in which an oxygen atom is bonded to the phosphorus.

Examples. Triphenylphosphine oxide, Fig. 4.1040*a*, and trimethylenebis(phosphine oxide), Fig. 4.1040*b*.

$Ph_3P^+-\ddot{O}^- \longleftrightarrow Ph_3P=\ddot{O}$

(*a*)

$H-P(=O)(H)-CH_2CH_2CH_2-P(=O)(H)-H$

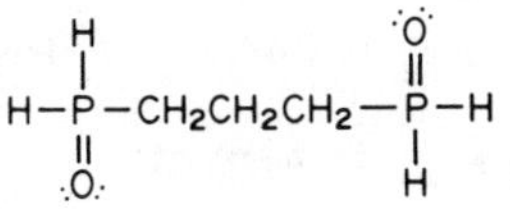

(*b*)

Figure 4.1040 (*a*) Triphenylphosphine oxide; (*b*) trimethylenebis(phosphine oxide).

4.1050 Phosphine Imides, R_3PNH

Analogous to phosphine oxides but with an imino (=NH) group rather than an oxygen atom attached to phosphorous.

Example. *P,P,P*-Triphenylphosphine imide, Fig. 4.1050.

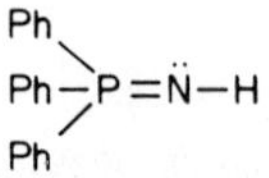

```
Ph
  \
Ph—P=N̈—H
  /
Ph
```

Figure 4.1050 *P,P,P*-Triphenylphosphine imide.

4.1060 Organophosphorus Acids

Organophosphorus compounds containing one or more hydroxy groups bonded to phosphorus.

4.1070 Phosphates $(RO)_3PO$

Derivatives of phosphoric acid, $P(O)(OH)_3$ (also called orthophosphoric acid) in which one (or more) of the hydrogens is replaced by R groups.

Examples. Trimethyl phosphate, Fig. 4.1070*a*; di-*n*-propyl hydrogen phosphate, 4.1070*b*; ethyl dihydrogen phosphate, 4.1070*c*.

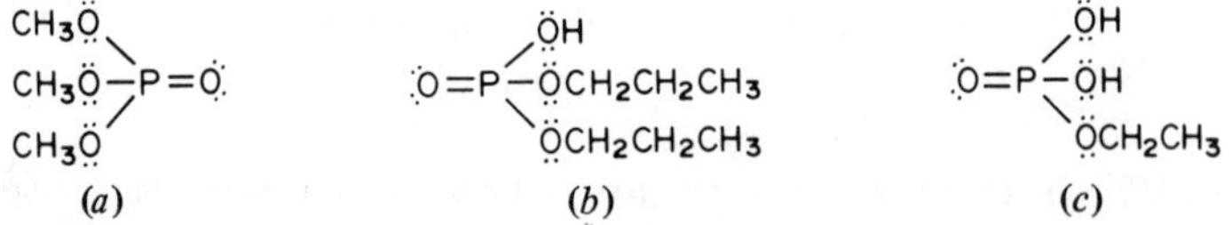

```
CH3O                      OH                       OH
     \                   /                        /
CH3O—P=O        O=P—OCH2CH2CH3          O=P—OH
     /                   \                        \
CH3O                      OCH2CH2CH3               OCH2CH3
  (a)                    (b)                      (c)
```

Figure 4.1070 (*a*) Trimethyl phosphate; (*b*) di-*n*-propyl hydrogen phosphate; (*c*) ethyl dihydrogen phosphate.

4.1080 Phosphinic Acid, $H_2P(O)OH$

A four coordinate phosphorus acid. This acid, Fig. 4.1080*a*, is in equilibrium with a three coordinate species, Fig. 4.1080*b*, called phosphonous acid.

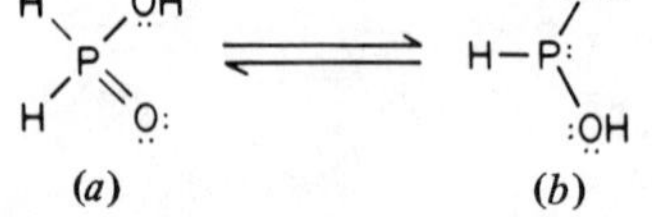

```
H   OH                    OH
 \ /                     /
  P        ⇌        H—P:
 / \\                    \
H   O                     OH
  (a)                   (b)
```

Figure 4.1080 (*a*) Phosphinic acid; (*b*) phosphonous acid.

4.1090 Hypophosphorous Acid, $H_2P(O)OH$

Synonymous with phosphinic acid; the latter is the preferred name.

4.1100 Phosphonous Acid, $HP(OH)_2$

A three coordinate phosphorus acid in equilibrium with phosphinic acid.

4.1110 Alkyl or Aryl Phosphonous Acids, $RP(OH)_2$

Compounds in which the hydrogen atom attached to phosphorus in phosphonous acid is replaced by R—.

Example. Ethylphosphonous acid, Fig. 4.1110.

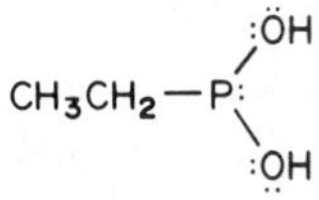

Figure 4.1110 Ethylphosphonous acid.

4.1120 Alkyl or Aryl Phosphinic Acids, RPH(O)OH

Compounds in which one of the hydrogen atoms attached to phosphorus in phosphinic acid is replaced by R—.

Examples. Phenylphosphinic acid, Fig. 4.1120.

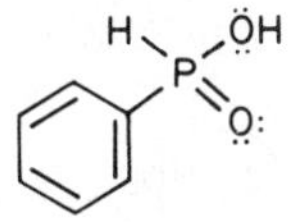

Figure 4.1120 Phenylphosphinic acid.

4.1130 Dialkyl or Diaryl Phosphinic Acids, $R_2P(O)OH$

Compounds in which both hydrogen atoms attached to phosphorus in phosphinic acid are replaced by R— groups.

Example. Ethylmethylphosphinic acid, Fig. 4.1130.

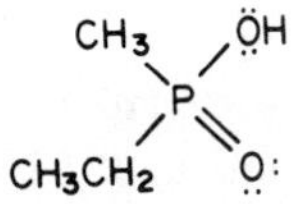

Figure 4.1130 Ethylmethylphosphinic acid.

4.1140 Phosphinous Acid, H_2POH

A three coordinate, monobasic phosphorus acid.

4.1150 Dialkyl or Diaryl Phosphinous Acids, R_2POH

Example. Dimethylphosphinous acid, Fig. 4.1150.

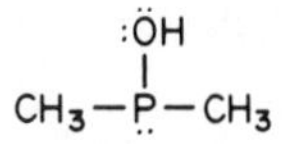

$$CH_3-\ddot{P}(\ddot{O}H)-CH_3$$

Figure 4.1150 Dimethylphosphinous acid.

4.1160 Phosphonic Acid, $HP(O)(OH)_2$

A four coordinate phosphorus acid. This acid, Fig. 4.1160*a*, is in equilibrium with the three coordinate species, Fig. 4.1160*b*, called phosphorous acid.

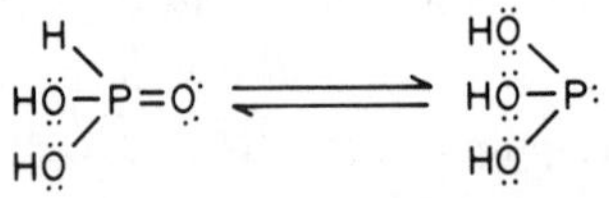

$$(HO)_2P(H)=O \rightleftharpoons (HO)_3P:$$

(*a*) (*b*)

Figure 4.1160 (*a*) Phosphonic acid; (*b*) phosphorous acid.

4.1170 Alkyl or Aryl Phosphonic Acids, $RP(O)(OH)_2$

Example. Ethylphosphonic acid, Fig. 4.1170.

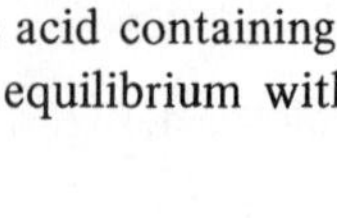

$$(HO)_2P(CH_3CH_2)=O$$

Figure 4.1170 Ethylphosphonic acid.

4.1180 Phosphorous Acid, $P(OH)_3$

A three coordinate phosphorus acid containing three hydroxy groups attached to phosphorus. This acid is in equilibrium with the four coordinated species, phosphonic acid.

4.1190 Phosphites, $(RO)_3P$

Derivatives of phosphorous acid. The trialkyl or triaryl phosphites are compounds in which the three hydrogen atoms in $P(OH)_3$ are replaced by R groups. The dialkyl phosphites are better formulated as four coordinate species, $HP(O)(OR)_2$, while the monoalkyl phosphites are monobasic, four coordinate acids (phosphonic acid structure) $HP(O)(OR)(OH)$.

Examples. Trimethylphosphite, Fig. 4.1190*a*. If the dialkyl phosphite were generically related to phosphorus acid, it would have the structure 4.1190*b* and

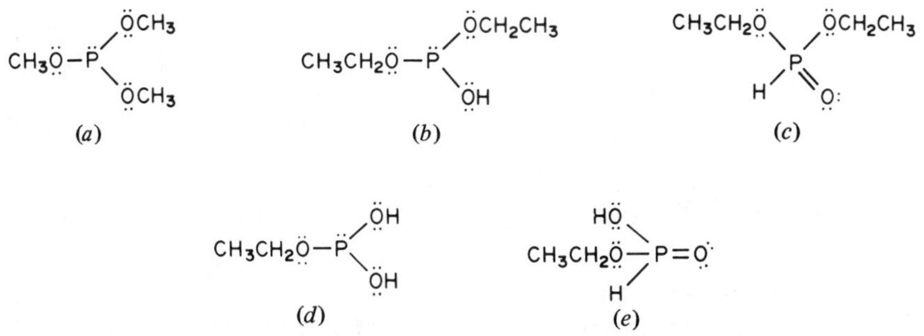

Figure 4.1190 (*a*) Trimethylphosphite; (*b*) diethyl hydrogen phosphite; (*c*) diethyl phosphite; (*d*) ethyl dihydrogen phosphite; (*e*) monoethyl phosphonate.

would be called diethyl hydrogen phosphite. Actually diethyl phosphite has the structure 4.1190*c* and is the diester of phosphonic acid. The monoalkyl phosphite, if it were related to phosphorous acid, would have the structure $ROP(OH)_2$, e.g., ethyl dihydrogen phosphite, Fig. 4.1190*d*. Actually the compound has structure 4.1190*e* and is a monoester of phosphonic acid.

4.1200 Heterocyclic Phosphorus Compounds

Organic compounds in which a phosphorus atom is a member of an organic ring system.

Examples. 2-Phosphanaphthalene, Fig. 4.1200*a*, and 1-phosphacyclopentane 1-oxide, Fig. 4.1200*b*.

Figure 4.1200 (*a*) 2-Phosphanaphthalene; (*b*) 1-phosphacyclopentane 1-oxide.

4.1210 Halophosphorus Compounds

Organophosphorus compounds that possess halogens attached to phosphorus.

Examples. Phenylphosphonous dichloride, Fig. 4.1210*a*. This compound is more commonly called dichlorophenylphosphine, but approved nomenclature recommends that when one or more halogens are attached to a phosphorus atom carrying no OH or OR groups, the compound should be named as an acid halide; in this example the corresponding acid is phenylphosphonous acid, $PhP(OH)_2$. Dimethylphosphinic chloride, Fig. 4.1210*b*.

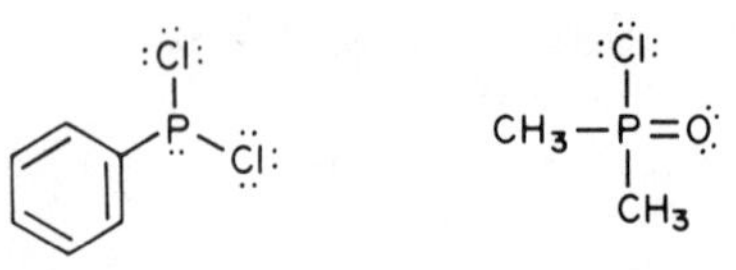

(a) (b)

Figure 4.1210 (*a*) Phenylphosphonous dichloride; (*b*) dimethylphosphinic chloride.

4.1220 Organophosphorus Acid Amides

Organophosphorus acids having one or more NH_2 groups attached to a phosphorus atom.

Examples. *P,P*-Diphenylphosphinous amide, Fig. 4.1220*a*, also called aminodiphenylphosphine but preferably named after diphenylphosphinous acid, $(Ph)_2P—OH$. Compounds carrying no OH or OR groups attached to phosphorus are preferably called amides. *P*-Methylphosphonic diamide, Fig. 4.1220*b*.

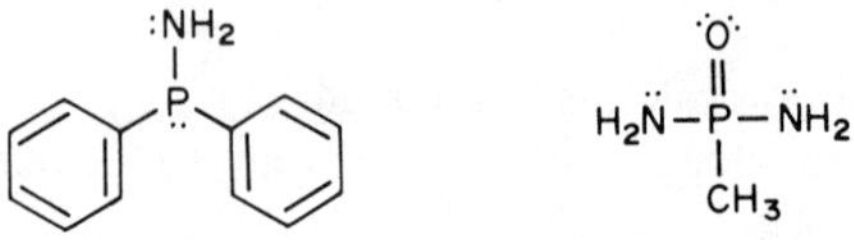

(a) (b)

Figure 4.1220 (*a*) *P,P*-Diphenylphosphinous amide (aminodiphenylphosphine); (*b*) *P*-methylphosphonic diamide.

4.1230 Organophosphorus Thioacids

Organophosphorus acids and esters in which one or more sulfur atoms are attached to phosphorus.

Examples. Dimethylphosphinothioic *O*-acid, Fig. 4.1230*a*, and methylphosphonomonothioic *S*-acid, Fig. 4.1230*b*.

(a) (b)

Figure 4.1230 (*a*) Dimethylphosphinothioic *O*-acid; (*b*) methylphosphonomonothioic *S*-acid.

4.1240 Phosphonium Salts, $R_4\overset{+}{P}\overset{-}{X}$

Tetracoordinated phosphorus cations.

Example. Ethyltrimethylphosphonium bromide, Fig. 4.1240.

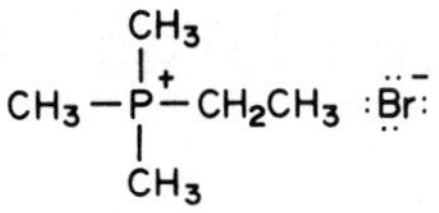

$$CH_3-\overset{+}{P}(CH_3)_2-CH_2CH_3 \quad :\ddot{Br}:^-$$

Figure 4.1240 Ethyltrimethylphosphonium bromide.

5 Stereochemistry and Conformational Analysis

CONTENTS

5.010 Stereochemistry
A description of the three-dimensional spatial relationships between atoms in a molecule.

5.020 Isomers
Two or more molecules that have identical molecular formulas (same number and kinds of atoms), but different arrangements of their atoms. There are three major classes of isomers: structural; stereochemical; and conformational.

5.030 Structural Isomers
Molecules having the same molecular formula, but different sequences of their atoms.

Examples. Ethanol (CH_3CH_2OH) and dimethyl ether (CH_3OCH_3) are structural isomers.

5.040 Constitutional Isomers

Synonymous with structural isomers.

5.050 Positional Isomers

Structural isomers that differ only with respect to the point of attachment of a substituent.

Examples. 1-Chloropropane ($ClCH_2CH_2CH_3$) and 2-chloropropane ($CH_3CHClCH_3$).

5.060 Tautomers

Structural isomers of different energies which are interconvertible via a low energy barrier; the isomerization involves atom or group migration.

5.070 Proton Tautomers (Prototropic Tautomers)

Structural isomers of differing energies that interconvert via migration of a proton; most commonly they have the general structures:

$$H{-}A{-}B{=}Y \rightleftharpoons A{=}B{-}Y{-}H$$

Examples. Enol-keto isomerization (Sect. 4.190) is an example of prototropy (change in position of a proton). The interconversion of the tautomers 2-hydroxypyridine, Fig. 5.070*a*, and pyridone, Fig. 5.070*b*.

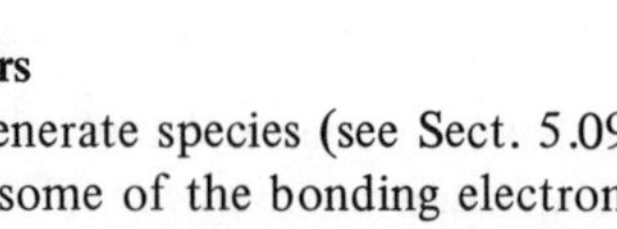

(*a*) (*b*)

Figure 5.070 Prototropic tautomers; (*a*) 2-hydroxypridine and (*b*) 2-pyridone.

5.080 Valence Tautomers

Structural isomers or degenerate species (see Sect. 5.090) that are interconvertible by reorganization of some of the bonding electrons. The interconversion is accompanied by atom movement but does not involve atom migration. (In this sense valence tautomers are not actually tautomers and therefore should preferably be called valence isomers.) Valence tautomers can be separately identifiable molecules or if they have the same structure (degenerate species) the individual atoms can be separately identified. Valence tautomers should not be confused with resonance structures (Sect. 3.360), which represent different electron distributions of the same molecule with no atom movement.

Example. The interconversion of cyclooctatetraene and its bicyclic isomer, Fig. 5.080.

Figure 5.080 Valence tautomers.

5.090 Fluxional Molecules

Molecules which undergo rapid degenerate rearrangement, that is, rearrangement into indistinguishable molecules; the rearrangement may involve either bond reorganization or atom (group) migration.

Examples. The tricyclic compound tricyclo[3.3.2.0^{4,6}]deca-2,7,9-triene, called bullvalene because of the disbelief with which its proposed structure was greeted, Fig. 5.090*a*. It has been calculated that, if each of the 10 carbon atoms of bullvalene were individually labelled, 1,209,600 structures of the labelled compound would be possible! Bullvalene is an example of degenerate valence tautomerism. The fluxionality of the sigma bonded metal cyclopentadienide, Fig. 5.090*b*, involves atom migration but because degeneracy is involved this is not considered to be a case of tautomerism. Fluxionality is most readily ascertained by means of nuclear magnetic resonance (nmr) spectroscopy. In examining a system of interconverting molecules, electronic and vibrational spectral probes detect "instantaneous structures," whereas the structure revealed by nmr at

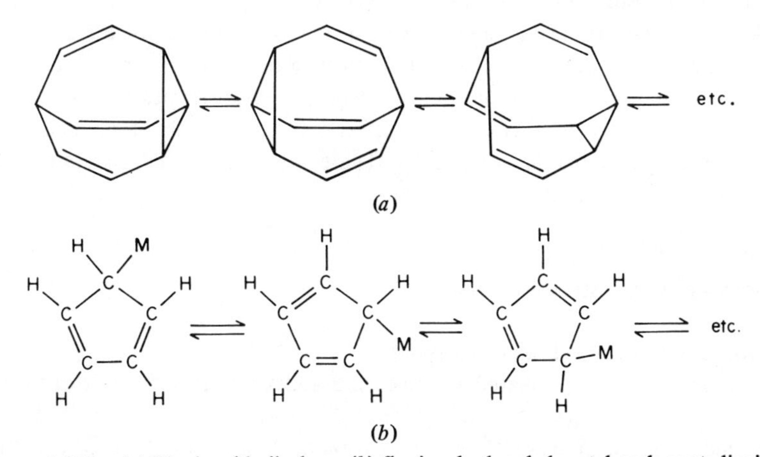

Figure 5.090 (*a*) Fluxional bullvalene; (*b*) fluxional σ bonded metal cyclopentadienide.

about room temperature is usually a "time averaged structure" because, unlike these other types of spectroscopy, the transition times in nmr are long compared to rapid chemical changes. Although the nmr spectrum of fluxional molecules changes with temperature, it is generally possible, with sufficient cooling, to obtain a limiting spectrum. When this occurs, the molecule has been frozen into its "instantaneous structure," which corresponds to any one of the indistinguishable fluxional structures shown in Fig. 5.090*a* or *b*.

5.100 Allylic Isomers

Isomers that result from a rearrangement involving the allyl (propenyl) group; sufficiently common in organic chemistry as to warrant separate recognition.

Example. Figure 5.100. Such rearrangements usually occur via delocalized ion or radical intermediates, which are written in resonance theory as shown in Fig. 5.100*b*. The loss of X at one end and its return to the opposite end of the allyl group results in the overall rearrangement, 5.100*a* → 5.100*c*. Allylic isomers result from a 1,3-rearrangement (see Sect. 10.1050).

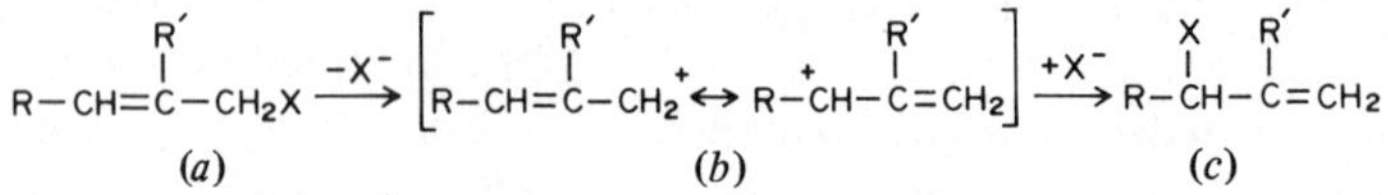

$$\underset{(a)}{R-CH=C(R')-CH_2X} \xrightarrow{-X^-} \underset{(b)}{\left[R-CH=C(R')-\overset{+}{C}H_2 \leftrightarrow R-\overset{+}{C}H-C(R')=CH_2\right]} \xrightarrow{+X^-} \underset{(c)}{R-CH(X)-C(R')=CH_2}$$

Figure 5.100 Rearrangement of allylic halides.

5.110 Configuration

The fixed relative spatial arrangement of atoms in a molecule.

5.120 Stereoisomers (Stereochemical Isomers)

Molecules having the same sequence of atoms and bonds, but differing in the fixed three-dimensional arrangement of these atoms (i.e., having different configurations, hence nonsuperimposable). Each stereoisomer has a unique configuration that can only be converted to a different configuration by chemical means (i.e., breaking and making of bonds). There are two distinct classes of stereoisomers: enantiomers and diastereomers.

5.130 Configurational Isomers

Synonymous with stereoisomers.

5.140 Enantiomers (Enantiomorphs)

Two molecules that are related as object and *nonsuperimposable* mirror image.

Examples. Enantiomers always occur in pairs. Both isomers of the pair shown in Fig. 5.140 have identical physical and chemical properties under symmetrical

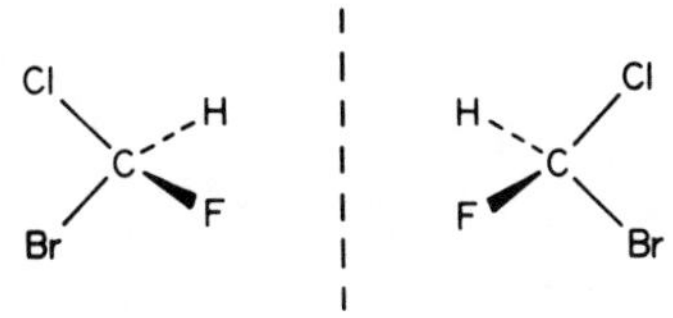

Figure 5.140 Enantiomers.

conditions. Molecules with an alternating axis of any fold [e.g., S_1 (a plane of symmetry) or S_2 (center of symmetry); see Sect. 1.070] have superimposable mirror images and cannot have enantiomers.

5.150 Chiral

Having a nonsuperimposable mirror image [i.e., dissymmetric (Sect. 1.190) or asymmetric (Sect. 1.200)]. Each enantiomer of an enantiomeric pair is chiral.

5.160 Achiral

Lacking a nonsuperimposable mirror image, i.e., nondissymmetric.

5.170 Chiral Center

A tetrahedral atom with four different groups attached to it. In the case of third row elements a nonbonding pair of electrons may count as one of the groups. A chiral center is often designated with an asterisk (⋆).

Example. The carbon atom in each enantiomer in Fig. 5.140 is a chiral center.

5.180 Asymmetric Center

Synonymous with chiral center.

5.190 Dissymmetric Grouping

An arrangement of atoms leading to dissymmetry (Sect. 1.190), hence chiral but lacking a specific chiral center.

Example. 1,3-Dimethylallene, Fig. 5.190, has a C_2 axis as its only symmetry element. The molecule is therefore chiral and exists as a pair of enantiomers.

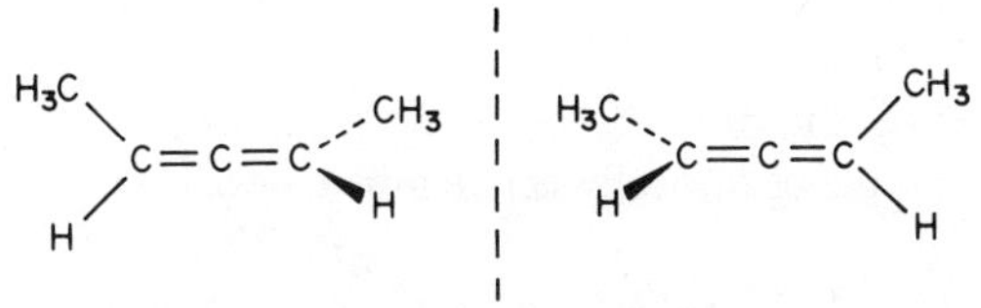

Figure 5.190 1,3-Dimethylallene.

5.200 Optical Isomers
Synonymous with enantiomers.

5.210 Optical Antipodes
Synonymous with enantiomers.

5.220 Plane-Polarized Light
Electromagnetic radiation in which the electric vector (**E**) oscillates in a single plane.

Example. Figure 5.220 shows the electric vector of a plane-polarized light wave.

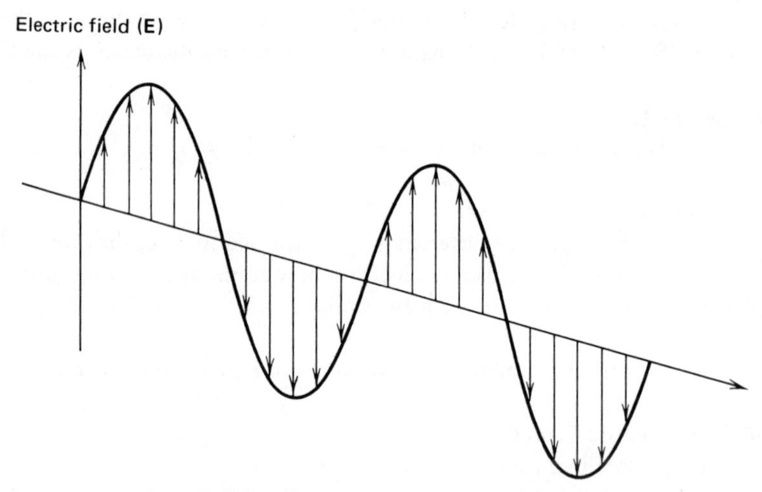

Figure 5.220 Electric vector of plane-polarized light.

5.230 Optical Activity
The property of the molecules of an enantiomer to rotate the polarization plane as plane-polarized light interacts with them. Such molecules (and the compounds they constitute) are said to be optically active. The words *optically active* and *chiral* are frequently used interchangeably even though a chiral molecule shows optical activity only when exposed to plane-polarized light.

5.240 Optical Rotation, α
The angle by which the polarization plane is rotated as plane-polarized light passes through a sample of optically active molecules.

Example. Figure 5.240; the angle α is -40° in this example.

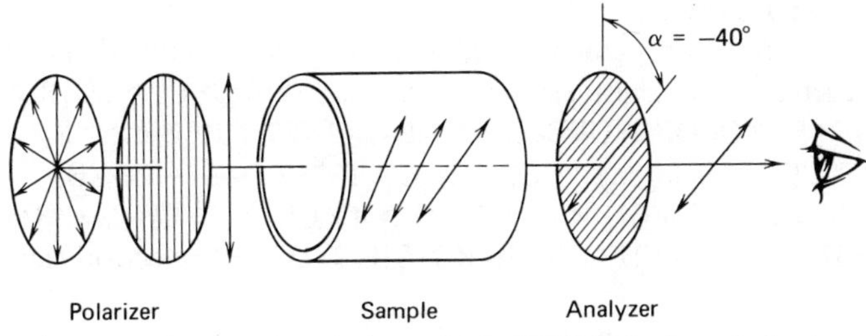

Figure 5.240 Rotation of plane of polarized light in the polarimeter.

5.250 Specific Rotation, [α]

Related to the optical rotation (α) by the formula:

$$[\alpha]_{\lambda}^{T} = \frac{100\alpha}{l \cdot c}$$

where T is the temperature
λ is the wavelength of plane-polarized light
l is the sample cell length in decimeters
c is the sample concentration in g/100 mL

The sign and magnitude of [α] are functions of these variables, as well as the nature of the solvent. Under a given set of conditions the specific rotation of one enantiomer has the same magnitude but opposite sign as that of its mirror image.

5.260 Polarimeter

The instrument that measures optical rotation, Fig. 5.240.

5.270 (+)-Enantiomer

The enantiomer that rotates the polarization plane in the clockwise direction, when viewed as shown in Fig. 5.240.

5.271 (−)-Enantiomer

The enantiomer that rotates the polarization plane in the counterclockwise direction, when viewed as shown in Fig. 5.240.

5.280 *d* (Dextrorotatory)

Synonymous with (+).

5.281 *l* (Levorotatory)

Synonymous with (−).

5.290 Group Priority

An ordering of functional groups based on their atomic number (heaviest isotope first). A partial list (highest priority first) has I > Br > Cl > SO_2R > SOR > SR > SH > F > OCOR > OR > OH > NO_2 > NO > NHCOR > NR_2 > NHR > NH_2 > CX_3 (X = halogen) > COX > CO_2R > CO_2H > $CONH_2$ > COR > CHO > CR_2OH > CH(OH)R > CH_2OH > C≡CR > C≡CH > C(R)=CR_2 > C_6H_5 > CH_2 > CR_3 > CHR_2 > CH_2R > CH_3 > D > H > electron pair.

5.300 (*R*) (*Latin* Rectus, Right)

That configuration of a chiral center with a clockwise relationship of group priorities (highest to second lowest) when viewed along the bond toward the lowest priority group.

Example. Figure 5.300*a*.

(R)

(*a*)

(S)

(*b*)

5.301 (*S*) (*Latin* Sinister, Left)

That configuration of a chiral center with a counterclockwise relationship of group priorities (highest to second lowest) when viewed along the bond toward the lowest priority group.

Example. Figure 5.300*b*. For molecules with a dissymmetric grouping the (*R*) or (*S*) configuration is found by assigning priority **1** to the higher priority group *in front*, **2** to the lower priority group *in front*, **3** to the higher priority group in back, etc., then examining the path **1** → **2** → **3** → **4**. The assignment for 1,3-dimethylallene enantiomers, Fig. 5.190, is shown in Fig. 5.300*c*.

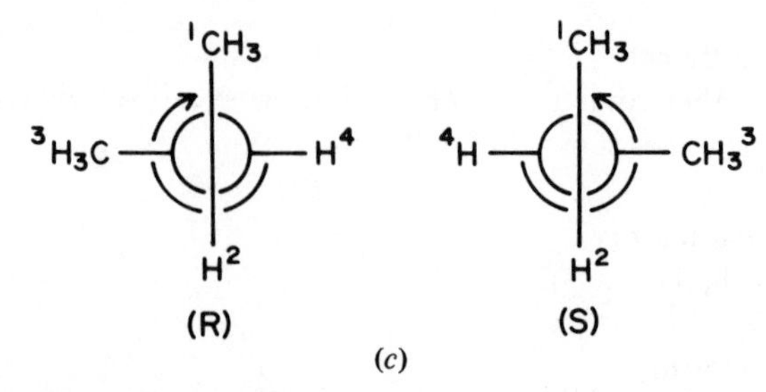

Figure 5.300 Assignment of absolute configuration: (*a*) chiral center with (*R*) configuration; (*b*) (*S*), its enantiomer; (*c*) dissymmetric group assignment.

5.310 Absolute Configuration

The specification of (*R*) or (*S*) at each chiral center (and dissymmetric grouping) in a molecule.

Example. Figure 5.310.

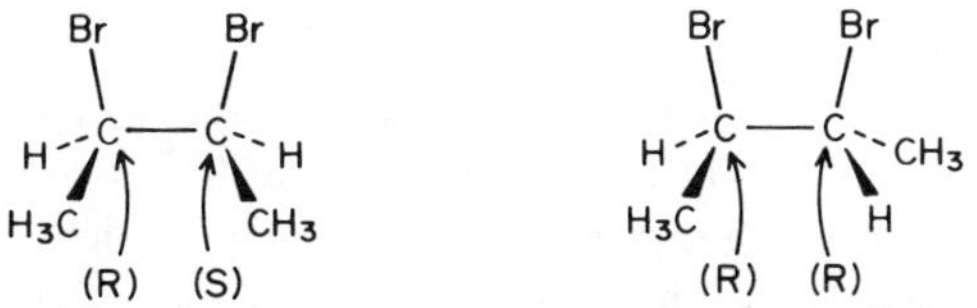

Figure 5.310 Absolute configuration at several chiral centers.

5.320 Relative Configuration

The relationship between the configurations of two chiral molecules. When two such molecules can be chemically interconverted (at least in principle) without breaking any bonds to the chiral center, they are said to have the same relative configuration independent of the direction of rotation of the plane of polarized light and independent of the (*R*) and (*S*) designation.

Example. Oxidation of (−)-2-methyl-1-butanol to (+)-2-methylbutanoic acid, Fig. 5.320, does not involve alteration of bonds to the chiral center. Thus these two molecules must have the same relative configuration at their chiral centers, even though one may not know their absolute configuration.

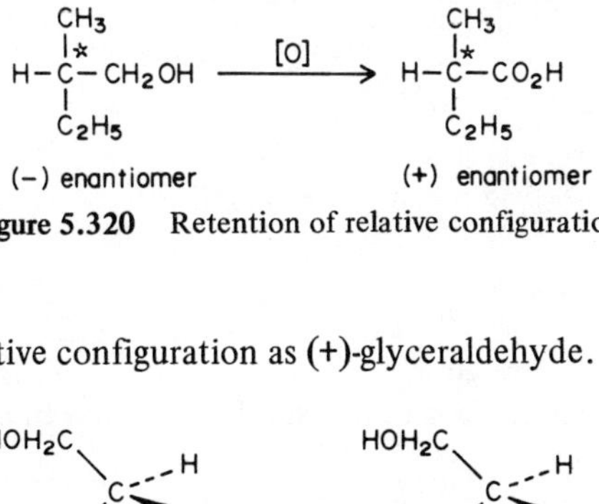

Figure 5.320 Retention of relative configuration.

5.330 D

Having the same relative configuration as (+)-glyceraldehyde.

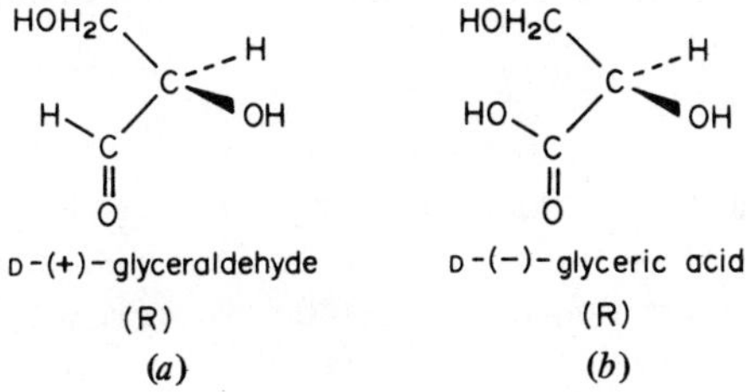

Figure 5.330 Assignments of D and L configurations: (*a*) D-glyceraldehyde; (*b*) D-glyceric acid.

Example. (+)-Glyceraldehyde is now known to have the (*R*) configuration, Fig. 5.330*a*. Conversion to glyceric acid does not involve breaking a bond to the chiral center and thus gives a compound with D configuration, Fig. 5.330*b*.

5.331 L

Having the same relative configuration as (−)-glyceraldehyde. Historically, glyceraldehyde was the reference compound for the stereochemical elaboration of the sugars. The simple sugars can actually be synthesized by starting with either (+) or (−) glyceraldehyde. Thus of the 16 possible aldohexoses (Sect. 6.240) half may be considered as being derived from D-(+)-glyceraldehyde and the other 8 from L-(−)-glyceraldehyde. The D and L in the names of these sugars specifies that configuration at only the starred chiral carbon, Fig. 5.331*a* and *b*. Usually a common name specifies the configuration at the other chiral centers. Thus D-(+)-glucose means that dextrorotatory glucose has the same configuration at the highest numbered chiral carbon (Fig. 5.331*c*) as that of D-glyceraldehyde, and the configuration at the other chiral centers must be memorized. In a common convention showing configurational relationships (the Fischer projection, Sect. 5.830, *vide infra*) the OH group is placed on the right-hand side of the plane formula to indicate the D configuration on the carbon related to D-glyceraldehyde.

```
      CHO              CHO              CHO
       |                |                |1
      CHOH             CHOH          H - C - OH
       |                |                |2
      CHOH             CHOH         HO - C - H
       |                |                |3
      CHOH             CHOH          H - C - OH
       | *              | *              |4
  H - C - OH      HO - C - H         H - C - OH
       |                |                |
      CH2OH            CH2OH            CH2OH
      (a)              (b)              (c)
```

Figure 5.331 Assignments of D and L configuration: (*a*) D-aldohexose; (*b*) L-aldohexose; (*c*) D-glucose.

5.340 Inversion (of Configuration)

The conversion of one molecule into another that has the opposite relative configuration.

Examples. Two reactions that proceed with inversion are shown in Fig. 5.340*a* and *b*. Inversion does not *require* a change in designated absolute configuration, nor does a change in designated absolute configuration require inversion. The process illustrated in Fig. 5.340*c* proceeds with inversion although (*S*)-reactant → (*S*)-product!

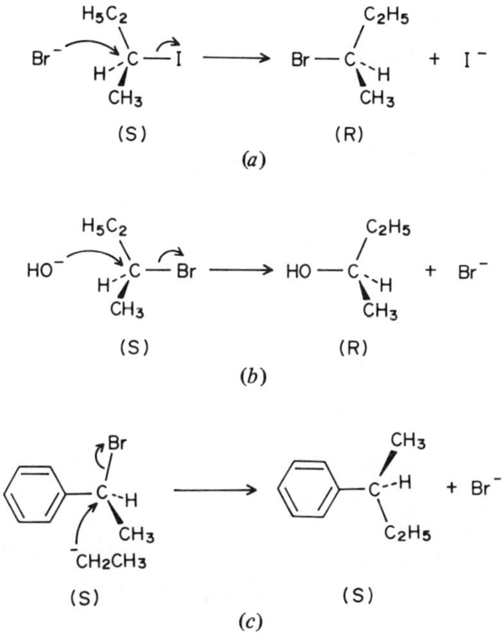

Figure 5.340 Inversion of relative configuration: (*a*, *b*) with change in absolute configuration (*c*) without change in absolute configuration.

5.350 Retention (of Configuration)

The conversion of one molecule into another with the same relative configuration; the opposite of inversion of configuration.

5.360 Racemic (Racemic Modification)

The generic term for a 50:50 combination of a pair of enantiomers, with the mixture having zero optical rotation.

5.370 *d*,*l*-Pair

Synonymous with racemic modification.

5.380 Conglomerate

An exactly 50:50 mixture of *crystals* of pure (+)-enantiomer with crystals of the pure (−)-enantiomer. Each individual crystal is a single enantiomer but the mixture, when well mixed, has a lower melting point than the pure enantiomers and shows zero optical rotation. A small amount of pure enantiomer added to the conglomerate always raises the melting point of a conglomerate.

5.390 Racemic Mixture

Synonymous with conglomerate.

5.400 Racemate

A *compound*, individual crystals of which contain equal numbers of *d* and *l* molecules. It has a sharp melting point that usually differs from that of the pure enantiomers, and the physical properties of the racemate differ from those of the pure enantiomers, but it has zero optical rotation. A small amount of pure enantiomer added to a racemate always lowers the melting point.

5.410 Racemize

To generate a racemic modification.

5.420 Resolution

The separation of a racemic modification into its pure enantiomers.

5.430 Enantiomeric Purity

The fraction or percentage of one enantiomer in a mixture. The enantiomeric purity of a racemic modification is 50% *d* and 50% *l*.

5.440 Optical Purity

The excess of one enantiomer over the *d*,*l* pair in a *d*,*l* mixture. The optical purity (OP) is related to the enantiomeric purity by the formula

$$\text{OP} = 2\ (\%\ \text{enantiomer that is in excess}) - 100\%$$

Example. The specific rotation of any *d*,*l* mixture is equal to the specific rotation of the pure enantiomer times the optical purity (expressed as a fraction). These relationships are illustrated in the following table for a compound whose maximum specific rotation is 150°:

Enantiomeric purity	Optical purity (%)	[α] (°)
100% *d*	100	+150
75% *d* (25% *l*)	50	+75
50% *d* (50% *l*)	0	0
75% *l* (25% *d*)	50	-75
100% *l*	100	-150

5.450 Enantiomeric Excess

Synonymous with optical purity.

5.460 Optical Yield

The optical purity of the product(s) from the reaction of a pure enantiomer, regardless of the chemical yield.

Example. If optically pure A gives a 100% yield of racemic B, the optical yield is 0%. If optically pure A gives a 66% chemical yield of optically pure B, the optical yield is 100%. If optically pure A gives a 50% chemical yield of 36% optically pure B, the optical yield is 36%.

Note that this definition is valid regardless of whether the process involves retention or inversion.

5.470 Optical Stability

A measure of how resistant a pure enantiomer is toward racemization under a given set of conditions.

5.480 Diastereomers (Diastereoisomers)

Stereoisomers that are not enantiomers.

Examples. An important example of one type is the relationship of the compounds shown in Fig. 5.310. The molecule with (*R*),(*S*) configuration is a diastereomer of the one with (*R*),(*R*) configuration. The (*S*),(*S*) configuration (not shown) is a diastereomer of the (*R*), (*S*) molecule but an enantiomer of the (*R*), (*R*) molecule. Other examples are shown in Fig. 5.620.

5.490 Homotopic Hydrogens

A set of hydrogen atoms that can be interchanged by a rotational axis of symmetry C_p ($\infty > p > 1$). The homotopic relationship can be ascertained by the atoms replacement test; i.e., if one hydrogen atom in a set of hydrogens under consideration is replaced by any other atom and this leads to a compound indistinguishable from one obtained by similar replacement of another hydrogen atom in the set, these hydrogens of the original molecule constitute a set of homotopic hydrogens.

Examples. The following are sets of homotopic hydrogen atoms: the 3H's of NH_3 or of CH_3Cl; the 4H's of $CH_2{=}CH_2$ or of CH_4; the 6H's of benzene or of CH_3CH_3. The 3H's of the methyl group in CH_3OH are homotopic (e.g., only one CH_2DOH is possible) because of local symmetry (Sect. 1.150) of the CH_3 group (local C_3 axis) and the free rotation of the CH_3 group around the C–O bond axis. However, the 2H's of CH_2ClF, Fig. 5.490*a*, are not homotopic because of the absence of a C_p, $p > 1$. The hydrogen atoms on C-1 and C-2 of *trans*-1,2-dichlorocyclopropane, Fig. 5.490*b*, are homotopic; in this example replacement

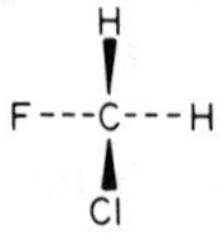

Figure 5.490 (*a*) CH_2ClF, with nonhomotopic hydrogen atoms; (*b*) homotopic hydrogen atoms.

of either hydrogen of this set by another atom (e.g., Br) leads to an identical (chiral) molecule.

5.500 Equivalent Hydrogens

Synonymous with homotopic hydrogens. This term is still widely used, but its popularity will probably decrease because it lacks precision. All homotopic (equivalent) hydrogens are necessarily symmetry equivalent (i.e., exchangeable by *any* symmetry operation, Sect. 1.250) but not all symmetry equivalent hydrogens are homotopic.

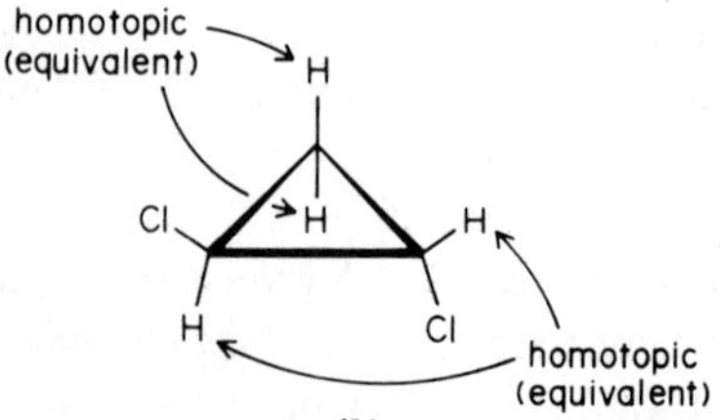

5.510 Equivalent Groups

Atoms (including H's) or groups of atoms that can be interchanged by an axis of rotation C_p ($\infty > p > 1$); analogous to homotopic or equivalent hydrogen atoms but generalized to cover other atoms.

Examples. *cis*-Dichloroethylene, shown in Fig. 5.510, has the following groups of equivalent atoms: 2 H's; 2 Cl's; 2C's; and 2 CHCl's.

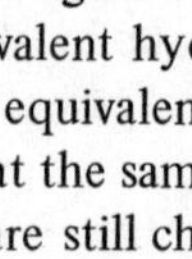

Figure 5.510 (*Z*)-Dichloroethylene, in which all like atoms and groups are homotopic.

5.520 Chemically Equivalent Hydrogens or Groups

Synonymous with symmetry equivalent hydrogens (Sect. 1.250). In the absence of a chiral environment symmetry equivalent atoms are chemically indistinguishable; e.g., they undergo reactions at the same rate. In the presence of a chiral environment homotopic hydrogens are still chemically indistinguishable, but enantiotopic hydrogens (Sect. 5.540) are chemically distinguishable, hence in such an environment are not chemically equivalent. The term chemically equivalent is somewhat ambiguous and is being abandoned by some authors.

5.530 Equivalent Faces

Opposite faces parallel to the nodal plane of a π orbital system that can be exchanged by an axis of rotation C_p ($\infty > p > 1$). Attack at either face of the π bond leads to the same species.

Examples. Cyclopropanone, Fig. 5.530*a*, and *trans*-2,3-dichlorocyclopropanone, Fig. 5.530*b*, both have equivalent faces, although the former molecule is achiral and the latter is dissymmetric and chiral. The opposite faces of (*Z*)-2-butene are equivalent but those of (*E*)-2 butene are not (see Sect. 5.570).

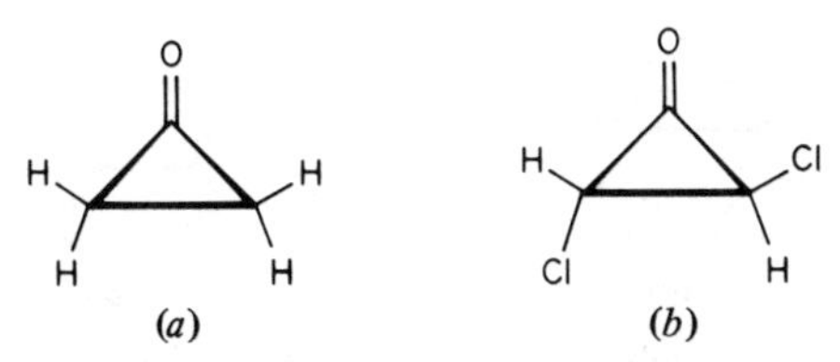

Figure 5.530 Molecules with equivalent faces: (*a*) cyclopropanone; (*b*) (*E*)-2,3-dichlorocyclopropanone.

5.540 Enantiotopic Hydrogens

A pair of hydrogen atoms in a molecule that are interchangeable by an S_p axis only, usually an S_1 axis equivalent to a plane of symmetry or S_2 equivalent to a center of inversion. Substitution of one enantiopic hydrogen by another atom or group leads to a dissymmetric molecule and similar substitution of the other hydrogen gives the enantiomer.

Examples. Most commonly, enantiotopic hydrogens are bonded to the same carbon atom and are symmetry equivalent by virtue of a plane of symmetry. Figure 5.540*a* shows the pair of enantiotopic H's of $ClCH_2F$ (the plane of symmetry lies in the plane of the paper) and the enantiomers formed by substitution. Enantiotopic hydrogens are pervasive in organic chemistry, since all compounds having the general structure shown in Fig. 5.540*b* possess them. It is not a necessary requirement of the enantiotopic relationship that the two symmetry equivalent hydrogen atoms be on the same carbon atom. Thus the hydrogen atoms on C-1 and C-2 in *cis*-1,2-dichlorocyclopropane, Fig. 5.540*c* (and on 3-chlorocyclopropene, Fig. 5.550*a*), are enantiotopic. Enantiotopic hydrogens exchangeable only by S_2 (center of inversion) are shown in Fig. 5.540*d*.

5.550 Prochiral Hydrogens

Synonymous with enantiotopic hydrogens (or other atoms).

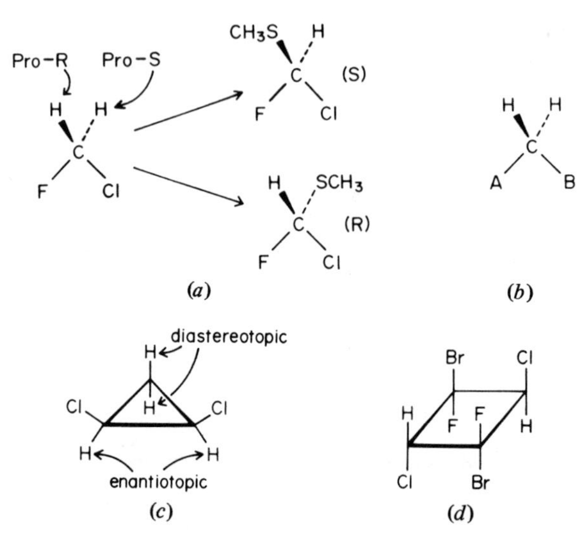

Figure 5.540 Enantiotopic hydrogens: (*a*) in CH_2ClF; (*b*) in a general case; (*c*) in *cis*-dichlorocyclopropane; (*d*) in a molecule with a center of inversion (S_2) only.

Examples. The enantiotopic hydrogens in Fig. 5.540*a* are prochiral atoms. They may be separately indicated as pro-(*R*) and pro-(*S*). The hydrogen under consideration is pro-(*R*) if in the orientation that places this hydrogen in front and the other hydrogen in the rear, one must move in the clockwise direction in going from the highest priority to the lower priority grouping. Note that in the example shown in Fig. 5.540*a* the replacement of the pro-(*R*) group leads to an (*S*) configuration; the configuration of the product depends on the nature of the group that replaces the pro-(*R*) hydrogen. In 3-chlorocyclopropene, Fig. 5.550*a*, the two hydrogen atoms on the olefinic carbons are enantiotopic and prochiral; the replacement of either one of them by another atom leads to chirality that is centered on the methine carbon and not on either of the olefinic carbons. Some authors refer to diastereotopic hydrogens (Sect. 5.580) on the same carbon atom in chiral molecules, Fig. 5.550*b*, as being prochiral because substitution of one of them leads to a **new chiral center.** Such hydrogens can be said to be prochiral with respect to the new chiral center, but they are not truly prochiral with respect to the entire molecule since it was chiral to begin with.

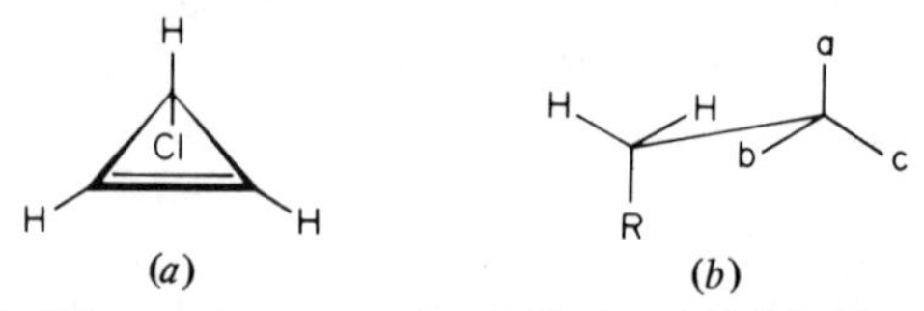

Figure 5.550 (*a*) 3-Chlorocyclopropene, in which the olefinic hydrogens are prochiral; (*b*) diastereotopic hydrogens that may be considered prochiral.

5.560 Enantiotopic Groups

Analogous to enantiotopic hydrogen atoms but generalized to cover other atoms or groups of atoms.

Examples. The two CHFCl groups in Fig. 5.560*a* are enantiotopic groups. In this molecule with the configuration as shown, the H's, F's, and Cl's of the CHFCl groups are also sets of enantiotopic atoms. In Fig. 5.560*b* the two $—CH_2CO_2H$ groups are enantiotopic.

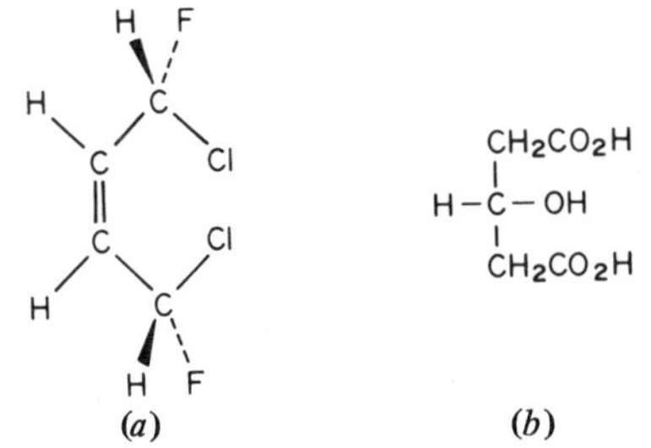

Figure 5.560 Enantiotopic groups: (*a*) CHFCl; (*b*) CH_2CO_2H.

5.570 Enantiotopic Faces

Opposite faces parallel to the nodal plane of a π orbital system that can be exchanged by an alternating axis (S_p) *only*. Attack of a reagent on one side of the π bond leads to one enantiomer, whereas attack on the opposite side leads to the other enantiomer.

Examples. A carbonyl compound of the general structure shown in Fig. 5.570*a* has enantiotopic faces (exchanged by a plane of symmetry equivalent to S_1). Re-

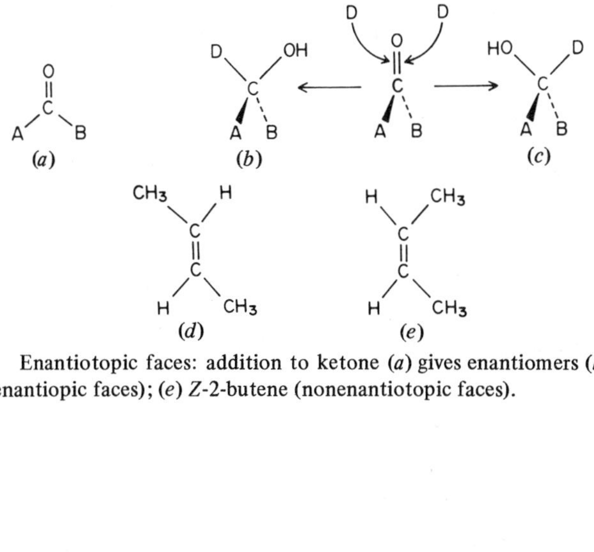

Figure 5.570 Enantiotopic faces: addition to ketone (*a*) gives enantiomers (*b*) and (*c*); (*d*) *E*-2-butene (enantiopic faces); (*e*) *Z*-2-butene (nonenantiotopic faces).

action of an achiral reagent at one face gives rise to a molecule that is the enantiomer of the molecule obtained by reaction at the other face, Fig. 5.570*b* and *c*. The two faces of (*E*)-2-butene, Fig. 5.570*d*, are enantiotopic (S_1 and S_2) in contrast to those of (*Z*)-2-butene, Fig. 5.570*e* (S_1 but C_2 also, hence homotopic). The difference in the faces of these isomers can be visualized by considering the structures of the corresponding epoxides (or analogous derivatives). The epoxide from (*E*)-2-butene is chiral (C_2 only), whereas the one from (*Z*)-2-butene (plane of symmetry) is achiral, hence the faces of the butenes from which they were formed are enantiotopic and homotopic, respectively. Enantiotopic faces are frequently called *prochiral faces* (see Sect. 5.620 for *E* and *Z*).

5.580 Diastereotopic Hydrogens

If substitution of one of the hydrogens of a pair under consideration by another atom gives a diastereomer of the molecule obtained by similar substitution of the other hydrogen atom, the hydrogens are diastereotopic; such hydrogen atoms cannot be interchanged by any symmetry operation.

Examples. The two H's on the same carbon of chloroethylene, Fig. 5.580*a*, constitute a set of diastereotopic hydrogens in this achiral molecule. The relationship between either of the diastereotopic hydrogens and the hydrogen on the chlorine-containing carbon is constitutional (positional) and not stereochemical; replacement of this hydrogen by another atom gives a positional isomer of the compound obtained by replacement of either diastereotopic hydrogen by the same atom. The methylene hydrogens of the chiral molecule, 1,2-fluorochlorocyclopropane, shown in Fig. 5.580*b*, are diastereotopic, as are those of *cis*-1,2-dichlorocyclopropane, Fig. 5.540*c*, an achiral molecule. The methylene hydrogens of the chiral molecule *trans*-1,2-dichlorocyclopropane, Fig. 5.490*b*, are equivalent (homotopic), not diastereotopic. Substitution of the diastereotopic methylene hydrogens alpha to the carbonyl group in the compound shown in Fig. 5.580*c* by OH leads to the diastereomers, erythrose and threose.

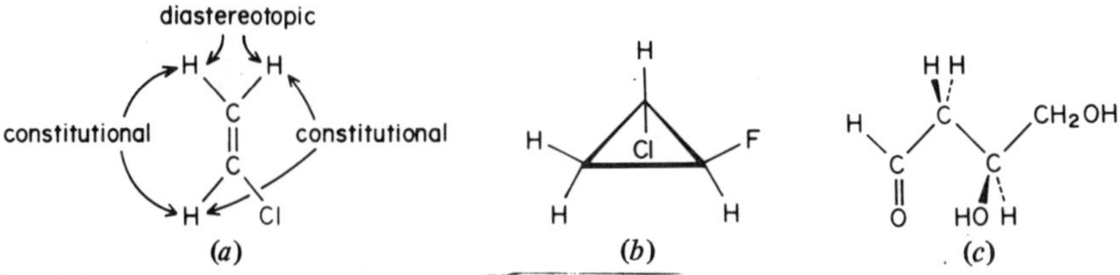

Figure 5.580 Relationship of hydrogen atoms: (*a*) in chloroethylene; (*b*), (*c*) diastereotopic methylene hydrogens.

5.590 Diastereotopic Groups

Analogous to diastereotopic hydrogens but generalized to cover other atoms or groups of atoms.

5.600 Diastereotopic Faces

Opposite faces parallel to the nodal plane of a π orbital system of a molecule that cannot be interchanged by any symmetry operation.

Examples. 5-Bromocyclopentadiene, Fig. 5.600*a*. The chiral molecule 3-chloro-2-butanone, Fig. 5.600*b*, has diastereotopic faces; reduction of the carbonyl to a carbinol (e.g., by $LiAlH_4$) gives rise to two diastereoisomers. The achiral molecule shown in Fig. 5.600*c* possesses diastereotopic faces (this is a *meso* compound, *vide infra*).

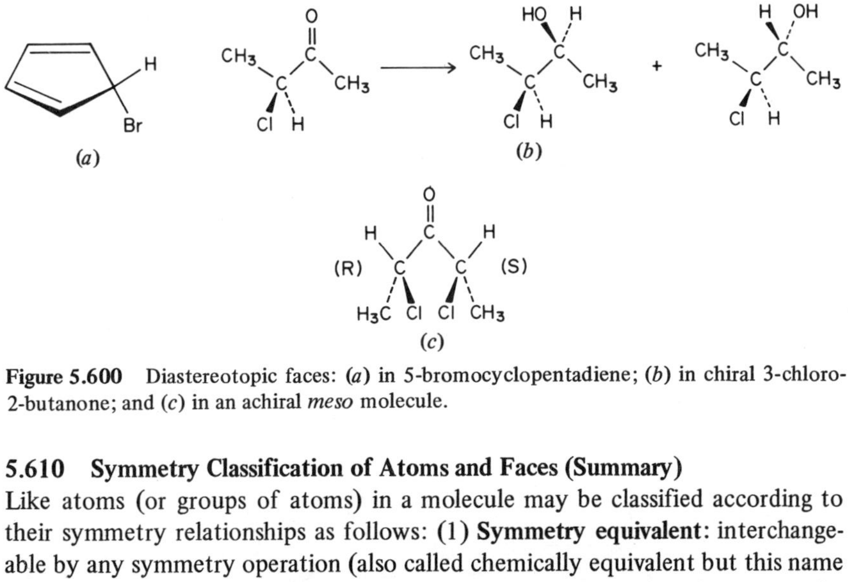

Figure 5.600 Diastereotopic faces: (*a*) in 5-bromocyclopentadiene; (*b*) in chiral 3-chloro-2-butanone; and (*c*) in an achiral *meso* molecule.

5.610 Symmetry Classification of Atoms and Faces (Summary)

Like atoms (or groups of atoms) in a molecule may be classified according to their symmetry relationships as follows: (1) **Symmetry equivalent**: interchangeable by any symmetry operation (also called chemically equivalent but this name is somewhat ambiguous). (2) **Homotopic**: symmetry equivalent by virtue of a rotation (proper) axis (also called equivalent). (3) **Enantiotopic (or prochiral)**: symmetry equivalent by virtue of an alternating (improper) axis (S_p) *only.* (4) **Diastereotopic**: not symmetry equivalent, i.e., noninterchangeable by any symmetry operation (but excluding, e.g., hydrogen atoms whose substitution by the same atom would lead to positional isomers rather than diastereomers). Analogously, the faces of a π system (planes on opposite sides of the nodal plane) are equivalent, enantiotopic, or diastereotopic depending on whether they can be exchanged by C_p, S_p (only), or are nonexchangeable, respectively.

5.620 *Z* (*G.* Zusammen, Together)

Part of the *E-Z* nomenclature for designating stereochemistry around a double bond (or ring). Priorities are assigned (Sect. 5.290) to the two groups attached to each atom of the double bond or ring. In double bond systems ($>C{=}C<$; $>C{=}\ddot{N}<$; $\ddot{N}{=}\ddot{N}<$) the arrangement in which the two higher priority groups eclipse each other is denoted by *Z*.

5.621 *E* (*G.* Entgegen, Opposite)

The arrangement in which the two higher priority or two lower priority groups are on opposite sides of a double bond. The *E-Z* nomenclature can also be used for cyclic compounds: a compound is *Z* when the two higher priority groups are on the same side of the ring and *E* when they are on opposite sides.

Examples. The *Z* and *E* isomers of various compounds are shown in Fig. 5.620*a–d*.

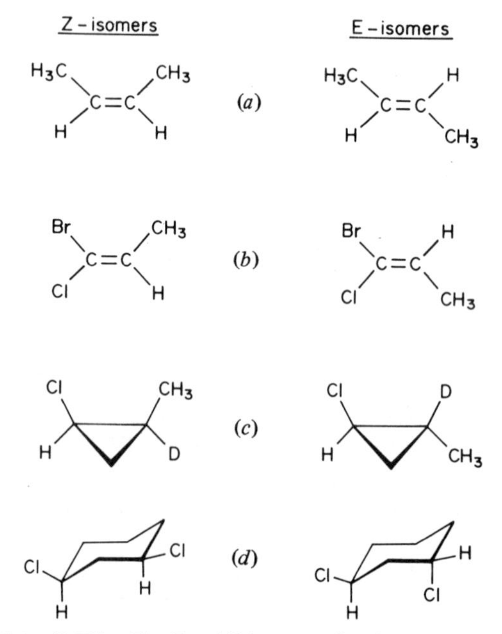

Figure 5.620 The *Z* and *E* isomers of various compounds.

5.630 *cis, c*

A term similar in meaning to *Z*, and referring to identical atoms on the same side of a double bond or ring. Because of possible ambiguity this term is gradually being abandoned.

5.631 *trans*, *t*

A term similar in meaning to *E*, and referring to identical atoms on opposite sides of a double bond or ring. Because of possible ambiguity the *cis*, *trans* prefixes, especially for acyclic olefins, are being abandoned in favor of the *E*-*Z* nomenclature.

Examples. The diastereomers in Fig. 5.620*a* and *d* are *cis*, *trans* pairs, whereas the pairs in Fig. 5.620*b* and *c* cannot be strictly classified as *cis* or *trans.* The *cis*-*trans* nomenclature of both cyclic and acyclic olefins and polyenes can be made unambiguous by using the main chain as a reference standard. If the main chain continues on the same side of a double bond, the configuration is *cis* at that double bond, and if it continues on opposite sides, it is *trans* at that double bond. Thus the [14] annulene shown in Fig. 5.630*a* is *cis*,*trans*,*trans*,*cis*,*trans*,*cis*,*trans*-1,3,5,7,9,11,12-cyclotetradecaheptaene, and Fig. 5.630*b* is *cis*-3,4-dimethyl-1,3-hexadiene (or (*Z*)-3,4-dimethyl-1,3-hexadiene). The terms are also used in designating configuration at a ring junction; Fig. 5.630*c* is a *trans* and Fig. 5.630*d* is a *cis* ring fusion.

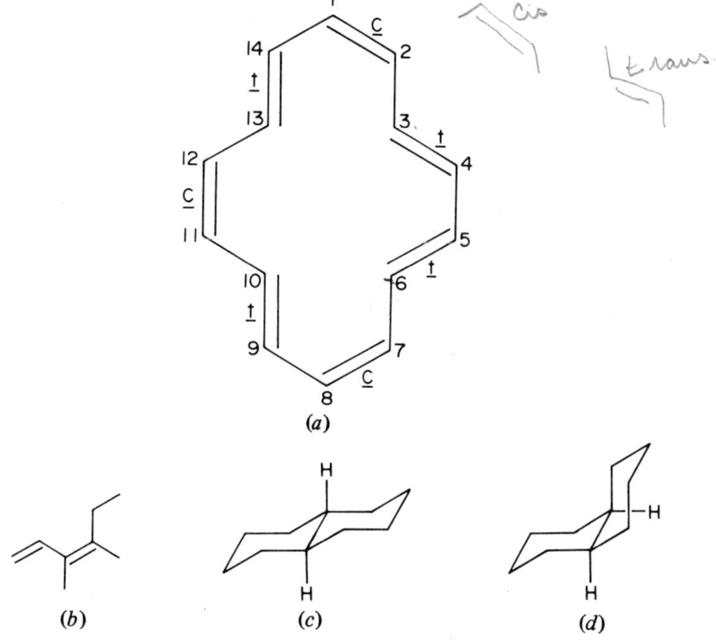

Figure 5.630 (*a*) [14] Annulene stereochemistry; (*b*) *cis* (or *Z*)-3,4-dimethyl-1,3-hexadiene; (*c*) *trans* ring fusion; (*d*) *cis* ring fusion.

5.640 *r* (Reference)

When there are more than two substituent-bearing atoms in a ring, the symbol *r* represents the substituent designates the substituent to which the others are referenced.

Examples. *t*-4-Ethyl-*c*-3-methyl-*r*-1-cyclohexanecarboxylic acid, Fig. 5.640*a*, and methyl 2-cyano-*t*-3-phenyl-*r*-2-oxiranecarboxylate, Fig. 5.640*b* (*t* and *c* stand for *trans* and *cis* in this nomenclature.)

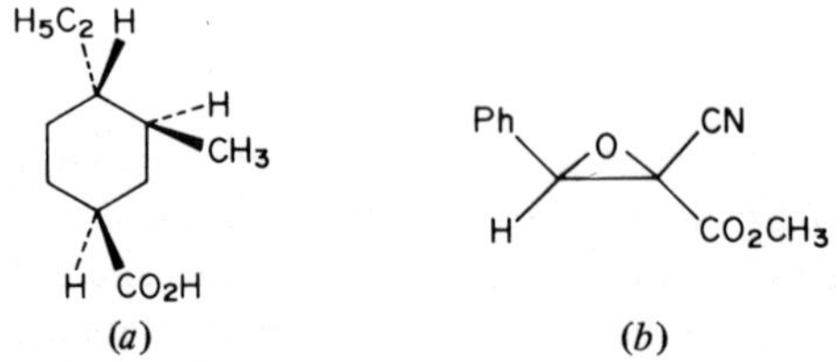

H_5C_2 H H CH_3 H CO_2H

(*a*)

Ph CN O H CO_2CH_3

(*b*)

Figure 5.640 (*a*) *t*-4-Ethyl-*c*-3-methyl-*r*-1-cyclohexanecarboxylic acid; (*b*) methyl 2-cyano-*t*-3-phenyl-*r*-2-oxiranecarboxylate.

5.650 *syn*

A term now falling into disuse, similar in meaning to *Z* (or *cis*), used in naming certain classes of compounds.

5.651 *anti*

A term now falling into disuse, similar in meaning to *E* (or *trans*), used in naming certain classes of compounds.

Examples. Four pairs of *syn-anti* compounds are shown in Fig. 5.650.

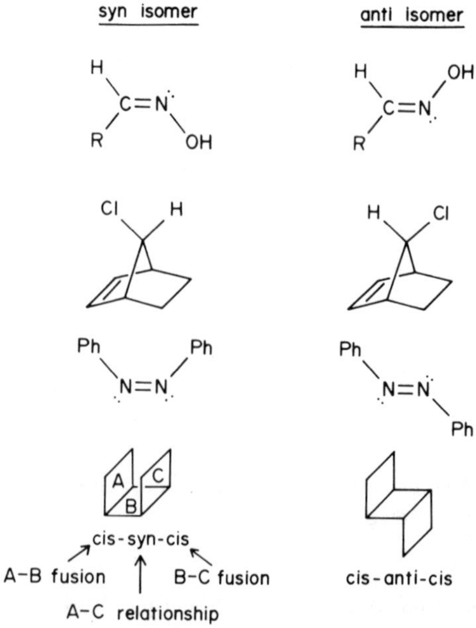

syn isomer

anti isomer

H C=N R OH

H OH C=N R

Cl H

H Cl

Ph Ph N=N

Ph N=N Ph

A C B

cis-syn-cis

A–B fusion

B–C fusion

A–C relationship

cis-anti-cis

Figure 5.650 *syn*- and *anti*-Isomers.

5.660 Geometrical Isomers

The class of diastereomers which includes E, Z, *cis*, *trans* as well as *syn* and *anti* isomers.

5.670 *exo*

That position in a bicyclic molecule nearer the main bridge. The exo substituent is often considered to be on the "outside" or less hindered position.

The main bridge is determined by applying the priorities (highest first): (1) bridge with hetero atoms; (2) bridge with fewest members; (3) saturated bridge; (4) bridge with fewest substituents; (5) bridge with lowest priority substituents.

5.671 *endo*

That position in a bicyclic molecule opposite the main bridge. The *endo* substituent is generally considered to be in the "inside" or more hindered position.

Examples. The *exo* and *endo* positions in a [2.2.1] and a generalized bicyclic system are given in Fig. 5.670*a* and *b*, respectively.

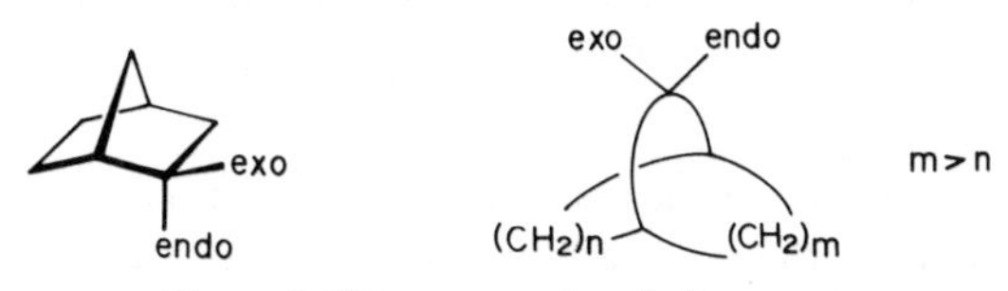

Figure 5.670 *exo*- and *endo*-Isomers.

5.680 α-Configuration

A term used primarily with steroid-type molecules (Sect. 13.590) to denote the downward substituent positions when the molecule is viewed as shown in Fig. 5.680*a*, or the back substituent when viewed as in Fig. 5.680*b*.

5.681 β-Configuration

A term used primarily with steroid-type molecules to denote the upward substituent positions when the molecule is viewed as shown in Fig. 5.680*a*, or the front position when viewed as in Fig. 5.680*b*.

Note: It may be a source of confusion that α, β, γ, etc., are also used to denote substitution positions with respect to a functional group. Thus $CH_3CHCl \cdot CH_2CO_2H$ is β-chlorobutyric acid.

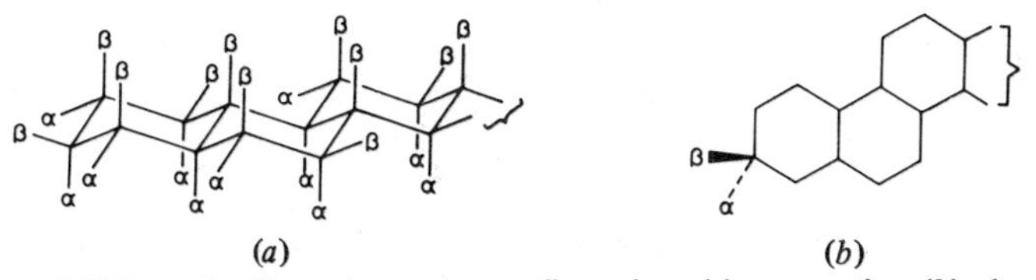

Figure 5.680 α-Configuration and β-configuration: (*a*) perspective; (*b*) planar.

5.690 *meso*

A nonplanar diastereomer that is optically inactive by virtue of a plane or center of symmetry in one of its possible conformations. Usually this term is applied to nondissymmetric molecules that possess multiple chiral centers. A *meso* form has different physical and chemical properties than any of its diastereomers.

Examples. Figure 5.690*a* shows a *meso* compound, whereas Fig. 5.690*b* shows a *d,l*-pair. The *meso* compound also has a conformation with a center of symmetry.

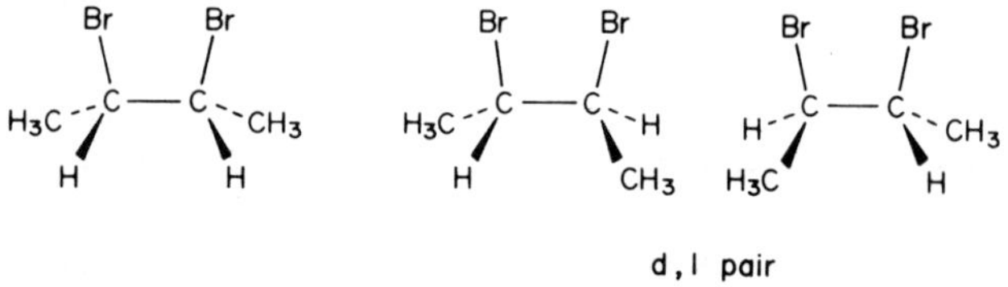

d, l pair

(*a*) (*b*)

Figure 5.690 (*a*) A meso compound; (*b*) a *d,l* pair.

5.700 *erythro*

A prefix given to the name of *d,l*-pair with configuration similar to the two adjacent chiral centers of the four carbon aldose erythrose, Fig. 5.700*a* (one enantiomer is shown). The *erythro* form is often described as "mesolike" because in the orientation shown there would be a plane of symmetry if the two dissimilar groups were equivalent.

Example. *erythro*-3-Bromo-2-butanol, Fig. 5.700*b*.

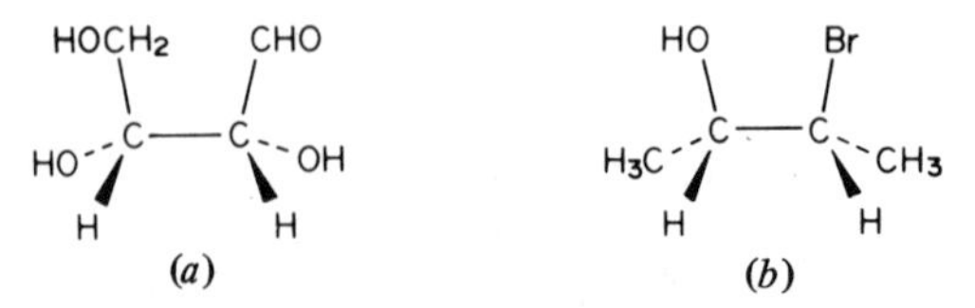

(*a*) (*b*)

Figure 5.700 (*a*) Erythrose; (*b*) *erythro*-3-bromo-2-butanol.

5.710 *threo*

A prefix given to the name of a *d,l*-pair with configuration similar to the two adjacent chiral centers of the four carbon aldose threose, Fig. 5.710*a* (one enantiomer is shown).

Example. *threo*-3-Bromo-2-butanol, Fig. 5.710*b*. The individual members of the two sets of equivalent groups are "*trans*" to each other in this orientation.

(a) (b)

Figure 5.710 (a) Threose; (b) *threo*-3-bromo-2-butanol.

5.720 Epimers

Diastereomers differing in configuration at only one (of several) chiral center.

Example. Figure 5.720, two epimers of 2,5-dimethylcyclohexanone.

(2R, 5S) (2S,5S)

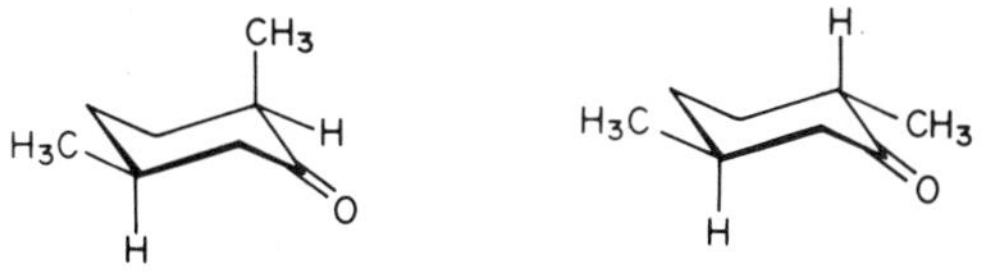

Figure 5.720 Epimers of 2,5-dimethylcyclohexanone.

5.730 Epimerize

To invert configuration at only one (of several) chiral center of a molecule.

Example. Figure 5.720 shows the inversion of configuration at the 2-position. The interconversion is an epimerization.

5.740 Anomers

Used primarily in carbohydrate chemistry to indicate the two epimers at the hemiacetal or acetal carbon of a sugar in its cyclic form.

Example. Figure 5.740 shows the two anomers of the aldohexose sugar D-(+)-glucose; their interconversion (anomerization) occurs through the open form shown in Fig. 5.740*b*.

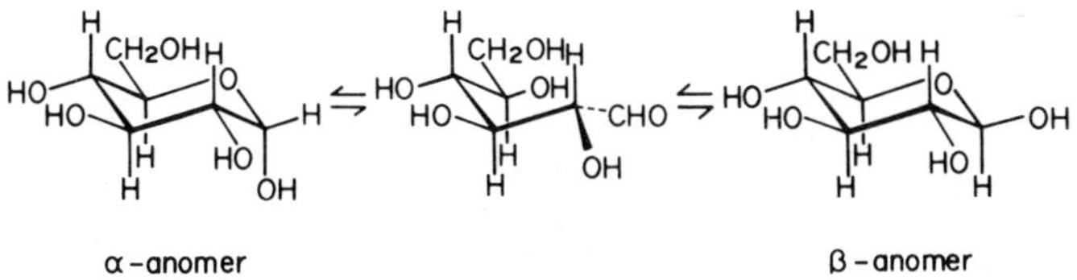

α-anomer β-anomer

(a) (b) (c)

Figure 5.740 Anomerization: (a) α-anomer; (b) open form; (c) β-anomer.

5.750 Mutarotation

The spontaneous change in optical rotation of a freshly prepared solution of a pure stereoisomer caused by epimerization or some other structural change. The equilibrium value of rotation is generally not zero.

Example. When the pure α-anomer (α-D-glucose) shown in Fig. 5.740*a* ($[\alpha]_D$ = +111°) is dissolved in water, the specific rotation gradually decreases to a value of +52.5°, corresponding to a mixture of 38% α and 62% β, Fig. 5.740*c* ($[\alpha]_D$ = +19°). The D refers to measurement with the sodium D line (λ = 5893 Å). The intermediate is the open chain aldehyde, Fig. 5.740*b*.

5.760 Pseudoasymmetric Center

A tetrahedral atom lying in the mirror plane of a *meso* form, substitution on which gives different meso epimers. Molecules with pseudosymmetric centers are achiral.

Example. Figure 5.760.

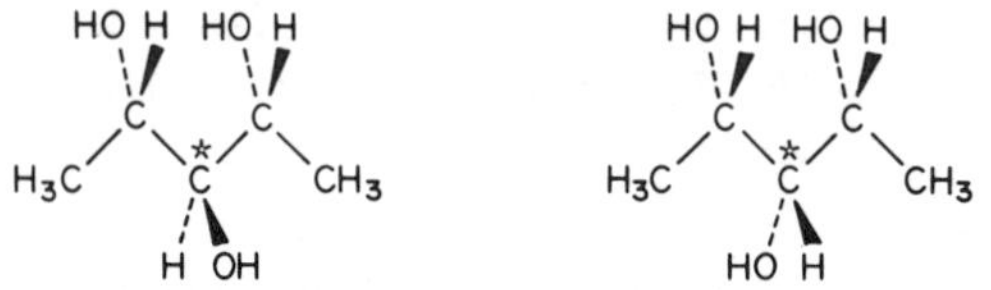

HO H HO H / H₃C C C* C CH₃ / H OH — HO H HO H / H₃C C C* C CH₃ / HO H

Figure 5.760 Pseudoasymmetric atoms (☆) in epimers.

5.770 Stereospecific

A process in which a particular stereoisomer *reacts* to give one specific stereoisomer (or *d,l*-pair) of product; accordingly starting compounds differing only in their stereoisomerism must be converted into stereoisomerically different products.

Example. The addition of Br_2 to an olefinic π bond is generally stereospecifically *trans*, as shown in Fig. 5.770*a*. A direct nucleophilic displacement generally proceeds with stereospecific inversion, as shown in Fig. 5.770*b*.

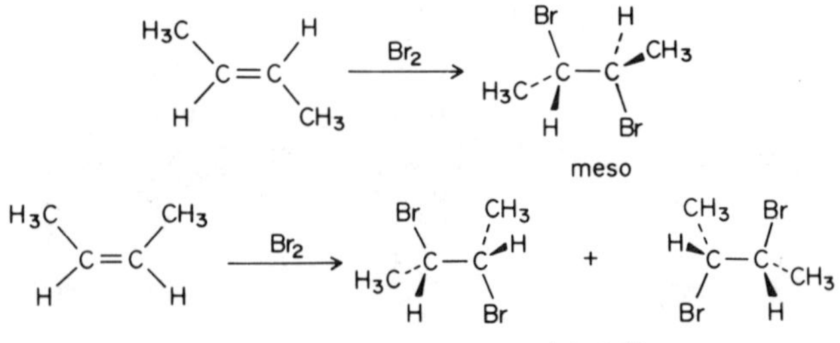

$H_3C(H)C{=}C(H)CH_3$ $\xrightarrow{Br_2}$ meso

$H_3C(H)C{=}C(H)CH_3$ $\xrightarrow{Br_2}$ + d,l pair

(*a*)

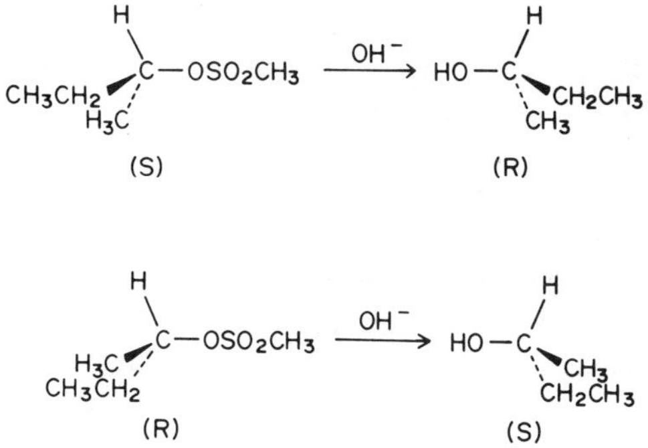

(*b*)

Figure 5.770 (*a*) Stereospecific addition; (*b*) stereospecific inversion.

5.780 Stereoselective

A reaction in which one stereoisomer in a mixture is *produced* (or *destroyed*) more rapidly than another, resulting in predominance of the favored stereoisomer in the mixture of products.

Example. The epimers shown in Fig. 5.780 react (are destroyed) at vastly different rates. The reactions shown in Fig. 5.770*a* are each stereoselective; *all* stereospecific reactions are necessarily stereoselective. However, a stereoselective reaction need not be stereospecific. Thus if conditions existed under which both (Z) and (E)-2-butene individually gave a reaction mixture of the same composition but one that was rich in one stereoisomer, the reactions would be stereoselective but not stereospecific.

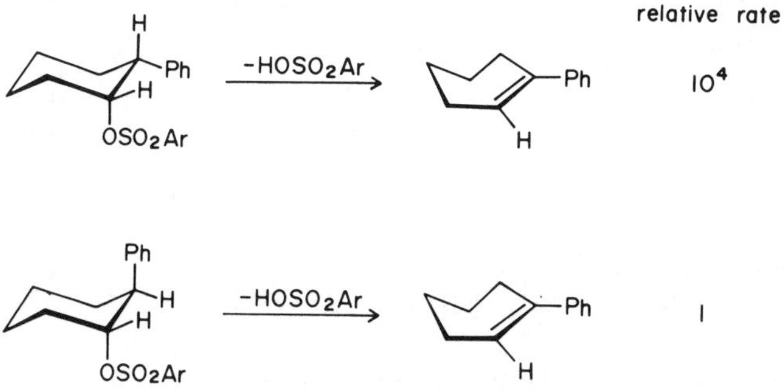

Figure 5.780 A stereoselective elimination.

5.790 Regioselective

A reaction in which one structural (or positional) isomer is favored over another, leading to its predominance in the mixture of products.

Examples. The fact that neither of the elimination reactions shown in Fig. 5.780 gives an appreciable amount of the nonconjugated olefin, Fig. 5.790*a*, suggests that this elimination is regioselective toward the phenyl-bearing carbon. The ionic addition of H—Cl to substituted olefins proceeds regioselectively to place the Cl on the more substituted carbon, Fig. 5.790*b*.

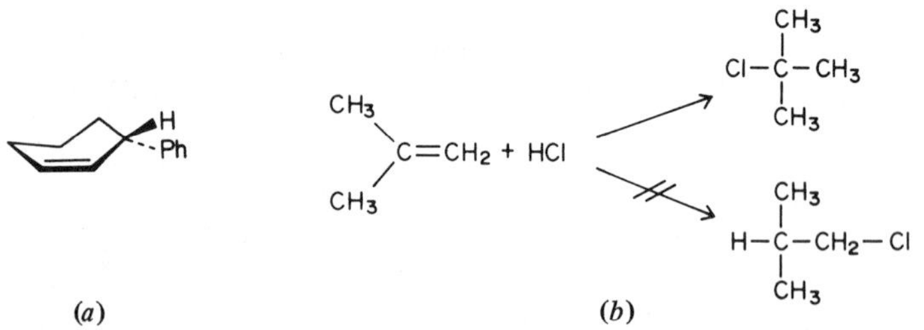

(*a*) (*b*)

Figure 5.790 (*a*) Nonconjugated olefin (not formed in the reaction shown in Fig. 5.780); (*b*) regioselective addition.

5.800 Regiospecific

A reaction in which one specific structural or positional isomer is formed when other isomers are possible. This term is also frequently used synonymously with regioselective.

5.810 Asymmetric Induction

Control of stereoselectivity exerted by an existing chiral center on the formation of a new chiral center. The new chiral center can be created in the same molecule as the existing one (intramolecular) or in a different one (intermolecular).

Example. Figure 5.810*a*, intramolecular and Fig. 5.810*b*, intermolecular.

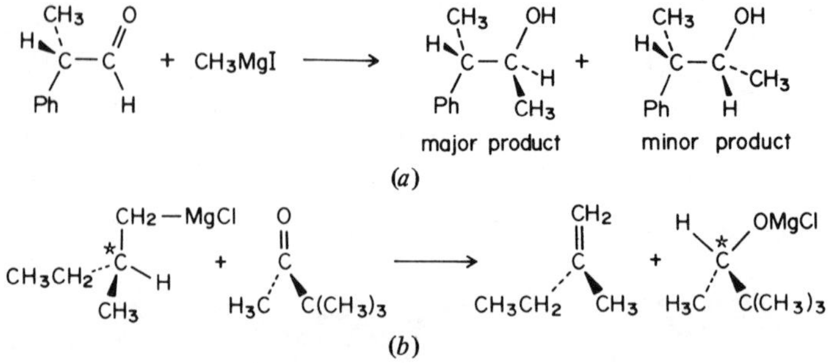

major product minor product

(*a*)

(*b*)

Figure 5.810 Asymmetric induction: (*a*) new chiral center on existing chiral molecule (intramolecular); (*b*) new chiral center introduced by chiral reagent (intermolecular).

5.820 Asymmetric Synthesis

A reaction that involves asymmetric induction.

5.830 Fischer Projection (E. Fischer, 1852-1919)

A convention for displaying the three-dimensional configurational relationship of chiral molecules in a planar representation. The two groups appearing on the left and right side of each chiral carbon are understood to have been in front of the carbon and projected backwards onto the plane; the two groups above and below the chiral carbon are understood to have been behind the projection plane.

Examples. The Fischer projection convention provides a method for ascertaining the absolute configuration from a plane projection drawing, as the relationship in Fig. 5.830*a* shows. This type of projection is used frequently in sugar chemistry, and it is customary to place the H and OH groups on the right and left (in front of the projection plane) of the plane projection formula, Fig. 5.830*b*.

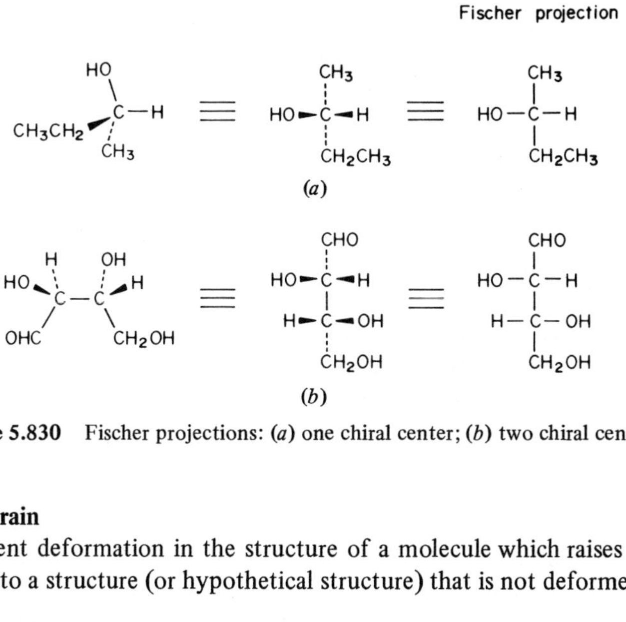

Figure 5.830 Fischer projections: (*a*) one chiral center; (*b*) two chiral centers.

5.840 Strain

A permanent deformation in the structure of a molecule which raises its energy compared to a structure (or hypothetical structure) that is not deformed.

5.850 Bond-Stretching Strain

A permanent deformation caused by the stretching or compression of a bond away from the lowest energy length of that bond.

Example. The preferred C_{sp^3}—C_{sp^3} bond length is 1.54 Å. The indicated bond in Fig. 5.850 is constrained by the molecular structure to be 1.75 Å, and as a result the bond is readily broken.

Figure 5.850 Bond-stretching strain.

5.860 Angle (or Baeyer) Strain (J. F. W. A. von Baeyer, 1835–1917)

Permanent deformation caused by the opening or closing of a bond angle away from the lowest energy angle of that bond sequence.

Example. Small rings (three- to five-membered, Fig. 5.860*a–c*) have internal bond angles smaller than the preferred tetrahedral angle of 109.5°. Cyclopropane, Fig. 5.860*a*, manifests this strain in the unusually high reactivity of its C—C bonds.

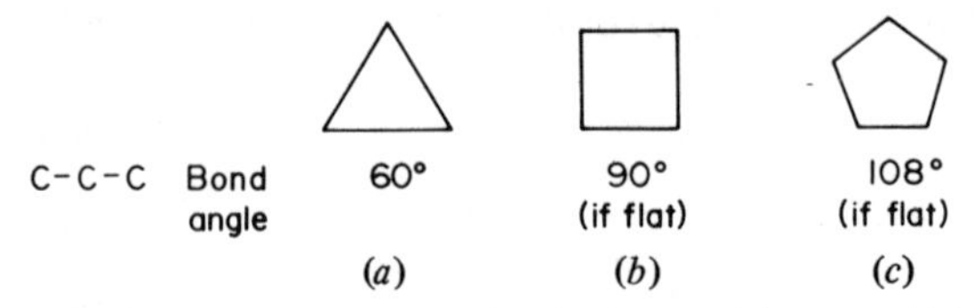

Figure 5.860 Angle strain: (*a*) in cyclopropane; (*b*) in cyclobutane; (*c*) in cyclopentane.

5.870 Steric Strain

Caused by the approach of two or more nonbonded atoms to a position where their electron clouds begin to repel each other.

5.880 Conformation (Conformational Isomer, Conformer)

A particular orientation of the atoms in a molecule, differing from other possible orientations by rotation(s) around single bonds (or in special cases by inversion, *vide infra* for invertomers). Unless it is held rigid by small rings or double bonds, a molecule can have an infinite number of conformations, but only one configuration.

Examples. Whether two stereoisomeric structures constitute different configurations or merely different conformations depends on how readily they interconvert. For example, the various *conformations* of ethane derivatives (see Sect. 5.940) interconvert millions of times per second at 25°C, while the *Z* and *E* *configurations* of ethylene derivatives (see Sect. 5.620) do not interconvert spontaneously even at 200°C. Since, in principle, it should be possible to stop conformational interconversion by sufficient cooling, the operational distinction between configurations and conformations usually is based on whether the structures interconvert at room temperature (25°C).

5.890 Rotamers (Rotational Isomers)

Synonymous with conformers.

5.900 Invertomers (Inversion Isomers)

Two conformers that interconvert by inversion at an atom possessing a nonbonding pair of electrons.

Examples. Ammonia and ammonia derivatives, Fig. 5.900.

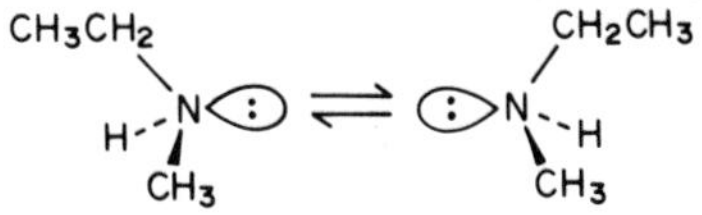

Figure 5.900 Invertomers.

5.910 *gem* (Geminal)

Like atoms or groups attached to the same atom in a molecule, i.e., separated by two bonds.

Example. Figure 5.910 shows geminal methyls.

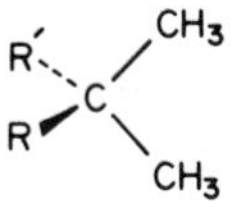

Figure 5.910 Geminal methyl groups.

5.920 *vic* (Vicinal)

Like atoms or groups attached to directly bonded atoms, i.e., separated by three bonds.

Example. Figure 5.920 shows vicinal chlorines.

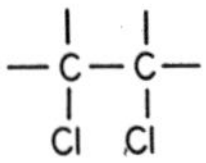

Figure 5.920 Vicinal chlorines.

5.930 Dihedral Angle

The angle between planes $\overline{ABC}$ and $\overline{BCD}$ in the nonlinear bond sequence A—B—C—D.

Examples. Figure 5.930 shows arrangements of atoms resulting in various dihedral angles.

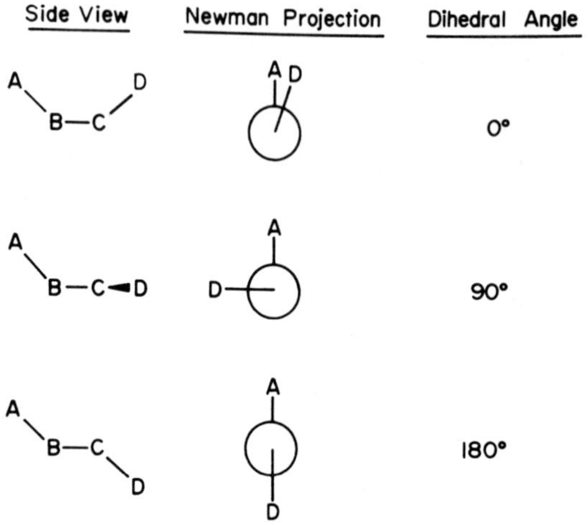

Side View	Newman Projection	Dihedral Angle
A–B–C–D		0°
A–B–C–D		90°
A–B–C–D		180°

Figure 5.930 Dihedral angles.

5.940 Eclipsed Conformation
A conformation with a 0° dihedral angle.

5.941 Staggered Conformation
A conformation with a 60° dihedral angle.

Examples. Figure 5.940*a* and *b*, eclipsed and staggered, respectively. There are an infinite number of conformations intermediate between these extremes; they have no special names. Of these (in acyclic molecules) the staggered conformation is the most stable, and the eclipsed least stable.

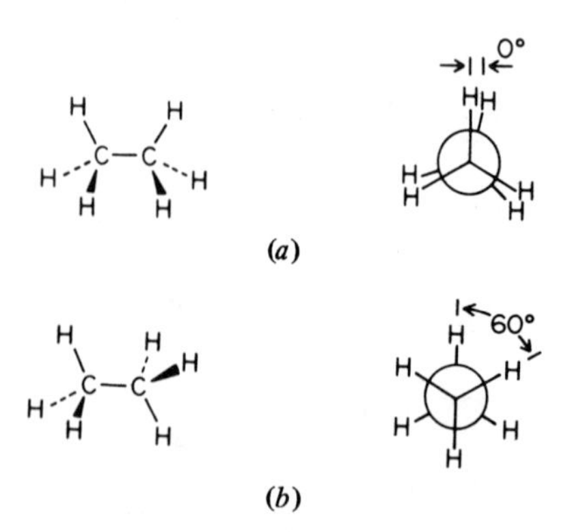

Figure 5.940 (*a*) Eclipsed conformation of ethane; (*b*) staggered conformation of ethane.

5.950 *cis* Conformation (Synperiplanar)
In molecules of the type X—CH_2—CH_2—Y the fully eclipsed conformation, with X—C—C—Y dihedral angle of 0°, Fig. 5.950*a*.

5.951 *gauche* (or *skew*) Conformation (Synclinal)
Staggered conformation with X—C—C—Y dihedral angle of 60°, Fig. 5.950*b*.

5.952 Partially Eclipsed Conformation (Anticlinal)
Eclipsed conformation with X—C—C—Y dihedral angle of 120°, Fig. 5.950*c*.

5.593 *anti* (or *trans*) Conformation (Antiperiplanar)
Staggered conformation with X—C—C—Y dihedral angle of 180°, Fig. 5.590*d*. The relative stability of these four conformations is $d > b > c > a$ in Fig. 5.950.

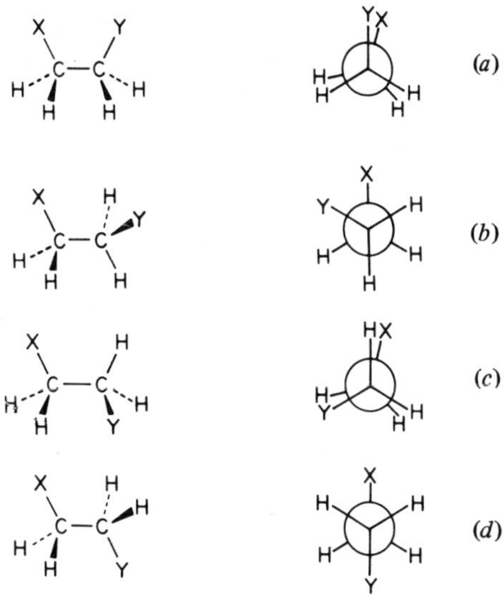

Figure 5.950 Forms of molecule X—C—C—Y: (*a*) *cis*; (*b*) *gauche*; (*c*) partially eclipsed; (*d*) *anti*.

5.960 *s-cis*
In molecules of the type A=CH—CH=B (or A=CH—$\ddot{B}$) the fully eclipsed conformation with a dihedral angle of 0°.

5.961 *s-trans*
In molecules of the type A=CH—CH=B (or A=CH—$\ddot{B}$) the conformation with a dihedral angle of 180°.

Examples. *s-cis*-1,3-Butadiene and *s-trans*-1,3-butadiene, Fig. 5.960*a* and *b*, respectively, and, less commonly, *s-cis* and *s-trans-N*-methylformamide, Fig. 5.960*c* and *d*, respectively. The *s* preceding the *cis* and *trans* refers to *cis* and *trans* with respect to a single bond possessing some double bond character that provides the torsional barrier to interconversion.

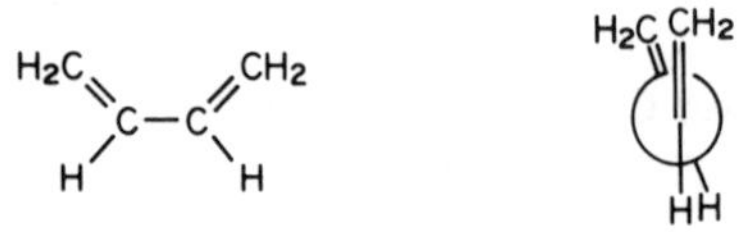

(*a*)

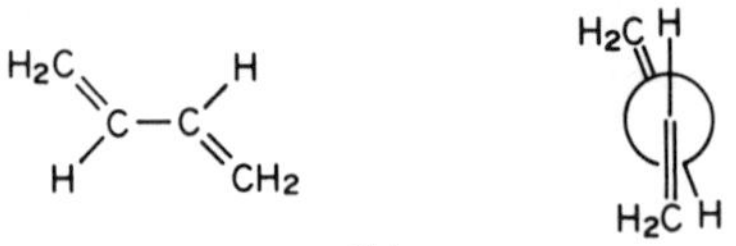

(*b*)

(*c*)

(*d*)

Figure 5.960 (*a*) *s-cis*-1,3-Butadiene; (*b*) *s-trans*-1,3-butadiene; (*c*) *s-cis-N*-methylformamide; (*d*) *s-trans-N*-methylformamide.

5.970 Chair Conformation

The all-staggered conformation of a six-membered ring.

Example. Chair cyclohexane; Fig. 5.970*a*, in perspective, and Fig. 5.970*b*, in Newman projection.

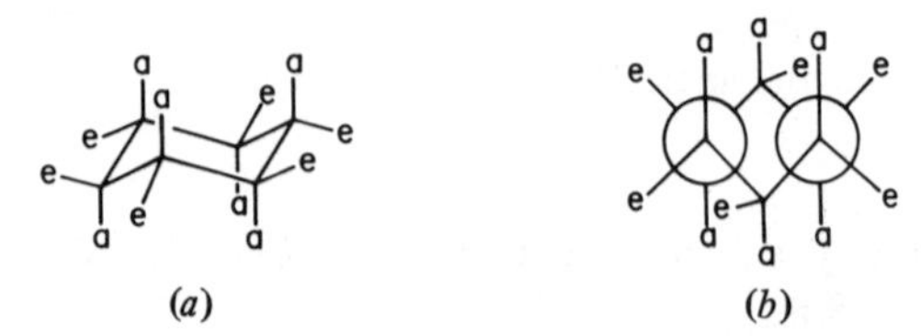

Figure 5.970 Chair form of cyclohexane; (*a*) in perspective; (*b*) in Newman projection.

5.980 Axial Positions

The six bond positions (three up and three down) labeled *a* in Fig. 5.970, which are parallel to the molecular C_3 axis. Any axial position is *gauche* to both vicinal equatorial positions and *anti* to both vicinal axial positions.

5.990 Equatorial Positions

The bond positions around the "equator" of the molecule, labeled *e* in Fig. 5.970. Any equatorial position is *gauche* to both adjacent axial and equatorial positions. By a series of connected rotations, one chair form can "flip" to another chair form, exchanging the axial and equatorial positions, Fig. 5.990*a*. Vicinal axial (and vicinal equatorial) positions are considered *trans* (or *E*), and this nomenclature can be extended as shown in Fig. 5.990*b*.

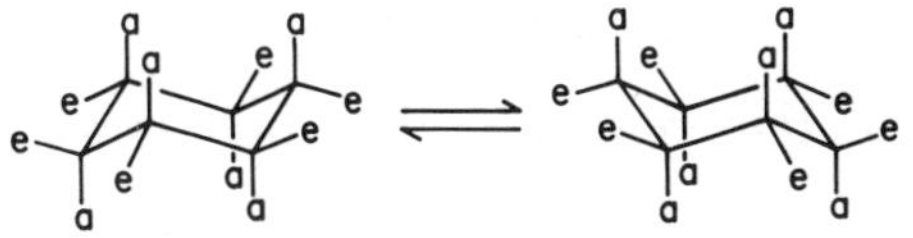

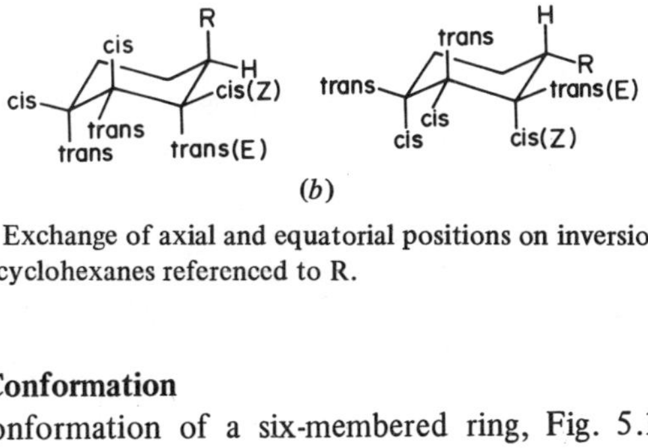

Figure 5.990 (*a*) Exchange of axial and equatorial positions on inversion; (*b*) stereochemistry of substituted cyclohexanes referenced to R.

5.1000 Boat Conformation

The eclipsed conformation of a six-membered ring, Fig. 5.1000*a*; drawn in Newman projection, Fig. 5.1000*b*.

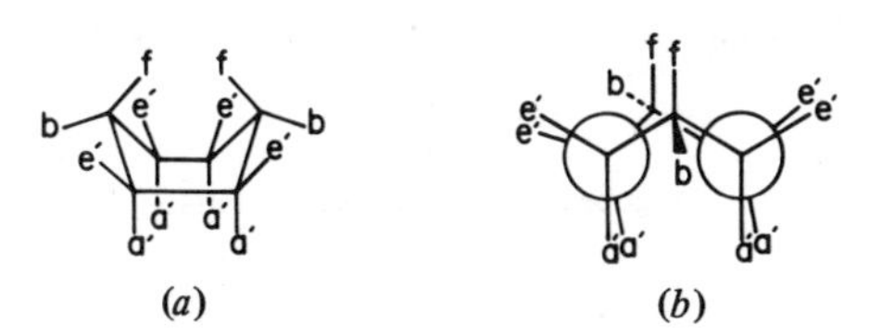

Figure 5.1000 Boat form of cyclohexane: (*a*) in perspective; (*b*) in Newman projection.

5.1001 Pseudoequatorial

Positions labeled e', Fig. 5.1000.

5.1002 Pseudoaxial
Positions labeled a', Fig. 5.1000.

5.10003 Flagpole
Positions labeled f, Fig. 5.1000.

5.1004 Bowsprit
Positions labeled b, Fig. 5.1000.

5.1010 Twist Boat (Skew Boat)
Another type of conformation of a six-membered ring, Fig. 5.1010. The order of stability of conformations is boat $<$ twist boat $<$ chair.

Figure 5.1010 Twist or skew boat form of cyclohexane.

5.1020 Torsional Strain
Strain caused by the repulsion between aligned electron pairs in eclipsed forms.

5.1030 Atropisomers
Two conformations of a molecule whose interconversion is sufficiently slow under a given set of conditions to allow separation and isolation.

Examples. The two enantiomeric conformations of the *ortho* tetrasubstituted biphenyl, Fig. 5.1030, exhibit hindered rotation about the bond joining the two rings. If the *ortho* groups are large enough, they inhibit rotation (and therefore interconversion) sufficiently to enable the two forms to be resolved into enantiomers.

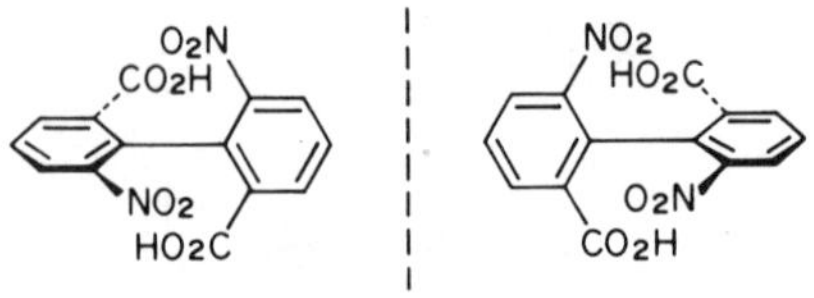

Figure 5.1030 Atropisomers.

6 Polyfunctional Compounds and Compounds of Mixed Function

CONTENTS

In this chapter we consider compounds of multiple functions, i.e., compounds having only one type of function although it may be repeated one or more times (polyfunctional), as well as compounds of mixed function, i.e., compounds that contain more than one function. Although the number and types of functions and their combinations that molecules may possess are almost limitless, we confine our vocabulary to the more common types of polyfunctional and mixed function compounds.

6.010 Polyhydric Alcohols

Compounds possessing two or more hydroxy groups.

Examples. 1,2,4-Butanetriol, Fig. 6.010*a*, and 1,2-dihydroxybenzene, Fig. 6.010*b*, also known as catechol.

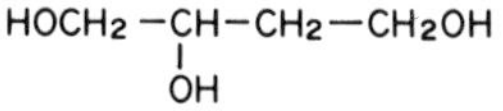

$HOCH_2-CH(OH)-CH_2-CH_2OH$

(*a*) 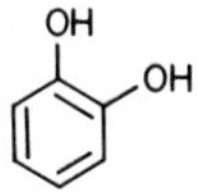 (*b*)

Figure 6.010 (*a*) 1,2,4-Butanetriol; (*b*) catechol (1,2-dihydroxybenzene).

6.020 Glycols

A class name for compounds containing two hydroxy groups.

Examples. Ethylene glycol, Fig. 6.020*a*, also called 1,2-ethanediol, and propylene glycol, Fig. 6.020*b*, also called 1,2-propanediol.

$HOCH_2CH_2OH$ (*a*)

$CH_3CH(OH)CH_2OH$ (*b*)

Figure 6.020 (*a*) Ethylene glycol (1,2-ethanediol); (*b*) propylene glycol (1,2-propanediol).

6.030 Polyols

Synonymous with polyhydric alcohols.

6.040 Ether-Alcohols

Compounds possessing both an ether and a hydroxy function.

Example. Diethylene glycol, Fig. 6.040, also called 3-oxapentane-1,5-diol. Poly(ethylene oxide), $H(OCH_2CH_2)_nOH$, is an important polymer of ethylene oxide.

$HOCH_2CH_2OCH_2CH_2OH$

Figure 6.040 Diethylene glycol (3-oxapentane-1,5-diol).

6.050 Glyme, $CH_3OCH_2CH_2OCH_3$

An acryonym for **gly**col **m**ethyl **e**ther.

6.060 Diglyme, $CH_3OCH_2CH_2OCH_2CH_2OCH_3$

An acryonym for **di**ethylene **gly**col **m**ethyl **e**ther.

6.070 Unsaturated Carbonyl Compounds, $R(CH_2)_aCH{=}CH(CH_2)_bCOZ$

Compounds containing an olefinic (or acetylenic) linkage as well as a carbonyl group.

6.080 α,β-Unsaturated Acids, $R(CH_2)_aCH{=}CHCO_2H$

Carboxylic acids having an unsaturated linkage on the carbon adjacent to the carboxy group.

Example. Acrylic acid, Fig. 6.080*a*, also called (IUPAC) propenoic acid, and 2-butynoic acid, Fig. 6.080*b*.

$$CH_2{=}CH{-}\overset{\overset{\displaystyle O}{\|}}{C}{-}OH \qquad CH_3{-}C{\equiv}C{-}CO_2H$$

(*a*) (*b*)

Figure 6.080 (*a*) Propenoic acid (acrylic acid); (*b*) 2-butynoic acid.

6.090 Substitutive Nomenclature

The most common scheme for naming organic compounds, consisting of giving a name to all functional groups, and placing this name as a prefix to the name of the compound (called the parent) to which it is attached; the functional group name may alternatively be attached to the parent as a suffix. Certain functional groups, e.g., halogens, nitro, nitroso, and hydroperoxy (—OOH), are always cited as prefixes, in which case the parent compound is the corresponding hydrocarbon. If one (or more) functional groups have suffix names, e.g., carboxylic acid, acyl halide, amide, alcohol, and amine, then that functional group which stands highest in a table of precedence (the principal group, IUPAC rules) is selected and used in deriving the name of the parent compound. The priority list is shown in Table 6.090.

Examples. 2-Bromo-1-chloro-4-nitrosobenzene, Fig. 6.090*a* (the substituents are ordered alphabetically); 3-amino-1-propanol, Fig. 6.090*b*; methyl 2-methoxy-6-methyl-2-cyclohexene-1-carboxylate, Fig. 6.090*c*; 4-amino-*N*-(2-hydroxyethyl)butanamide, Fig. 6.090*d*.

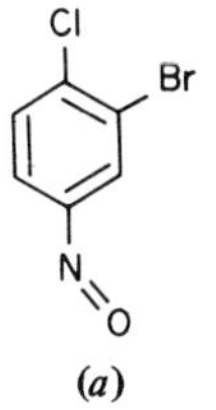

Cl, Br, N, O

(*a*)

$HOCH_2CH_2CH_2NH_2$

(*b*)

OCH_3, O, C, OCH_3, CH_3

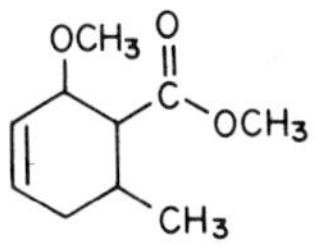

(*c*)

$H_2NCH_2CH_2CH_2CONHCH_2CH_2OH$

(*d*)

Figure 6.090 (*a*) 2-Bromo-1-chloro-4-nitrosobenzene; (*b*) 3-amino-1-propanol; (*c*) methyl 2-methoxy-6-methyl-3-cyclohexene-1-carboxylate; (*d*) 4-amino-*N*-(2-hydroxyethyl)butanamide.

Table 6.090 Partial list of functional suffixes in descending order of preference

Structure	Name
Cations	-onium
Anions	-ate, -ide
—COOH	-oic acid[a]
$—SO_2OH$	-sulfonic acid
—S(O)OH	-sulfinic acid
—SOH	-sulfenic acid
—COX	-oyl halide
$—CON_3$	-oyl azide
$—CONH_2$	-amide[b]
—CHO	-al
>C=O	-one
>C=S	-thione
—OH	-ol
—SH	-thiol
$—NH_2$	-amine

[a]When this suffix name is used, the carbon of the carboxy group is included in the root name, e.g., propanoic acid ($CH_3CH_2CO_2H$), as contrasted with the suffix name carboxylic acid, as in cyclohexanecarboxylic acid ($C_6H_{11}CO_2H$), where the carboxy carbon is named separately.

[b]For example, $CH_3CH_2CONH_2$ is propanamide or ethanecarboxamide, analogous to suffix names -oic acid and carboxylic acid. The convenience of the suffix name carboxamide is apparent, e.g., in 1,1-butanedicarboxamide, [$CH_3CH_2CH_2CH(CONH_2)_2$].

6.100 Keto Acids, $R(CH_2)_aCO(CH_2)_bCO_2H$

The class of carboxylic acids in which another carbonyl group is also present; the position of the carbonyl or keto group is usually indicated by the Greek letter α, β, or γ, referring respectively to whether the doubly bonded oxygen is on the carbon adjacent (b = 0), once (b = 1), or twice (b = 2) removed from the carboxyl function.

Example. γ-Ketovaleric acid, Fig. 6.100.

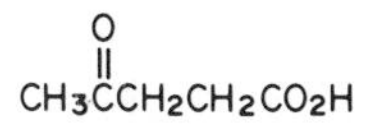

$$CH_3\overset{\overset{\displaystyle O}{\|}}{C}CH_2CH_2CO_2H$$

Figure 6.100 γ-Ketovaleric acid (levulinic acid).

6.110 Dicarboxylic Acids

Hydrocarbons in which two of the hydrogen atoms have been replaced by carboxy groups.

Examples. 1,2-Benzenedicarboxylic acid, Fig. 6.110*a*, also known as phthalic acid, and 1,6-hexanedioic acid, Fig. 6.110*b*, also known as adipic acid.

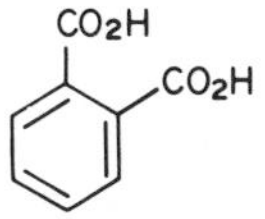

CO_2H, CO_2H (benzene ring)

$HO_2CCH_2CH_2CH_2CH_2CO_2H$

(*a*) (*b*)

Figure 6.110 (*a*) 1,2-Benzenedicarboxylic acid (phthalic acid); (*b*) 1,6-hexanedioic acid (adipic acid).

6.120 Conjunctive Names

A type of nomenclature sometimes employed when the principal function occurs in a side chain attached to a ring; the name is formed by connecting the name of the acyclic component to the cyclic component.

Examples. 2-Naphthaleneacetic acid, Fig. 6.120*a*, and β-chloro-α-methyl-β-3-pyridineethanol, Fig. 6.120*b*.

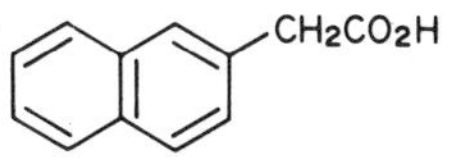

CH_2CO_2H

CHCHOH, Cl CH_3, N

(*a*) 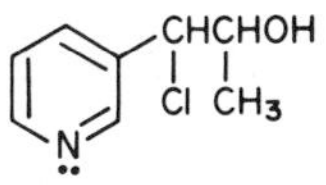 (*b*)

Figure 6.120 (*a*) 2-Naphthaleneacetic acid; (*b*) β-chloro-α-methyl-3β-pyridineethanol.

6.130 Symmetrical Polyfunctional Compounds

Compounds in which the structural unit carrying the functional group occurs more than once, giving rise to a plane (or other element) of symmetry. These compounds are named to emphasize the unit being repeated.

Examples. 1,4-Bis(2-chloroethyl)benzene, Fig. 6.130*a*; nitrilotriacetic acid, Fig. 6.130*b*; tris(2,3-dibromopropyl)phosphate, Fig. 6.130*c*, a flame retardant, now banned, commonly known simply as tris.

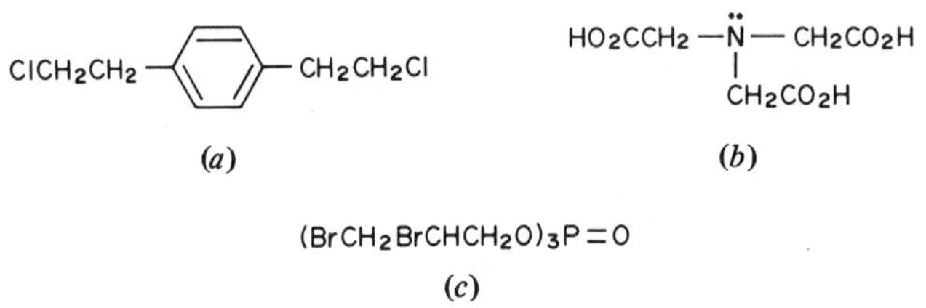

$ClCH_2CH_2-C_6H_4-CH_2CH_2Cl$

(*a*)

$HO_2CCH_2-\ddot{N}(CH_2CO_2H)-CH_2CO_2H$

(*b*)

$(BrCH_2BrCHCH_2O)_3P=O$

(*c*)

Figure 6.130 (*a*) 1,4-Bis(2-chloroethyl)benzene; (*b*) nitrilotriacetic acid; (*c*) tris(2,3-dibromopropyl)phosphate.

6.140 Isotopically Labeled Compounds

Compounds containing isotopic atoms in place of the corresponding atoms having the usual mass number.

Examples. Ethylene-*cis*-1,2-d_2, Fig. 6.140*a*; ethanol-O-*t*, Fig. 6.140*b*; [2-^{13}C]-propanoic acid, Fig. 6.140*c*.

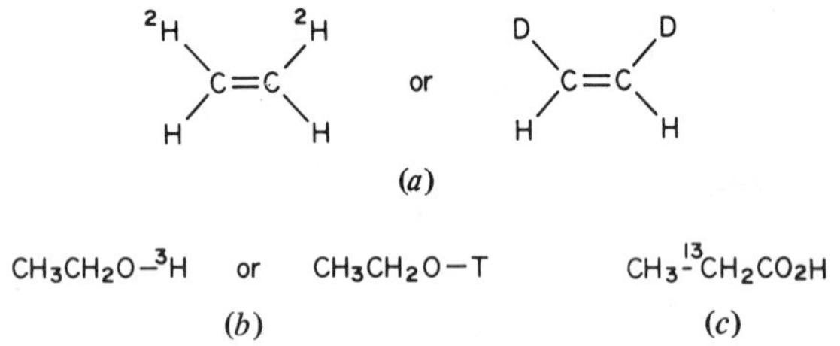

$(^2H)HC=CH(^2H)$ or $(D)HC=CH(D)$

(*a*)

$CH_3CH_2O-{}^3H$ or CH_3CH_2O-T

(*b*)

$CH_3-{}^{13}CH_2CO_2H$

(*c*)

Figure 6.140 (*a*) Ethylene-*cis*-1,2-d_2; (*b*) ethanol-O-*t*; (*c*) [2-^{13}C]propanoic acid.

6.150 Hydroxyaldehydes

Compounds possessing both a hydroxy group(s) and a formyl (—CH=O) group(s). Countless compounds of this type occur in nature.

6.160 Carbohydrates

Polyhydroxy aldehydes or ketones, or substances that yield such compounds on treatment with water (hydrolysis); many of them possess the empirical formula $C_n(H_2O)_n$, hence the name hydrate of carbon or carbohydrate.

6.170 Sugars

Essentially synonymous with carbohydrates but generally used to indicate the less complex, lower molecular weight, water-soluble carbohydrates.

Examples. Glucose, $C_6H_{12}O_6$, and maltose, $C_{12}H_{22}O_{11}$, are sugars but starch $(C_6H_{10}O_5)_n$ is a carbohydrate and not a sugar.

6.180 Pentoses

Sugars possessing five carbon atoms.

Example. D-(−)-Ribose, whose Fischer projection is shown in Fig. 6.180.

```
    CHO
     |
  H—C—OH
     |
  H—C—OH
     |
  H—C—OH
     |
    CH2OH
```

Figure 6.180 D-(−)-Ribose.

6.190 Hexoses

Sugars possessing six carbon atoms.

Example. D-(+)-Glucose, whose Fischer projection is shown in Fig. 6.190.

```
     1CHO
      2|
   H—C—OH
      3|
  HO—C—H
      4|
   H—C—OH
      5|
   H—C—OH
      6|
     CH2OH
```

Figure 6.190 D-(+)-Glucose.

6.200 Aldoses

Sugars containing an aldehyde function.

Examples. Ribose and glucose, Figs. 6.180 and 6.190, are aldoses.

6.210 Ketoses

Sugars containing a keto group.

Example. D-(−)-Fructose, Fig. 6.210.

```
        CH2OH
         |
         C=O
         |
   HO—C—H
         |
    H—C—OH
         |
    H—C—OH
         |
        CH2OH
```

Figure 6.210 D-(−)-Fructose.

6.220 Aldopentose

A five-carbon sugar possessing a formyl group.

6.230 Ketopentose

A five-carbon sugar possessing a keto group.

6.240 Aldohexose

A six-carbon sugar possessing a formyl group.

6.250 Ketohexose

A six-carbon sugar possessing a keto group.

6.260 D- and L-Configuration

(See also Sect. 5.330.) The configurational notation, attributed to Emil Fischer, used to designate the configurations of various sugars relative to the enantiomeric (+)- and (−)-glucoses as reference.

Examples. Fischer originally used the designations *d* and *l* but, to avoid some confusion between configurational relationship and observed signs of rotation, it was actually Rosanoff who suggested the use of the small capital letters D (dee) and L (ell) to indicate configuration relative to (+)- and (−)-glucose. After it was shown that (+)-glucose had the same relative configuration at the chiral atom, C-5, Fig. 6.190, as that of the most simple aldose, (+)-glyceraldehyde, Fig. 6.260*a*, the latter was adopted as the reference compound. D-(+)-glucose is actually (2*R*, 3*S*, 4*R*, 5*R*)-2,3,4,5,6-pentahydroxyhexanal. Carbohydrate or sugar chemists find it convenient to retain the D and L configurational notation. All sugars whose Fischer projection formula shows the OH group on the carbon

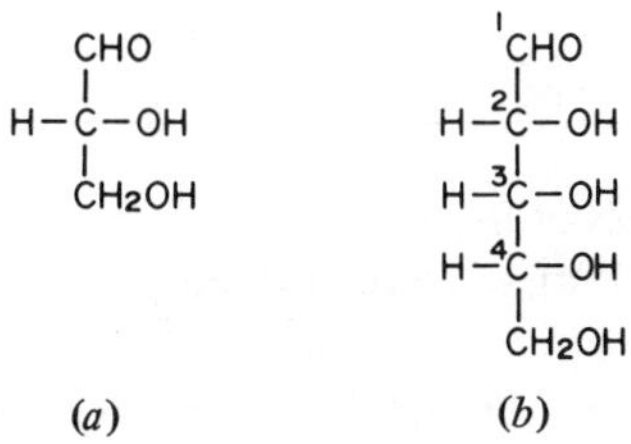

CHO
H—C—OH
CH₂OH

(a)

¹CHO
H—²C—OH
H—³C—OH
H—⁴C—OH
CH₂OH

(b)

Figure 6.260 (*a*) D-(+)-Glyceraldehyde; (*b*) D-(−)-ribose.

atom adjacent to the terminal CH_2OH group on the right-hand side belong to the D series. Thus Fig. 6.260*b* is D-(−)-ribose, which means that the levorotatory enantiomer of ribose has the same relative configuration at C-4 as that of the chiral atom of (+)-glyceraldehyde.

6.270 Monosaccharides

As distinguished from di- and polysaccharides; carbohydrates that do not undergo cleavage on hydrolysis (treatment with water) to smaller sugar molecules.

Examples. Ribose, glucose, and fructose, Figs. 6.180, 6.190, and 6.210, respectively.

6.280 Disaccharides

Compounds that yield two molecules of a monosaccharide on hydrolysis; those that yield two to about eight units are classed as **oligosaccharides.**

Example. Sucrose, Fig. 6.280. On hydrolysis, sucrose gives D-(+)-glucose, Fig. 6.190, and D-(−)-fructose, Fig. 6.210.

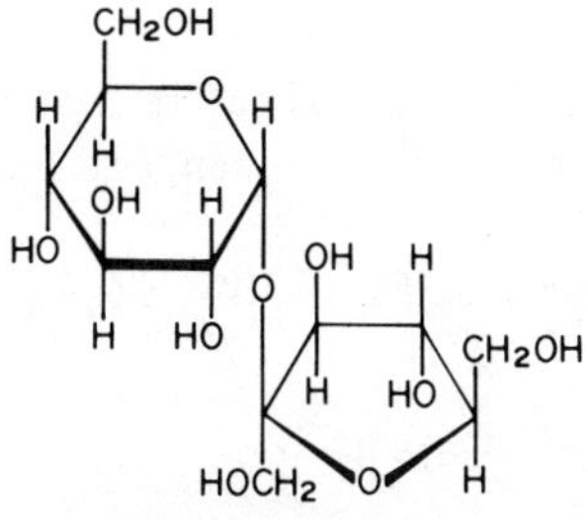

Figure 6.280 Sucrose.

6.290 Polysaccharides

Carbohydrates that yield a large number of monosaccharides on hydrolysis.

6.300 Cyclic Hemiacetals of Sugars

Intramolecular hemiacetals of an aldose or ketose.

Examples. D-(+)-Glucose, Fig. 6.300*b*, forms cyclic hemiacetals; the most stable is the one involving the —OH group on the carbon attached to the $-CH_2OH$ group. This particular hydroxy group (on the C-5 carbon) is favored because the cyclic hemiacetal that results is a six-membered, chair-like ring with minimal steric strain. The open chain form of glucose is in equilibrium, in aqueous solution, with the cyclic hemiacetal; because a new chiral center is formed in the process (the anomeric carbon atom), two epimers are possible, Fig. 6.300*a*, *c*. This equilibrium is responsible for the mutarotation of (+)-glucose (Sect. 5.750). The two cyclic isomers are designated α and β. The designation α is given to the form in which the hydroxy group on the anomeric carbon is on the right in the Fischer projection formula for members of the D family and on the left for members of the L family.

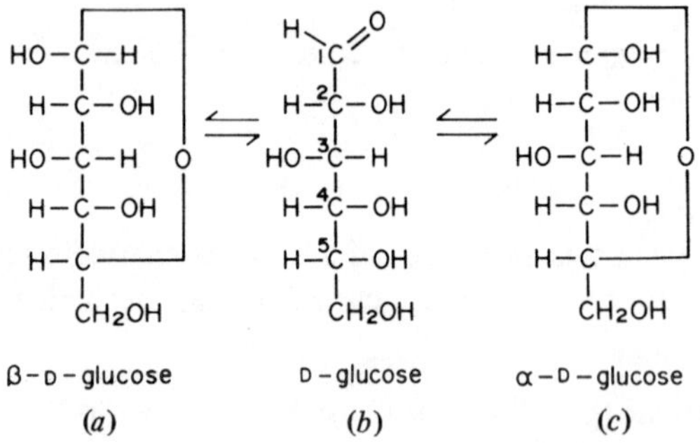

β-D-glucose (*a*) D-glucose (*b*) α-D-glucose (*c*)

Figure 6.300 Cyclic hemiacetals and the mutarotation of glucose.

6.310 Pyranoses

Six-membered ring hemiacetals or acetal sugars containing one oxygen atom as part of the ring; named after the parent, pyran, Fig. 6.310.

Example. The cyclic hemiacetal, Fig. 6.300*c* is α-D-glucopyranose.

Figure 6.310 Pyran.

6.320 Furanoses

Five-membered ring hemiacetals or acetal sugars containing one oxygen atom as part of the ring; named after the parent, furan, Fig. 6.320*a*.

(*a*)

Example. α-D-glucofuranose, Fig. 6.320*b*.

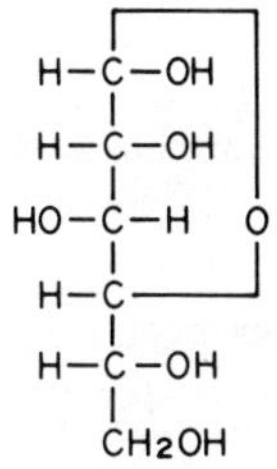

```
  ┌─────────┐
H—C—OH      │
  |         │
H—C—OH      │
  |         │
HO—C—H      O
  |         │
H—C─────────┘
  |
H—C—OH
  |
  CH2OH
```

(*b*)

Figure 6.320 (*a*) Furan; (*b*) α-D-glucofuranose.

6.330 Glycosides

The generic name for the cyclic acetals of the aldoses.

Example. Methyl β-L-glucopyranoside, Fig. 6.330; also called methyl β-L-glucoside.

```
┌──────┐
│  H—C—OCH3
│    |
│ HO—C—H
│    |
O  H—C—OH
│    |
│ HO—C—H
│    |
└────C—H
     |
     CH2OH
```

Figure 6.330 Methyl β-L-glucopyranoside (methyl β-L-glucoside).

6.340 Haworth Formulas (W. N. Haworth, 1883–1950)

Perspective formulas of the sugar hemiacetals and acetals showing the five- or six-membered ring in a plane oriented perpendicular to the plane of the writing surface with substituent groups above and below this plane.

Examples. The Haworth formulas for α-D-glucose and β-D-glucose, Fig. 6.340*a* and *b*, respectively. These correspond to the Fischer projection formulas for the same anomers, Fig. 6.300*c* and *a*, respectively.

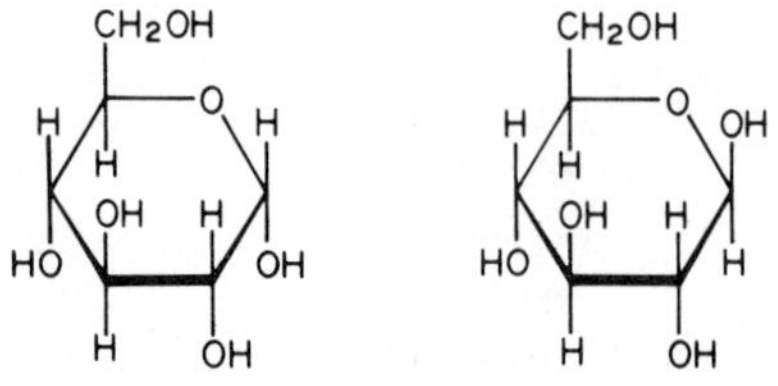

(*a*) (*b*)

Figure 6.340 Haworth representations of (*a*) α-D-glucose; (*b*) β-D-glucose.

6.350 Strainfree Formulas of Sugars

The cyclic structures of pyranoses written in the chair (and if necessary, other) conformation.

Example. Methyl α-D-glucopyranoside, Fig. 6.350. This structure is related to the Haworth formula, Fig. 6.340*a*, and the Fischer projection formula, Fig. 6.300*c*. It will be noted that in methyl α-D-glucoside all the bulky groups, with the exception of the CH_3O— on the anomeric carbon, are equatorial; in methyl β-D-glucoside all bulky groups are equatorial.

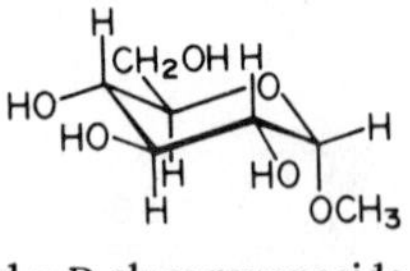

Figure 6.350 Methyl α-D-glucopyranoside (strainfree formula).

6.360 Glycaric Acids

Dicarboxylic acids resulting from the oxidation of aldoses at the two terminal positions.

Example. Galactaric acid, Fig. 6.360*a*; this acid is optically inactive and is the nitric acid oxidation product from either D-(+)-galactose or L-(−)-galactose, Fig. 6.360*b* and *c*, respectively.

6.370 Saccharic Acids

Synonomous with gylcaric acids but falling into disuse.

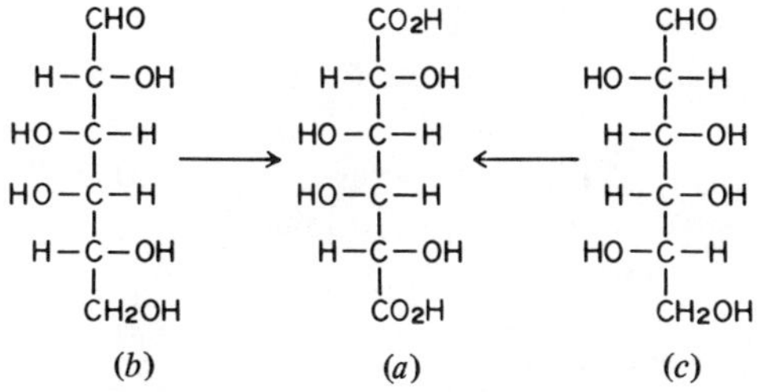

```
     CHO                 CO2H                 CHO
      |                   |                    |
   H—C—OH              H—C—OH              HO—C—H
      |                   |                    |
  HO—C—H    ——→       HO—C—H     ←——       H—C—OH
      |                   |                    |
  HO—C—H              HO—C—H               H—C—OH
      |                   |                    |
   H—C—OH              H—C—OH              HO—C—H
      |                   |                    |
    CH2OH                CO2H                CH2OH
     (b)                 (a)                  (c)
```

Figure 6.360 (*a*) Galactaric acid; (*b*) D-(+)-galactose; (*c*) L-(−)-galactose.

6.380 Glyconic (or Aldonic) Acids

Monobasic acids in which the —CHO function of a sugar is replaced by the carboxy function.

Example. D-Gluconic acid, Fig. 6.380.

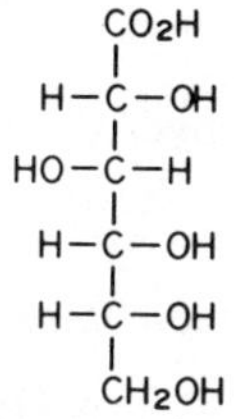

```
     CO2H
      |
   H—C—OH
      |
  HO—C—H
      |
   H—C—OH
      |
   H—C—OH
      |
    CH2OH
```

Figure 6.380 D-Gluconic acid.

6.390 Deoxysugars

Sugars in which an OH group is replaced by hydrogen; the prefix *de* is a generic prefix that is followed by the name of the atom which is deleted.

Examples. 2-Deoxy-D-ribose, Fig. 6.390*a*, and 2,6-dideoxy-D-allose, Fig. 6.390*b*.

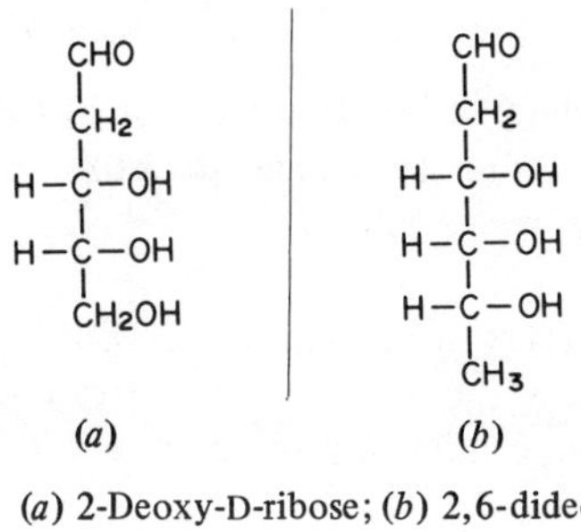

```
     CHO            CHO
      |              |
     CH2            CH2
      |              |
   H—C—OH         H—C—OH
      |              |
   H—C—OH         H—C—OH
      |              |
    CH2OH         H—C—OH
                     |
                    CH3
     (a)            (b)
```

Figure 6.390 (*a*) 2-Deoxy-D-ribose; (*b*) 2,6-dideoxy-D-allose.

6.400 Anhydrosugars

Sugars that have undergone intramolecular loss of the elements of water.

Example. 3,6-Anhydro-β-D-glucopyranose, Fig. 6.400*a*, Fischer projection, and Fig. 6.400*b*, chair form perspective formula.

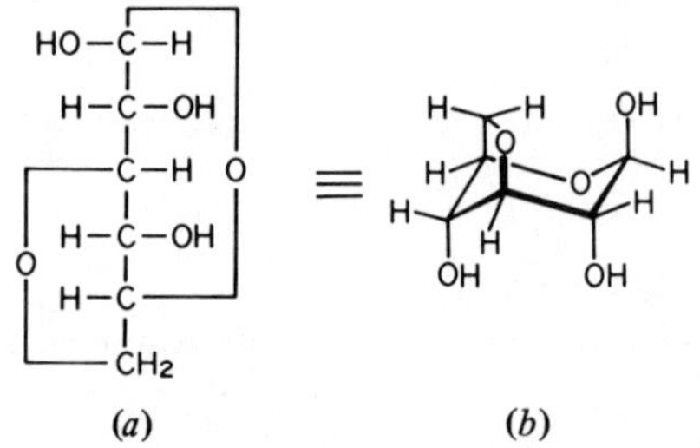

(*a*) (*b*)

Figure 6.400 3,6-Anhydro-β-D-glucopyranose: (*a*) Fischer projection; (*b*) chair form perspective drawing.

6.410 Amino Sugars

Aldoses in which one (or more) hydroxy group(s) is replaced by an amino group(s).

Example. 2-D-Glucosamine (common name), Fig. 6.410.

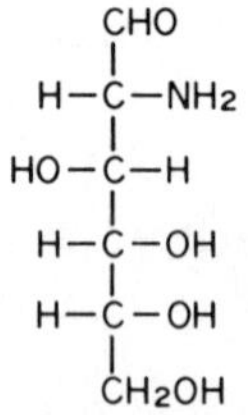

Figure 6.410 2-D-Glucosamine.

6.420 Reducing Sugars

Sugars that possess an aldehyde function (free or in the form of a cyclic hemiacetal) or an α-hydroxy ketone group, and therefore reduce very mild oxidizing agents such as Cu^{2+} or Ag^{+}.

6.430 Fehling Solution (H. von Fehling, 1812-1885)

An alkaline solution of Cu^{2+} containing the tartrate ion to complex with and keep the Cu^{2+} in solution; a mild oxidant used to test for the presence of a reducing sugar.

6.440 Benedict Solution, (S. R. Benedict, 1884–1936)

An alkaline solution of Cu^{2+} containing the citrate ion to complex with and keep the Cu^{2+} in solution; a mild oxidant used to test for the presence of a reducing sugar.

6.450 Tollens Reagent (B. C. Tollens, 1841–1918)

A solution containing $[Ag(NH_3)_2]^+$; a mild oxidant used to test for the presence of a reducing sugar.

6.460 Osazone

The product, containing two phenylhydrazine residues, that results from the reaction between phenylhydrazine (or substituted hydrazines) and a reducing sugar.

Example. Phenylosazone of D-glucose, Fig. 6.460.

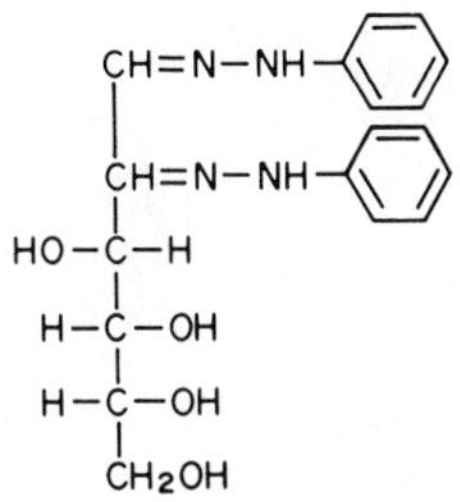

Figure 6.460 D-Glucose phenylosazone.

6.470 Reducing Disaccharides

Disaccharides that are readily oxidized by mild oxidizing agents such as Cu^{2+} (Fehling or Benedict solution) to give the reduced form of copper.

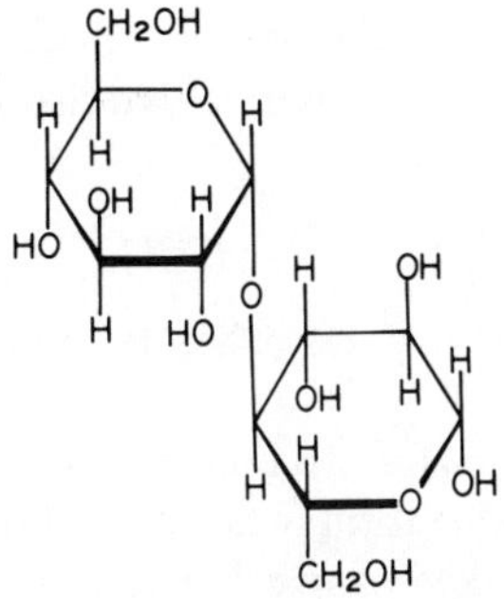

Figure 6.470 4-O-(α-D-Glucopyranosyl)-β-D-glucopyranose (α-maltose).

Example. α-Maltose [4-O-(α-D-glucopyranosyl)-β-D-glucopyranose], Fig. 6.470, shown in the Haworth formula. In this disaccharide, it is the hemiacetal linkage (the potential aldehyde function) in the ring on the right and not the acetal linkage that is the site of oxidation.

6.480 Nonreducing Disaccharides

Disaccharides that are not readily oxidized by mild oxidizing agents such as Cu^{2+}.

Example. Sucrose, Fig. 6.280; such sugars do not possess either an aldehyde or a hemiactal function.

6.490 Cellobiose

4-O-(β-D-Glucopyranosyl)-β-D-glucopyranose, Fig. 6.490. A disaccharide consisiting of two units of D-glucose united by a β-glycoside linkage of one, to the 4-hydroxy position of the other; a product of the enzymatic hydrolysis of cellulose.

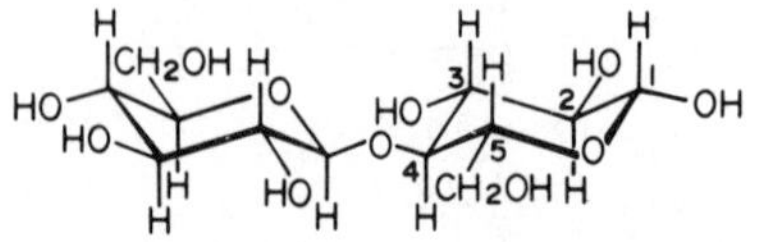

Figure 6.490 4-O-(β-D-Glucopyranosyl)-β-D-glucopyranose (cellobiose).

6.500 Cellulose

A high molecular weight polymer consisting of long chains of D-glucose units linked as in cellobiose; the principal constituent of the cell walls of plants. Cotton cellulose consists of about 3000 glucose units per molecule.

6.510 Hemicelluloses

The major constituents, besides cellulose, of the cell walls of plants; they are not related to cellulose but consist of other polysaccharides, many of which contain aldopentose units.

6.520 Starch

A naturally occurring polymeric polysaccharide composed of D-glucose units.

Examples. There are two major structures of starch and the composition depends on the plant source. One form is called amylose and is composed of about 250–300 units of glucose linked as in cellulose except that the glycoside linkage is α rather than β. Enzymatic hydrolysis (diastase) leads to the disaccharide maltose, 4-O-(α-D-glucopyranosyl)-β-D-glucopyranose, Fig. 6.470 and in perspective, Fig. 6.520. The other form of starch is called amylopectin and contains about 1000 units of glucose in a structure that is branched because of some linkage through the $—CH_2OH$ group.

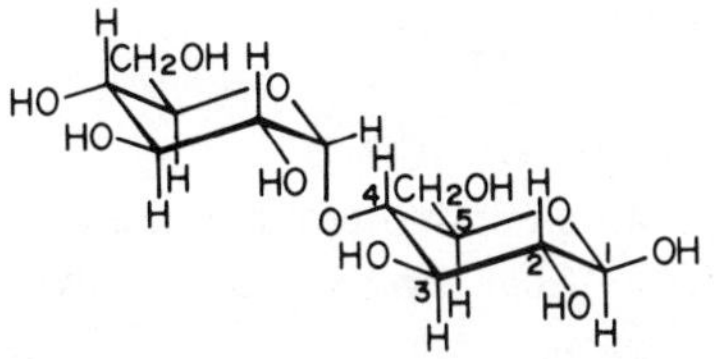

Figure 6.520 4-O-(α-D-Glucopyranosyl)-β-D-glucopyranose (α-maltose).

6.530 Glycogen

An extremely high polymer (~10^6 units) of D-glucose which occurs throughout the protoplasm of animals and forms their reserve carbohydrate.

6.540 Sugar Alcohols

Polyhydric alcohols closely related to the hexoses, hence most are hexahydric.

Example. Sorbitol, the reduction product of D-glucose, Fig. 6.540*a*; this is also one of the reduction products of the ketohexose, L-sorbitose, Fig. 6.540*b*.

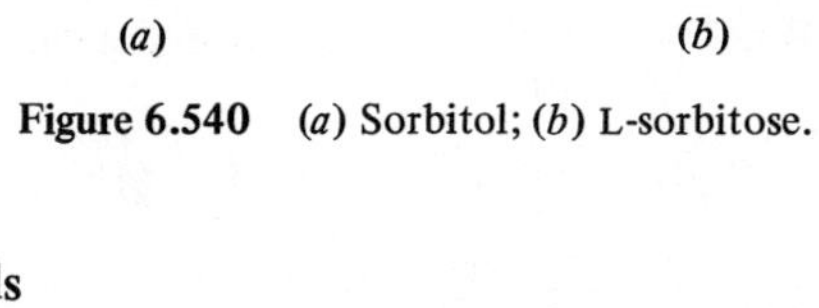

Figure 6.540 (*a*) Sorbitol; (*b*) L-sorbitose.

6.550 Amino Acids

Compounds possessing both an amino group and a carboxy function.

Examples. Anthranilic acid (2-aminobenzoic acid), Fig. 6.550*a*, and 3-aminobutanoic acid, Fig. 6.550*b*.

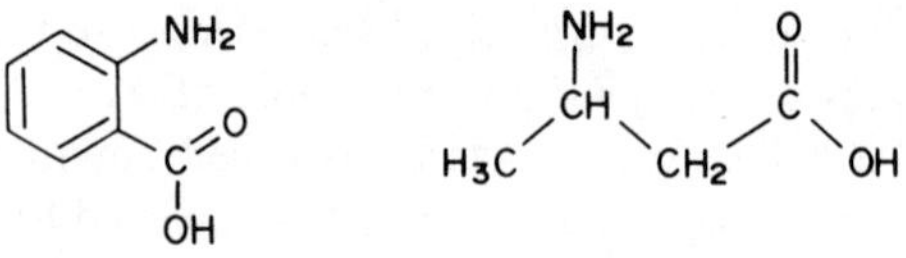

(a) (b)

Figure 6.550 (*a*) Anthranilic acid (2-aminobenzoic acid); (*b*) 3-aminobutanoic acid.

6.560 α-Amino Acids, $RCH(NH_2)CO_2H$

Compounds possessing both an amino group and a carboxy function bonded to the same carbon atom; the final products of the hydrolysis of proteins (*vide infra*).

Examples. There are only about 25 common amino acids that make up the proteins which occur naturally in living matter. Eight amino acids cannot be synthesized by the body and must be added to the diet to preserve the health of human beings. These are listed in Table 6.560 with names, three letter abbreviations, which is common practice in amino acid designation, and structures.

Table 6.560 Essential amino acids

Structure	Name	Abbreviation of residue
$(CH_3)_2CH—CH(NH_2)CO_2H$	L-(+)-Valine	Val
$(CH_3)_2CHCH_2—CH(NH_2)CO_2H$	L-(−)-Leucine	Leu
$CH_3CH_2CH(CH_3)—CH(NH_2)CO_2H$	L-(+)-Isoleucine	Ile
$CH_3CH(OH)—CH(NH_2)CO_2H$	L-(−)-Threonine	Thr
$CH_3S(CH_2)_2—CH(NH_2)CO_2H$	L-(−)-Methionine	Met
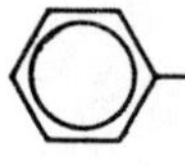 $—CH_2—CH(NH_2)CO_2H$	L-(−)-Phenylalanine	Phe
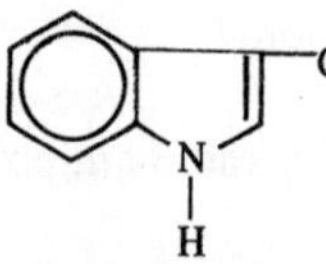 $—CH_2—CH(NH_2)CO_2H$ (N—H)	L-(−)-Tryptophan	Trp
$H_2N(CH_2)_4—CH(NH_2)CO_2H$	L-(+)-Lysine	Lys

6.570 Neutral α-Amino Acids

α-Amino acids possessing an equal number of acid (carboxy) and basic (amino or substituted amino) groups.

Examples. The first seven amino acids listed in Table 6.560.

6.580 Basic Amino Acids

α-Amino acids containing more basic groups than acidic groups.

Examples. Lysine (Lys), Table 6.560; histidine (His), Fig. 6.580.

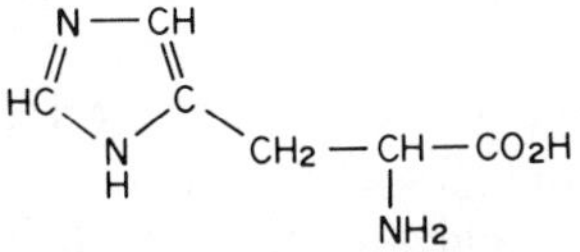

Figure 6.580 Histidine.

6.590 Acidic Amino Acids

α-Amino acids containing more acidic groups than basic groups.

Examples. Glutamic acid (Glu), Fig. 6.590*a*, and aspartic acid, (Asp), Fig. 6.590*b*.

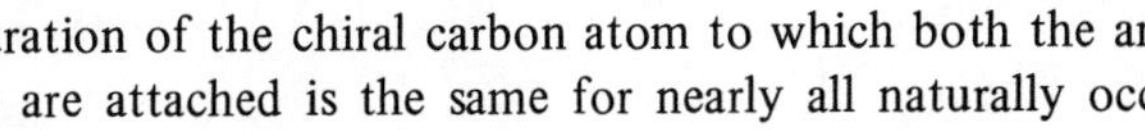

$HO_2C{-}CH_2{-}CH_2{-}CH(NH_2){-}CO_2H$

(*a*)

$HO_2C{-}CH_2{-}CH(NH_2){-}CO_2H$

(*b*)

Figure 6.590 (*a*) Glutamic acid; (*b*) aspartic acid.

6.600 α-Amino Acid Configuration

The configuration of the chiral carbon atom to which both the amino and carboxy group are attached is the same for nearly all naturally occurring amino acids and is (*S*).

Examples. If in the formula $RCH(NH_2)CO_2H$ the carboxy, amino, hydrogen, and R groups are considered to correspond to the aldehyde, hydroxy, hydrogen, and hydroxymethyl groups of glyceraldehyde, then the (*S*)-amino acids have the same configuration as (−)-glyceraldehyde, hence the amino acids were originally called members of the L-family. This designation is no longer in widespread use and instead the absolute configuration around each chiral atom is specified by the (*R*) and (*S*) nomenclature: Fig. 6.600*a* is (*S*)-alanine and Fig. 6.600*b* is (2*S*, 3*R*)-2-amino-3-hydroxybutanoic acid (L-threonine).

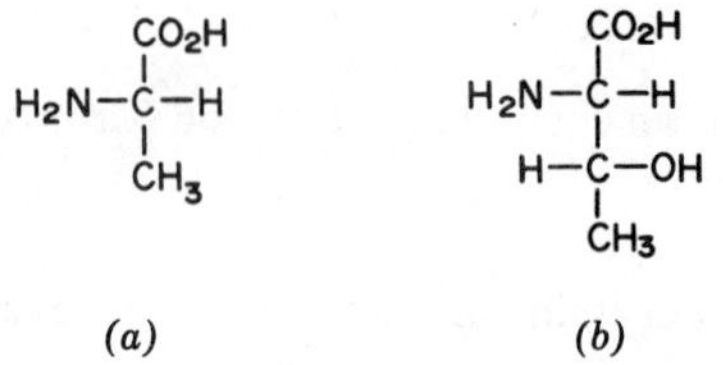

$$\begin{array}{c} CO_2H \\ | \\ H_2N-C-H \\ | \\ CH_3 \end{array} \qquad \begin{array}{c} CO_2H \\ | \\ H_2N-C-H \\ | \\ H-C-OH \\ | \\ CH_3 \end{array}$$

(*a*) (*b*)

Figure 6.600 (*a*) (*S*)-Alanine; (*b*) (2*S*,3*R*)-2-amino-3-hydroxybutanoic acid (L-threonine).

6.610 Dipolar Ion (or Zwitterion) of Amino Acids

The charge separated species that results from the intramolecular transfer of a proton from the acidic to the basic site in an amino acid.

Example. The dipolar ionic form of an amino acid, Fig. 6.610*a*, is in equilibrium with the nonpolar form Fig. 6.610*b*. The concentration of the dipolar form in solution in the absence of added acid or base depends on the relative acidity of the carboxy group and the relative basicity of the amino group, both of which are influenced by the nature of the R group. The addition of acid or base results in a shift in the equilibria toward either structures Figs. 6.610*c* or *d*, respectively.

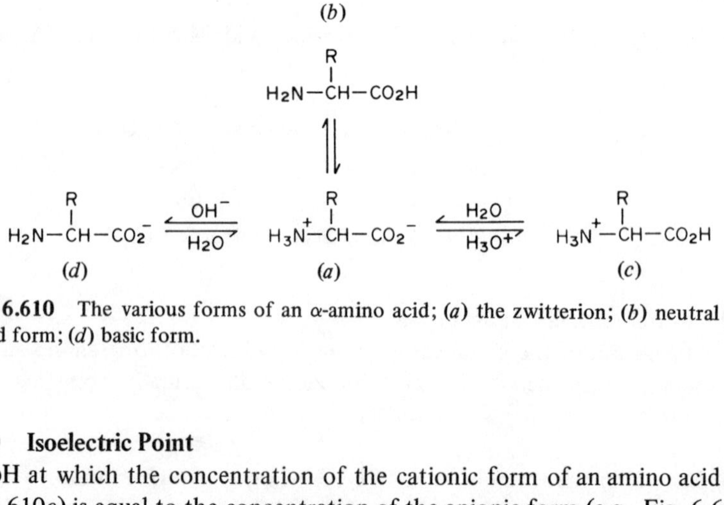

(*b*)

$$H_2N-CH(R)-CO_2H$$

$$\updownarrow$$

$$\underset{(d)}{H_2N-CH(R)-CO_2^-} \underset{H_2O}{\overset{OH^-}{\rightleftharpoons}} \underset{(a)}{H_3\overset{+}{N}-CH(R)-CO_2^-} \underset{H_3O^+}{\overset{H_2O}{\rightleftharpoons}} \underset{(c)}{H_3\overset{+}{N}-CH(R)-CO_2H}$$

Figure 6.610 The various forms of an α-amino acid; (*a*) the zwitterion; (*b*) neutral form; (*c*) acid form; (*d*) basic form.

6.620 Isoelectric Point

The pH at which the concentration of the cationic form of an amino acid (e.g., Fig. 6.610*c*) is equal to the concentration of the anionic form (e.g., Fig. 6.610*d*). At this pH there is no net migration of the amino acid when it is placed in a solution containing electrodes. The concentration of the dipolar ion (e.g., Fig. 6.610*a*) is at a maximum at the isoelectric point. Amino acids can be characterized by their isoelectric point, which depends on the nature of the R group, Fig. 6.610.

6.630 Peptides

Polymers of amino acids which may be considered to have been formed by the loss of water between the $—NH_2$ group of one amino acid and the $—OH$ part of the carboxy group of a second amino acid.

Examples. Glycylalanine (Gly · Ala), Fig. 6.630*a*, a common dipeptide; on hydrolysis it yields two amino acids. It is customary to write the structure of peptides with the $H_2N—$ terminal end to the left of the chain and $—CO_2H$ terminal group to the right of the chain. Alanylcysteinylserine (Ala · CySH · Ser), Fig. 6.630*b*, is a tripeptide.

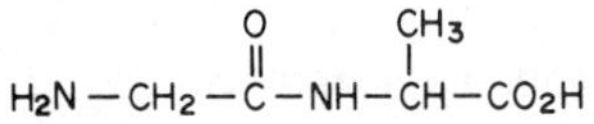

$$H_2N-CH_2-\overset{\displaystyle O}{\overset{\|}{C}}-NH-\overset{\displaystyle CH_3}{\overset{|}{C}H}-CO_2H$$

(*a*)

$$H_2N-\underset{\displaystyle CH_3}{\underset{|}{C}H}-\overset{\displaystyle O}{\overset{\|}{C}}-NH-\underset{\displaystyle \underset{\displaystyle SH}{\underset{|}{CH_2}}}{\underset{|}{C}H}-\overset{\displaystyle O}{\overset{\|}{C}}-NH-\underset{\displaystyle CH_2OH}{\underset{|}{C}H}-CO_2H$$

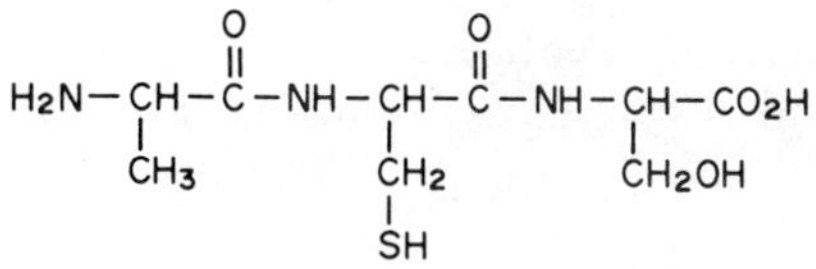

(*b*)

Figure 6.630 (*a*) The dipeptide Gly·Ala; (*b*) the tripeptide Ala·CySH·Ser.

6.640 Peptide Linkage

The amide grouping, $—\overset{\displaystyle H}{\overset{|}{N}}—\overset{\displaystyle O}{\overset{\|}{C}}—$, which constitutes linkage of the amino acid residues in peptides.

6.650 *N*-Terminal Amino Acid Residue

The terminal amino acid residue in a peptide that contains the free amino group.

6.660 *C*-Terminal Amino Acid Residue

The terminal amino acid residue in a peptide that contains the free carboxy group.

6.670 Oligopeptides

An inexact name denoting peptides that contain more than two (dipeptides) or three (tripeptides) amino acid units but fewer than perhaps about 10.

6.680 Polypeptides

Peptides containing many (perhaps more than about 10) amino acid residues but usually having a molecular weight less than 5000.

6.690 Proteins

Polypeptides ranging in molecular weight from about 6000 (insulin, which consists of 51 amino acids) to 40,000,000; they constitute about 75% of the dry weight of most living systems.

6.700 Conjugated Proteins

Proteins associated with nonprotein material (prosthetic group, see Sect. 13.050).

6.710 Conjugated Metalloproteins

Proteins associated with a metal.

Examples. Hemoglobin and myoglobin contain the iron atom in the heme prosthetic group.

6.720 Primary Structure of a Peptide

The sequence of amino acid residues making up the chain or backbone of the peptide.

6.730 Secondary Structure of a Peptide

The structure resulting from the backbone interacting with itself and giving rise to a helical structure.

Example. The helical structure of many polypeptides is due to hydrogen bonding between the hydrogen atom on the amide linkage with the oxygen of the carbonyl group of a neighboring amide linkage in the so-called α-helix arrangement.

6.740 Tertiary Structure of a Peptide

The structure resulting from the interactions of groups off the main backbone of the polypeptide with other such groups. Such interaction determines the gross shape or way the polypeptide helix is folded.

Examples. Myoglobin and hemoglobin have a tertiary structure that has been determined by x-ray analysis, which shows their three-dimensional structure. Covalent S—S bonding and van der Waal interactions help determine the particular folding patterns.

6.750 Quaternary Structure of a Peptide

The fourth level of organization of a peptide or protein which describes the way subunits aggregate to form large complexes.

6.760 Lipoproteins

Conjugate proteins in which the prosthetic group is a lipid or fat, typically a trigylceride involving a fatty acid such as stearic or palmitic.

6.770 Mucoproteins

Conjugated proteins in which the prosthetic group is a sugar moiety.

7 Separation Techniques and Physical Properties

CONTENTS

7.010 Refluxing

The process of converting a liquid into its vapor, condensing the vapor, and continuously returning the condensate to the vessel from which it was vaporized.

7.020 Distillation

The process of converting a liquid to its vapor in one vessel, usually called the pot, and collecting the condensate in a second vessel, called the receiver.

7.030 Raoult's Law (F. M. Raoult, 1830-1901)

The partial pressure P_A of a particular component A over a liquid mixture is equal to the mole fraction X_A of A in the liquid times the vapor pressure P_A° of pure A at the temperature of interest:

$$P_A = X_A P_A^\circ$$

7.040 Ideal Solution

A solution that obeys Raoult's law. Such solutions have properties that are the weighted averages of the properties of the components, and the interactions between molecules of each individual component are unaffected by the presence of the other components.

7.050 Relative Volatility, α

The ratio of vapor pressures at a particular temperature of two components of a solution; by convention the vapor pressure of the more volatile pure component P_A° is the numerator, i.e., $\alpha > 1$.

$$\alpha = \frac{P_A^\circ}{P_B^\circ} = \frac{P_A/X_A}{P_B/X_B} = \frac{P_A \cdot X_B}{P_B \cdot X_A}$$

Because the partial pressures P_A and P_B are directly proportional to the moles of A and B in the vapor phase, these terms can be replaced by the mole fractions of the components in the vapor phase, Y_A and Y_B, hence

$$\alpha = \frac{Y_A \cdot X_B}{Y_B \cdot X_A}$$

7.060 Vapor-Liquid Equilibrium

In a mixture of liquids, the composition of the vapor phase relative to the composition of the liquid phase in equilibrium with it at any particular temperature.

Example. Figure 7.060 shows a typical vapor-liquid equilibrium diagram for an ideal two-component mixture of A and B.

7.070 Fractional Distillation

A distillation performed on a mixture of liquids such that the most volatile component is separated first, followed by those with successively higher boiling points.

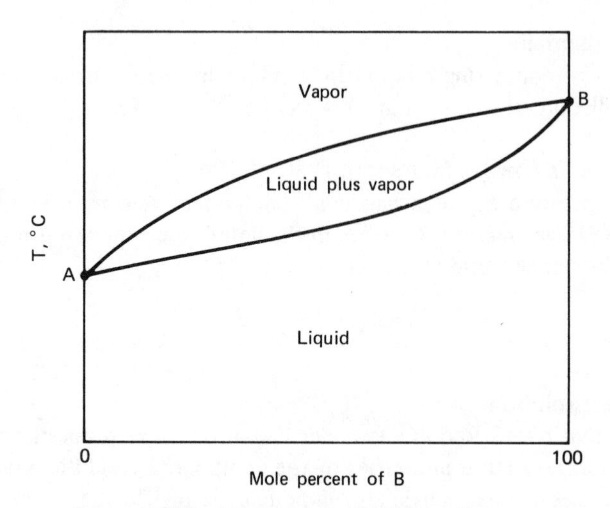

Figure 7.060 Liquid-vapor equilibrium diagram.

Examples. Consider the binary mixture shown in Fig. 7.060 and redrawn in Fig. 7.070, and suppose the mixture to be distilled has the composition represented by X_1. On bringing this mixture to its boiling point t_1, and then condensing the first small quantity of vapor in equilibrium with the liquid at t_1, the condensate would have the composition X_2 and boil at t_2. At the boiling point of liquid X_2, namely t_2, the vapor in equilibrium would have the composition X_3. These hypothetical successive distillations could be repeated until a very small quantity of pure A with boiling point t_A is obtained. To obtain good liquid-vapor equilibrium during a distillation an inert packing material is placed in a column above the distilling flask. Many expensive laboratory and many commercial distilling columns are filled with perforated trays or plates to enhance vapor-liquid equilibrium.

7.080 Theoretical Plate

A plate in a distilling column on which complete liquid-vapor equilibrium is obtained. Since in practice this is impossible, such a plate is a theoretical plate and even under the best of distilling conditions several plates are required to obtain equilibrium.

Examples. In Fig. 7.070 the dotted lines within the liquid-vapor curve represented by the step t_1X_1, t_1X_2, t_2X_2 shows the change in liquid composition obtained by one equilibration with vapor, i.e., one equilibrium stage, hence this change is equivalent to that which can be achieved by one theoretical plate. The distillation column that is capable of achieving an enrichment of the more vola-

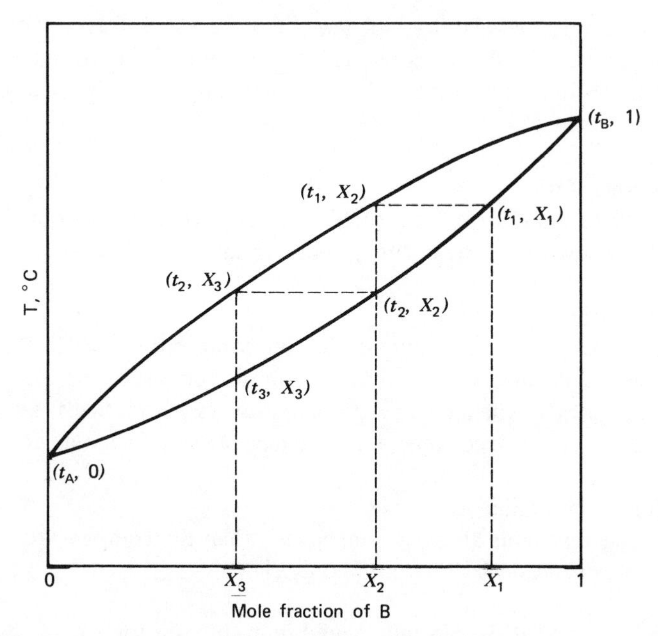

Figure 7.070 Temperature-composition curve showing the effect of two theoretical plates.

tile component from the liquid composition represented by X_1 to that represented by X_3 in Fig. 7.070 is said to have an efficiency of two theoretical plates.

7.090 Height Equivalent to a Theoretical Plate (HETP)

The height of a column packing which can produce a separation equivalent to one theoretical plate.

Examples. The HETP of a particular column is evaluated by determining the total number of theoretical plates divided by the height. To determine the number of theoretical plates the following are required: (*a*) a standard binary test mixture whose vapor-liquid diagram (Fig. 7.060), hence its relative volatility α is known; (*b*) a method for determining the composition of this mixture in the distilling flask and the composition of the condensate under particular operating conditions. The distilling apparatus is operated under total reflux until everything connected with the column operation appears to be in a steady state, then a small amount of condensate is removed for analysis. When these data are obtained, the number of plates n may be determined from the formula

$$n = \frac{1}{\log \alpha} \log \frac{X_D^A/X_D^B}{X_S^A/X_S^B}$$

where α is the relative volatility, X_D^A and X_D^B are the mole fractions of A and B, respectively, in the distillate or overhead, and X_S^A and X_S^B are the mole fractions of A and B, respectively, in the still pot or distilling flask. This equation was developed by Fenske.

7.100 Reflux Ratio

The quantity of liquid being condensed in a distillation apparatus and returned to the column divided by the quantity collected in the receiver.

7.110 Azeotrope

A mixture of two (or more) liquids of definite composition that distills at a constant temperature. Such systems deviate from Raoult's law; i.e., the total pressure is not the sum of the individual partial pressures at a particular temperature. The components of an azeotrope cannot be separated by fractional distillation.

7.120 Minimum Boiling Azeotrope

An azeotrope that boils at lower temperature than the components; such solutions show a positive deviation from Raoult's law.

Examples. A typical liquid-vapor equilibrium diagram for a minimum boiling azeotrope is shown in Fig. 7.120. The azeotrope of water and alcohol containing 96.0% ethanol and 4.0% water by weight boils at 78.17°C, whereas pure ethanol boils at 78.3°C and water at 100.0°C. Water, benzene, and ethanol form a ternary azeotrope that boils at 64.9°C (at 760 mm).

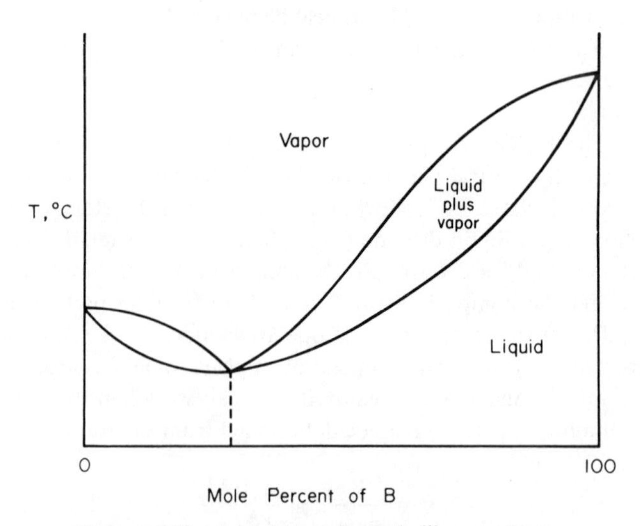

Figure 7.120 A typical minimum boiling azeotrope.

7.130 Maximum Boiling Azeotrope

An azeotrope that boils at higher temperature than any of its components; such solutions show a negative deviation from Raoult's law.

Examples. The azeotrope of HCl and water consists of 20.2% HCl and 79.8% water and boils at 108.6°C , whereas HCl gas has a boiling point of −80°C.

7.140 Molal Boiling Point Elevation Constant, K_b

The increase in boiling point of a pure liquid that is 1 molal (1 mole of solute per 1000 g of solvent) in nonvolatile solute; the value is a constant, i.e., independent of the nature of the solute.

Examples. The boiling point of a 1 molal aqueous solution of ethylene glycol, $(CH_2OH)_2$, is 100.512°C, hence the molal boiling point constant, K_b, for water is 0.512°C m^{-1}.

7.150 Ebullioscopic Constant

Synonymous with molal boiling point elevation constant.

7.160 Extractive Distillation

A distillation process in which a solvent that is higher boiling than any of the components of the sample is added to the distillation column containing the vapor-liquid mixture of sample, in order to enhance the relative volatility of the components to be separated. The extractive solvent functions analogously to the stationary phase (*vide infra*) of a chromatographic column by interacting to a differing extent with the individual sample components and thus facilitating their separation.

7.170 Steam Distillation

The distillation of water mixed with a higher boiling, immiscible liquid that enables the high boiling liquid to be relatively rapidly distilled at temperatures near that of boiling water. The process is made possible by the fact that in such a system the total vapor pressure at any temperature is the sum of the vapor pressures of water and the liquid at that temperature. The steam helps to transport the higher boiling substance from the pot to the receiver.

Example. Bromobenzene boils at 157°C but steam distills at 95°C because at this temperature the vapor pressure of bromobenzene is 120 mm Hg and that of water is 640 mm, and thus the total pressure is one atmosphere (760 mm Hg).

7.180 Flash Point

The temperature to which a substance must be heated in air before its vapor can be ignited by a free flame; a measure of a substance's flammability.

Examples. The flash point of a typical gasoline is about -45°C and that of a typical lubricating oil is about 230°C.

7.190 Ignition Temperature (Spontaneous Ignition Temperature)
The temperature at which a combustible mixture of vapor and air ignites in the absence of a flame.

Examples. A mixture of pentane and air must be heated to about 300°C before it will ignite spontaneously. Carbon disulfide has an ignition temperature of about 100°C; heating it on a steam bath may ignite it.

7.200 Explosive Limits
The range of concentration of a vapor of interest that, when mixed with air (or oxygen) and exposed to a spark or other source of ignition, results in an explosion.

Examples. Mixtures of pentane and air are explosive in the range of 1.5 to 7.5% by volume of pentane in air; if less or more pentane is present, an explosion will not occur. Hydrogen in air is explosive in the range of 4 to 74% and of course in the presence of pure oxygen this range is considerably expanded so that mixtures of oxygen and hydrogen in any proportions are extremely hazardous.

7.210 Chromatography
A separation procedure that depends on the partitioning of the components of a mixture between a stationary phase and a moving phase.

7.220 Adsorption Chromatography
Chromatography in which the stationary phase, the absorbent, is a solid. Separation of the components in a mixture depends on their relative affinities toward the absorbent; historically, the first chromatographic technique.

Example. In one of the earliest examples the separation of chlorophyll from other colored organic plant material was achieved by passing a mixture of plant extracts through a column of alumina and washing with solvents such as benzene. Because separate colored bands developed on the column, the separation process was called chromatography.

7.230 Column Chromatography
A type of chromatography in which the stationary phase is placed in a column and a mixture of substances to be separated is placed onto the column. Various solvents are then passed through the column. The components in the mixture be-

come distributed between the stationary phase and the solvent phase by a combination of adsorption and desorption processes. The least strongly adsorbed and most easily desorbed substance, under the condition of operation, gets eluted from the column first.

Example. The experiment with chlorophyll described in Sect. 7.220 is one example of column chromatography. The technique is also used in liquid-liquid (Sect. 7.260) and gas-liquid (Sect. 7.270) partition chromatography.

7.240 Partition Coefficient

The ratio of the quantities of a pure material which appear in each of two immiscible phases when the material is equilibrated between them.

7.250 Thin-Layer Chromatography (tlc)

A special kind of adsorption chromatography that utilizes a thin-layer of solid adsorbent on a glass or plastic plate onto which is placed a tiny amount of the solution of components (solutes) to be separated. The bottom of the plate is then immersed in an appropriate solvent or mixture of solvents in a closed container. The solvent then diffuses up the plate through the sample to be resolved. The various components move up the plate in accordance with their partition coefficients.

Examples. A typical arrangement is shown in Fig. 7.250.

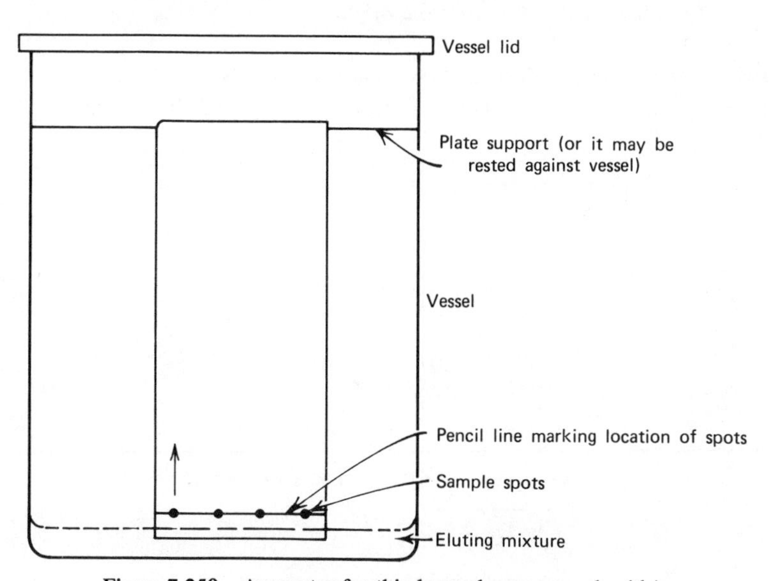

Figure 7.250 Apparatus for thin-layer chromatography (tlc).

7.260 Liquid-Liquid Partition Chromatography

Chromatography in which the stationary phase is a liquid supported on the surface of an inert solid.

7.270 Gas-Liquid Partition Chromatography (glpc, vpc or glc)

Partition chromatography in which the sample mixture is vaporized by passing it in a hot stream of inert gas over a partitioning liquid (the stationary phase) dispersed as a thin film over the surface of a porous inert support material. The technique, frequently called glc, has revolutionized the separation of volatile materials.

7.280 Paper Chromatography

Liquid-liquid partition chromatography in which the components of a mixture of solutes are separated through their differing partition coefficients between a water phase adsorbed on the paper's surface and an eluting solvent that moves the components at differing rates up the paper.

Example. A typical setup is similar to that shown in Fig. 7.250 except that the glass plate is replaced by a strip of paper.

7.290 R_f Value

In tlc or paper chromatography the distance the solute moves divided by the distance traveled by the solvent front; the measurement is made when the solvent front has moved quite far from the origin.

Examples. Typically, a mixture of compounds A and B is spotted on a tlc plate or paper very close to the bottom, as shown in Fig. 7.290, and the paper or

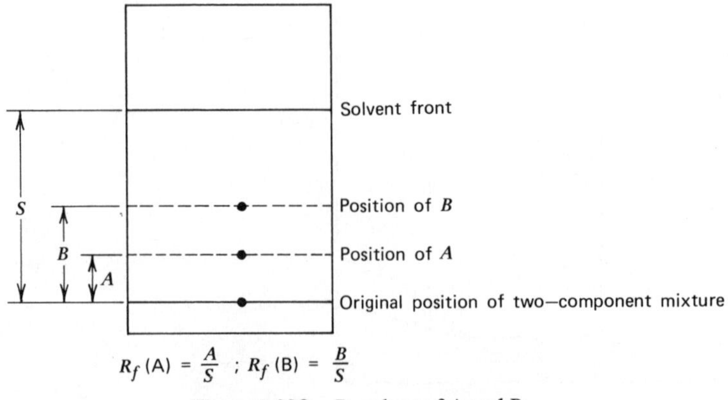

$$R_f\,(A) = \frac{A}{S}\ ;\ R_f\,(B) = \frac{B}{S}$$

Figure 7.290 R_f values of A and B.

plate is placed in a vessel with an eluting solvent. After an appropriate length of time the eluting solvent can be seen at the position depicted in Fig. 7.290. The plate or paper is removed and the spots developed appropriately, whereupon the R_f values for A and B may be calculated and preferably compared with the R_f values of authentic A and B, each determined as above.

7.300 Electrophoresis

A technique that utilizes an electric field to effect separation of high molecular weight charged ions. Most frequently it is combined with paper chromatography. As the ions move in a vertical direction up the paper as in chromatography, they are spread in a horizontal direction by the influence of the electrical voltage placed on the paper. The positive ions migrate horizontally toward the cathode, and the negative ions move in the opposite direction.

7.310 Molecular Sieves

Those used in chromatography are polymeric carbohydrates or polymeric acrylamides (monomer, $CH_2{=}CH{-}CONH_2$) that have an open network formed by cross-linking. They are hydrophilic, adsorb water, and swell, causing the development of "holes" of various diameters. The resulting structures are *gels.*

7.320 Gel Permeation Chromatography

A chromatography procedure using gels in which the separation depends on the ability of relatively small molecules to be trapped in the holes in the gel while the large molecules pass freely through the medium. High swelling gels are used to fractionate high molecular weight substances, whereas the denser and less highly swollen gels are used for the separation of relatively low molecular weight substances.

7.330 Molecular Exclusion Chromatography

A more elegant name for gel permeation chromatography.

7.340 Reverse-Phase Chromatography

A type of partition chromatography in which hydrocarbons as well as polar samples are partitioned between a nonpolar stationary phase and a polar eluting phase. Under these conditions the most polar substances elute most rapidly. This is the reverse of the more common partition chromatography such as glpc where the stationary phase is polar and the least polar substances elute most rapidly with the nonpolar eluting phase. The stationary phase often consists of a chain of atoms chemically bonded to an inert surface such as glass and the eluting phase is frequently aqueous methanol or aqueous acetonitrile in reverse-phase chromatography.

Liquid partition chromatography techniques were vastly improved by the introduction of the use of small packing spheres (5–10 μm in diameter) in small diameter (5 mm) columns operating under high pressures (1000–5000 psi) with the stationary phase chemically bonded to the solid packing. Chromatography performed with such columns (often not more than 10–25 cm long) is known as **high performance liquid chromatography** or **high pressure liquid chromatography (hplc)**; reverse phase chromatography is one of the most widely used variations of hplc.

7.350 Thermal Conductivity Detector

Used in many glpc instruments, it depends on the fact that when a gas is passed over a heated thermister, the temperature, hence the resistance, of the thermister varies with the thermal conductivity of the gas. The difference in thermal conductivity, measured as a function of the resistance, between pure carrier gas (usually helium because of its low thermal conductivity) and a similar stream of carrier gas carrying a minute quantity of one of the components, is detected by means of a Wheatstone bridge circuit. The change in resistance occasioned by the presence of one of the components of the sample in the carrier gas, which is proportional to its concentration, is registered on a recorder.

7.360 Hydrogen Flame Detector

Used in some glpc instruments with hydrogen as the carrier gas, this type of detector is based on the difference in luminosity between a flame produced by burning pure hydrogen and a flame when the same quantity of hydrogen is diluted with a small amount of organic material present in the sample. The change in flame temperature can also be measured by a thermocouple.

7.370 Flame Ionization Detector

A detector that is almost 10^3 times more sensitive than the thermal conductivity detector; its functioning depends on the fact that most organic compounds form ions in a hydrogen flame. The ions are measured by a pair of oppositely charged electrodes.

7.380 Electron Capture Detector

Extremely sensitive for compounds that contain electronegative atoms. The cathode in the detector cell consists of a metal foil impregnated with an element that emits β-rays (usually tritium or ^{63}Ni). On application of a potential to the cell, the β-rays emitted at the cathode migrate to the anode, generating a current. If a compound with a high electron affinity is introduced into the cell, it captures electrons and becomes a negative ion, decreasing the electric field; the resulting change is suitably recorded. Such a detector is very selective for compounds containing halogen, carbonyl, and nitro groups, but has a low sensitivity for hydrocarbons other than aromatics.

7.390 Ion Exchange Chromatography

A chromatography technique useful for the separation of inorganic ions and amino acids based on the exchange of ions in the sample with ions in a stationary phase. The stationary phase is usually a polystyrene resin in which some of the aromatic hydrogens have been replaced by appropriate functional groups.

7.400 Cation Exchange Resin

The exchange of cations in solution for protons provided by the exchange resin. The exchange resin is a cross-linked polystyrene, in the form of beads, that has been sulfonated (or carboxylated) and thus has $-SO_3H$ (or $-COOH$) functions on the aromatic rings of the polymer or resin.

Examples. The exchange of metal cations, M^{n+}, with a strong acid cation exchange resin depends on the equilibrium

$$n[\text{Ⓟ}-SO_3H] + M^{n+} \rightleftharpoons (\text{Ⓟ}-SO_3)_nM + nH^+$$

With a weak carboxylic acid cation exchange resin there is a similar equilibrium:

$$n[\text{Ⓟ}-CO_2H] + M^{n+} \rightleftharpoons (\text{Ⓟ}-CO_2)_nM + nH^+$$

where P — represents the polymeric polystyrene. The weaker acid ion exchange resin has a greater affinity for protons, and therefore requires a higher pH for a given ion to exchange.

7.410 Anion Exchange Resin

A resin that allows the exchange of anions in solution for OH^- anions provided by the exchange resin.

Examples. The equilibrium with a strong base exchange resin (alkyl quaternary ammonium salt) is

$$n[\text{Ⓟ}-NR_3(OH)] + A^{n-} \rightleftharpoons (\text{Ⓟ}-NR_3)_nA + nOH^-$$

and with a weaker base exchange resin (ammonium salt) there is a similar equilibrium:

$$n[\text{Ⓟ}-NH_3(OH)] + A^{n-} \rightleftharpoons (\text{Ⓟ}-NH_3)_nA + nOH^-$$

The weaker base ion exchange resin has greater affinity for hydroxide groups, and a lower pH is therefore required for a given ion to exchange.

7.420 Molal Freezing Point Depression Constant, K_f

The decrease in freezing point of a pure liquid that is 1 molal (1 mole of solute per 1000 g of solvent) in a nondissociating solute; the value is characteristic of each solvent and is a constant independent of the nature of the solute (providing that there is no dissociation of the solute).

Examples. Pure benzene has a freezing point of 5.5°C. When 2.40 g of biphenyl (mol. wt. 154) is dissolved in 75.0 g of benzene, the freezing point of the solution is lowered 1.06°C (ΔT). The molal freezing point depression constant K_f of benzene can then be calculated from the equation

$$K_f = \frac{\Delta T}{\text{molality}} = \frac{1.06}{.208} = 5.1°\text{C}\, m^{-1}$$

Thus a 1 molal solution of any organic solute in benzene would freeze at 0.4°C.

7.430 Cryoscopic Constant
Synonomous with molal freezing point depression constant.

7.440 Van't Hoff Factor, *i* (J. H. van't Hoff, 1852-1911)
The ratio of the observed freezing point depression (ΔT) to that expected on the basis of no dissociation of the solute ($m \cdot K_f$):

$$i = \frac{\Delta T}{mK_f}$$

where m is the molality of the solution and K_f the cryoscopic constant of the solvent.

Example. 100% H_2SO_4 has a convenient freezing point (-12°C). The fact that a solution of *o*-benzoylbenzoic acid in this solvent gives a van't Hoff factor of 4 is evidence that the reaction shown in Fig. 7.440 occurs giving 4 particles.

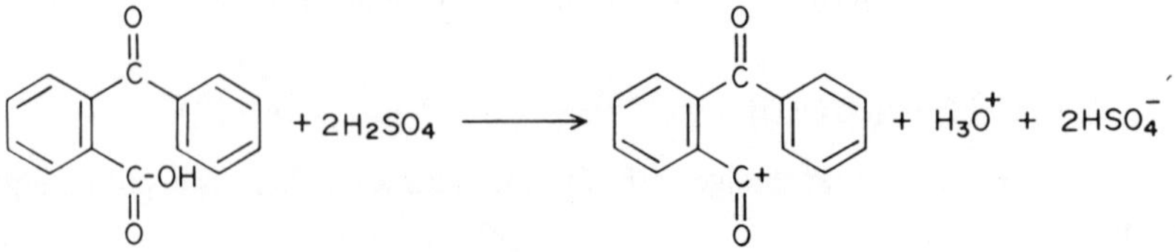

Figure 7.440 The ionization of benzoylbenzoic acid.

7.450 Viscosity (Coefficient of Viscosity), η
A measure of the resistance to flow of any fluid; the greater the resistance the higher the viscosity. The coefficient of viscosity, η, a term used interchangeably with viscosity, is given in units of poise (J. L. M. Poiseuille, 1799-1869) and has the dimension of 10^{-1} kg m^{-1} s^{-1}. η is frequently determined from the rate of flow of a liquid through cylindrical tubes.

7.460 Viscosity Index
A measure of the change in viscosity as a function of temperature. Good lubricants have a low viscosity index.

7.470 Viscometer
An instrument for measuring viscosity. These are of various types. Three common methods of measurement are: the determination of the rate of flow of the sample through a calibrated tube; the rate of settling of a ball in the liquid to be measured; and measurement of the force required to turn one of two concentric cylinders, filled with liquid, at a certain angular velocity.

7.480 Refractive Index, *n*
A measure of the apparent speed of light (c') passing through matter relative to the speed of light passing through a vacuum (c): $n = c/c'$. The change in refractive index of a substance as a function of the wavelength of light is known as **dispersion.** The density and refraction index of a liquid are useful in characterizing a pure liquid by means of its **molecular refractivity, [*R*]**, calculated from the equation

$$[R] = \frac{n^2 - 1}{n^2 + 2} \cdot \frac{M}{d}$$

where M is the molecular weight of the substance, n is its refraction index, and d is its density (determined at the same temperature as n). Refractive index measurements are commonly carried out using the sodium D line.

7.490 Colligative Properties
Properties of solutions that depend principally on the concentration of dissolved particles and not on the characteristics of these particles, e.g., molal freezing point depression and boiling point elevation.

7.500 Semipermeable Membrane
A membrane that permits some types of molecules but not others to pass through it.

Examples. When an aqueous solution of sucrose is placed in one arm of a U tube the bottom of which contains a semipermeable membrane (e.g., cellophane) and pure water is placed in the other arm, the water will diffuse into the solution of sucrose because the concentration of water on one side of the membrane is greater than that on the other and only the water but not the sugar can diffuse through the membrane.

7.510 Osmosis

The process by which one type of molecule in a solution will pass through a semipermeable membrane but another type will not.

7.520 Osmotic Pressure, π

The hydrostatic pressure that develops at equilibrium on one side of a semipermeable membrane as a result of selective diffusion.

Examples. The U-tube shown in Fig. 7.520 has unequal hydrostatic pressure on the two sides of the membrane shown. If the pure solvent were water and the solution an aqueous solution of sucrose, the difference in heights of the two legs of the tubes would be a measure of the osmotic pressure. The osmotic pressure π is very similar to p (pressure) in the gas equation ($pV = nRT$):

$$\pi V = nRT$$

where π is the osmotic pressure (in atmospheres), n equals the number of moles of solute dissolved in volume V (in liters), R is the gas constant (0.08206 L atm K^{-1} mol^{-1}), and T is the absolute temperature.

7.530 Reverse Osmosis

An osmosis process in which external pressure is applied to force a solvent to flow in the direction opposite to that normally expected.

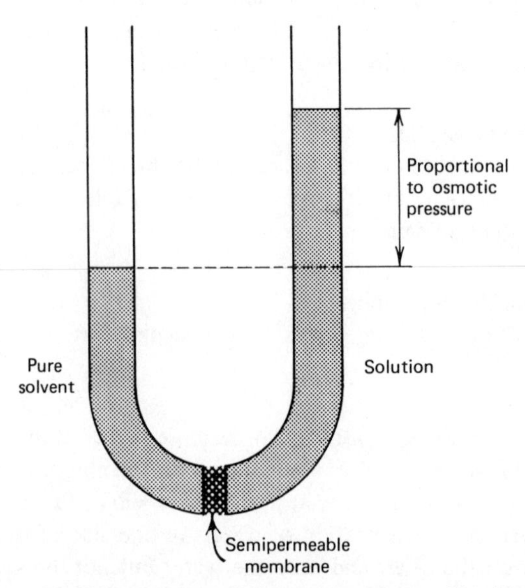

Figure 7.520 Measurement of osmotic pressure with a U-tube.

Examples. If, at equilibrium, the right arm of the U-tube shown in Fig. 7.520 is subjected to pressure, water can be made to flow through the membrane to the left arm of the U-tube. This is one of the most important processes used in water desalination.

7.540 Polarized Bond

The bond that results when two different atoms are bonded and the electrons bonding them are unequally shared, e.g., $\overset{+\!\!\longrightarrow}{\text{H—Cl}}$.

7.550 Electric Dipole

The dipole that characterizes a polarized bond; i.e., the positive and negative charges do not coincide.

7.560 Electric Dipole Moment, μ

A directed (vector) quantity (+→) that is the product of the charge (q) times the distance R separating $+q$ and $-q$: $\mu = qR$. In molecules a typical distance is 1 Å (10^{-8} cm) and a typical charge is that associated with an electron, namely 4.80×10^{-10} electrostatic units (esu).

7.570 Debye Unit, D (P. Debye, 1884-1966)

Equal to 1×10^{-18} esu cm.

7.580 Bond Dipole Moment

The dipole moment associated with a particular bond independent of the other bonds present in the molecule.

Examples. Typical bond moments (in Debye units) of some important C—Z bonds in organic chemistry are: C—H(0.4); C—F(1.41); C—Cl(1.46); C—N(0.22); C—O(0.74); C=O(2.3). These values may then be substituted in other real or theoretical compounds having C—Z bonds, and the dipole moments for such compounds calculated by vectorial addition.

7.590 Vectorial Addition of Dipole Moments

The vectorial sum of individual bond dipole moments of a molecule. Each group in a substituted molecule has an associated directed dipole moment that is (only) approximately independent of the nature of the rest of the molecule. The resultant of the vectorial addition of all group moments yields the overall dipole moment.

Example. Chlorobenzene has a measured dipole moment of 1.70 D, Fig. 7.590*a*. The dipole moment of 1,2-dichlorobenzene can be calculated from the dipole moment of the monochlorobenzene by vector addition as shown in Fig. 7.590*b*.

The vector OB is simply equal to 1.70 cos 30° or 1.47, and the desired dipole moment (O*b*, Fig. 7.590*b*) is twice this quantity or 2.94 D. However the experimental value for the dipole moment of 1,2-dichlorobenzene is 2.25 D. If one uses this experimental value as the vector sum and calculates the individual vectors,

$$\text{OCl} = \frac{1}{2}\,\frac{\text{Ob}}{\cos 30^\circ} = \frac{2.25}{2\cos 30^\circ} = 1.30\ \text{D}$$

these are found to be considerably less than the single vector in monochlorobenzene. This indicates that the dipole vectors in dichlorobenzene interact with each other, resulting in a vector sum less than that calculated on the basis of no interaction. In these simplified calculations the bond moments of other bonds in the molecule have been neglected.

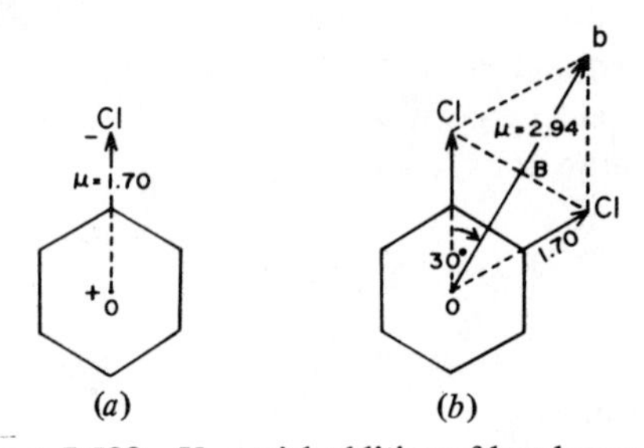

Figure 7.590 Vectorial addition of bond moments.

7.600 Van der Waals Forces (J. D. van der Waals, 1837-1923)
Any attractive force between uncharged molecules.

7.610 London Dispersion Forces (F. London, 1900-1954)
The interaction between uncharged molecules which arises because of the transient noncoincidence of the centers of negative and positive charge; these transient dipoles interact with similar transient dipoles in neighboring molecules. The attraction caused by London forces varies with the seventh power of the distance separating the molecules involved. Van der Waals forces and London forces are frequently used interchangeably, but the latter are really a specific type of van der Waals interaction.

7.620 Dipole-Dipole Interaction
An interaction between molecules having permanent electric dipole moments.

7.630 Ion-Dipole Interaction

The interaction or solvation of ions by molecules with electrical dipoles; the solubility of most inorganic salts in water is due to such interactions.

7.640 Polarizability

The susceptibility of an uncharged molecule to the development of an electric dipole under the influence of a transient or permanent dipole usually provided by a neighboring molecule. The polarizability of a molecule increases as the number of electrons increases and as their distance from the nucleus increases.

7.650 Dielectric Constant, ϵ

A measure of the relative effect of a solvent on the force with which two opposite charges attract each other; it is determined by measuring the electrical capacitance of a condenser when empty and when filled with the solvent.

Examples. Solvents with high dielectric constants such as water ($\epsilon = 80$) and acetonitrile ($\epsilon = 39$) are much better solvents for ions than solvents with low dielectric constants such as acetone ($\epsilon = 21$) and benzene ($\epsilon = 2.3$).

7.660 Surface Tension

A measure of the force acting on surface molecules by the molecules immediately below this surface; units are dyn cm^{-1}.

Example. The surface tension of a liquid may be measured by observing liquid in a capillary tube. Surface tension is responsible for the formation of spherical droplets because the forces involved result in surface contraction to a minimum area.

7.670 Adsorption

A process in which molecules (either gas or liquid) adhere to the surface of a solid.

7.680 Physical Adsorption

An adsorption process that depends on the van der Waals force of interaction between molecules (gas or liquid) and the solid surface on which they are being adsorbed.

7.690 Chemisorption (Chemical Adsorption)

An adsorption process that depends on chemical interaction (weak chemical bonds) between molecules (gas or liquid) and a solid surface.

Example. The chemisorption of molecules on a solid catalyst is presumed to be the first step in heterogeneous catalysis.

7.700 Surface Area

The area of an adsorbent surface that can be covered completely with a unimolecular layer of gas. It is calculated from the volume of gas adsorbed.

Example. A typical silica gel may have a surface area of ~500 $m^2 g^{-1}$.

7.710 B.E.T. Equation

An equation named after S. Brunauer, P. Emmett, and E. Teller, used in the experimental determination of surface areas of solids.

7.720 Absorption

The intimate mixing of the atoms or molecules of one phase with the atoms or molecules of a second phase; essentially the same as solution.

7.730 Interfacial Tension

The surface tension at the interface between two mutually saturated liquids which tends to contract the interface area.

7.740 Colloid

A two-phase system consisting of very fine particles (in the 10–10,000 Å range) dispersed in a second phase. The very large surface area of the dispersed phase is responsible for the unique properties of a colloidal system.

7.750 Emulsion

A colloidal system of two immiscible liquids, one of which is dispersed throughout the other in small drops.

7.760 Micelle

The stable, ordered aggregation of molecules that constitutes the dispersed phase of an emulsion.

Examples. In water, soap molecules, e.g., sodium oleate [$CH_3(CH_2)_7CH{=}CH{=}(CH_2)_7CO_2^-\overset{+}{Na}$], orient themselves in the micelle form depicted in Fig. 7.760.

7.770 Hydrophilic and Hydrophobic

Water attractive and water repellant, respectively.

Examples. Soaps such as sodium oleate have a hydrophilic site (the $-CO_2^-$) and a hydrophobic site (the hydrocarbon chain). The emulsifying properties of soaps

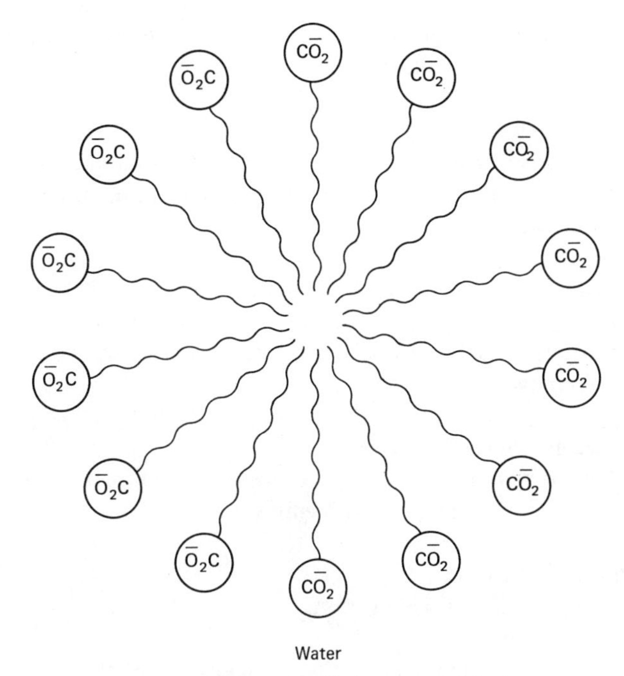

Figure 7.760 A micelle (a soap emulsion).

depend on their hydrophobic chains being immersed in the hydrocarbon or oil phase and the carboxylate site being solvated (immersed) in the aqueous phase, thus allowing the hydrocarbon to be dispersed in the water as an emulsion.

7.780 Emulsifying Agent

A substance that, when added to an emulsion that tends to separate, keeps the system more or less permanently emulsified.

Examples. The emulsifying action of soap depends on its ability to reduce the interfacial tension between a liquid hydrocarbon layer and water. The interfacial tension of benzene-water (35 dyn cm^{-1}) is reduced to 2 dyn cm^{-1} by the addition of sodium oleate. The emulsion that results may have as much as 100 parts of benzene spread out as drops throughout only 1 part of water.

7.790 Detergents

Compounds that act as emulsifying agents.

Examples. Soaps (sodium salts of long chain fatty acids) are detergents, but other types of molecules have been synthesized that are even better detergents than soaps. One group of commercially important synthetic detergents are the sodium salts of alkylbenzenesulfonic acids, $RC_6H_4S\bar{O}_3\overset{+}{Na}$.

7.800 Wetting Agents

A substance adsorbed on the surface of a solid which interacts more strongly with water than does the solid surface, thus permitting the otherwise inactive surface to become wet with water.

Examples. The effectiveness of detergents in removing soil from fabrics depends on wetting and emulsification. The adsorption of the detergent at the solid surface allows the fabric to be wet by water and allows the soil (oil) to be emulsified and removed.

7.810 Surfactants (Surface Active)

A general term for substances that reduce the surface tension between immiscible liquids; essentially synonymous with emulsifiers.

7.820 Cationic Surfactants

Surfactants whose polar group is a cation.

Examples. Quaternary ammonium salts having as one of the R groups a long chain alkyl group, e.g., $C_{16}H_{33}\overset{+}{N}(CH_3)_3\bar{C}l$. Compounds such as $RC_6H_4S\bar{O}_3\overset{+}{Na}$ are **anionic surfactants.**

7.830 Invert Soaps

Synomymous with cationic surfactants; so called because the polar group is a cation rather than an anion, as in the more conventional soaps and detergents.

7.840 Nonionic Surfactants

Surfactants that are uncharged (neutral) species without cationic or anionic sites.

Examples. Alkyl aryl ethers of poly(ethylene glycol) prepared from alkylphenols and ethylene oxide: $RC_6H_4O(CH_2CH_2O)_xCH_2CH_2OH$. This type of detergent is used in formulating low-suds products for automatic washers.

7.850 Aerosols

A colloidal system in which a solid or a liquid is dispersed in a gas.

7.860 Phase-Transfer Catalysts

Compounds whose addition to a two-phase organic–water system helps transfer a water soluble reactant across the interface to the organic phase where a homogeneous reaction can occur, thus enchancing the rate of reaction.

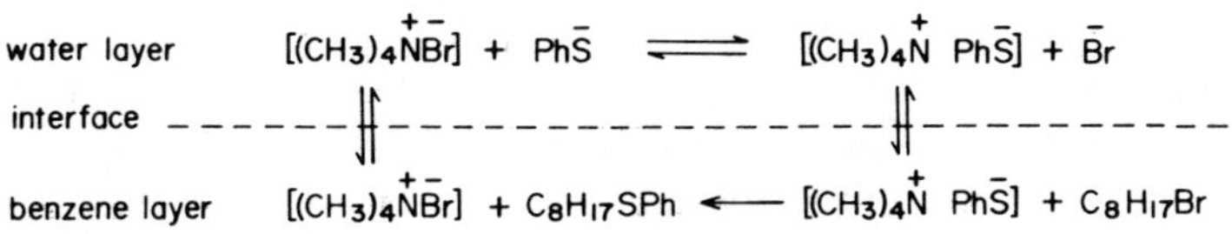

water layer $[(CH_3)_4\overset{+}{N}\overset{-}{Br}] + Ph\overset{-}{S} \rightleftharpoons [(CH_3)_4\overset{+}{N}\ Ph\overset{-}{S}] + \overset{-}{Br}$

interface ⇅ ⇅

benzene layer $[(CH_3)_4\overset{+}{N}\overset{-}{Br}] + C_8H_{17}SPh \longleftarrow [(CH_3)_4\overset{+}{N}\ Ph\overset{-}{S}] + C_8H_{17}Br$

Figure 7.860 Phase-transfer catalysis for $PhS^- + RX \rightarrow PhSR + X^-$.

Example. A quaternary salt that is a phase-transfer catalyst functions as shown in Fig. 7.860 for the reaction between 1-bromooctane and the sodium thiophenoxide.

7.870 Crown Ethers

Cyclic polyethers. The structure of these compounds permits a conformation possessing certain sized "holes" in which cations are trapped by coordination with the lone pair electrons on the oxygen atoms.

Examples. Figure 7.870 shows an 18-crown-6 ether; it is given this name because there are a total of 18 atoms in the cyclic crown, 6 of which are oxygen atoms. These compounds dissolve metal salts in nonpolar solvents. Thus KF dissolves in a benzene solution containing 18-crown-6 which complexes K^+ and leaves an extremely reactive "naked" fluoride ion. The crown ether forms a hydrophilic cavity and a hydrophobic shell. The crown ethers can be used as phase-transfer catalysts, since they can transfer ionic materials from an aqueous phase to an organic one.

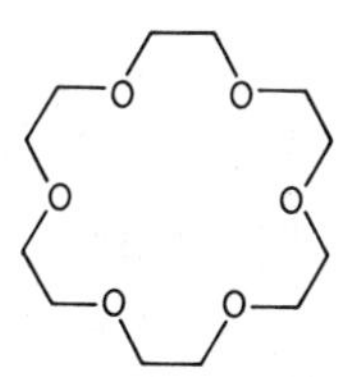

Figure 7.870 A crown ether (18-crown-6).

7.880 Liquid Crystals

Rodlike molecules that undergo transformations into stable, intermediate semifluid states before passing into the liquid state.

7.890 Mesophases

The stable phases exhibited by the liquid crystal molecules as they pass from the crystalline solid to the liquid state.

7.900 Anisotropic

The property of molecularly oriented substances to exhibit variations in physical properties along different axes of the substance.

7.910 Isotropic

Lack of anisotropy; the property of having identical properties in all directions. All liquids are isotropic because there is complete randomness in the arrangements the molecules may assume.

7.920 Thermotropic Liquid Crystals

Liquid crystals that are formed by heating. One of the two main classes of liquid crystals.

7.930 Lyotropic Liquid Crystals

A second main class of liquid crystals; these are formed by mixing two components, one of which is highly polar such as water. Such liquid crystals are found in living systems.

7.940 Smectic Mesophase

The mesophase of liquid crystals which shows more disorder than the crystalline phase but which still possesses a layered structure. In the smectic mesophase the layers of molecules have moved relative to each other, but the molecules retain their parallel orientation with respect to one another.

Example. Figure 7.940.

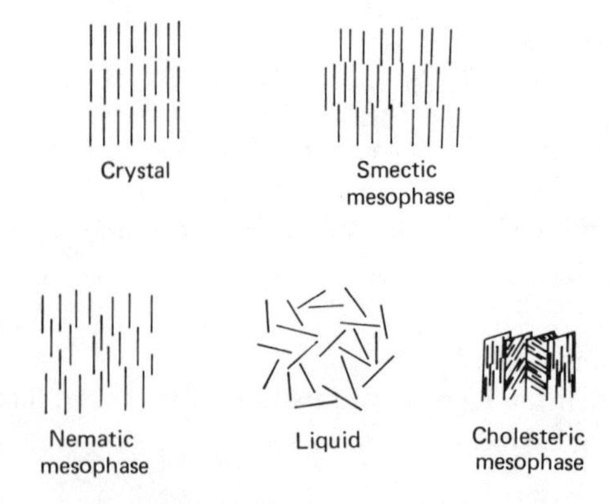

Figure 7.940 The various forms and phases of a liquid crystal.

7.950 Nematic Mesophase

In this phase the liquid crystal molecules are still constrained to be parallel to each other, but they are no longer separated into the layers that characterize the smectic mesophase (see Fig. 7.940).

Example. Butyl *p*-(*p*-ethoxyphenoxycarbonyl)phenyl carbonate, Fig. 7.950, is a crystalline material that, on heating to ~55°C, goes over to the nematic mesophase and at ~87°C becomes an isotropic liquid.

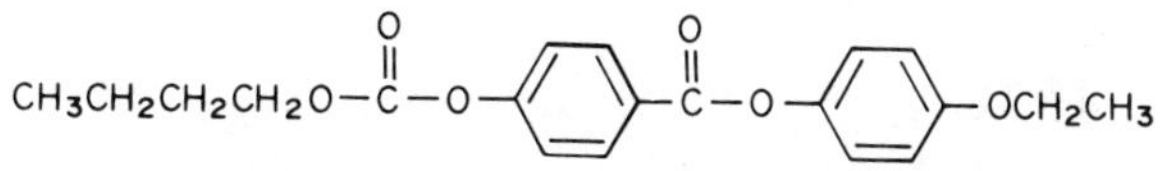

$$CH_3CH_2CH_2CH_2O-\overset{\overset{O}{\|}}{C}-O-C_6H_4-\overset{\overset{O}{\|}}{C}-O-C_6H_4-OCH_2CH_3$$

Figure 7.950 Butyl *p*-(*p*-ethoxyphenoxycarbonyl)phenyl carbonate.

7.960 Cholesteric Mesophase

In this phase the liquid crystal molecular arrangement resembles a twisted structure. The direction of the long axis of the molecules in the nematic phase is displaced from the long axis in an adjacent sheet of nematic phase molecules. The overall displacement results in a helical arrangement.

Example. See Fig. 7.940.

8 Ions and Electron Deficient Species

CONTENTS

8.010 Species

A very loose but convenient term used to denote atoms, molecules, ions, radicals, or other chemical entities; alternatively, **molecular particles.**

8.020 Charged Species

A species in which the total number of electrons and the summed over all nuclei number of protons is unequal. The number of protons is always equal to the sum of the atomic numbers of the constituent atoms of the species.

Examples. Methyl cation, Fig. 8.020*a*, and methyl anion, Fig. 8.020*b*, possess 8 and 10 electrons, respectively, but both species possess 9 protons associated with the three hydrogen and one carbon nuclei (sum of the atomic numbers: $1 + 1 + 1 + 6 = 9$). In order to avoid confusion, a positive electrical charge is sometimes represented by ⊕ and a negative electrical change by ⊖; although the circle is usually omitted, in this chapter only, all charges will be circled for emphasis.

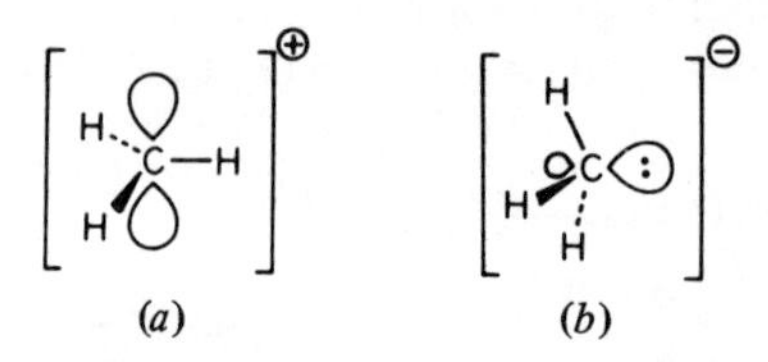

Figure 8.020 (*a*) Methyl cation; (*b*) methyl anion.

8.030 Ion

Synonymous with **charged species.**

8.040 Cation

A positively charged atom or group of atoms.

Examples. Benzyl cation, Fig. 8.040*a*; methylene dication, Fig. 8.040*b*; trimethyloxonium ion, Fig. 8.040*c*.

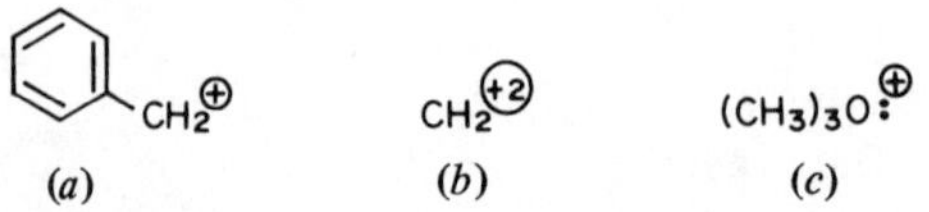

$C_6H_5CH_2^{\oplus}$ (*a*) $CH_2^{(+2)}$ (*b*) $(CH_3)_3O{:}^{\oplus}$ (*c*)

Figure 8.040 (*a*) Benzyl cation; (*b*) methylene dication; (*c*) trimethyloxonium ion.

8.050 Anion

A negatively charged atom or group of atoms.

Examples. 2-Naphthyl anion, Fig. 8.050*a*, and phenylmethylene dianion, Fig. 8.050*b*.

Figure 8.050 (*a*) 2-Naphthyl anion; (*b*) phenylmethylene dianion.

8.060 Uncharged Species

A species that has neither a net positive nor a net negative charge.

Examples. Atomic carbon, C, and methane, CH_4; the total number of electrons equals the total number of protons in the nuclei.

8.070 Neutral Species

Synonymous with uncharged species; not to be confused with the word neutral when it is used in the description of acid-base properties.

8.080 Electron Deficient Species

A species that possesses fewer than the maximum possible number of electrons in the valence shells of its atoms.

Examples. Methyl radical, Fig. 8.080*a*; methylene, Fig. 8.080*b*; and methyl cation, Fig. 8.080*c*; possess seven, six, and six valence electrons, respectively; the full complement of valence electrons around carbon and all other second row elements is eight. Under this definition Lewis acids (Sect. 9.520) such as boron trifluoride (BF_3) as well as compounds containing multicenter bonds, e.g., diborane (B_2H_6), which contains two two-electron, three-center bonds, are electron deficient.

$CH_3\cdot$ (*a*) $CH_2:$ (*b*)

$CH_3^{\oplus}$ (*c*)

Figure 8.080 (*a*) Methyl radical; (*b*) methylene; (*c*) methyl cation.

8.090 Free Radical

An uncharged species possessing one or more unpaired electrons. If the species bears a charge, it is known as a **radical-ion**. Transition metal complexes that have one or more unpaired electrons associated with the metal are not generally referred to as free radicals. If a radical is associated with another radical counterpart in a solvent cage, the system is referred to as a **radical pair**, or **geminate pair**.

Examples. Benzyl radical, Fig. 8.090*a*, and methoxyl radical, Fig. 8.090*b*. In the standard convention a dot (·) or single arrow (↑) represents an unpaired electron. Two dots (:) or two arrows facing in opposite directions (↑↓) represent two paired electrons. Two arrows facing in the same direction (↑↑) represent two unpaired electrons but these cannot, of course, occupy the same orbital.

$C_6H_5CH_2\cdot$ (*a*) $CH_3\ddot{\underset{..}{O}}\cdot$ (*b*)

Figure 8.090 (*a*) Benzyl radical; (*b*) methoxyl radical.

8.100 Radical

Synonymous with free radical. In early radicofunctional nomenclature, prefix names such as methyl and ethyl were called radicals, but these are now referred to as groups.

8.110 Diradical

An uncharged species possessing two unpaired electrons not on the same atom and being able to exist in two electronic states with different **electron spin multiplicity**.

Examples. Methylethylene diradical, Fig. 8.110*a*, and dimethylenediaminyl diradical, Fig. 8.110*b*. Compounds such as triplet methylene, Fig. 8.110*c*, have been called diradicals but are better referred to as triplet carbenes. (See Sect. 8.180.)

$CH_3\dot{C}H\dot{C}H_2$ (*a*) $H\dot{\underset{..}{N}}CH_2CH_2\dot{\underset{..}{N}}H$ (*b*) $\dot{C}H_2\cdot$ (*c*)

Figure 8.110 (*a*) Methylethylene diradical; (*b*) dimethylenediaminyl diradical; (*c*) triplet methylene.

8.120 Biradical

Synonymous with diradical.

8.130 Divalent (or Dicovalent) Carbon

A carbon atom having only two covalent bonds attached to it; it may be charged or uncharged.

Examples. Methylene, Fig. 8.130*a*; methylmethylene radical-cation, Fig. 8.130*b*; and phenylmethylene radical-anion, Fig. 8.130*c*. The underlined carbons are divalent.

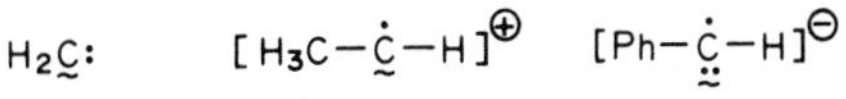

$H_2\underset{\sim}{C}{:}$ (*a*) $[H_3C-\dot{\underset{\sim}{C}}-H]^{\oplus}$ (*b*) $[Ph-\dot{\underset{\sim}{\underset{\cdot\cdot}{C}}}-H]^{\ominus}$ (*c*)

Figure 8.130 (*a*) Methylene; (*b*) methylmethylene radical-cation; (*c*) phenylmethylene radical-anion.

8.140 Carbene, R_2C:

An uncharged species containing a dicovalent carbon atom. This carbon atom is surrounded by a sextet of valence electrons, hence it is electron deficient.

Examples. Dichlorocarbene, Fig. 8.140*a*, and methylenecarbene, Fig. 8.140*b*.

$Cl-\ddot{C}-Cl$ (*a*) $H_2C{=}C{:}$ (*b*)

Figure 8.140 (*a*) Dichlorocarbene; (*b*) methylene carbene.

8.150 Methylene

The carbene that has the molecular formula $:CH_2$.

8.160 Electron Spin Multiplicity

The number of spin orientations observed when the species in question is placed in a magnetic field. The multiplicity is determined by the number of unpaired electron spins n and the spin angular momentum number s ($\pm\frac{1}{2}$); it is equal to $2|S| + 1$ where $|S|$ is equal to $|ns|$.

Examples. CH_4, $S = 0$, multiplicity = 1 (singlet); $CH_3\uparrow$, $S = \frac{1}{2}$, multiplicity = 2 (doublet); $CH_2\uparrow\uparrow$, $S = 1$, multiplicity = 3 (triplet).

8.170 Singlet Carbene

A carbene that has an electron spin multiplicity of 1 (no unpaired electrons).

Examples. Singlet methylene, Fig. 8.170*a*, and singlet phenylcarbene, Fig. 8.170*b*. The pair of nonbonding electrons occupy a single orbital lying in the H—C—H plane.

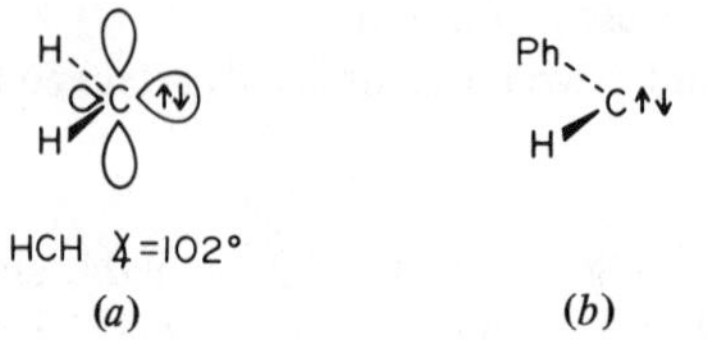

HCH ∡=102°

(*a*) (*b*)

Figure 8.170 (*a*) Singlet methylene; (*b*) singlet phenylcarbene.

8.180 Triplet Carbene

A carbene that has an electron spin multiplicity of 3 (two unpaired electrons).

Examples. Triplet methylcarbene, Fig. 8.180*a*, and triplet methylene, Fig. 8.180*b*. The unpaired electrons occupy separate orbitals, one perpendicular to, and the other one in, the H—C—H plane.

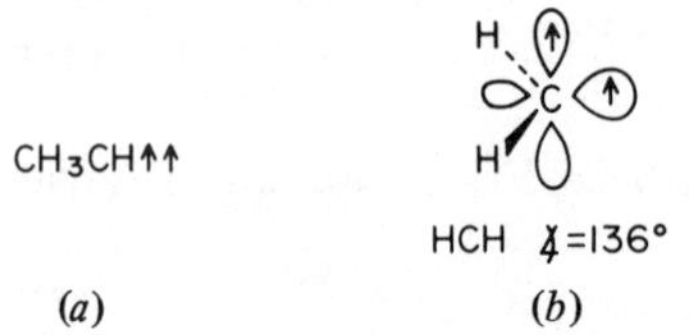

$CH_3CH\uparrow\uparrow$

HCH ∡=136°

(*a*) (*b*)

Figure 8.180 (*a*) Triplet methylcarbene; (*b*) triplet methylene.

8.190 Carbenoid Species

A species that, when undergoing reaction, has the characteristics of a carbene, but in fact is not a free carbene.

Example. Iodomethylzinc iodide, Fig. 8.190, reacts with ethylene to yield the same product, cyclopropane, as the reaction of ethylene with methylene, a free carbene.

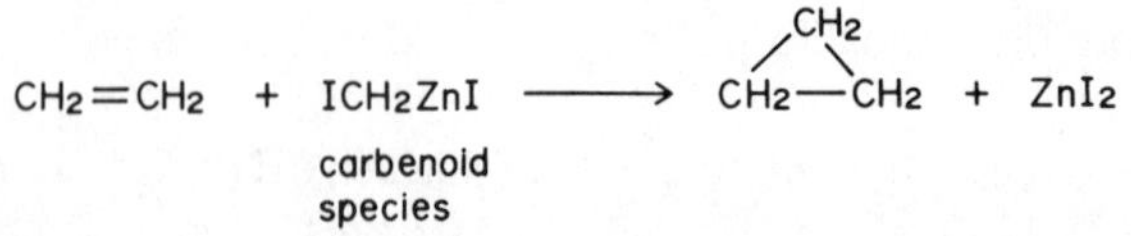

$$CH_2{=}CH_2 + \underset{\text{carbenoid species}}{ICH_2ZnI} \longrightarrow \text{cyclo-}(CH_2)_3 + ZnI_2$$

Figure 8.190 The reaction of iodomethylzinc iodide with ethylene.

8.200 α-Ketocarbene, RCOC̈R

A carbene in which at least one of the groups attached to the dicovalent carbon atom is an acyl group.

Example. Acetylcarbene, Fig. 8.200.

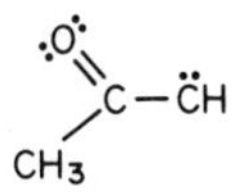

Figure 8.200 Acetylcarbene.

8.210 Acylcarbene

Synonymous with α-ketocarbene.

8.220 Nitrene RN̈:

An uncharged species containing a monocovalent nitrogen atom; this nitrogen is surrounded by a sextet of valence electrons and hence is electron deficient.

Example. Phenylnitrene, Fig. 8.220.

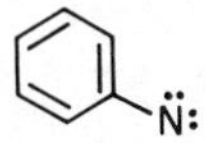

Figure 8.220 Phenylnitrene.

8.230 α-Ketonitrene, RCON̈:

A nitrene in which the group attached to the monocovalent nitrogen atom is an acyl group; analogous to an α-ketocarbene.

Example. Benzoylnitrene, Fig. 8.230.

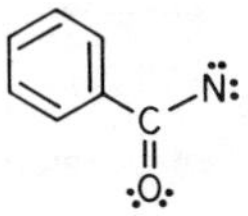

Figure 8.230 Benzoylnitrene.

8.240 Acylnitrene

Synonymous with α-ketonitrene.

8.250 Charge Localized (or Localized) Ion

An ion in which the charge resides principally on an individual atom and is so represented.

Examples. *t*-Butyl cation, Fig. 8.250*a*, and the cyclohexyl anion, Fig. 8.250*b*; the electrical charges on these ions are represented as being on the underlined carbon atoms.

(a) (b)

Figure 8.250 (a) t-Butyl cation; (b) cyclohexyl anion.

8.260 Charge Delocalized (or Delocalized) Ion

An ion in which the charge is predominantly distributed over more than one atom; these ions are usually written to reflect the delocalization.

Examples. Benzyl anion, Fig. 8.260*a*, and acetyl cation, Fig. 8.260*b*.

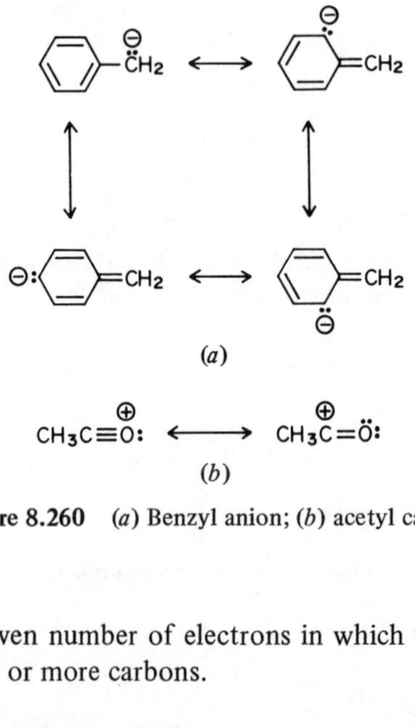

(a)

(b)

Figure 8.260 (a) Benzyl anion; (b) acetyl cation.

8.270 Carbocation

Any cation with an even number of electrons in which the charge may be formally localized on one or more carbons.

Examples. *t*-Butyl cation, Fig. 8.270*a*; protonated acetone, Fig. 8.270*b*; methylidyne cation, Fig. 8.270*c*. All carbenium ions (Sect. 8.330) and all carbonium ions (Sect. 8.310) are carbocations.

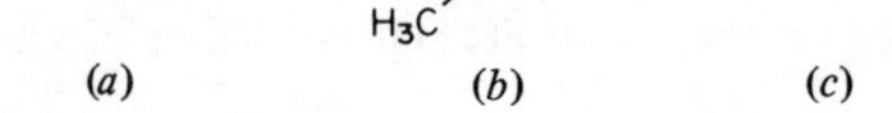

(a) (b) (c)

Figure 8.270 (a) t-Butyl cation; (b) protonated acetone; (c) methylidyne cation.

8.280 Nonclassical Cation

A delocalized cation in which the charge is distributed by means of *closed* multicenter bonding (Sect. 2.830). The most common case involves three carbon atoms, two of which are bonded to each other by a σ bond, and the third is bonded to the other two by a two-electron three-center bond. The third atom can, alternatively, be a hydrogen atom.

Example. Norbornyl cation, Fig. 8.280*a*. The figure shows two main resonance structures and *not* two carbocations in equilibrium with each other. The delocalization of charge represented by Fig. 8.280*a* can be displayed by one structure, Fig. 8.280*b* which shows C-6 as having a coordination number of 5 and C-1 and C-2 with a coordination number of 4. The usual coordination number for a posi-

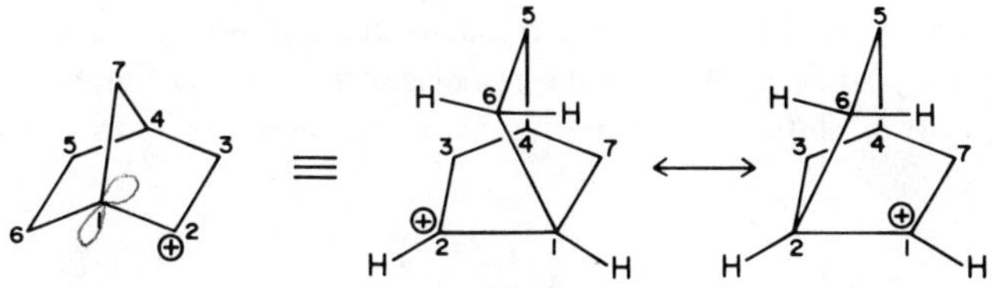

(*a*)

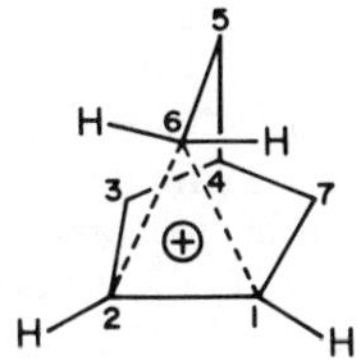

(*b*)

tively charged carbon atom without a multicenter bond is 3. If the coordination number of positively charged carbon is greater than 3, it is considered **hypervalent** (and is called a carbonium ion, Sect. 8.310). The structure shown in Fig. 8.280*b* is intended to indicate that there is $\frac{1}{2}$ plus charge on both of C-1 and C-2. If some of the charge is also on C-6, a double bond between C-1 and C-2 is required, Fig. 8.280*c*. The most accurate single structure showing the distribution of charge would then be Fig. 8.280*d*. From a molecular orbital point of view the carbon sp^3 orbital on C-6 combines with the two carbon p orbitals on C-1 and C-2, Fig. 8.280*e*, to generate three molecular orbitals; in order of increasing en-

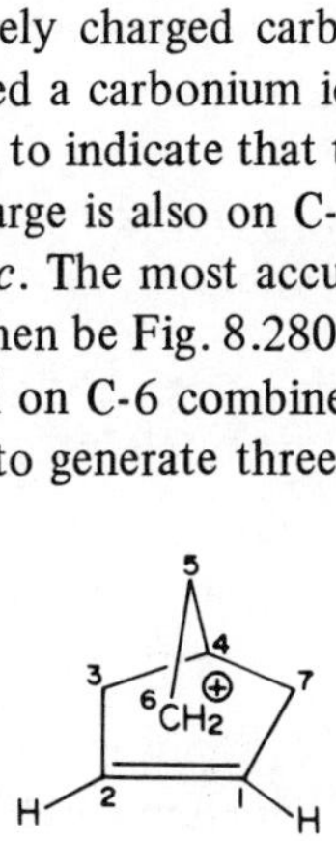

(*c*)

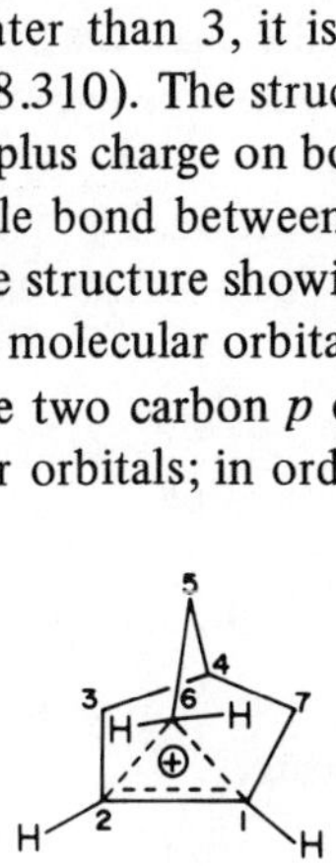

(*d*)

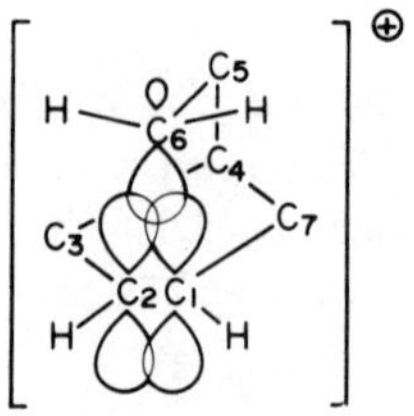

(*e*)

ergy these are bonding, weakly antibonding, and strongly antibonding orbitals. Because only two electrons are involved, these occupy the bonding orbital. Essentially the same type of two-electron three-center bonding is involved at the transition state in a typical 1,2-methyl migration of a carbocation, Fig. 8.280*f*. The important distinction is that the nonclassical norbornyl cation, Fig. 8.280*e*, has a finite existence, i.e., it is an intermediate corresponding to a minimum on a potential energy surface, whereas the nonclassical cation of Fig. 8.280*f* is a postulated transition state corresponding to a maximum on a potential energy surface.

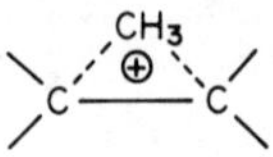

(*f*)

Figure 8.280 (*a*) Two resonance forms of the norbornyl cation; (*b*) a single structure representing Fig. 8.280*a*; (*c*) a third resonance form of the norbornyl cation; (*d*) a single representation of the three resonance forms of the norbornyl cation; (*e*) a representation of the orbital overlap in the norbornyl cation; (*f*) a representation of the transition state in a 1,2-methyl migration of a carbocation.

8.290 Classical Cation

A cation in which the charge is not distributed by means of a closed multicenter bond, i.e., a cation other than a nonclassical cation.

Examples. Methyl cation, Fig. 8.290*a*, and 1-hydroxyethyl cation, Fig. 8.290*b*. All carbenium ions (Sect. 8.330) are classical cations.

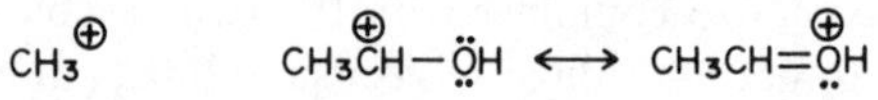

$CH_3^{\oplus}$

(*a*)

$CH_3\overset{\oplus}{C}H-\ddot{O}H \longleftrightarrow CH_3CH=\overset{\oplus}{O}H$

(*b*)

Figure 8.290 (*a*) Methyl cation; (*b*) 1-hydroxyethyl cation.

8.300 Bridged Cation

Synonymous with nonclassical cation. The bromonium ion shown in Fig. 8.300*a*, and the phenonium ion shown in Fig. 8.300*b*, are examples of classical ions

(Sect. 8.290) but they are sometimes (ambiguously) referred to as bridged ions; neither involves three-center two-electron bonding. In the bromonium ion there are as many valence electrons available as are required for separate localized bonds.

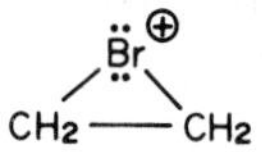

(*a*)

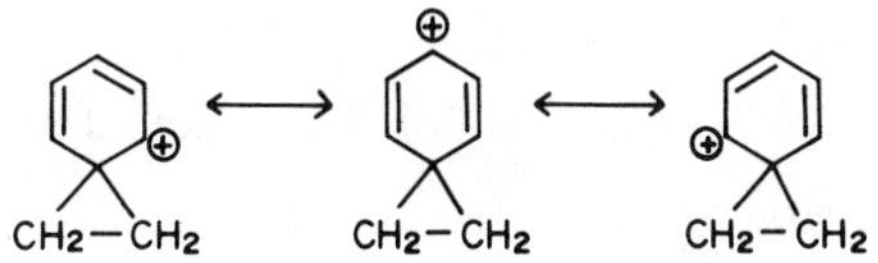

(*b*)

Figure 8.300 (*a*) A bromonium ion; (*b*) a phenonium ion.

8.310 Carbonium Ion

A hypervalent carbocation, i.e., a positively charged carbon having a valence (coordination number) greater than 3. Unfortunately in some literature *carbonium ion* is used as a generic term for the species we should call carbocations.

Examples. The nonclassical norbornyl cation, Fig. 8.280*b*, in which C-1, C-2, and C-6 each have a coordination number greater than 3.

8.320 Enium (en-ee-um) Ion (Ylium Ion)

A generic name for the cationic portion of an ionic species in which the positively charged nonmetallic atom has two electrons less than the normal complement in its valence shell. The charged atom thus has one less covalent bond than the corresponding uncharged species; the name is usually attached as a suffix to a root name.

8.330 Carbenium Ion, $R_3C^{\oplus}$

An enium ion in which the charged atom is carbon.

Examples. Methylcarbenium ion (ethyl cation), Fig. 8.330*a*, and trimethylcarbenium ion (*t*-butyl cation), Fig. 8.330*b*, have six electrons in the valence shell of the charged atom. Although the suffix -enium should be used in the systematic naming of the cationic portion of an ionic compound, its use in describing only the cationic component as a single species has become widespread and is advantageous. This practice is used for -ium (Sect. 8.350) and -onium (Sect. 8.360) ions.

(a) (b)

Figure 8.330 (a) Ethyl cation (methylcarbenium ion); (b) t-Butyl cation (trimethylcarbenium ion).

The effect of structure on the relative stabilities of carbenium ions can be evaluated from standard heat of formation data, e.g.:

$$\Delta H_f^\circ\,(CH_3CH_2\oplus) - \Delta H_f^\circ\,(CH_3CH_3) = 240 \pm 4\ \text{kcal mol}^{-1}\ (1004 \pm 17\ \text{kJ mol}^{-1})$$

$$\Delta H_f^\circ\,[(CH_3)_3C\oplus] - \Delta H_f^\circ\,[(CH_3)_3CH] = 202 \pm 4\ \text{kcal mol}^{-1}\ (845 \pm 17\ \text{kJ mol}^{-1})$$

These data are in accordance with the general statement that tertiary carbenium ions are more stable than secondary ones, which are more stable than primary ones.

8.340 Nitrenium ($R_2N\oplus$), Oxenium ($RO\oplus$), and Halenium ($X\oplus$) Ions

Enium ions in which the charged atoms are nitrogen, oxygen, and halogens, respectively.

Examples. Diphenylnitrenium ion, Fig. 8.340*a*; methyloxenium ion, Fig. 8.340*b*; bromenium ion, Fig. 8.340*c*. A set of nonbonding electrons in each of these three species may be either paired (singlet) or unpaired (triplet) and it is likely that the ground states at least in *a* and *b* are triplets. However when such species are generated experimentally they are usually initially singlets.

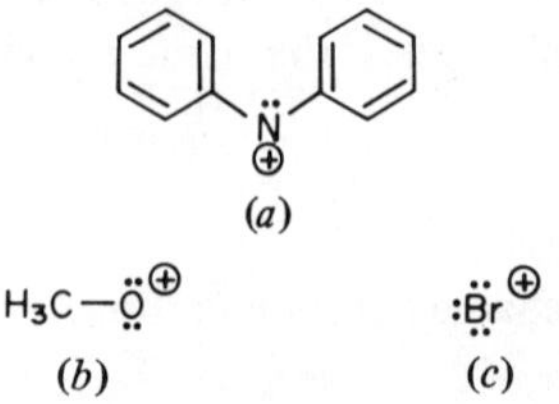

(a)

(b) (c)

Figure 8.340 (a) Diphenylnitrenium ion; (b) methyloxenium ion; (c) bromenium ion.

8.350 Ium (ee-um) Ion

A generic name for a cation in which a charged nonmetallic atom other than carbon or silicon possesses a closed shell (rare gas) configuration; usually attached as a suffix to a root name.

Examples. 1-Methylpyridinium ion, Fig. 8.350*a*; diphenylammonium ion, Fig. 8.350*b*; diethyloxonium ion (protonated diethyl ether), Fig. 8.350*c*.

$(C_6H_5)_2\overset{\oplus}{N}H_2$ $CH_3CH_2-\overset{\oplus}{\ddot{O}}(H)-CH_2CH_3$

(*a*) (*b*) (*c*)

Figure 8.350 (*a*) 1-Methylpyridinium ion; (*b*) diphenylammonium ion; (*c*) diethyloxonium ion (protonated diethyl ether).

8.360 Onium Ion: Ammonium ($R_4N^{\oplus}$), Oxonium ($R_3O^{\oplus}$), and Halonium ($R_2X^{\oplus}$) Ions

An ium ion named so as to emphasize the identity of the charged hetero atom and the groups attached to it rather than the name of the uncharged parent compound; usually attached as a suffix to a root name.

Examples. Tetraethylammonium ion, Fig. 8.360*a*; triethyloxonium ion, Fig. 8.360*b*; diphenyliodonium ion, Fig. 8.360*c*; these all have eight electrons in the valence shell of the charged atom. $Ph\overset{\oplus}{N}H_3$ may be named either anilinium ion or phenylammonium ion.

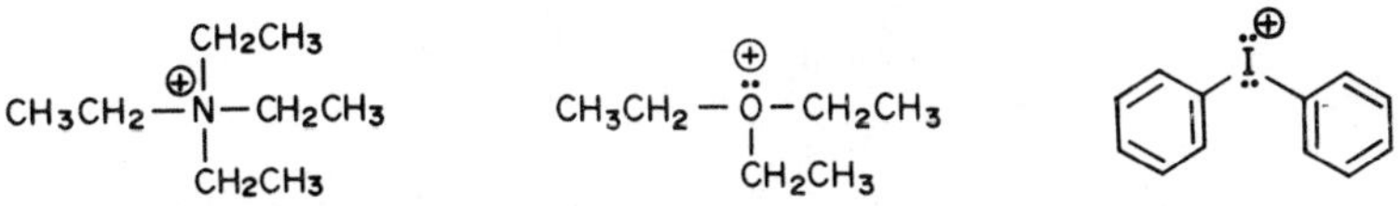

$(CH_3CH_2)_4\overset{\oplus}{N}$ $CH_3CH_2-\overset{\oplus}{\ddot{O}}(CH_2CH_3)-CH_2CH_3$ $Ph-\overset{\oplus}{\ddot{I}}-Ph$

(*a*) (*b*) (*c*)

Figure 8.360 (*a*) Tetraethylammonium ion; (*b*) triethyloxonium ion; (*c*) diphenyliodonium ion.

8.370 Aminium Ion, $R_3\overset{\oplus}{N}H$

An ium ion in which the charge resides on a tetracovalent nitrogen atom bonded to at least one hydrogen atom.

Example. 1-Propanaminium ion, Fig. 8.370, more commonly, *n*-propylammonium ion.

$CH_3CH_2CH_2\overset{\oplus}{N}H_3$

Figure 8.370 *n*-Propylammonium ion (1-propanaminium ion).

8.380 Iminium Ion, $R_2C{=}\overset{\oplus}{N}RR'$
An ium ion in which a charge resides primarily on a tetracovalent nitrogen atom connected to a carbon atom by a double bond.

Example. Cyclohexaniminium ion, Fig. 8.380.

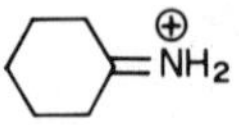

$$C_6H_{10}{=}\overset{\oplus}{N}H_2$$

Figure 8.380 Cyclohexaniminium ion.

8.390 Diazonium Ion, $R{-}\overset{\oplus}{N}{\equiv}N$
An onium ion in which the charge resides primarily on a tetracovalent nitrogen atom that is part of a nitrogen-nitrogen triple bond and that is bonded to a carbon atom.

Example. Phenyldiazonium ion, Fig. 8.390. It should be appreciated that the positive charge is partially delocalized over the second nitrogen atom as well, as the resonance structure shown in Fig. 8.390*b* illustrates. However, Fig. 8.390*a* represents a much more important contributing structure than Fig. 8.390*b*, since in the former both nitrogens are surrounded by an octet of electrons.

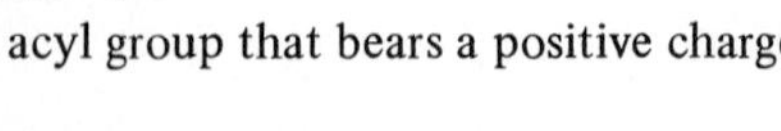

$$C_6H_5{-}\overset{\oplus}{N}{\equiv}N{:} \longleftrightarrow C_6H_5{-}\ddot{N}{=}\overset{\oplus}{N}{:}$$

(*a*) (*b*)

Figure 8.390 (*a, b*) Resonance forms of phenyldiazonium ion.

8.400 Acylium Ion, $RC{\equiv}O^{\oplus}$
A generic name for an acyl group that bears a positive charge.

Example. Acetyl cation, Fig. 8.400.

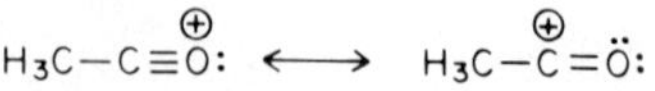

$$H_3C{-}C{\equiv}\overset{\oplus}{O}{:} \longleftrightarrow H_3C{-}\overset{\oplus}{C}{=}\ddot{O}{:}$$

Figure 8.400 Resonance forms of acetyl cation.

8.410 Third Row Enium Ions: Silicenium ($R_3Si^{\oplus}$), Phosphenium ($R_2P^{\oplus}$), Sulfenium ($RS^{\oplus}$), and Chlorenium ($Cl^{\oplus}$) Ions
Enium ions of silicon, phosphorus, sulfur, and chlorine, respectively.

Examples. Methylsilicenium ion, Fig. 8.410*a*; dimethylphosphenium ion (dimethylphosphino cation), Fig. 8.410*b*; methylsulfenium ion (methylsulfanyl cation), Fig. 8.410*c*; chlorenium ion, Fig. 8.410*d*.

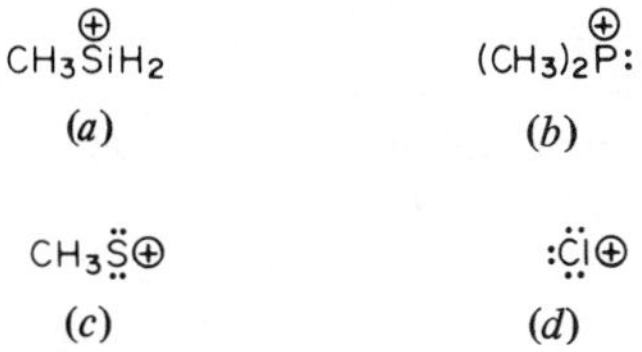

$CH_3\overset{\oplus}{Si}H_2$ (*a*) $(CH_3)_2\overset{\oplus}{P}{:}$ (*b*)

$CH_3\ddot{\underset{\cdot\cdot}{S}}\oplus$ (*c*) $:\ddot{\underset{\cdot\cdot}{Cl}}\oplus$ (*d*)

Figure 8.410 (*a*) Methysilicenium ion; (*b*) dimethylphosphenium ion (dimethylphosphino cation); (*c*) methylsulfenium ion (methylsulfanyl cation); (*d*) chlorenium ion.

8.420 Siliconium Ion, $R_3Si^{\oplus}$

Used synonymously with silicenium ion. Unfortunately this term implies that the species is an onium ion, which it is not.

8.430 Third Row Onium Ions: Phosphonium ($R_4P^{\oplus}$), Sulfonium ($R_3S^{\oplus}$), and Chloronium ($R_2Cl^{\oplus}$) Ions

Onium ions of phosphorus, sulfur, and chlorine, respectively.

Examples. Trimethylphenylphosphonium ion, Fig. 8.430*a*; trimethylsulfonium ion, Fig. 8.430*b*; dimethylchloronium ion, Fig. 8.430*c*.

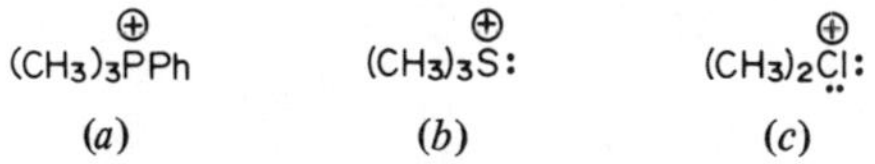

$(CH_3)_3\overset{\oplus}{P}Ph$ (*a*) $(CH_3)_3\overset{\oplus}{S}{:}$ (*b*) $(CH_3)_2\overset{\oplus}{\underset{\cdot\cdot}{Cl}}{:}$ (*c*)

Figure 8.430 (*a*) Trimethylphenylphosphonium ion; (*b*) trimethylsulfonium ion; (*c*) dimethylchloronium ion.

8.440 Carbanion

A generic name for an anion with an even number of electrons in which the charge formally resides on one or more carbon atoms.

Examples. Isopropyl anion, Fig. 8.440*a*, and allyl anion, Fig. 8.440*b*.

$H_3C{-}\overset{\ominus}{\ddot{C}}(H){-}CH_3$ (*a*)

$\overset{\ominus}{\ddot{C}}H_2{-}CH{=}CH_2 \longleftrightarrow CH_2{=}CH{-}\overset{\ominus}{\ddot{C}}H_2$ (*b*)

Figure 8.440 (*a*) Isopropyl anion; (*b*) allyl anion.

8.450 Nitrogen Anion (Nitranion)

An anion with an even number of electrons in which the charge formally resides on a nitrogen atom. Nitranion is a term that is not presently in use, but such a name would be systematic and convenient.

Examples. Dimethylaminyl anion, Fig. 8.450. It is common practice to name certain compounds with nitrogen anions as amides, e.g., sodium amide, $Na^{\oplus}\,{}^{\ominus}NH_2$, and lithium dimethylamide, $Li^{\oplus}\,{}^{\ominus}N(CH_3)_2$.

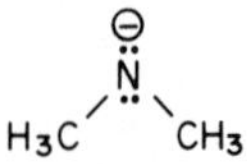

$H_3C-\ddot{N}^{\ominus}-CH_3$

Figure 8.450 Dimethylaminyl anion.

8.460 Oxygen Anion (Oxyanion)

An anion with an even number of electrons in which the charge formally resides on an oxygen atom. Oxyanion is a term that is not presently in use, but such a name would be systematic and convenient.

Examples. Methoxyl anion, Fig. 8.460*a*, and acetate anion, Fig. 8.460*b*.

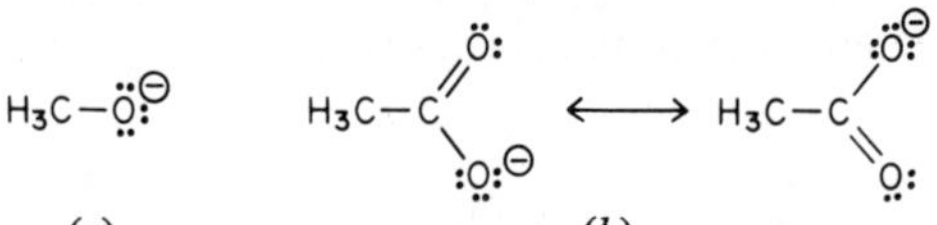

(*a*) $H_3C-\ddot{O}:^{\ominus}$

(*b*) $H_3C-C(=\ddot{O}:)-\ddot{O}:^{\ominus} \longleftrightarrow H_3C-C(-\ddot{O}:^{\ominus})=\ddot{O}:$

Figure 8.460 (*a*) Methoxyl anion; (*b*) acetate anion.

8.470 Alkoxides, $M^{\oplus}\,{}^{\ominus}OR$

Salts in which the negatively charged oxygen is bonded to an alkyl group.

Examples. Sodium methoxide, $NaOCH_3$, and aluminum isopropoxide, $Al[OCH(CH_3)_2]_3$.

8.480 Enolate Anion

The delocalized anion remaining after removal of a proton from an enol or the carbonyl compound in equilibrium with it.

Example. Acetone enolate anion, Fig. 8.480, can be generated by the loss of a proton from the methyl group adjacent to the carbonyl group or by loss of a proton from the —OH group of the tautomeric enol.

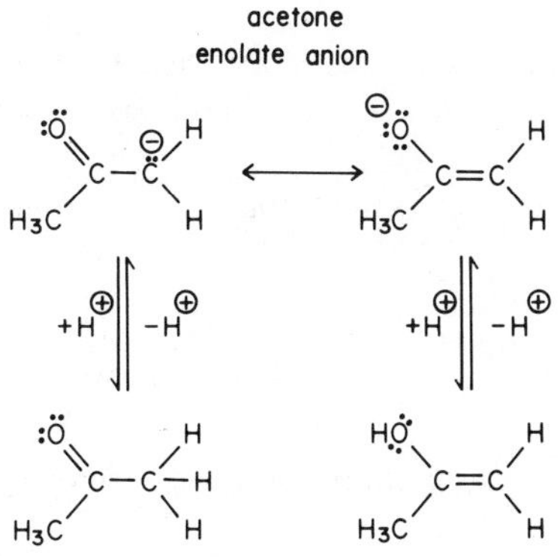

acetone enolate anion

Figure 8.480 Keto-enol equilibrium via the acetone enolate anion.

8.490 Radical-Ion

An ion with an unpaired electron.

8.500 Radical-Cation

A cation with an unpaired electron.

Examples. Methane radical-cation, Fig. 8.500*a*, and *N*,*N*,*N*-trimethylamine *N*-radical-cation, Fig. 8.500*b*. A dot (·) with a charge (⊕ or ⊖) adjacent to the molecular formula or outside a bracket represents an ion with an odd electron.

$[CH_4]^{\oplus\cdot}$ $(CH_3)_3N^{\oplus\cdot}$

(*a*) (*b*)

Figure 8.500 (*a*) Methane radical-cation; (*b*) *N*,*N*,*N*-trimethylamine *N*-radical-cation.

8.510 Radical-Anion

An anion with an unpaired electron.

Example. Benzene radical-anion, Fig. 8.510.

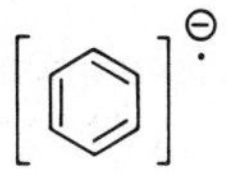

Figure 8.510 Benzene radical anion.

8.520 Ion Pair

A general term used to describe a closely associated cation and anion that behave as a single unit.

8.530 Tight Ion Pair

An ion pair in which individual ions retain their individual stereochemical configuration. No solvent molecules separate the cation from the anion.

Example. An asymmetric carbenium ion with an anion in a nonpolar solvent is shown in Fig. 8.530.

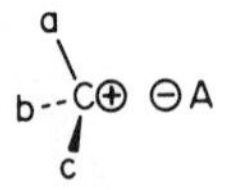

Figure 8.530 A tight ion pair.

8.540 Intimate Ion Pair

Synonymous with tight ion pair.

8.550 Contact Ion Pair

Synonymous with tight ion pair.

8.560 Solvent Separated (or Loose) Ion Pair

An ion pair in which the individual ions are separated by one or more solvent molecules. The ions may or may not retain their individual stereochemistry, depending on reaction conditions. The symbol generally used to represent this type of ion pair is $R^{\oplus}||X^{\ominus}$.

8.570 Counterion

An ion associated with another ion of opposite charge.

Example. In sodium acetate, Fig. 8.570, the sodium cation is the counterion of the carboxylate anion and vice versa.

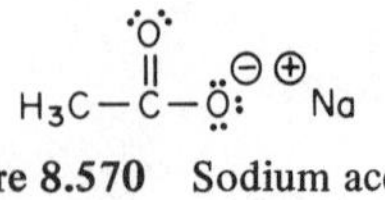

Figure 8.570 Sodium acetate.

8.580 Gegenion

From the German meaning counter; synonymous with counterion.

8.590 Zwitterion (Dipolar Ion)

An uncharged species that has separate cationic and anionic sites.

Example. The zwitterion (dimethylammonio)acetate, Fig. 8.590*a*, in equilibrium in an aqueous solution with the corresponding α-amino acid, Fig. 8.590*b*.

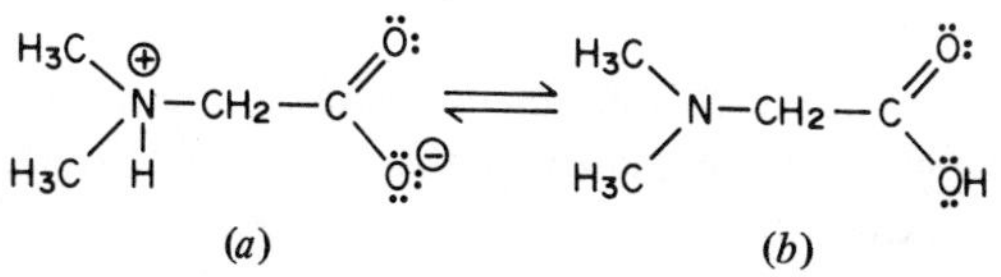

Figure 8.590 (*a*) The zwitterion (dimethylammonio)acetate; (*b*) *N*,*N*-Dimethylamino-acetic acid.

8.600 Betaine (bāta-ene)

An uncharged species having isolated, nonadjacent cationic and anionic sites and not possessing a hydrogen atom bonded to the cationic site; a special class of zwitterions.

Examples. (Dimethylsulfonio)acetate, Fig. 8.600*a*. (Trimethylammonio)acetate, Fig. 8.600*b*, has the common name betaine, which has now taken on a generic meaning.

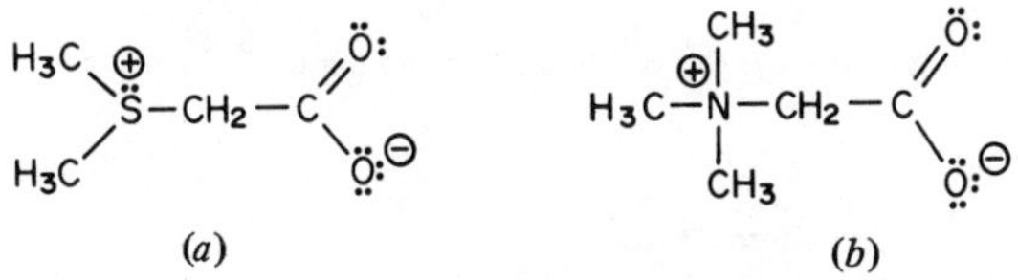

Figure 8.600 (*a*) (Dimethylsulfonio)acetate: (*b*) betaine [(trimethyammonio)acetate].

8.610 Ylide

An uncharged species whose structure may be represented as a 1,2-dipole with a negatively charged carbon or other atom with a lone pair of electrons bonded to a positively charged *hetero* atom. The term is usually preceded by the name of the hetero atom.

Examples. The phosphorus ylide methylenetriphenylphosphorane, Fig. 8.610*a*, and the nitrogen ylide (trimethylammonio)methylide, Fig. 8.610*b*.

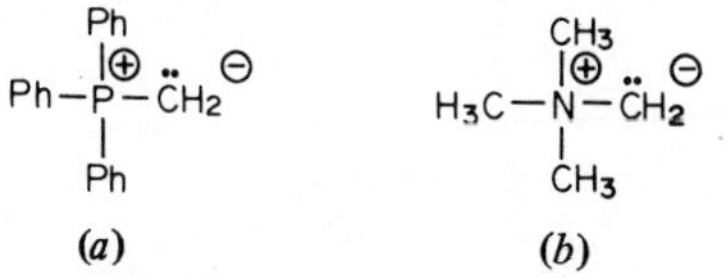

Figure 8.610 (*a*) The phosphorus ylide methylenetriphenylphosphorane; (*b*) the nitrogen ylide (trimethylammonio)methylide.

8.620 Aryne

An uncharged species in which two adjacent atoms of an aromatic ring lack substituents, thus leaving two orbitals (perpendicular to the aromatic π system) to be occupied by two electrons.

Examples. Singlet benzyne, Fig. 8.620*a*, and triplet 2,3-pyridyne, Fig. 8.620*b*.

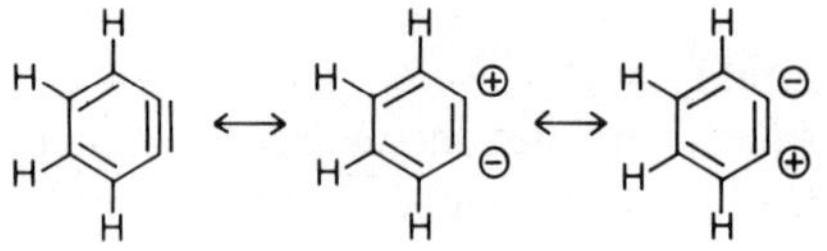

(*a*)

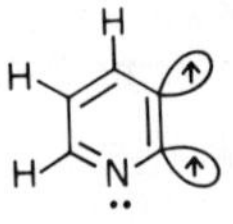

(*b*)

Figure 8.620 (*a*) Singlet benzyne; (*b*) triplet 2,3-pyridyne.

9 Thermodynamics, Acids and Bases, and Kinetics

CONTENTS

9.010 Thermodynamics
The scientific discipline that deals with energy changes accompanying chemical and physical transformations.

9.020 Thermochemistry
The branch of chemistry dealing with the quantity of heat that accompanies chemical and physical transformations.

9.030 Equilibrium
In the general reaction

$$A \rightleftarrows B$$

equilibrium is achieved when the rate of the forward reaction equals the rate of the reverse reaction. Two arrows facing in opposite directions ($\rightleftarrows$) are used to represent an equilibrium. At equilibrium there is no net change in the properties of the **system** and its surroundings.

9.040 System
A defined region of the universe under investigation. Boundaries, real or imaginary, delineate a system, and the region outside the boundaries is known as the **surroundings.** An **isolated system** is not affected by changes in the surroundings and does not exchange energy (heat or work) or mass with its surroundings. A **closed system** is one in which there is no transfer of mass across the boundaries; energy may, however, be exchanged.

9.050 Reversible Process

A process (chemical or physical change) in which a system undergoes change at a sufficiently slow rate such that at any instant the system and its surroundings are in a state of equilibrium (no net change). In theory a reversible process is achievable but in practice it is not. If a change can occur in both directions under a given set of conditions, for example, A to B and B to A, then the individual processes are referred to as being reversible, even though in reality they are not.

9.060 Irreversible Process

A process (chemical or physical change) in which a system undergoes change such that at any instant the system is not in a state of equilibrium. All transformations are irreversible in practice, even those that can proceed in opposite directions. If a transformation occurs only in one direction, A to B, under a given set of conditions, the organic chemist refers to the process as being irreversible. On the other hand, if reactions occur in opposite directions, A to B and B to A, under the same conditions, the individual reactions are said to be reversible.

9.070 Microscopic Reversibility

At equilibrium (see Sect. 9.030) individual molecular processes and the exact reverse of these processes have an equal probability of occurring. If a certain number of molecules follow one path (of many possible paths) in the forward direction, the same number follow that path in the reverse direction. Microscopic reversibility is also known as the **Principle of Detailed Balancing.**

Example. In the reaction

$$A \rightleftarrows B \rightleftarrows C$$

if B is an intermediate in going from A to C, then under the same conditions B must be an intermediate in going from C to A.

9.080 State

The specific set of conditions that completely describe the properties of the system. The quantity of each chemical species present and independent physical variables such as temperature, pressure, and magnetic field strength must be specified.

Example. One mole of acetic acid, at 1 atm pressure and at 25°C. The description of the state must enable one to reproduce the system.

9.090 State Function (State Variable)

A function that is dependent only on the state of the system. When a system undergoes a change from one state to another, the change in state function is independent of the path taken.

Examples. The internal energy, enthalpy, and entropy of a system are each state functions.

9.100 Internal Energy, U (or E)

A state function that describes the sum of the kinetic and potential energy components of a system. Changes in internal energy ΔU for closed systems usually result in a transfer of work or heat between the system and its surroundings:

$$\Delta U = q + w$$

where q is heat *gained* by the system and w is work *done* on the system.

9.110 Kinetic Energy, E_k

The energy that a system possesses by virtue of the motion of its constituent particles. Translational energy (arising from motion of the center of gravity of a species), rotational energy (arising from motion that results in angular momentum about the center of gravity), and vibrational energy (arising from motion that results in changes in bond angles and distances) have kinetic components. The temperature of a system is a measure of its average kinetic energy.

Examples. The translational kinetic energy of 1 mole of an ideal gas is equal to $\frac{3}{2}RT$ where R is the gas constant and T is the absolute temperature. At 25°C the translational kinetic energy is 3715 J mol^{-1}. Thus one molecule has an average kinetic energy of 6.17×10^{-21} J. The mass m and the velocity v of a molecule are related to its kinetic energy by the equation

$$E_k = \frac{1}{2}mv^2$$

The average velocity of a gaseous water molecule at 25°C is 1214 mph.

9.120 Potential Energy, E_p (or V)

The energy a system or body possesses by virtue of its relationship (position) to a factor that affects internal energy. Kinetic energy may be converted to potential energy and vice versa but their sum is constant for an isolated system.

Examples. A plot showing the relationship between the potential energy of a chemical bond between two atoms as a function of the internuclear distance is shown in Fig. 9.120; such a plot is called a **Morse curve** (Philip M. Morse, 1903-). Note: D_0 in this figure represents the experimental dissociation energy, 432 kJ mol^{-1} (103 kcal mol^{-1}), usually determined spectroscopically, hence frequently called the **spectroscopic dissociation energy**; D_e represents the energy required to separate the atoms to an infinite distance from their equilibrium position where the potential energy is at the (unobtainable) minimum; and the difference $D_0 - D_e$ is the zero point energy 25.9 kJ mol^{-1} (6.2 kcal mol^{-1}). **A potential energy profile**, Fig. 9.650*a*, is a plot of potential energy as a function of reaction coordinate.

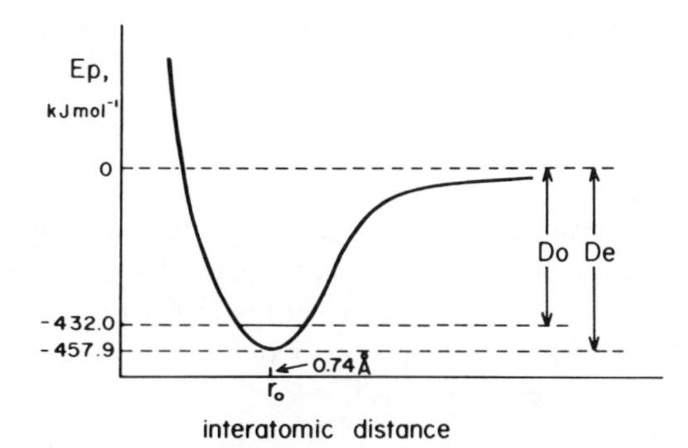

Figure 9.120 A Morse curve for the hydrogen molecule.

9.130 Standard State

An arbitrary reference state. For solids and liquids it is the pure substance in its most stable form under a pressure of 1 atm. For gases the standard state is that of the hypothetical ideal gas at 1 atm pressure; in this state the enthalpy is the same as that of the real gas at zero pressure. Although the temperature of the standard state is arbitrary, the one most frequently used is 298.15 K (25.15°C). A superscript of zero is attached to a state function (enthalpy, entropy, free energy, etc.) to indicate that it represents a system in a standard state. The temperature of the state is indicated by a subscript.

9.140 Enthalpy (Heat Content), H

A state function defined by

$$H = U + PV$$

where U is the internal energy, P is the pressure, and V is the volume of the system. The **standard enthalpy** H° is the enthalpy of a system in its standard state.

9.150 Enthalpy of Reaction, ΔH_r

The change in enthalpy after a system has undergone a chemical or physical transformation:

$$\Delta H_r = \Delta U + \Delta(PV)$$

where U is the internal energy, P is pressure, and V is volume. For reactions occurring at a constant pressure (most solution reactions) the change in enthalpy is the heat of reaction q_p:

$$\Delta H_r = q_p = \Delta U + P\Delta V$$

If heat is absorbed by the system, q_p has a positive value, and if heat is given off by the system, q_p is negative. If the volume change is negligible (no gases involved), the difference between ΔH_r or q_p and ΔU is of the order of only 1 J (0.2 cal) mol^{-1}.

9.160 Heat of Reaction, q

The heat absorbed or given off as the result of a system undergoing a chemical or physical transformation. For reactions performed at a constant pressure, the heat of reaction q_p is identical with the enthalpy of reaction ΔH_r. For reactions in which the volume remains constant (reactions run in a closed container such as a bomb calorimeter), the heat of reaction is

$$q_v = \Delta H - V\Delta P = \Delta U$$

where H is enthalpy at a constant volume, V is volume, P is pressure, and U is internal energy. If the system absorbs heat from the surroundings, q is positive; if heat is given off by the system, q is negative.

9.170 Standard Enthalpy (or Standard Heat) of Reaction, ΔH_r°

The change in enthalpy after a system in its standard state has undergone a transformation and the final system is also in its standard state. Since pressure is constant for standard states (1 atm), the enthalpy change is equal to the heat of reaction. A positive value of ΔH_r° indicates heat is absorbed by the system, and a negative value indicates heat is given off.

Example. The overall equation for photosynthesis:

$$6CO_2(g) + 6H_2O(l) \longrightarrow C_6H_{12}O_6(aq) + 6O_2(g)$$

$$\Delta H_{298}^\circ = 2900 \text{ kJ mol}^{-1} \text{ (693 kcal mol}^{-1}\text{)}$$

9.180 Standard Heat (or Standard Enthalpy) of Formation, ΔH_f°

The change in enthalpy when 1 mole of a substance in its standard state is formed from its elements in their standard states. For all elements in their standard state ΔH_f° is defined as zero.

Example. At 25°C (298 K) and 1 atm

$$C(s) + 2Cl_2(g) \longrightarrow CCl_4(l)$$

The heat evolved in this reaction is 139.3 kJ mol^{-1} (33.3 kcal mol^{-1}):

$$\Delta H_f^\circ(\mathrm{CCl_4,l}) = -139.3 \text{ kJ mol}^{-1}$$

9.190 Hess's Law of Constant Heat Summation (H. Hess, 1802–1850)
Enthalpy changes are independent of pathways taken (enthalpy is a state function). Therefore it is possible to calculate the enthalpy change of a reaction from related reactions whose enthalpy changes are known.

Example

$$(1)\ 2\mathrm{C(s)} + 2\mathrm{O_2(g)} \longrightarrow 2\mathrm{CO_2(g)} \qquad \Delta H_r = -787.4 \text{ kJ } (-188.2 \text{ kcal})$$

$$(2)\ 2\mathrm{CO(g)} + \mathrm{O_2(g)} \longrightarrow 2\mathrm{CO_2(g)} \qquad \Delta H_r = -568.6 \text{ kJ } (-135.9 \text{ kcal})$$

Subtracting equation 2 from equation 1 yields

$$2\mathrm{C(s)} + \mathrm{O_2(g)} \longrightarrow 2\mathrm{CO(g)} \qquad \Delta H_r = -218.8 \text{ kJ } (-52.3 \text{ kcal})$$

9.200 Adiabatic Process
A process during which the system neither gains nor loses heat.

9.210 Isothermal Process
A process during which the system remains at a constant temperature and therefore at a constant internal energy.

9.220 Heat Capacity, *C*
That quantity of heat q required to increase the temperature of a system or substance one degree of temperature:

$$C = \frac{q}{\Delta T}$$

At constant volume

$$C_v = \left(\frac{\partial U}{\partial T}\right)_v$$

where U is the internal energy. At a constant pressure

$$C_p = \left(\frac{\partial H}{\partial T}\right)_p$$

where H is the enthalpy. The difference between heat capacities at constant pressure and constant volume is given by

$$C_p - C_v = \left[P + \left(\frac{\partial U}{\partial V}\right)_T\right]\left(\frac{\partial V}{\partial T}\right)_p$$

Specific heat capacity c (often referred to simply as **specific heat**) is heat capacity divided by mass: $c = C/m$ i.e., the heat capacity per gram.

Examples

Compound	C_p (J K^{-1} mol^{-1})	C_v (J K^{-1} mol^{-1})
CS_2	75.7	47.1
$CHCl_3$	116.3	72.8
CCl_4	131.7	89.5
C_6H_6	134.3	91.6

9.230 Exothermic Reaction

A transformation accompanied by the evolution and transfer of heat from the system to the surroundings. ΔH_r and q have negative values for an exothermic reaction. When any form of energy is evolved in a reaction, the reaction is **exoergic.**

9.240 Endothermic Reaction

A transformation accompanied by the absorption of heat by the system from the surroundings. ΔH_r and q have positive values for an endothermic reaction. When any form of energy is absorbed in a reaction, the reaction is **endoergic.**

9.250 Heat of Hydrogenation

The heat of reaction (enthalpy of reaction at a constant pressure) that accompanies the addition of molecular hydrogen to a specified amount (usually 1 mole) of a reactant to form a particular product.

Examples

$$CH_3CH{=}CH_2 + H_2 \xrightarrow{\text{catalyst}} CH_3CH_2CH_3$$

$$\Delta H_r = -125.9 \text{ kJ mol}^{-1} \ (-30.1 \text{ kcal mol}^{-1})$$

$$CH_2{=}C{=}CH_2 + H_2 \xrightarrow{\text{catalyst}} CH_3CH{=}CH_2$$

$$\Delta H_r = -172.4 \text{ kJ mol}^{-1} \ (-41.2 \text{ kcal mol}^{-1})$$

$$CH_2{=}C{=}CH_2 + 2H_2 \xrightarrow{\text{catalyst}} CH_3CH_2CH_3$$

$$\Delta H_r = -298.3 \text{ kJ mol}^{-1} \ (-71.3 \text{ kcal mol}^{-1})$$

9.260 Heat of Combustion

The heat of reaction (enthalpy of reaction at a constant pressure) that results when a specified amount of substance, usually 1 mole reacts with gaseous oxygen to completely convert all the carbon to carbon dioxide. All other elements are formally converted to specified compounds even though these compounds may not be formed in the actual reaction. For example, hydrogen is converted to H_2O, nitrogen to N_2 (even though HNO_3 may be produced), and iodine to I_2. Thermochemical adjustments are made on the experimental data to reflect the formal production of specific compounds.

Examples

$$CH_4(g) + 2O_2(g) \longrightarrow CO_2(g) + 2H_2O(l)$$

$$\Delta H_r = -890.4 \text{ kJ mol}^{-1} \ (-212.8 \text{ kcal mol}^{-1})$$

$$CH_3I(l) + 1.75O_2(g) \longrightarrow CO_2(g) + 1.5H_2O(l) + 0.5I_2(s)$$

$$\Delta H_r = -807.9 \text{ kJ mol}^{-1} \ (-193.1 \text{ kcal mol}^{-1})$$

9.270 Heat of Fusion (or Melting), ΔH_m

The enthalpy or heat change associated with the conversion of a specified quantity of substance, usually 1 mole **(molar heat of fusion)** or 1 g **(specific heat of fusion)**, from the solid to the liquid phase at a specified pressure, usually 1 atm, and a specified temperature, the melting point if the pressure is specified as 1 atm. The process is isothermal.

Examples

Substance	Temperature (°C)	ΔH_m (kJ mol^{-1})[a]
CCl_4	-22.9	2.51
CH_3OH	-97.9	3.17
C_6H_6	5.5	10.58

[a] At 1 atm pressure.

9.280 Heat of Vaporization, ΔH_{vap}

The enthalpy or heat change associated with the conversion of a specified quantity of substance, usually 1 mole **(molar heat of vaporization)** or 1 g **(specific heat of vaporization)**, from the liquid to the vapor phase at a specified temperature, usually the boiling point, and a specified pressure, usually 1 atm. The process is isothermal.

Examples

Substance	Temperature (°C)	ΔH_{vap} (kJ mol^{-1})[a]
CH_3OH	64.7	35.3
CH_3NH_2	-6.2	25.8
C_6H_6	80.1	30.8

[a] At 1 atm pressure

9.290 Heat of Sublimation, ΔH_s

The enthalpy or heat change associated with the conversion of a specified quantity of substance, usually 1 mole **(molar heat of sublimation)** or 1 g **(specific heat of sublimation)**, from the solid phase directly to the vapor phase at a specified temperature and pressure. The process is isothermal.

Examples

Substance	ΔH_s (kJ mol^{-1})[a]
Naphthalene	72.8
Azulene	95.4

[a] At 1 atm pressure and 278 K.

9.300 Bond Dissociation Energy (Bond Strength, Bond Enthalpy) $D°$ (AB)

The enthalpy change that accompanies the homolytic cleavage of a bond under standard state conditions at 298.15 K (25°C):

$$\text{A—B} \longrightarrow \text{A}\cdot + \cdot\text{B}$$

The more precise term for this quantity is **standard bond dissociation energy**.

Examples

$$CH_3\text{—}CH_3 \longrightarrow 2CH_3\cdot$$

$$D°(H_3C\text{—}CH_3) = 368 \text{ kJ mol}^{-1} \text{ (88 kcal mol}^{-1})$$

$$CH_3COO\text{—}H \longrightarrow CH_3COO\cdot + \cdot H$$

$$D°(CH_3COO\text{—}H) = 469 \text{ kJ mol}^{-1} \text{ (112 kcal mol}^{-1})$$

$$H_2C\text{=}\ddot{O}: \longrightarrow H_2C: + :\dot{\underset{\cdot\cdot}{O}}\cdot$$

$$D°(H_2C\text{=}O) = 732 \text{ kJ mol}^{-1} \text{ (175 kcal mol}^{-1})$$

9.310 Average (or Mean) Bond Dissociation Energy, D^a

The average of enthalpy changes for the homolytic cleavage of more than one bond of a set of equivalent bonds of a molecule. For example,

$$AB_n \longrightarrow \text{A (with } n \text{ electrons)} + nB\cdot$$

$$D^a(\text{A—B}) = \frac{\Delta H_r}{n}$$

Example

$$CH_4 \longrightarrow \cdot\dot{\underset{\cdot}{C}}\cdot + 4H\cdot$$

$$\Delta H_r = 1662 \text{ kJ mol}^{-1} \text{ (397 kcal mol}^{-1})$$

Four carbon-hydrogen bonds are broken:

$$D^a(\text{C—H}) = \frac{1662}{4}\ \text{kJ mol}^{-1} = 415\ \text{kJ mol}^{-1}$$

although

$$CH_4 \longrightarrow \cdot CH_3 + H\cdot \qquad \Delta H_r = 435\ \text{kJ mol}^{-1}$$

$$\cdot CH_3 \longrightarrow \cdot\dot{C}H_2 + H\cdot \qquad \Delta H_r = 444\ \text{kJ mol}^{-1}$$

$$\cdot\dot{C}H_2 \longrightarrow \cdot\dot{\underset{\cdot}{C}}H + H\cdot \qquad \Delta H_r = 444\ \text{kJ mol}^{-1}$$

$$\cdot\dot{\underset{\cdot}{C}}H \longrightarrow \cdot\dot{\underset{\cdot}{C}}\cdot + H\cdot \qquad \Delta H_r = 339\ \text{kJ mol}^{-1}$$

The term **bond energy** is loosely used for either the average bond dissociation energy or an average of specific bond dissociation energies obtained from different molecules of a given type, for example,

Bond	Bond energy	
	kJ mol^{-1}	kcal mol^{-1}
C—H	414	99
C—C	347	83
C—N	305	73
C—O	356	85
C—F	485	116
C=C	611	146
C=N	615	147
C=O	749	179

9.320 Heterolytic Bond Dissociation Energy, $D(A^+B^-)$

The enthalpy change that accompanies the heterolytic cleavage of a bond. For example,

$$\text{A—B} \longrightarrow A^+ + B^-$$

$$D(A^+B^-) = \Delta H_r = \Delta H_f(A^+) + \Delta H_f(B^-) - \Delta H_f(\text{A—B})$$

Examples

$$(CH_3)_3\text{C—Br(g)} \longrightarrow (CH_3)_3C^+(g) + Br^-(g)$$

$$\Delta H_f\ [(CH_3)_3CBr)] = -117\ \text{kJ mol}^{-1}\ (-28\ \text{kcal mol}^{-1})$$

$$\Delta H_f\ [(CH_3)_3C^+] = 736\ \text{kJ mol}^{-1}\ (176\ \text{kcal mol}^{-1})$$

$$\Delta H_f\ (Br^-) = -234\ \text{kJ mol}^{-1}\ (-56\ \text{kcal mol}^{-1})$$

$$D\ [(CH_3)_3C^+Br^-] = 619\ \text{kJ mol}^{-1}\ (148\ \text{kcal mol}^{-1})$$

It is also possible to calculate heterolytic bond dissociation energies from **bond dissociation energies** ($D°$), adiabatic **ionization energies** (IE), and **electron affinities** (EA). Consider the following reactions:

$$(1) \quad A\text{—}B \longrightarrow A\cdot + B\cdot$$

$$(2) \quad A\cdot \longrightarrow A^+ + e^{\cdot\ominus}$$

$$(3) \quad B\cdot + e^{\cdot\ominus} \longrightarrow B^-$$

$$(1 + 2 + 3) \quad A\text{—}B \longrightarrow A^+ + B^-$$

$$D(A^+B^-) = D°(A\text{—}B) + IP\,(A\cdot) - EA\,(B\cdot)$$

For example,

$$CH_3Cl(g) \longrightarrow CH_3^{\cdot+}(g) + Cl^-(g)$$

$$D°(H_3C\text{—}Cl) = 351 \text{ kJ mol}^{-1} \ (84 \text{ kcal mol}^{-1})$$

$$IP\,(CH_3\cdot) = 950 \text{ kJ mol}^{-1} \ (227 \text{ kcal mol}^{-1})$$

$$EA\,(Cl\cdot) = 368 \text{ kJ mol}^{-1} \ (88 \text{ kcal mol}^{-1})$$

$$D(CH_3^{\cdot+}Cl^-) = 933 \text{ kJ mol}^{-1} \ (223 \text{ kcal mol}^{-1})$$

9.330 Ionization Energy (Ionization Potential), *IE*

The energy required to remove an electron from a molecule (or atom) in its gaseous state. If such an electron is removed from the molecule in its lowest vibrational and rotational state to produce a radical-cation in its lowest vibrational and rotational state the process is more explicitly referred to as the **adiabatic ionization energy** Fig. 9.330, process 1. The **vertical ionization energy** is the en-

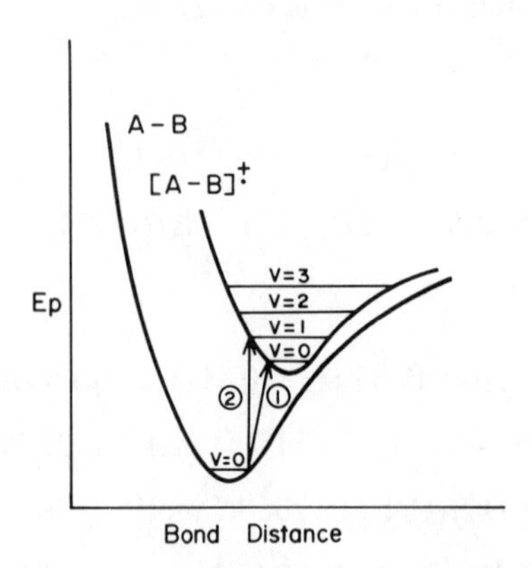

Figure 9.330 Morse curves for A–B and [A–B]$^{\cdot+}$ showing adiabatic (1) and vertical (2) ionization processes.

ergy associated with an ionization process that occurs in accordance with the Franck-Condon principle (James Franck, 1882–1946, and E. U. Condon, 1902–1974), i.e., without a change in nuclear geometry or momentum (Fig. 9.330, process 2). This process usually requires more energy than an adiabatic ionization owing to the excess vibrational and rotational energy imparted to the ionized molecule (radical-cation). Although ionization energy is most frequently used to indicate the energy required to remove an electron from the highest occupied orbital, electrons in different orbitals each have a characteristic ionization energy.

9.340 Electron Affinity, *EA*

The energy associated with the capture of an electron by an atom, ion, or molecule in the gas phase:

$$M + e^{\cdot\ominus} \longrightarrow M^{\cdot\ominus}$$

If energy is released, then the electron affinity has a positive value. This sign convention runs counter to most energy conventions, where a negative value indicates that energy is released.

Examples

Atom	Electron affinity (eV)[a]
C	1.25 ± 0.03
O	1.465 ± 0.005
F	3.448 ± 0.005

[a] 1 eV = 96.49 kJ mol^{-1} = 23.06 kcal mol^{-1}

9.350 Entropy, S

A state function that is a measure of the degree of disorder or randomness of a system. For a change from one state to another

$$\Delta S = \frac{q_{\text{rev}}}{T}$$

where q_{rev} is the heat change along a reversible pathway and T is the absolute temperature, which is constant. For an irreversible pathway

$$\Delta S > \frac{q_{\text{irrev}}}{T}$$

An increase in disorder results in an increase in entropy, so that ΔS is positive.

Example

$$C_2H_4(g) + H_2O(g) \longrightarrow C_2H_5OH(g)$$

Less ordered system ⟶ More ordered system

At 298 K (25°C)

$$\Delta S = -125.6 \text{ J deg}^{-1} \text{ mol}^{-1} \ (-30.02 \text{ cal deg}^{-1} \text{ mol}^{-1})$$

9.360 Standard Entropy, $S°$

The entropy of a substance in the state specified, based on $S°$ being equal to zero for a perfect crystal of that substance at 0 K.

Examples

$$C_2H_5OH(l) \quad S^\circ_{298} = 160.7 \text{ J deg}^{-1} \text{ mol}^{-1} \ (38.4 \text{ cal deg}^{-1} \text{ mol}^{-1})$$

$$C_2H_5OH(g) \quad S^\circ_{298} = 282.0 \text{ J deg}^{-1} \text{ mol}^{-1} \ (67.4 \text{ cal deg}^{-1} \text{ mol}^{-1})$$

9.370 First Law of Thermodynamics

The equality

$$\Delta U = q + w$$

where ΔU is a change in internal energy, q is heat gained by the system, and w is work done on the system. Thus any change in the energy content of a system must be accounted for in terms of heat and work flowing in and out of the system. For an isolated system where q and w must equal zero, no net change in energy content can take place, thus the energy of the system is said to be conserved.

9.380 Second Law of Thermodynamics

The mathematical statement

$$\Delta S > \frac{q_{irrev}}{T}$$

where ΔS is the entropy change, q_{irrev} is the heat change along an irreversible path, and T is the absolute temperature. For an **isolated system** such as the universe, q is equal to zero, hence $\Delta S > 0$. The sum of all processes that occur in our universe results in an increase in entropy. However, systems that are not isolated (all chemical reactions) may either increase or decrease in entropy when they undergo a change.

9.390 Third Law of Thermodynamics

All perfect crystals of any pure substance have the same entropy at the absolute zero of temperature. The value of the standard entropy $S°$ at the absolute zero is zero.

9.400 Gibbs Free Energy, G (J. W. Gibbs, 1839-1903)
A state function defined by

$$G = H - TS$$

where H is the enthalpy, T is the absolute temperature, and S is the entropy. For a change between two states at a constant temperature

$$\Delta G = \Delta H - T\Delta S$$

For spontaneous reactions $\Delta G < 0$, at equilibrium $\Delta G = 0$, and for nonspontaneous reactions $\Delta G > 0$.

Example. In the reaction

$$A \rightleftarrows B + C$$

the free energy change ΔG can be related to the standard free energy change ΔG° and concentrations of reactants and products by the equation

$$\Delta G = \Delta G^\circ + RT \ln \frac{[B]\,[C]}{[A]} \cdot \frac{[A^\circ]}{[B^\circ]\,[C^\circ]}$$

where R is the gas constant, T is the absolute temperature, [A], [B], and [C] are concentrations of products and reactants *in the reacting mixture* (not equilibrium concentrations), and $[A^\circ]$, $[B^\circ]$, and $[C^\circ]$ are concentrations in a reference standard state. The values of $[A^\circ]$, $[B^\circ]$, and $[C^\circ]$ are generally chosen to be unity and seldom appear in the equation. Near the start of the reaction the concentrations of B and C are small and

$$\Delta G = \Delta G^\circ + RT \text{ times a large negative number}$$

Therefore

$$\Delta G < 0$$

At equilibrium $\Delta G = 0$ and then

$$\Delta G^\circ = -RT \ln K$$

9.410 Standard Free Energy Change, ΔG°
The free energy change between two systems in their standard states. At a constant temperature

$$\Delta G^\circ = \Delta H^\circ - T\Delta S^\circ = -RT \ln K$$

where ΔH° is the standard enthalpy of reaction, T is the absolute temperature, ΔS° is the standard entropy of reaction, R is the gas constant, and K is a unitless "thermodynamic" equilibrium constant. The thermodynamic equilibrium constant represents a ratio of two constants, one using concentrations at equilibrium and the other using concentrations of a reference standard state. The latter (sometimes represented by the symbol Q) is generally chosen so that its value is one.

Example. For

$$C_2H_4(g) + H_2O(g) \rightleftarrows C_2H_5OH(g)$$

$$K\,(\text{atm}^{-1}) = \frac{P_{C_2H_5OH}}{P_{C_2H_4} \cdot P_{H_2O}}; \qquad Q\,(\text{atm}^{-1}) = \frac{P^{\circ}_{C_2H_5OH}}{P^{\circ}_{C_2H_4} \cdot P^{\circ}_{H_2O}} = 1$$

where P is the partial pressure of the gas at equilibrium and P° is the partial pressure of the gas in a standard reference state.

	298 K	500 K
ΔH°, kcal[a]	-10.89	-11.23
ΔS°, cal deg^{-1}	-30.02	-30.98
$T\Delta S^{\circ}$, kcal	- 8.95	-15.49
ΔG°, kcal	- 1.94	+ 4.26
K, atm^{-1}	26.42	0.0137

[a] 1 kcal = 4.18 kJ.

Ethanol, the compound on the right-hand side of the equation, is favored at 298 K, and ethylene and water are favored at 500 K.

9.420 Standard Free Energy of Formation, ΔG_f°

The change is free energy when 1 mole of a substance in its standard state is formed from its elements in their standard states. ΔG_f° of all elements in their standard states is by convention equal to zero.

Examples. At 298 K

$$C(\text{graphite}) \longrightarrow C(\text{gas})$$

$$\Delta G_f^{\circ}\,(C, \text{gas}) = 697.9 \text{ kJ mol}^{-1} \;(166.8 \text{ kcal mol}^{-1}).$$

$$C(\text{graphite}) + 2Cl_2(\text{gas}) \longrightarrow CCl_4(\text{liquid})$$

$$\Delta G_f^{\circ}\,(CCl_4, \text{liquid}) = -68.6 \text{ kJ mol}^{-1} \;(-16.4 \text{ kcal mol}^{-1})$$

9.430 Helmholtz Free Energy, A (H. L. von Helmholtz, 1821-1894)

A state function defined by

$$A = U - TS$$

where U is the internal energy, T is the absolute temperature, and S is the entropy. The difference between Gibbs free energy G and Helmholtz free energy A

is given by

$$G - A = PV$$

but more importantly, the thermodynamic relationships developed from the formal definitions are used in different situations. Expressions involving G are generally more applicable to constant pressure processes, whereas those involving A are generally more applicable to constant volume processes.

9.440 Equilibrium Constant, K

A quantity calculated from the relative amounts of products and reactants of a reaction present at equilibrium. For the reaction

$$a\mathrm{A} + b\mathrm{B} \underset{k_{-1}}{\overset{k_1}{\rightleftharpoons}} c\mathrm{C}$$

$$K = \frac{[C]^c}{[\mathrm{A}]^a[\mathrm{B}]^b} = \frac{k_1}{k_{-1}}$$

where a, b, and c are the relative number of moles of each corresponding species in the equilibrium equation, [A], [B], and [C] are the effective concentrations (partial pressures, mole fractions, moles per liter, activities, fugacities etc.) of those species, and k_1 and k_{-1} are rate constants for the forward and reverse reactions, respectively. Equilibrium constants may or may not have units associated with them. The equilibrium constant can be related to the standard free energy change ΔG° by the equation

$$\Delta G^\circ = -RT \ln \left(\frac{K}{Q}\right)$$

where R is the gas constant, T is the absolute temperature, and Q is a constant with a value of one associated with a standard reference state (see Sect. 9.130, 9.400 and 9.410)

Examples. For

$$C_2H_6(g) \rightleftharpoons C_2H_4(g) + H_2(g)$$

at 298 K, $\Delta G^\circ = 100.8 \text{ kJ mol}^{-1}$ ($24.10 \text{ kcal mol}^{-1}$)

$$K = 2.187 \times 10^{-18} \text{ atm} = \frac{P_{C_2H_4} \cdot P_{H_2}}{P_{C_2H_6}}$$

where P is the partial pressure of the respective species. For

$$CH_3COOC_2H_5(l) + H_2O(l) \rightleftharpoons CH_3COOH(l) + C_2H_5OH(l)$$

at 298 K, $\Delta G° = 3.72$ kJ mol^{-1} (0.89 kcal mol^{-1})

$$K = 2.2 \times 10^{-1} = \frac{[CH_3COOC_2H_5]\,[H_2O]}{[CH_3COOH]\,[C_2H_5OH]}$$

9.450 Fugacity, f

A term that replaces pressure so that the nonideal behavior of a gas is adjusted to correspond to ideal behavior. It is defined by the equation

$$\Delta G_p - \Delta G° = RT \ln\left(\frac{f_p}{f_s}\right)$$

where ΔG_p and $\Delta G°$ are molar free energies at a specified and at a standard state pressure, respectively, R is the gas constant, T is the absolute temperature, and f_p and f_s are the fugacities at the specified and standard state pressures, respectively. The ratio of fugacity to pressure defines a **fugacity coefficient** γ_i:

$$\frac{f_i}{p_i} = \gamma_i$$

Fugacity has the units of pressure. Since all gases tend to become ideal as the pressure approaches zero, $f_i/p_i \rightarrow 1$.

9.460 Activity, a

A term that replaces concentration so that the nonideal behavior of a condensed phase species is adjusted to correspond to ideal behavior. It is defined by the equation

$$a_i = \gamma_i x_i$$

where a_i is the activity, γ_i is the **activity coefficient**, and x_i is the concentration of a given species. Activities are unitless parameters and activity coefficients have the inverse units of concentration. The activity coefficient in essence is a correction factor.

9.470 Concentration Constant, K_c

An equilibrium constant calculated for a reaction in which the relative amounts of products and reactants present at equilibrium are expressed in concentration units.

9.480 Ionization Constant

An equilibrium constant that measures the degree of dissociation of a species into ions.

Example. For water at 298 K

$$H_2O(l) \rightleftharpoons H^+(aq) + {}^-OH(aq)$$

$$K_{ion} = \frac{[H^+][{}^-OH]}{[H_2O]} = \frac{[1.004 \times 10^{-7}][1.004 \times 10^{-7}]}{[55.5]}$$

$$= 1.816 \times 10^{-16} \text{ mol L}^{-1}$$

In most aqueous solutions the concentration of water is not significantly changed by the small quantity that dissociates, and its value is combined with the equilibrium constant to form a new constant. For pure water the new constant is given the symbol K_w and is called the ion-product constant. At 298 K

$$K_w = [H_2O]K_{ion} = [H^+][{}^-OH] = 1.008 \times 10^{-14} \text{ mol}^2 \text{ L}^{-2}$$

It is rather rare to see textbooks in which K_w is expressed in units, but it is correct to do so.

9.490 Instability Constant (Dissociation Constant), K_d

An equilibrium constant that expresses the degree of dissociation of a particular species.

Examples. In pyridine (py) at 25°C

$$Ag(py)^+ \rightleftharpoons Ag^+ + py$$

$$K_d = \frac{[Ag^+][py]}{[Ag(py)^+]} = 1.0 \times 10^{-2} \text{ mol L}^{-1}$$

and

$$Ag(py)_2^+ \rightleftharpoons Ag^+ + 2py$$

$$K_d = \frac{[Ag^+][py]^2}{[Ag(py)_2^+]} = 7.8 \times 10^{-5} \text{ mol}^2 \text{ L}^{-2}$$

9.500 Stability Constant (Formation Constant)

An equilibrium constant that expresses the propensity of a species to form from its component parts. The stability constant is the reciprocal of the instability constant. The larger the stability constant, the more stable is the species.

Example. In pyridine (py) at 25°C the stability constants for

$$Ag^+ + py \rightleftharpoons Ag(py)^+$$

and

$$Ag^+ + 2py \rightleftharpoons Ag(py)_2^+$$

are 1.0×10^2 L mol^{-1} and 1.3×10^4 L^2 mol^{-2}, respectively, the reciprocals of the instability constants.

9.510 Brønsted Acid and Its Conjugate Base; Brønsted Base and Its Conjugate Acid

A substance that can act as a proton donor in a chemical reaction is a Brønsted acid (J. N. Brønsted, 1879-1947) and the species remaining after the loss of the proton is the conjugate base of that acid. The species that accepts the proton is a Brønsted base and upon protonation becomes the conjugate acid of that base. The conjugate relationship was developed by Lowry (T. M. Lowry, 1878-1936) and the proton donor-acceptor concept is frequently referred to as the Brønsted-Lowry theory of acids and bases.

Examples. In the equilibrium mixture

$$CH_3COOH + H_2O \rightleftarrows CH_3COO^- + H_3O^+$$

CH_3COOH and H_3O^+ are Brønsted acids (proton donors) and H_2O and CH_3COO^- are Brønsted bases (proton acceptors). The conjugate relationships are

$$\begin{cases} CH_3COOH, \text{ Brønsted acid (or conjugate acid)} \\ CH_3COO^-, \text{ conjugate base (or Brønsted base)} \end{cases}$$

$$\begin{cases} H_2O, \text{ Brønsted base (or conjugate base)} \\ H_3O^+, \text{ conjugate acid (or Brønsted acid)} \end{cases}$$

In the equilibrium mixture

$$CH_3COOH + CF_3COOH \rightleftarrows CH_3COOH_2^+ + CF_3COO^-$$

CF_3COOH and $CH_3COOH_2^+$ are Brønsted acids, and their conjugate bases are CF_3COO^- and CH_3COOH, respectively. In this mixture CH_3COOH is a Brønsted base and not an acid. In the CH_3COOH-H_2O mixture CH_3COOH is a Brønsted acid. Table 9.510 lists some additional conjugate relationships.

Table 9.510 Conjugate relationships

Brønsted acid (Conjugate acid)	Conjugate base (Brønsted base)
H_2O	HO^-
H_3O^+	H_2O
CH_5^+	CH_4
$CH_3CH_2^+$	$CH_2{=}CH_2$
CH_3COCH_3	$CH_3COCH_2^-$
$(CH_3)_2C{=}OH^+$	$(CH_3)_2C{=}O$

9.520 Lewis Acid and Lewis Base (G. N. Lewis, 1875-1946)

A substance that acts as an electron pair acceptor is a Lewis acid, and a substance that acts as an electron pair donor is a Lewis base. The species formed by the reaction of a Lewis acid with a Lewis base is a **Lewis adduct.**

Examples

Lewis acid	Lewis base		Lewis adduct
$(CH_3)_3C^+$	+ Cl^-	$\rightleftarrows$	$(CH_3)_3CCl$
$(CH_3)_3B$	+ $:NH_3$	$\rightleftarrows$	$(CH_3)_3\bar{B}\overset{+}{N}H_3$

The Brønsted-Lowry theory of acids and bases involves the centrality of the proton, whereas the Lewis theory involves the centrality of the electron pair.

9.530 Hard and Soft Acids and Bases, HSAB

A qualitative classification of acidity and basicity of Lewis acids and bases. Lewis acids and bases may be classified as being hard, soft, or borderline. Hard acids (electron pair acceptors) generally have a small electron acceptor site of high positive charge and do not possess unshared pairs of electrons in their valence shells. Hard acids are characterized by high electronegativity and low polarizability. Hard bases (electron pair donors) are generally difficult to oxidize and have no empty low energy orbitals available. Hard bases are characterized by a highly electronegative donor atom of low polarizability. Soft acids are characterized by a large electron acceptor atom of high polarizability. Soft bases generally have electrons that are easily removed by oxidizing agents and have empty orbitals of low energy. Soft bases are characterized by a polarizable donor atom. Hard acids prefer to combine with hard bases, and soft acids prefer to bond to soft bases.

Table 9.530 Classification of acids and bases

Acids		Bases	
Hard	Soft	Hard	Soft
H^+, Li^+, Na^+	Cu^+, Ag^+, Au^+	H_2O	R_2S
Mg^{2+}, Mn^{2+}	Pd^{2+}, Pt^{2+}	HO^-	RS^-
Al^{3+}, Sc^{3+}	Tl^{3+}, $Tl(CH_3)_3$	F^-	CN^-
Cr^{3+}, Co^{3+}	RS^+	AcO^-	C_2H_4
Si^{4+}, Ti^{4+}	I^+, Br^+	ROH	C_6H_6
BF_3, $B(OR)_3$	BH_3	R_2O	H^-
RSO_2^+, SO_3	I_2, Br_2	RO^-	R^-
RCO^+, CO_2		RNH_2	CO

Examples. Table 9.530 lists some hard and soft acids and bases. It is instructive to see that BF_3 is a hard acid and BH_3 is a soft acid. BH_3 forms more stable complexes with soft bases like carbon monoxide and olefins than with hard bases. On the other hand, the hard acid-hard base complex $BF_3 \cdot OR_2$ is more stable than $BF_3 \cdot SR_2$, a hard acid-soft base complex. In BF_3 boron is largely B^{3+} because of the electronegative fluorines, hence BF_3 is hard; in BH_3 boron is largely neutral, hence BH_3 is soft.

9.540 Superacid

An acidic medium that has a proton donating ability greater than 100% sulfuric acid. Superacids are generally mixtures of fluorosulfonic acid (FSO_2OH) and Lewis acids such as SO_3 and SbF_5. An equimolar mixture of FSO_2OH and SbF_5 is known as **magic acid.**

9.550 Acid Dissociation Constant, K_a

A constant that characterizes the ability of a Brønsted acid to donate a proton to a specific reference base. In dilute aqueous solution where water is the reference base

$$\text{AH (Brønsted acid)} + H_2O \text{ (reference base)} \rightleftharpoons H_3O^+ + A^-$$

$$K_a(\text{AH}) \text{ in mol L}^{-1} = [H_2O]K = \frac{[A^-][H_3O^+]}{[\text{AH}]}$$

where K is the equilibrium constant for the system. At 50% ionization K_a is equal to the hydrogen ion concentration. In other solvents

$$\text{AH} + \text{solv} \rightleftharpoons \text{solv—H}^+ + A^-$$

$$K_a(\text{AH})_{\text{solv}} = [\text{solv}]K = \frac{[A^-][\text{solv—H}^+]}{[\text{AH}]}$$

Examples. At 25°C for a dilute solution of acetic acid in water

$$CH_3COOH + H_2O \rightleftharpoons CH_3COO^- + H_3O^+$$

$$K_a(CH_3COOH) = 1.76 \times 10^{-5} \text{ mol L}^{-1}$$

For pure water at 25°C

$$H_2O + H_2O \rightleftharpoons H_3O^+ + {}^-OH$$

$$K_a(H_2O) = [H_2O]K = \frac{[H_3O^+][{}^-OH]}{[H_2O]} = 1.8 \times 10^{-16} \text{ mol L}^{-1}$$

9.560 Hydride Affinity

The negative of the standard enthalpy of reaction of a species with hydride ion. For

$$A^+ + H^- \longrightarrow AH$$

$$\text{Hydride affinity } (A^+) = \Delta H_f^\circ(A^+) + \Delta H_f^\circ(H^-) - \Delta H_f^\circ(AH)$$

Hydride affinity, equivalent to the **heterolytic bond dissociation energy** $D(A^+—H^-)$, is one quantitative measure of the gas phase acidity of Lewis acids. The higher the hydride affinity, the stronger is the acid.

Examples. The gas phase hydride affinities of cyclopentyl cation, Fig. 9.560*a*, and the norbornyl cation, Fig. 9.560*b*, are 245 and 234 kcal mol^{-1} (1025 and 979 kJ mol^{-1}), respectively. Cyclopentyl cation is a stronger Lewis acid than norbornyl cation.

Figure 9.560 (*a*) Cyclopentyl cation; (*b*) norbornyl cation.

9.570 Acidity Function, H

A parameter H that quantitatively measures the proton donating ability of a solvent system. It is defined by the equation

$$H = pK_a(BH^+) + \log \frac{[B]}{[BH^+]}$$

The pK_a of an acid BH^+ is generally known and the relative concentrations of BH^+ and its conjugate base B are measured.

Example. The **Hammett acidity function** H_0 measures the acidity of solvent systems using substituted anilines as proton acceptors B. The concentrations of BH^+ and B are conveniently measured spectroscopically and the pK_a's of BH^+ are known. The solvent system consisting of 60 wt % of sulfuric acid in water has an H_0 value of −4.32. 2,4-Dinitroaniline [$pK_a(BH^+) = -4.38$] was used as an indicator base to determine this value. Once an H_0 value is known for a solvent system, that system may be used to experimentally determine an unknown pK_a.

9.580 Acid Strength (Acidity)

For Brønsted acids acidity is expressed quantitatively by acid dissociation constants. For Lewis acids acidity is expressed quantitatively by stability constants with respect to specific Lewis adducts or qualitatively by hardness or softness. Gas phase acidities are generally expressed in terms of hydride affinities. For protolytic (acidic) solvent systems, acidity is quantitatively expressed in terms of acidity functions.

9.590 Base Dissociation Constant, K_b

A constant that characterizes the ability of a Brønsted base to accept a proton from a reference acid. In dilute aqueous solution water is the reference acid

$$\text{B (Brønsted base)} + \text{H}_2\text{O (reference acid)} \rightleftharpoons \text{BH}^+ + {}^-\text{OH}$$

$$K_b(\text{B}) \text{ in mol L}^{-1} = [\text{H}_2\text{O}]K = \frac{[\text{BH}^+][{}^-\text{OH}]}{[\text{B}]}$$

where K is the equilibrium constant for the system. The larger K_b is, the stronger is the base. The product of the aqueous acid and base dissociation constants of conjugate pairs is equal to K_w, the ion product of water.

$$K_a(\text{BH}^+) \cdot K_b(\text{B}) = K_w$$

Thus if $K_a(\text{BH}^+)$ is large (a strong acid), then $K_b(\text{B})$ is small (a weak base).

Examples

Base	K_b[a]	Conjugate acid	K_a[a]
$(CH_3)_2NH$	$10^{-3.27}$	$(CH_3)_2\overset{+}{N}H_2$	$10^{-10.73}$
HO^-	$10^{+1.74}$	H_2O	$10^{-15.74}$
H_2O	$10^{-15.74}$	H_3O^+	$10^{+1.74}$
CH_3O^-	$10^{+1.5}$	CH_3OH	$10^{-15.5}$

[a]In dilute aqueous solution, mol L^{-1}; $-\log$ of these numbers are, of course, the pK values.

9.600 Proton Affinity, *PA*

The negative of the standard enthalpy of reaction of a species with a proton. For

$$\text{B} + \text{H}^+ \longrightarrow \text{BH}^+$$

$$PA\,(\text{B}) = \Delta H_f^\circ(\text{B}) + \Delta H_f^\circ(\text{H}^+) - \Delta H_f^\circ(\text{BH}^+)$$

Proton affinity is a quantitative measure of the gas phase basicity of both Lewis and Brønsted bases. The higher the proton affinity, the stronger is the base.

Examples

Species	Proton affinity	
	kcal mol^{-1}	kJ mol^{-1}
CH_3NH_2	211.3	884.0
$CH_3COOCH_2CH_3$	198.0	828.4
CH_3COCH_3	194.6	814.2
$CH_3C{\equiv}N$	187.4	784.1
CH_3CH_2OH	186.8	781.6
Benzene	183.4	767.3
$CH_2{=}CHCH_3$	181.0	757.3
H_2O	168.9	706.7

9.610 Base Strength (Basicity)

For Brønsted bases basicity is expressed quantitatively by base dissociation constants or in terms of a related quantity, the acid dissociation constant of the conjugate acid of the Brønsted base. In aqueous solution

$$K_b(\mathrm{B}) \cdot K_a(\mathrm{BH^+}) = K_w$$

For Lewis bases acidity is expressed quantitatively by stability constants with respect to specific Lewis adducts or qualitatively by hardness or softness. Gas phase basicities are expressed in terms of proton affinities. Base strengths of solvent systems are expressed in terms of acidity functions.

9.620 Autoprotolysis

The reaction between two molecules of solvent, one acting as a Brønsted acid and the other as a Brønsted base.

Example.

$$CH_3OH + CH_3OH \rightleftharpoons CH_3OH_2^+ + CH_3O^-$$

9.630 Aprotic Solvent

A solvent that does not function as a proton donor under a specified set of reaction conditions.

Examples. Solvents such as acetonitrile, dimethylsulfoxide, and hexamethylphosphoramide (HMPA) do not act as proton donors under most reaction conditions.

9.640 Leveling Effect of Solvent

No acid stronger than the conjugate acid of the solvent can exist to an appreciable extent in that solvent. Likewise, no base stronger than the conjugate base of a solvent can exist to an appreciable extent in that solvent.

Example. If equal concentrations of HCl, HBr, and HI are placed separately into water, the three resulting solutions have the same concentration of H_3O^+. All three hydrohalic acids are stronger than H_3O^+, the conjugate acid of H_2O, and thus all available protons of each of these acids protonate H_2O:

$$HX + H_2O \rightleftharpoons H_3O^+ + X^-$$

9.650 Reaction Coordinate Diagram

A plot of potential energy changes as a function of changes in atomic bond distances in reacting molecules as they proceed from reactants through transition states and/or intermediates to products. In the two-dimensional plot called the **potential energy profile,** the potential energy of the most favorable pathway is plotted as the ordinate and the abscissa is a parameter called the reaction coordinate. The **reaction coordinate** represents changes in the coordinates defining the location of the reacting atoms, and these changes can be taken as a measure of the **progress of the reaction.**

Example. The displacement of chloride ion in methyl chloride by iodide ion in the reaction

$$I^- + CH_3Cl \longrightarrow \left[\begin{array}{c} \quad H \\ \quad | \\ I\text{-------}C\text{-------}Cl \\ \quad /\backslash \\ \quad H\ \ H \end{array}\right]^- \longrightarrow ICH_3 + Cl^-$$

can be represented by the potential energy profile shown in Fig. 9.650*a*. The maximum in the curve is called the transition state (Sect. 9.660) and is represented by the bracketed structure in the reaction equation. Reaction coordinate diagrams represent **microscopic phenomena.** The difference in potential energy between the starting system and the transition state, ϵ_a, represents the energy of activation for a given set of reacting species, a microscopic quantity. If this quantity is averaged over the range of values and is multiplied by Avogardro's number N_0

$$\bar{\epsilon}_a \times N_0 = E_a$$

then a macroscopic quantity, the energy of activation per mole E_a, is obtained (see Sect. 9.750). Macroscopic quantities such as free energies, enthalpies, and Arrhenius energies of activation are usually presented (incorrectly) on plots where

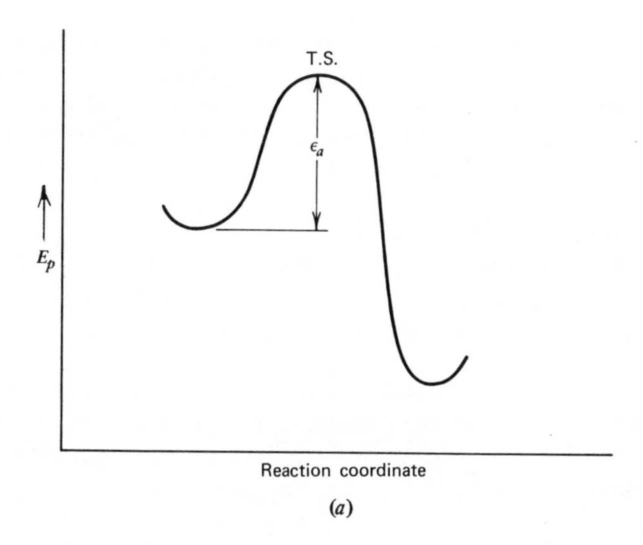

(*a*)

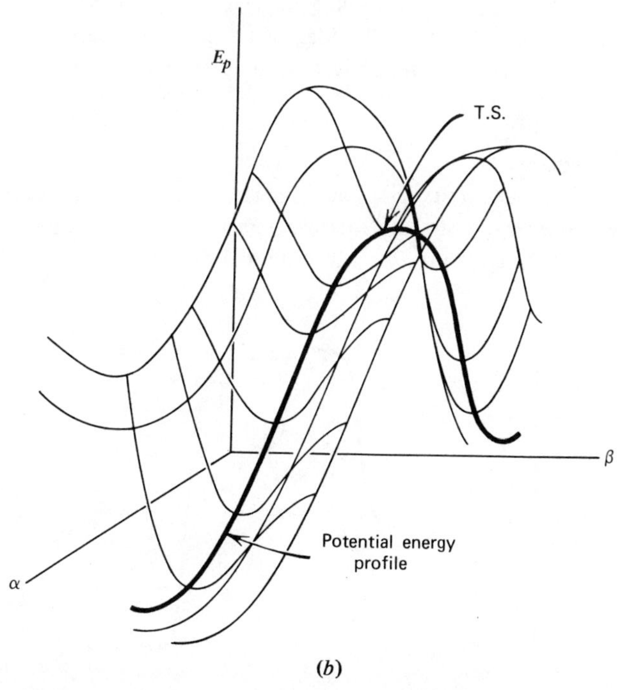

(*b*)

Figure 9.650 (*a*) A two-dimensional plot of potential energy E_p versus reaction coordinate; (*b*) a three-dimensional plot of potential energy E_p versus reaction coordinates α and β.

the reaction coordinate, a microscopic representation, is shown as the abscissa (see Sect. 9.710). Figure 9.650*b* shows a three-dimensional plot of two reaction coordinates against potential energy, giving rise to a **potential energy surface.** The heavy line represents the lowest energy pathway and is a representation of what is plotted in the two-dimensional energy profile.

9.660 Transition State

A state along a defined reaction path, characterized by a higher Gibbs free energy or potential energy than immediately adjacent states lying on that path. This state corresponds to a maximum in a potential energy profile.

Examples. In Figs. 9.650*a*, 9.680, and 9.710 transition states are represented by the abbreviation T.S. Although reactions are represented as proceeding in a specific direction, species in transition states have an equal probability of forming reactants or products.

9.670 Activated Complex

The hypothetical species at a transition state. Because such complexes cannot by definition be isolated, the bonds being made and broken are represented by dotted or broken lines to convey the idea of partial bond character in such species. The structures representing activated complexes are frequently placed in brackets.

9.680 Intermediate

A transient species that exists in a state characterized by a lower Gibbs free energy or potential energy than immediately adjacent states. Intermediates correspond to a minima between maxima in potential energy profiles. An intermedi-

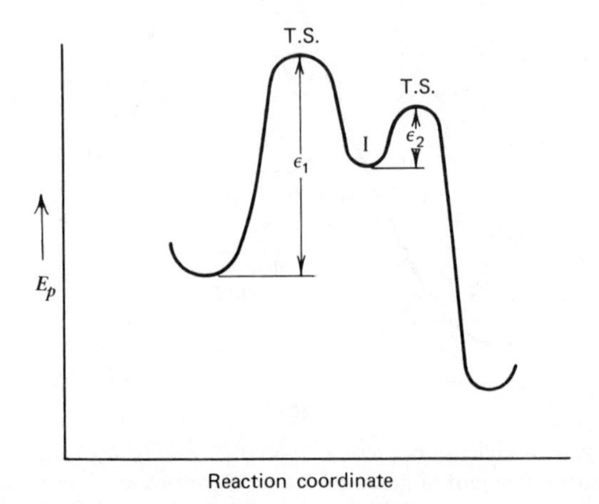

Figure 9.680 A potential energy profile of a two-step reaction.

ate has a finite lifetime that is appreciably greater than the period of a molecular vibration (>10^{-12} s).

Example. In a typical two-step reaction

$$Y^- + R_3CX \xrightarrow{\text{step 1}} Y^- + R_3C^+ + X^- \xrightarrow{\text{step 2}} R_3CY + X^-$$

R_3C^+ is an intermediate and is represented as being at point I in Fig. 9.680.

9.690 Reaction Steps

A sequence of **elementary reactions** (a reaction proceeding through a single transition state) in a multistep reaction (see Sect. 10.040).

Examples. A one-step reaction involves one transition state and no intermediates, whereas a two-step reaction has two transition states and one intermediate. An n-step reaction has n transition states and $n - 1$ intermediates.

9.700 Hammond Postulate (George S. Hammond, 1921-)

In any individual step in a reaction, the geometry and structure of the transition state resemble either the reactants or the products (or intermediate state if one is present) depending on which of these is closer in energy to the transition state. This implies that in an endothermic step the transition state resembles the products and that in an exothermic step the transition state resembles the reactants.

Example. Figure 9.700 shows the potential energy profile of the two-step reaction described by the equation

$$\underset{\text{Reactants}}{Y^- + R_3CX} \xrightarrow[\text{step 1}]{\text{slow}} \underset{\text{Intermediate state}}{Y^- + R_3C^+ + X^-} \xrightarrow[\text{step 2}]{\text{fast}} \underset{\text{Products}}{R_3CY + X^-}$$

The first step is endothermic and therefore the transition state resembles the intermediate state I to a greater extent than the starting reactants. The second step

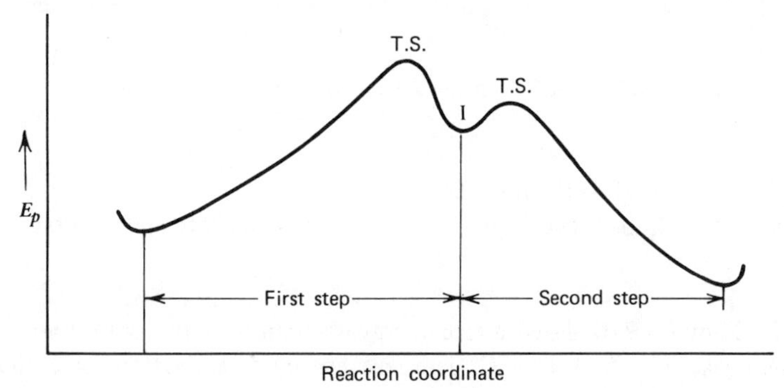

Figure 9.700 A potential energy profile illustrating the Hammond postulate.

is exothermic and the transition state for this step resembles the intermediate state more than the final products.

9.710 Free Energy Diagram

A diagram that shows the relative Gibbs free energies of reactants, transition states, intermediates, and products that occur during the course of a reaction.

Example. For the two-step reaction

$$A \longrightarrow B \longrightarrow C$$

the free energy diagram, Fig. 9.710 shows the relative free energies of five states, those that would correspond to points at either minima or maxima in a potential energy profile. The points are separated for clarity and are generally placed in the sequential order in which the state that they represent occurs during the reaction. The points should not be connected by a continuous curve and the abscissa is not defined. Point A represents the starting system, point B an intermediate system, and point C the final system. Two transition states (T.S.) are also shown. This type of diagram is used to represent macroscopic properties and not the microscopic properties of individual reacting molecules, which are represented in reaction coordinate diagrams (see Sect. 9.650).

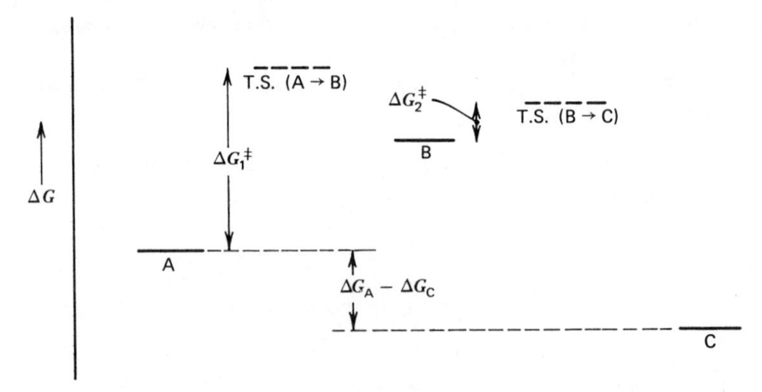

Figure 9.710 A free energy diagram of a two-step reaction.

9.720 Free Energy of Activation, $\Delta G^{\ddagger}$

The difference in Gibbs free energy between a transition state and the state of its reactants.

Example. Figure 9.710 shows a free energy diagram for a two-step (two transition states) reaction, A → B → C. The free energy of activation for each step may be calculated from the absolute rate equation

$$\Delta G^{\ddagger} = RT\left(\ln \frac{kT}{hk_r}\right)$$

where R is the gas constant, T is the absolute temperature, k is Boltzmann's constant, h is Planck's constant, and k_r is an experimental rate constant. For reactions in solution at constant pressures, the free energy of activation is related to the Arrhenius energy of activation E_a by the equation

$$\Delta G^{\ddagger} = E_a - RT - T\Delta S^{\ddagger}$$

where $\Delta S^{\ddagger}$ is the entropy of activation.

9.730 Enthalpy (or Heat) of Activation, $\Delta H^{\ddagger}$
The difference in enthalpy between a transition state and the state of its reactants:

$$\Delta H^{\ddagger} = \Delta G^{\ddagger} + T\Delta S^{\ddagger}$$

where $\Delta G^{\ddagger}$ is the free energy of activation, $\Delta S^{\ddagger}$ is the entropy of activation, and T is the absolute temperature. For reactions in solution at a constant pressure, the enthalpy of activation is related to the Arrhenius energy of activation E_a by the equation

$$\Delta H^{\ddagger} = E_a - RT$$

where R is the gas constant. At ordinary temperatures, 25°C, the difference between E_a and $\Delta H^{\ddagger}$ is approximately 0.6 kcal mol^{-1}.

9.740 Entropy of Activation, $\Delta S^{\ddagger}$
The difference in entropy between a transition state and the state of its reactants:

$$\Delta S^{\ddagger} = \frac{\Delta H^{\ddagger} - \Delta G^{\ddagger}}{T}$$

where $\Delta H^{\ddagger}$ is the enthalpy of activation, $\Delta G^{\ddagger}$ is the free energy of activation, and T is the absolute temperature. Entropies of activation are usually calculated from experimental rate data and the equation

$$\Delta S^{\ddagger} = \frac{E_a - RT - \Delta G^{\ddagger}}{T}$$

where E_a is the Arrhenius energy of activation and R is the gas constant. $\Delta G^{\ddagger}$ is obtained from the Eyring equation (see Sect. 9.1000).

9.750 Arrhenius Energy of Activation, E_a (Svante Arrhenius, 1859-1927)
An operationally defined quantity that relates rate constants to temperature by

the equation (see also Sect. 9.1010)

$$k = Ae^{-E_a/RT}$$

where k is a rate constant, A is a constant known as the **preexponential or frequency factor,** R is the gas constant, and T is the absolute temperature. At a constant pressure and assuming only a small change in A with respect to temperature,

$$E_a = RT^2 \left(\frac{\partial \ln k}{\partial T}\right)_p = -R \left(\frac{\partial \ln k}{\partial(1/T)}\right)_p$$

For solution reactions at constant pressures, the free energy of activation $\Delta G^{\ddagger}$ and the enthalpy of activation $\Delta H^{\ddagger}$ are related to the Arrhenius energy of activation by the equation

$$E_a = \Delta G^{\ddagger} + T\Delta S^{\ddagger} + RT = \Delta H^{\ddagger} + RT$$

where $\Delta S^{\ddagger}$ is the entropy of activation. At ordinary temperatures, 25°C, the difference between $\Delta H^{\ddagger}$ and E_a is ~0.6 kcal mol^{-1}. If energy of activation E_a is substituted for **free energy of activation** $\Delta G^{\ddagger}$ on free energy diagrams (see Sect. 9.710), then it is assumed that entropy factors and RT are not significant.

9.760 Thermodynamic (or Equilibrium) Control

Under reversible reaction conditions the ratio of possible products is determined by the relative stability of each product, measured by its standard free energy ($\Delta G°$). The composition of the equilibrium mixture does not depend on how fast ($\Delta G^{\ddagger}$) each product is formed in the reaction.

Example. The addition of HCl to 1,3-butadiene can lead to two major products, both derived from a common intermediate, as shown in Fig. 9.760*a*; Fig. 9.760*b* shows the potential energy profile for the second step in the addition. The more stable 1,4-addition product predominates at higher temperatures, where equilibrium is more rapidly attained, even though it is formed more slowly than the 1,2-addition product. At lower temperatures, where the equilibrium is not attained very rapidly, the 1,2-addition product predominates. In

+ HCl

Cl

1,2-addition intermediate 1,4-addition

(*a*)

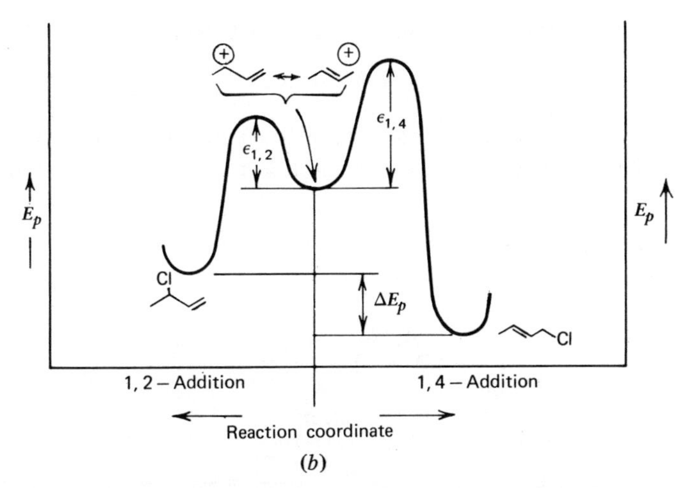

Figure 9.760 (*a*) The 1,2- and 1,4-addition of HCl to 1,3-butadiene; (*b*) a potential energy profile for the second step in the addition of HCl to 1,3-butadiene.

many cases the most stable product **(thermodynamic product)** may also be the one formed fastest **(kinetic product).**

9.770 Kinetic Control

The quantity of each possible product is determined by how fast each product is formed and is not a function of the relative stability ΔG° of each product.

Example. See thermodynamic control, Sect. 9.760.

9.780 Kinetics

The scientific discipline that deals with rates of reactions.

9.790 Extent of Reaction, ξ (xi)

A useful term in expressing the rate of reaction, which describes how far a reaction has proceeded. For a general reaction

$$a_1 A_1 + a_2 A_2 \longrightarrow a_3 A_3 + a_4 A_4$$

$$\sum a_i A_i = 0$$

where a_i is the stoichiometric coefficient, positive for products and negative for reactants, and A_i is the chemical symbol of the ith substance. The extent of reaction is defined by

$$d\xi = a_i^{-1}\, dn_i$$

where n_i is the amount of substance i. In integrated form

$$n_i = n_{io} + a_i \xi$$

where n_{io} is the initial amount of substance. The extent of reaction is defined to be zero at the beginning of the reaction and it increases as the reaction proceeds.

9.800 Rate of Reaction, v

A quantitative measure of how fast a reaction occurs. For the elementary (one-step) reaction

$$a\mathrm{A} + b\mathrm{B} \longrightarrow c\mathrm{C}$$

the rate of reaction v is given by the equation

$$v = -\frac{1}{a}\frac{d[\mathrm{A}]}{dt} = -\frac{1}{b}\frac{d[\mathrm{B}]}{dt} = \frac{1}{c}\frac{d[\mathrm{C}]}{dt}$$

where a, b, and c are stoichiometric numbers (coefficients) and [A], [B], and [C] represent concentrations of the respective substances. Another less frequently used definition for rate of reaction is the rate of extent of reaction:

$$\dot{\xi} = \frac{d\xi}{dt} = -\frac{1}{a}\frac{dn_a}{dt} = -\frac{1}{b}\frac{dn_b}{dt} = \frac{1}{c}\frac{dn_c}{dt}$$

where $\dot{\xi}$ is the rate of reaction, ξ is the extent of reaction, and n_a, n_b, and n_c are amounts of substances. In a one-step reaction the **rate of disappearance**, $-d[\mathrm{A}]/dt$ or $-d[\mathrm{B}]/dt$, or the **rate of appearance**, $d[\mathrm{C}]/dt$, is equal to the rate of reaction only when the stoichiometric number of the species in question is one.

9.810 Rate Equation (Rate Law)

An equation that describes the dependence of the rate of reaction on the concentrations of species present in the reaction mixture. The rate equation is frequently used to substantiate or negate a proposed reaction sequence (mechanism).

Example. For a one-step reaction

$$1\mathrm{A} + 1\mathrm{B} \longrightarrow 1\mathrm{C}$$

the rate equation is

$$\text{rate} = v = -\frac{d[\mathrm{A}]}{dt} = -\frac{d[\mathrm{B}]}{dt} = \frac{d[\mathrm{C}]}{dt} = k[\mathrm{A}][\mathrm{B}]$$

where k is the rate constant and symbols placed in brackets represent concentrations.

For a two-step reaction, A + B + D → E, where

$$1A + 1B \longrightarrow 1C \qquad \text{(Step 1)}$$

$$1C + 1D \longrightarrow 1E \qquad \text{(Step 2)}$$

in which step 1 is slow compared to step 2, the rate equation is

$$\text{rate} = v = k_1 [A] [B]$$

where k_1 is the rate constant for the first step. The rate equation shows that the rate of reaction is not dependent on the concentration of D; it is not involved in the slow step of the reaction.

9.820 Order of Reaction (Kinetic Order)

A number (or index) expressing the experimentally determined dependence of the rate of reaction on the concentration of a reacting species.

Example. If the rate equation for a reaction is

$$\text{rate} = k \frac{[A]^1 [B]^2}{[C]^2} = k [A]^1 [B]^2 [C]^{-2}$$

then the kinetic order with respect to A is one (first-order), with respect to B is two (second-order), and with respect to C is −2. The overall kinetic order of reaction for a reaction whose rate can be expressed as

$$\text{rate} = k [A]^a [B]^b \cdots$$

is the sum of the superscript numbers (indices). In the example given above the order of the reaction is one.

9.830 Molecularity

A description of the number of species involved in an elementary (one-step) reaction. Molecularity is not used to describe an overall reaction that has multiple steps.

Example. In the two-step reaction

$$A + B \longrightarrow C \longrightarrow D$$

the molecularity of the first step is two **(bimolecular)**, and that of the second step one **(unimolecular)**. The molecularity of any individual elementary reaction is the same as the order of reaction.

9.840 Rate Constant, k

A proportionality constant in rate equations which relates concentrations to the rate of reaction. The rate constant is independent of concentration, but is temperature dependent and characteristic for a specific reaction.

Example. In the rate equation for

$$A + B \longrightarrow C$$

$$\text{rate} = -\frac{d[A]}{dt} = -\frac{d[B]}{dt} = k[A][B]$$

the rate constant k has the units of L mol^{-1} s^{-1}. But in the rate equation for

$$A \longrightarrow B$$

$$\text{rate} = -\frac{d[A]}{dt} = k[A]$$

the rate constant k has the units of s^{-1}.

9.850 Rate-Determining (or Rate-Limiting) Step

In a multiple step reaction the step that is slower than any of the other steps.

Example. In the two-step reaction

$$RY \longrightarrow R^+ + Y^- \qquad \text{(Step 1)}$$

$$R^+ + X^- \longrightarrow RX \qquad \text{(Step 2)}$$

the first step is generally much slower than the second step. The rate of the overall reaction is essentially the same as that of the first step and the first step is said to be rate-determining.

9.860 Reactivity

A kinetic term that involves a comparison of rate constants for related reactions of different substances, one involving a chemical species of interest and the other a reference species.

Example. The rate constant for the reduction of cyclopentanone by sodium borohydride in isopropyl alcohol at 0°C is 7×10^{-4} L mol^{-1} s^{-1}, compared to a rate constant of 161×10^{-4} L mol^{-1} s^{-1} for cyclohexanone. Cyclohexanone is said to be more reactive or to have a higher reactivity than cyclopentanone in this reaction.

9.870 Relative Rate

The rate constant of a particular reaction compared to the rate constant of a related reference reaction whose rate is arbitrarily assigned a value of unity.

Example. Table 9.870 shows the relative rates of hydrolysis of alkyl bromides in water at 50°C. The hydrolysis of ethyl bromide is used as the reference reaction.

Table 9.870 Relative rates of hydrolysis of alkyl bromides in water at 50°C

Alkyl bromide	Relative rate
CH_3Br	1.05
C_2H_5Br	1.00
$(CH_3)_2CHBr$	11.6
$(CH_3)_3CBr$	1.2×10^6

9.880 Kinetic Stability

The reluctance of a substance to undergo a specified reaction; that is, the rate constant for the reaction is very small.

Example. Ethane is unreactive toward bromine in carbon tetrachloride, whereas under the same conditions ethylene is very reactive. Kinetic stability of a species should not be confused with stability or thermodynamic stability, which refers to the standard heat of formation ΔH_f° of the substance. The kinetic stability is related to an energy of activation.

9.890 Thermodynamic Stability

A measure of the magnitude of either the heat of formation (ΔH_f°) of a substance or the standard Gibbs free energy (ΔG°) of a system. The smaller (or larger negative) ΔH_f° or ΔG°, the more stable is the substance or system. **Stability** is synonymous with thermodynamic stability.

Examples. Ethane with a ΔH_f° of -84.5 kJ mol^{-1} (-20.2 kcal mol^{-1}) is more stable than ethylene, whose ΔH_f° is $+52.3$ kJ mol^{-1} ($+12.5$ kcal mol^{-1}). Likewise *trans*-2-butene ($\Delta H_f^\circ = -12.1$ kJ mol^{-1}, -2.9 kcal mol^{-1}) is more stable than *cis*-2-butene ($\Delta H_f^\circ = -7.1$ kJ mol^{-1}, -1.7 kcal mol^{-1}). At 0°C for the equilibrium

$$C_2H_4(g) + H_2O(g) \rightleftharpoons C_2H_5OH(g)$$

the standard Gibbs free energy difference is -8.12 kJ mol^{-1} (-1.94 kcal mol^{-1}); hence C_2H_5OH is more stable than C_2H_4 and H_2O, its component parts.

9.900 Pseudo Kinetics

A type of behavior observed when the concentration of a species normally appearing in the rate equation is essentially held constant so that the observed rate equation does not reflect its involvement.

Example. For the elementary reaction

$$A + B \longrightarrow C$$

the rate equation is given by

$$\text{rate} = v = -\frac{d[A]}{dt} = -\frac{d[B]}{dt} = k[A][B]$$

The reaction is said to have second-order kinetics. If the concentration of B is very large compared to that of A, then the rate equation would be

$$\text{rate} = -\frac{d[A]}{dt} = k'[A]$$

where $k' = k[B]$. The reaction is then said to follow pseudo first-order kinetics.

9.910 Diffusion (or Encounter) Controlled Rate

A bimolecular reaction that occurs with 100% efficiency every time the reaction partners undergo collision, hence a bimolecular reaction that proceeds with the maximum rate constant k_d possible under a given set of reaction conditions:

$$k_d = \frac{8RT}{3000\eta}$$

where R is the gas constant, T is the absolute temperature, and η is the viscosity of the medium in poise. These reactions have negligible or nonexistent free energies of activation and are exceedingly fast, typically $k_d = 10^9$ L mol^{-1} s^{-1}.

9.920 Half-Life, $t_{1/2}$

The time required for the concentration of a reactant to fall to one-half its initial value.

Examples. In a first-order reaction where the rate equation is given by

$$\text{rate} = k[A] = -\frac{d[A]}{dt}$$

the half-life of A, $t_{1/2}$, is given by the equation

$$t_{1/2} = \frac{\ln \frac{[A]}{[A/2]}}{k} = \frac{.0693}{k}$$

The half-life is independent of the initial concentration in a first-order reaction. This is not true for reactions of higher order.

9.930 Common Ion Effect

The effect produced by addition of an ion that is being formed in the reaction; the effect is a depression in the rate.

Example. The addition of chloride ion decreases the rate of solvolysis of diphenylmethyl chloride. The first step in the reaction,

$$RCl \rightleftarrows R^+ + Cl^-$$

is rate-determining. The presence of Cl^- from a second source has a mass-law effect on the equilibrium and thus on the overall rate of reaction.

9.940 Salt Effect

A change in the rate of a reaction caused by the addition of a salt (electrolyte) that generally does not have an ion in common with reactants or products. The addition of a salt changes the ionic strength of the solution, which in turn affects the free energies of ions that are either being formed or destroyed.

Example. Addition of $LiClO_4$ generally increases the rate of an S_N1 reaction. Ions are being formed in the rate-determining step,

$$RX \rightleftarrows R^+ + X^-$$

and an increase in the ionic strength of the solution favors their formation.

9.950 Steady (or Stationary) State

A state in which the concentration of a particular species does not change during the course of a reaction. The amounts of material entering that state are balanced by the amounts of material leaving that state.

Examples. In the two-step reaction

$$A + B \xrightarrow{k_1} C \qquad \text{(Step 1)}$$

$$C \xrightarrow{k_2} D \qquad \text{(Step 2)}$$

the rate equations for steps 1 and 2 are

$$-\frac{d[\mathrm{A}]}{dt} = k_1 [\mathrm{A}][\mathrm{B}]$$

and

$$\frac{d[\mathrm{D}]}{dt} = k_2 [\mathrm{C}]$$

respectively. If C is reacting as fast as it is formed, it reaches a steady state concentration:

$$\frac{d[\mathrm{C}]}{dt} = 0 \qquad \text{and} \qquad \frac{d[\mathrm{D}]}{dt} = k_1 [\mathrm{A}][\mathrm{B}]$$

Reactive intermediates generally are assumed to reach steady state concentrations. If a preequilibrium is involved,

$$\mathrm{A} + \mathrm{B} \underset{k_{-1}}{\overset{k_1}{\rightleftharpoons}} \mathrm{C} \qquad \text{(Step 1)}$$

$$\mathrm{C} \xrightarrow{k_2} \mathrm{D} \qquad \text{(Step 2)}$$

and a steady state of C is achieved, then

$$\frac{d[\mathrm{C}]}{dt} = k_1 [\mathrm{A}][\mathrm{B}] - k_{-1} [\mathrm{C}] - k_2 [\mathrm{C}] = 0$$

and

$$[\mathrm{C}] = \frac{k_1 [\mathrm{A}][\mathrm{B}]}{k_{-1} + k_2}$$

9.960 Preequilibrium (Prior Equilibrium)
A situation in a reaction sequence in which an equilibrium is established prior to the rate-determining step.

Example. In the two-step reaction

$$\mathrm{A} + \mathrm{B} \underset{k_{-1}}{\overset{k_1}{\rightleftharpoons}} \mathrm{C} \qquad \text{(Step 1)}$$

$$\mathrm{C} + \mathrm{D} \xrightarrow{k_2} \mathrm{E} \qquad \text{(Step 2)}$$

if the second step is slow compared to the first step, then a preequilibrium can be established and C will reach a **steady state** concentration:

$$\frac{d[\mathrm{C}]}{dt} = 0 = k_1[\mathrm{A}][\mathrm{B}] - k_{-1}[\mathrm{C}] - k_2[\mathrm{C}][\mathrm{D}]$$

If $k_2[\mathrm{C}][\mathrm{D}]$ is small so that it can be neglected,

$$[\mathrm{C}] = \frac{k_1[\mathrm{A}][\mathrm{B}]}{k_{-1}} = K[\mathrm{A}][\mathrm{B}]$$

where K is the equilibrium constant. Then

$$\frac{d[\mathrm{E}]}{dt} = k_2 K[\mathrm{A}][\mathrm{B}][\mathrm{D}]$$

9.970 General Acid or Base Catalysis

The enhancement of the rate of a reaction by the presence of any Brønsted acid (or base); each acid (base) present separately enhances the rate. The acid (or base) is unchanged by the overall reaction.

Example. If an aqueous reaction is catalyzed by both H_3O^+ and undissociated acids HA_i, the reaction is subject to general acid catalysis. For a reaction involving a two-step sequence,

$$\mathrm{Y} + \mathrm{HA}_i \overset{\text{slow}}{\rightleftharpoons} \mathrm{YH}^+ + \mathrm{A}_i^-$$

$$\mathrm{YH}^+ \xrightarrow{\text{fast}} \text{products}$$

where HA_i represents various acids, HA_1, HA_2 and so forth, the rate equation is

$$\text{rate} = \{k[\mathrm{H_3O^+}] + k_1[\mathrm{HA}_1] + k_2[\mathrm{HA}_2] + \cdots\}[\mathrm{Y}]$$

Each acid independently affects the overall rate. General base catalysis involves Brønsted bases in a similar fashion.

9.980 Specific Acid or Base Catalysis

The rate of a reaction is accelerated by one specific acid or base rather than by a group of acids (or bases). The acid or base is unchanged by the overall reaction.

Example. In a two-step, acid-catalyzed, aqueous reaction represented by

$$\mathrm{Y} + \mathrm{HA} \overset{\text{fast}}{\rightleftharpoons} \mathrm{YH}^+ + \mathrm{A}^-$$

$$\mathrm{YH}^+ \xrightarrow[k_2]{\text{slow}} \text{products}$$

the rate of the reaction is

$$\text{rate} = k_2 K \frac{[\text{Y}][\text{HA}]}{[\text{A}^-]}$$

If additional acids are present in the medium, the strongest acid present will always be H_3O^+ (see Sect. 9.640) and only its concentration appears in the above rate equation. Specific base catalysis involves a Brønsted base in a similar fashion. General acid catalysis can be experimentally distinguished from specific acid catalysis by observing the reaction rate as a function of buffer concentration. If the reaction rate increases at constant pH with an increase in acid concentration, such a reaction exhibits general acid catalysis.

9.990 Transition State Theory

A theory of rates of reaction which assumes that in an elementary reaction the reactants must pass through a transition state. The reactants are in equilibrium with an activated complex. For a bimolecular reaction

$$\text{A} + \text{B} \rightleftharpoons \text{AB}^{\ddagger} \longrightarrow \text{products}$$

where $AB^{\ddagger}$ is an activated complex at the transition state. The Eyring equation is developed from transition state theory.

9.1000 Eyring Equation (Henry Eyring, 1901-)

The equation

$$k_r = \frac{kT}{h} e^{-\Delta G^{\ddagger}/RT}$$

where k_r is a rate constant, k is the Boltzmann constant, h is Planck's constant, T is the absolute temperature, $\Delta G^{\ddagger}$ is the Gibbs free energy of activation, and R is the gas constant. The equation is used most frequently to determine free energies of activation and thus enthalpies ($\Delta H^{\ddagger}$) and entropies ($\Delta S^{\ddagger}$) of activation from rate data. For solution reactions at constant pressures

$$\Delta G^{\ddagger} = \Delta H^{\ddagger} - T\Delta S^{\ddagger} = E_a - RT - T\Delta S^{\ddagger}$$

where E_a is the Arrhenius energy of activation.

9.1010 Arrhenius Equation (Svante Arrhenius, 1859-1927)

The equation

$$k = Ae^{-E_a/RT}$$

where k is a rate constant, A is a constant known as the **preexponential factor**, E_a is the **energy of activation**, R is the gas constant, and T is the absolute temperature. The equation is used most frequently to determine energies of activation from rate data. According to the above equation a plot of ln k against $1/T$ gives a straight line whose slope is $-E_a/R$ and whose intercept is ln A (see Sect. 9.750).

9.1020 Kinetic Isotope Effect

A change of rate that occurs upon isotopic substitution, generally expressed as a ratio of rate constants, k_{light}/k_{heavy}. A **normal isotope effect** is one in which the ratio of k_{light} to k_{heavy} is greater than one; in an **inverse isotope effect,** the ratio is less than one. A **primary isotope effect** is one that results from the making or breaking of a bond to an isotopically substituted atom; this must occur in the rate-determining step. A **secondary isotope effect** is attributable to isotopic substitution of an atom not involved in bond making or breaking in the rate-determining step.

Examples. In the free radical bromination of toluene

$$C_6H_5CH_2Z + Br\cdot \longrightarrow C_6H_5CH_2\cdot + ZBr$$

where Z = H or D, k_H/k_D is 4.6. Primary isotope effects for k_H/k_D generally are between 2 and 7. In the solvolysis of isopropyl bromide

$$(CZ_3)_2CHBr + H_2O \longrightarrow (CZ_3)_2CHOH + HBr$$

where Z = H or D, k_H/k_D is 1.34. Secondary isotope effects for k_H/k_D are generally between 0.6 and 2.

9.1030 Migratory Aptitude

A term used to compare the relative rates with which different atoms or groups (usually one of two or possibly three similarly situated groups) migrate to another atom during the course of a reaction. Migratory aptitudes are a function of the type of reaction and the conditions of the reaction.

Examples. In a pinacol rearrangement of the type

$$R_2C(OH)(R_1)-C(OH)(R_1)R_2 \longrightarrow R_2-C(=O)-C(R_1)_2-R_2 \quad \text{or} \quad R_1-C(=O)-C(R_1)(R_2)-R_2$$

relative migratory aptitudes are *p*-methoxyphenyl, 500; *p*-tolyl, 15.7; *m*-tolyl, 1.95; *m*-methoxyphenyl, 1.6; phenyl, 1.0; *p*-chlorophenyl, 0.66. Electron releasing groups have a favorable influence on migratory aptitudes in this reaction.

9.1040 Substituent Effect

A change in the rate constant or equilibrium constant of a reaction caused by the replacement of a hydrogen atom by another atom or group of atoms. Such substituent effects result from the influence of the size of the substituent (steric effect) and/or its influence on the availability of electrons (electronic effect) on the reaction site. The electronic effect of the substituent may be either **electron releasing** or **electron withdrawing.** Such electronic effects can be subdivided into an inductive and a resonance (mesomeric) effect. The **inductive effect** depends on a substituent's intrinsic ability to supply or withdraw electrons, i.e., its electronegativity. This effect is transmitted through σ bonds or through space and weakens as the distance between the substituent and reactive center increases. The **resonance effect** involves delocalization of electrons through resonance via the π system (see Sect. 3.360).

Example. An electrophilic aromatic substitution reaction (see Sect. 10.590) on anisole (methoxybenzene) can lead to *ortho*, *meta*, and *para* monosubstitution, Fig. 9.1040*a*. The presence of the methoxy group influences the relative energies

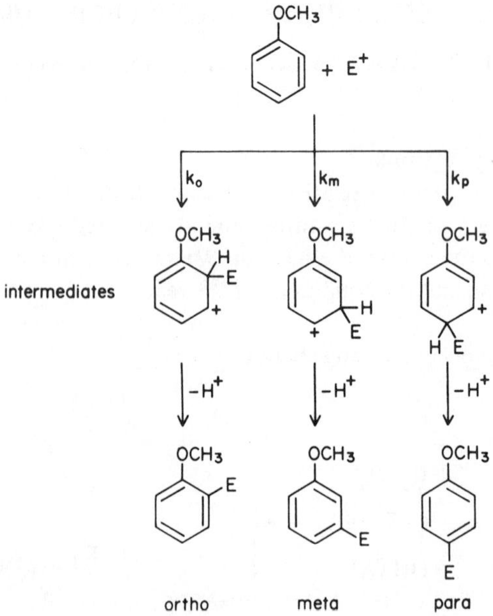

(*a*)

of the three possible transition states leading to the intermediates. In practice such energies are assessed by examining the structure of the intermediates (see Sect. 9.700). The methoxy substituent exerts an electron-releasing resonance effect (+*R* effect) and allows delocalization of the positive charge. In the *ortho*- and *para*-substituted intermediates, such charge can be delocalized over the substituent but this is not possible to any appreciable extent in the *meta* case, Fig. 9.1040*b*. The methoxy group, however, also exerts an electron-withdrawing in-

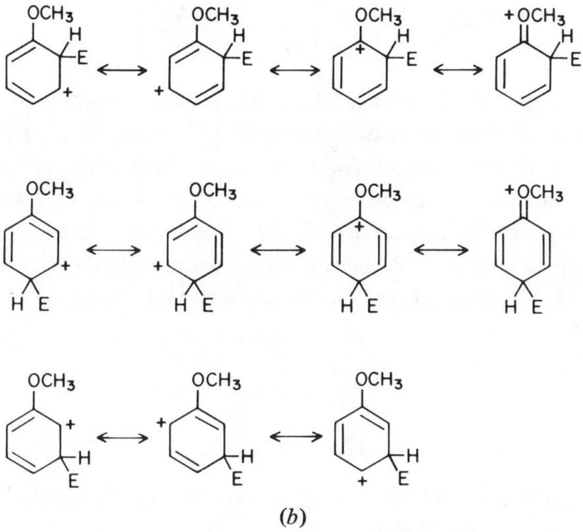

(*b*)

Figure 9.1040 (*a*) The electrophilic aromatic substitution of anisole; (*b*) resonance stabilization of the *ortho*, *para*, and *meta* intermediates.

ductive (–*I*) effect in all three intermediates. In the *ortho* and *para* cases, the result of the +*R* and –*I* effects of methoxy results in a net electron releasing effect, hence stabilization of these intermediates. In the *meta* case, the electron-withdrawing –*I* effect dominates and the intermediate has a considerably higher energy than the *ortho* and *para* intermediates. The substituent also creates a much more unfavorable steric effect in the *ortho* intermediate. The distribution of final products in the reaction is *para* > *ortho* >>>> *meta* ($k_p > k_o >>>> k_m$). Anisole undergoes electrophilic aromatic substitution 1000 times faster than benzene. The rate constants for individual positions in anisole can be expressed as **partial rate factors.**

9.1050 Partial Rate Factor

The rate of substitution at one particular position in a benzene derivative relative to the rate of similar substitution at *one* position in benzene.

Example. For monosubstituted benzenes ϕz the partial rate factors f for *para*, *meta*, and *ortho* positions are given by

$$f_p^{\phi z} = \left(\frac{k_{\phi z}/1}{k_{\phi \mathrm{H}}/6}\right) \cdot \left(\frac{\%\, para}{100}\right)$$

$$f_m^{\phi z} = \left(\frac{k_{\phi z}/2}{k_{\phi \mathrm{H}}/6}\right) \cdot \left(\frac{\%\, meta}{100}\right)$$

$$f_o^{\phi z} = \left(\frac{k_{\phi z}/2}{k_{\phi \mathrm{H}}/6}\right) \cdot \left(\frac{\%\, ortho}{100}\right)$$

For *para* substitution there is only one possible position available. Since there are six possible substitution sites in benzene, $k_{\phi \mathrm{H}}$ is divided by 6 in the $f_p^{\phi z}$ equation. For *meta* and *ortho* substitution two equivalent sites are available compared to 6 in benzene. Therefore $k_{\phi z}$ is divided by 2 and $k_{\phi \mathrm{H}}$ by 6 for these cases. Bromination of toluene proceeds 605 times faster than bromination of benzene. Three products are obtained: 66.8% *p*-bromotoluene; 0.3% *m*-bromotoluene; and 32.9% *o*-bromotoluene. The partial rate factors are

$$f_p^{\phi z} = 2425$$

$$f_m^{\phi z} = 5.4$$

$$f_o^{\phi z} = 597$$

If $f > 1$ for a given position, that position is activated compared to benzene; if $f < 1$, that position is deactivated.

9.1060 Hammett Equation (Louis P. Hammett, 1894-)

An equation of the form

$$\log \frac{k_z}{k_o} = \rho\sigma \qquad (9.1060a)$$

where k_z is the rate constant of a reaction for a species carrying a substituent, k_o is the rate constant for the reaction of the unsubstituted species, ρ (rho) is a number that characterizes the sensitivity of the reaction to substituent effects, and σ (sigma) is a constant expressing the electronic effects of the substituent. The value of σ is ascertained from the acid dissociation constants of benzoic acid K_o and a substituted benzoic acid K_z, from the following relationship

$$\sigma = \log \frac{K_z}{K_o} \qquad (9.1060b)$$

A positive value for σ indicates that the substituent is electron withdrawing ($K_z > K_o$), and a negative value indicates that it is electron releasing ($K_z < K_o$).

When log k_z/k_o is plotted against σ, according to equation 9.1060*a*, the result should be a straight line with a slope ρ. Reactions that are assisted by high electron density at the reaction site have negative ρ values, whereas reactions that are favored by withdrawal of electrons from the reaction site have positive ρ values. The ρ values for many types of equilibria can be ascertained by substituting equilibrium constants, K_o' and K_z', for rate constants in equation 9.1060*a*. The Hammett equation is subject to certain limitations and has been improved by various modifications of the choice of σ values.

Examples. Table 9.1060*a* gives some values of Hammett σ constants, and Table 9.1060*b* values of ρ for several equilibria and reaction rates.

Table 9.1060*a* Hammett substituent constants, σ

Substituent	*meta*	*para*
—OH	+0.12	-0.37
—OCH_3	+0.12	-0.27
—NH_2	-0.16	-0.66
—CH_3	-0.07	-0.17
—H	0.00	0.00
—Cl	+0.37	+0.23
—CF_3	+0.43	+0.54
—NO_2	+0.71	+0.78

Table 9.1060*b* Rho values for *meta*- and *para*-substituted benzene derivatives

	ρ
Equilibria	
$RC_6H_4COOH + H_2O \rightleftharpoons RC_6H_4COO^- + H_3O^+$	1.00
$RC_6H_4NH_3^+ + H_2O \rightleftharpoons RC_6H_4NH_2 + H_3O^+$	2.77
$RC_6H_4CHO + HCN \underset{}{\overset{EtOH}{\rightleftharpoons}} RC_6H_4CH(OH)CN$	-1.49
Rates	
$RC_6H_4COOC_2H_5 + {}^-OH \xrightarrow[30^\circ]{EtOH} RC_6H_4CO_2^- + C_2H_5OH$	2.43
$RC_6H_4COOH + CH_3OH \xrightarrow[25^\circ]{H^+} RC_6H_4CO_2CH_3 + H_2O$	-0.23

9.1070 Linear Free Energy Relationships
A mathematical expression of the form

$$-\Delta G_z^\circ = -\Delta G_o^\circ + 2.303RT\rho\sigma$$

which relates standard free energies of substituted (ΔG_z°) and unsubstituted (ΔG_o°) species with a constant ρ that is characteristic of an equilibrium's sensitivity to substituent effects and a constant σ that reflects the electronic effects of a specific substituent. Here σ and ρ have the same meaning as in the Hammett equation (see Sect. 9.1060). Standard free energies of activations, $\Delta G_z^\ddagger$ and $\Delta G_o^\ddagger$, may be substituted for standard free energies when the sensitivity of reaction rate constants, rather than equilibria, are determined. Since ΔG° is related to K by the equation

$$\Delta G^\circ = -RT \ln K$$

and $\Delta G^\ddagger$ is related to a rate constant k by the Eyring equation (see Sect. 9.1000), the linear free energy relationships are most frequently used in the form

$$\log\left(\frac{K_z}{K_o}\right) = \rho\sigma$$

and

$$\log\left(\frac{k_z}{k_o}\right) = \rho\sigma$$

Examples. The Hammett and Taft equations.

9.1080 Taft Equation (Robert Taft, 1922-)
A linear free energy relationship

$$\log\left(\frac{k_z}{k_o}\right) = \sigma_I \rho_I$$

σ_I is designed to more accurately take into account inductive and/or field effects of substituents, and its values are obtained from the equation

$$\sigma_I = 0.262 \log\left(\frac{K_z}{K_o}\right)$$

where K_z and K_o are acid dissociation constants of substituted acetic acids and acetic acid, respectively.

9.1090 Catalysis
The alteration of the rate of a reaction achieved by adding to it, in much less than stoichiometric quantities, a substance (the catalyst) that can (theoretically)

be quantitatively recovered. Catalysts are almost always employed to enhance the rate of reactions by lowering the energy of activation from that which prevails in the absence of a catalyst. The catalyst changes the mechanistic pathway.

9.1100 Mechanism of a Reaction

A complete detailed description of a reaction which includes characterization of all intermediates and transition states with respect to their composition, structure (including geometry), and relative free energies. The mechanism must be consistent with all experimental data. Studies of reaction rates and stereochemistry are two of the powerful tools used to acquire data. Experimental evidence can easily disprove a particular mechanism but only rarely can it prove one. Equations including detailed geometries, electron-pushing arrows to represent bond making and breaking processes, and energy diagrams are the main tools used by the organic chemist to describe mechanisms.

10 Types of Organic Reaction Mechanisms

CONTENTS

One of the greatest advances in modern organic chemistry has been the discovery that, although several million organic compounds have been prepared using thousands of "different" reactions, the microscopic details of these reactions (i.e., their mechanisms, Sect. 9.1100) can be viewed as slight variations of a very few basic processes. The recognition and exploitation of these patterns is, to many organic chemists, the most important aspect of the field. In this chapter we briefly describe these basic mechanisms, but we have had to forego a description of the methods used to investigate reaction mechanisms.

10.010 Electron-Pushing

A formalism used to picture the movement of electrons. Movement of single electrons is shown with a curved fishhook arrow ($\curvearrowright$); movement of electron pairs is shown with a normal curved arrow ($\curvearrowright$). Care must be taken to avoid attaching too much literal significance to the formalism. The Uncertainty Principle prevents one from knowing the exact location of the electrons, but the electron-pushing formalism allows one to rationalize (i.e., approximate) their movement.

Electron-pushing is used extensively in this and succeeding chapters to follow electronic movements associated with chemical reactions. The technique is also useful in showing the relationship between resonance structures (Sect. 3.360).

In depicting the movement of electrons, it is important to remember that the number of electrons around an atom can never exceed two for H or eight for B, C, N, O, or F.

Examples. Resonance structures for a variety of unsaturated species are shown in Fig. 10.010.

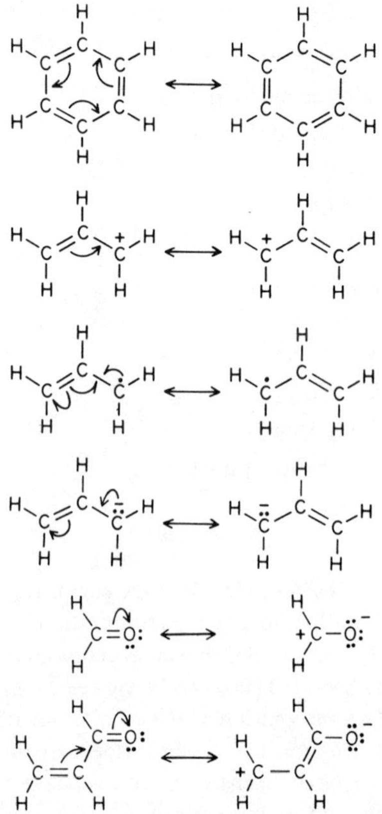

Figure 10.010 Resonance structures of various chemical species.

10.020 Homolytic Cleavage

The breakage of a single (two electron) bond, leaving one of the electrons on each of the fragments.

Examples. Figure 10.020.

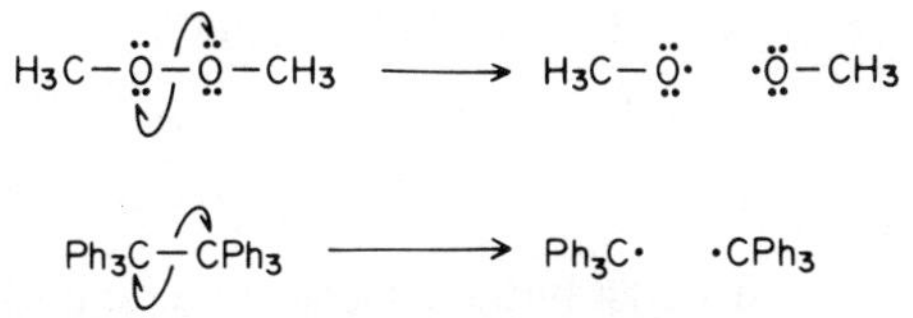

$$H_3C-\ddot{O}-\ddot{O}-CH_3 \longrightarrow H_3C-\ddot{O}\cdot \quad \cdot\ddot{O}-CH_3$$

$$Ph_3C-CPh_3 \longrightarrow Ph_3C\cdot \quad \cdot CPh_3$$

Figure 10.020 Homolytic cleavage.

10.030 Heterolytic Cleavage

The breakage of a single (two electron) bond, leaving both the electrons on one of the fragments.

Examples. Figure 10.030.

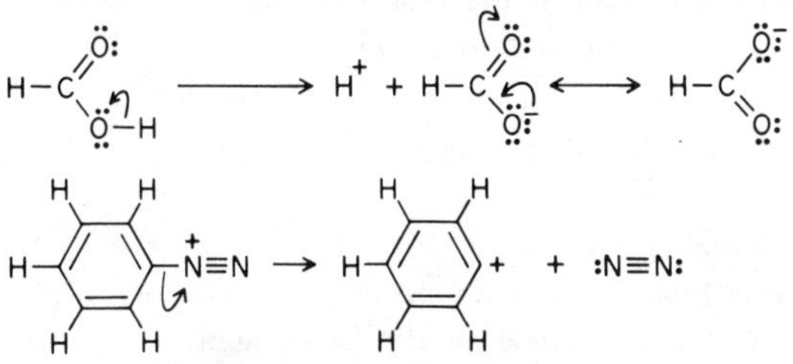

$$HCOOH \longrightarrow H^+ + HCOO^- \longleftrightarrow HCOO^-$$

$$C_6H_5-\overset{+}{N}\equiv N \rightarrow C_6H_5^+ + :N\equiv N:$$

Figure 10.030 Heterolytic cleavage.

10.040 Elementary Reaction (Elementary Process; Step)

A reaction involving only one transition state and no intermediates. For such a reaction the stoichiometric coefficients, molecularity, and kinetic order are all identical.

10.050 Concerted Reaction

A reaction in which all bond-making and bond-breaking occurs simultaneously (in the same step).

10.060 Nucleophile (Literally, Nucleus-Loving)

An electron pair donor, i.e., a **Lewis base.**

Examples. $R:\ddot{O}:^-$, $H-\ddot{O}-R$, $:\ddot{Cl}:^-$, $R_3N:$, the π electron pair of a multiple bond.

10.070 Nucleophilicity

The relative (kinetic) reactivity of a nucleophile. In general, nucleophilicity increases with atomic number as one goes down a column in the periodic table. For nucleophiles with the same attacking atom, nucleophilicity increases with basicity and decreases with increasing solvation. Although any relative order of nucleophilicity depends on conditions and substrate, a typical order is: $RS^- > R_3P > I^- > CN^- > R_3N > RO^- > Br^- > PhO^- > Cl^- > RCO_2^- > F^- > CH_3OH > H_2O$

10.080 Leaving Group

The group of atoms that departs during a substitution or displacement reaction. The group can be charged or uncharged.

10.090 Nucleofuge

The leaving group in a nucleophilic substitution or displacement reaction that departs *with* the electron pair that originally bonded the group to the remainder of the molecule.

10.100 Electrophile (Literally, Electron-Loving)

An electron *pair* acceptor, i.e., a **Lewis acid.**

Examples. H^+, R_3B, $AlCl_3$, Ag^+, $:\ddot{\underset{..}{Br}}^+$.

10.110 Electrophilicity

The relative (kinetic) reactivity of an electrophile. The electrophilicity of a series of electrophiles generally increases as the Lewis acidity increases and as the solvation (or coordination to other molecules) decreases.

10.120 Electrofuge

The leaving group in an electrophilic substitution reaction that departs *without* the electron pair that originally bonded the group to the remainder of the molecule.

10.130 Reagent

The compound that serves as the source of the molecule, ion, or free radical that is arbitrarily regarded as the attacking species in a reaction.

10.140 Substrate

The compound that is regarded as being attacked by the reagent.

Example. Figure 10.140.

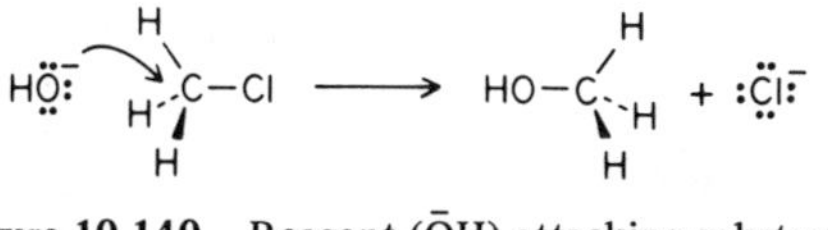

Figure 10.140 Reagent ($\bar{O}H$) attacking substrate.

10.150 Substitution

A reaction in which an attacking species of one type (nucleophile, electrophile, or free radical) replaces another group of the *same type.*

10.160 S_N2 (Substitution Nucleophilic Bimolecular)

The concerted displacement of one nucleophile by another. The site of substitution is usually sp^3 hybridized. This mechanism involves stereospecific backside approach by the attacking nucleophile relative to the nucleofuge, causing inversion of configuration at the reaction site. The reaction is kinetically first order in both attacking nucleophile and substrate, and thus second order overall. The S_N2 reaction is highly sensitive to steric effects, and thus the order of substrate reactivity is methyl > primary > secondary >> tertiary.

Example. Figure 10.160*a* shows a typical example, together with its potential energy profile, Fig. 10.160*b*.

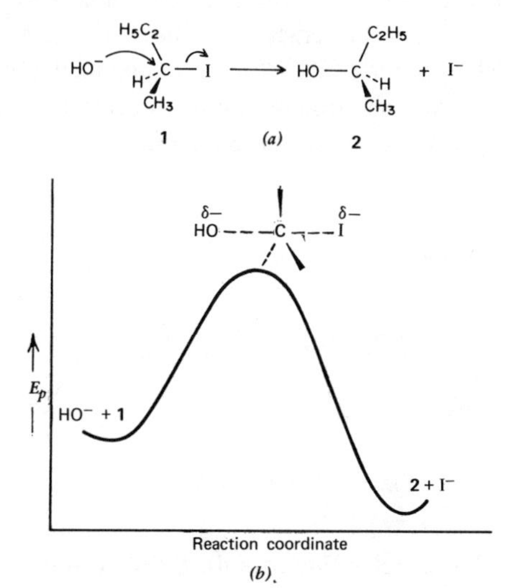

Figure 10.160 (*a*) An S_N2 reaction; (*b*) its potential energy profile.

10.170 Ambident (Literally, Two-sided Tooth)

A reagent with two or more possible attacking sites, or a substrate with two or more possible sites of attack.

Examples. Figure 10.170; arrows indicate reactive sites.

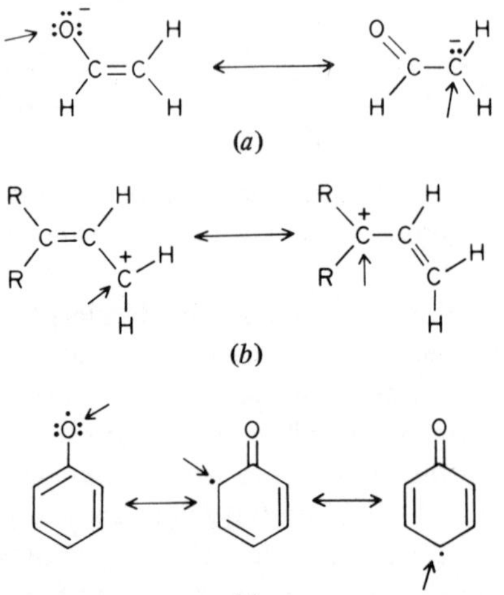

(*a*)

(*b*)

(*c*)

Figure 10.170 Ambident species: (*a*) nucleophile; (*b*) electrophile; (*c*) radical.

10.180 S_N2' (Substitution Nucleophilic Bimolecular with Rearrangement)

A concerted nucleophilic displacement in which the site of attack is at an atom other than the original point of attachment of the nucleofuge (usually one multiple bond separated from the original point of attachment). This mechanism is kinetically indistinguishable from the S_N2 reaction, but the attack/departure stereochemistry is *cis.*

Example. Figure 10.180.

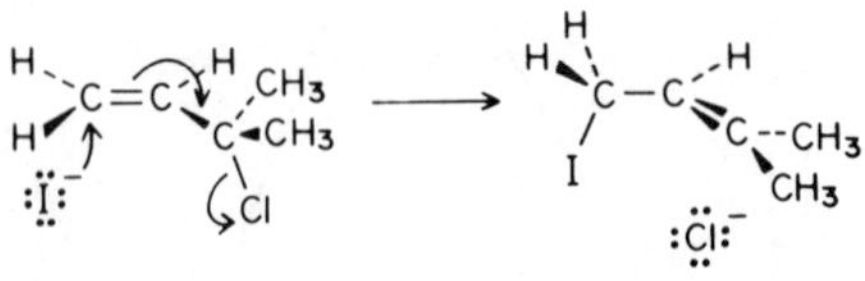

Figure 10.180 An S_N2' reaction.

10.190 Stepwise Bimolecular Nucleophilic Substitution

A multistep nucleophilic interchange in which bond-making precedes bond-breaking. This mechanism occurs at unsaturated carbon (see nucleophilic substitution at C=O, Sect. 10.740) or at saturated third row elements (Si, P, S), but it is unknown at saturated carbon. This mechanism is kinetically indistinguishable from the S_N2 mechanism under conditions where the intermediate is highly reactive. The stereochemistry of the reaction depends on the details of how the nucleofuge leaves the intermediate.

Examples. Substitution at carbon is shown in Fig. 10.190*a*; Fig. 10.190*b* substitution at phosphorus, and Fig. 10.190*c* the potential energy profile for substitution at phosphorus.

(*a*)

(*b*)

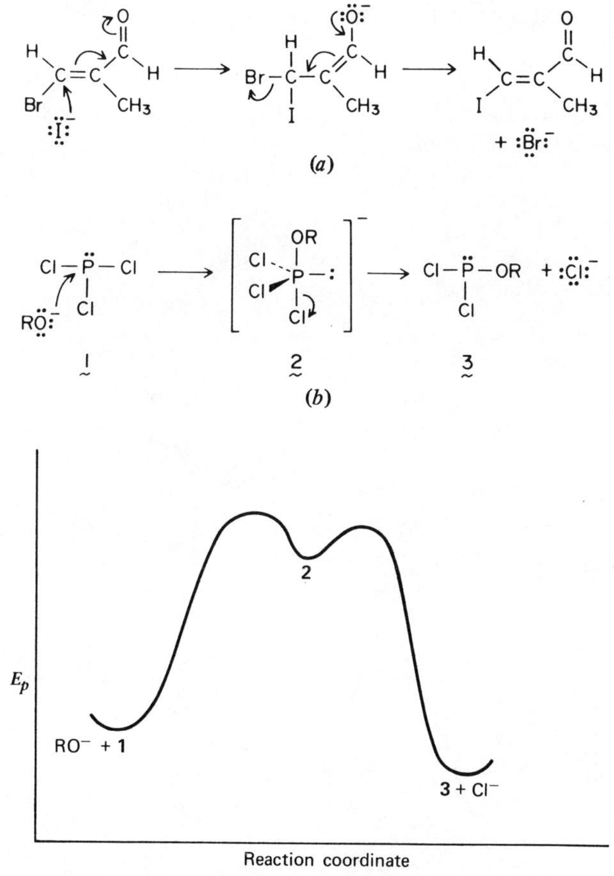

(*c*)

Figure 10.190 (*a*) Substitution at carbon; (*b*) substitution at phosphorus; (*c*) the potential energy profile for *b*.

10.200 S_N1 (Substitution Nucleophilic Unimolecular)

A multistep nucleophilic interchange with bond-breaking (the **rate-limiting step**) preceding bond-making. The reaction is first order in substrate, but zero order in attacking nucleophile:

$$\mathrm{RX} \xrightarrow{\text{slow}} \mathrm{R}^+ + \mathrm{X}^- \xrightarrow[\mathrm{Y}^-]{\text{fast}} \mathrm{RY}$$

The intermediate (R^+) is a carbocation; attack by Y^- can usually occur from either side of the carbocation. Only substrates that can lead to relatively stable carbocations react by this mechanism, the relative reactivity being tertiary (or resonance stabilized ions such as benzyl or allyl) >> secondary >> primary > methyl.

Example. A typical example is shown in Fig. 10.200*a* together with its potential energy profile, Fig. 10.200*b*.

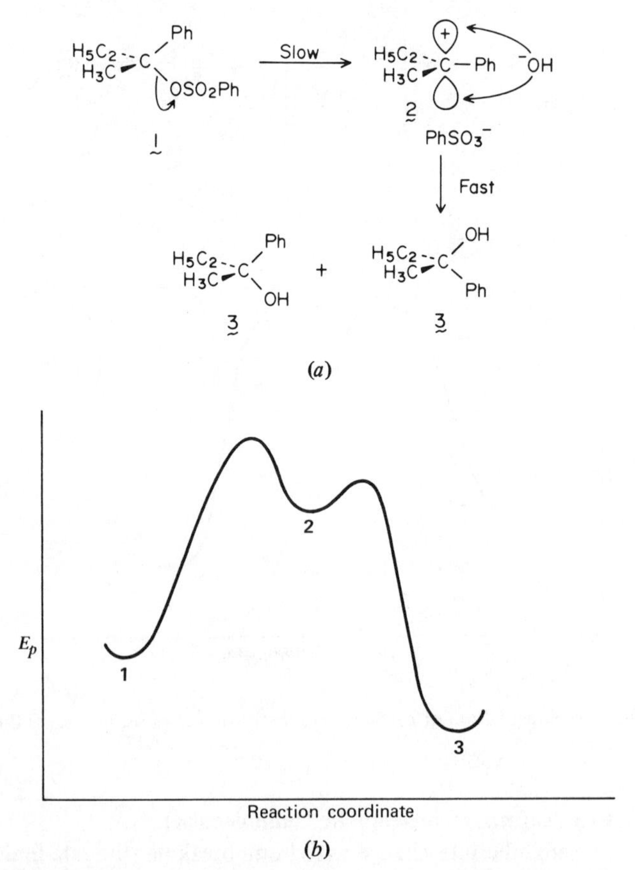

Figure 10.200 (*a*) An S_N1 reaction; (*b*) its potential energy profile.

10.210 S_N1' (Substitution Nucleophilic Unimolecular with Rearrangement)

Kinetically and stereochemically similar to the S_N1 mechanism, except that the attacking nucleophile becomes attached to a different atom than the one originally attached to the nucleofuge. Either a carbenium ion rearrangement or an allylic ion rearrangement is often involved in this mechanism.

Example. Figure 10.210.

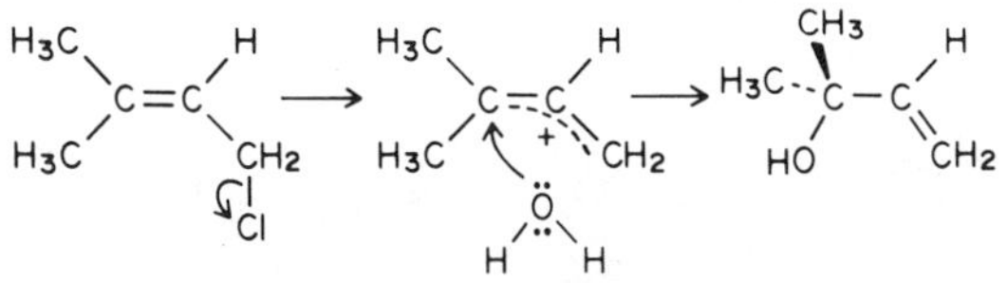

$$(H_3C)_2C{=}CH{-}CH_2Cl \longrightarrow (H_3C)_2C{\overset{+}{=}}CH{=}CH_2 \;(+\; H_2\ddot{O}) \longrightarrow H_3C{-}C(CH_3)(HO){-}CH{=}CH_2$$

Figure 10.210 An S_N1' reaction.

10.220 Solvolysis (Literally, a Cleavage (Lysis) with Solvent)

A nucleophilic substitution in which the solvent serves as the attacking reagent. Even if the reaction is bimolecular (e.g., S_N2), its kinetic order in solvent is indistinguishable from zero (see Sect. 9.900).

Example. Figure 10.220.

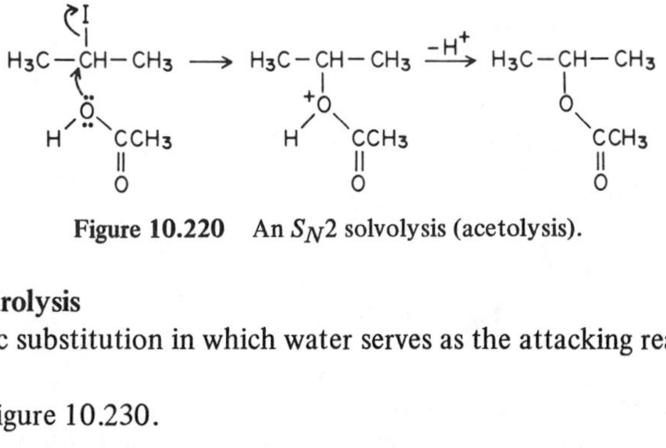

$$H_3C{-}CH(I){-}CH_3 + H{-}\ddot{O}{-}C(=O)CH_3 \longrightarrow H_3C{-}CH(\overset{+}{O}(H)C(=O)CH_3){-}CH_3 \xrightarrow{-H^+} H_3C{-}CH(OC(=O)CH_3){-}CH_3$$

Figure 10.220 An S_N2 solvolysis (acetolysis).

10.230 Hydrolysis

A nucleophilic substitution in which water serves as the attacking reagent.

Example. Figure 10.230.

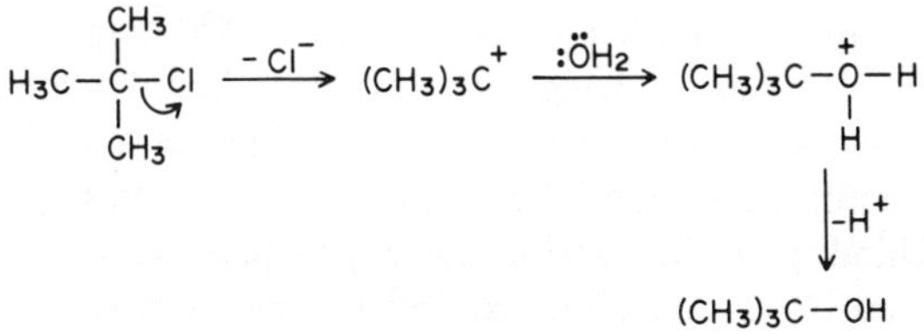

$$H_3C{-}C(CH_3)_2{-}Cl \xrightarrow{-Cl^-} (CH_3)_3C^+ \xrightarrow{:\ddot{O}H_2} (CH_3)_3C{-}\overset{+}{O}H{-}H \xrightarrow{-H^+} (CH_3)_3C{-}OH$$

Figure 10.230 Hydrolysis by an S_N1 mechanism.

10.240 $S_N i$ (Substitution Nucleophilic Internal)

An intramolecular (unimolecular) nucleophilic interchange in which the attacking nucleophile is part of the substrate. The reaction resembles an S_N1 process, except that migration of the nucleophile occurs with retention of configuration at the reaction site. Ion pairs are often implicated in this type of mechanism. The *concerted* "$S_N i$" is generally described as a sigmatropic shift (Sect. 10.1040).

Example. Figure 10.240.

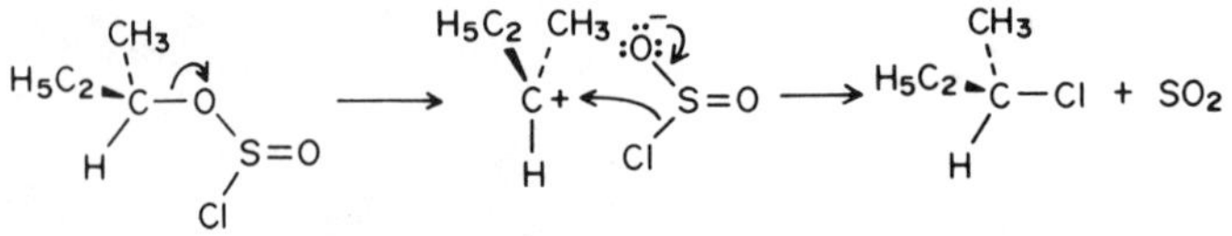

Figure 10.240 An $S_N i$ reaction.

10.250 $S_N i'$ (Substitution Nucleophilic Internal with Rearrangement)

Similar to the $S_N i$ mechanism, except that the internal nucleophile becomes attached to a different atom (usually two atoms removed) from the original point of attachment. The stereochemical path involves a *cis* departure/attack relationship:

Example. Figure 10.250.

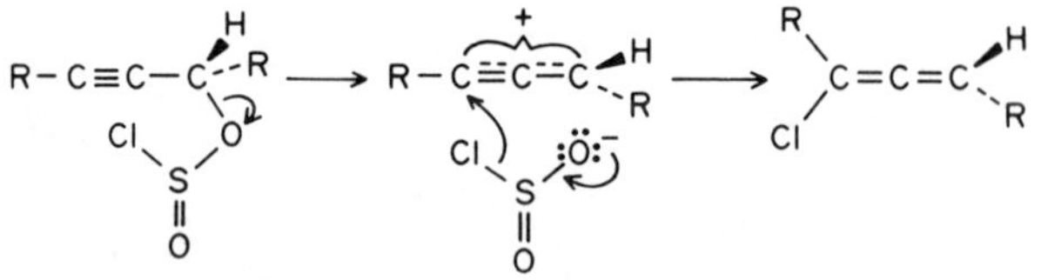

Figure 10.250 An $S_N i'$ reaction.

10.260 Neighboring Group Participation

The intramolecular involvement of one functional group (with its n, π, or σ electrons) in the reaction at another functional group. The term is most commonly encountered in reactions involving carbocations, where the neighboring group shares a pair of electrons with the positively charged carbon. Such interactions can take place before, during, or after the rate-determining step, and can involve retention of chirality in the unrearranged products. Rearranged products, if formed as a result of participation, can involve inversion at both the carbenium carbon and the neighboring group-bearing atom.

Examples. Figure 10.260*a*, *b* and *c* shows *n*, π, and σ electron participation, respectively.

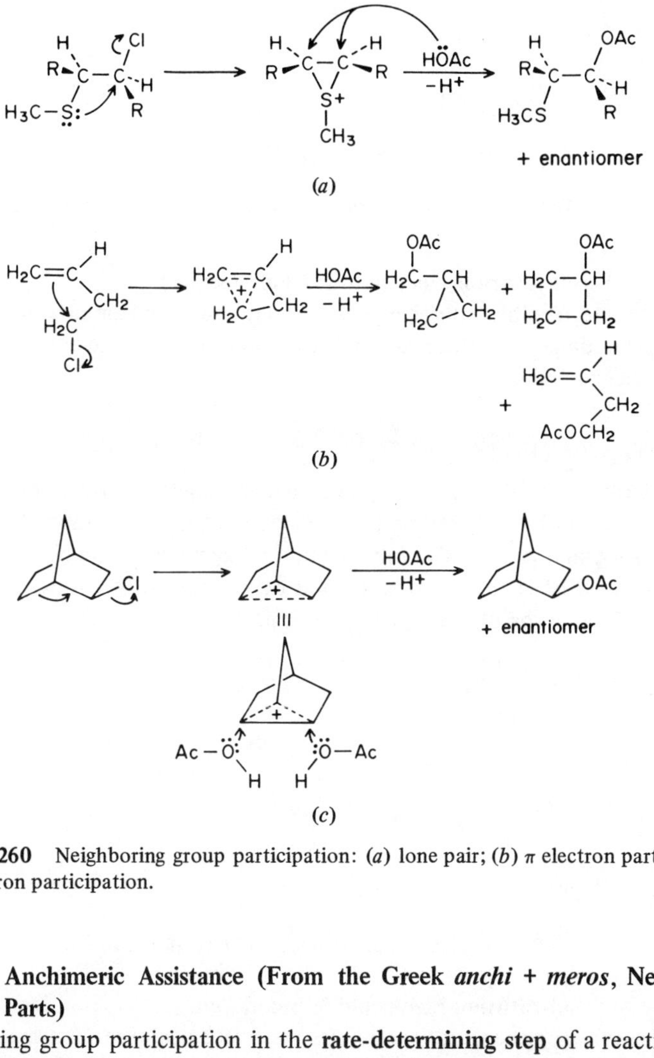

(*a*)

(*b*)

(*c*)

Figure 10.260 Neighboring group participation: (*a*) lone pair; (*b*) π electron participation; (*c*) σ electron participation.

10.270 Anchimeric Assistance (From the Greek *anchi* + *meros*, Neighboring Parts)

Neighboring group participation in the **rate-determining step** of a reaction. This term is most often encountered in reactions involving carbocation intermediates, where **neighboring group participation** occurs in the ionization step that generates this ion. Such assistance causes an increase in the reaction rate compared to a model reaction in which participation is absent.

Example. Figure 10.270.

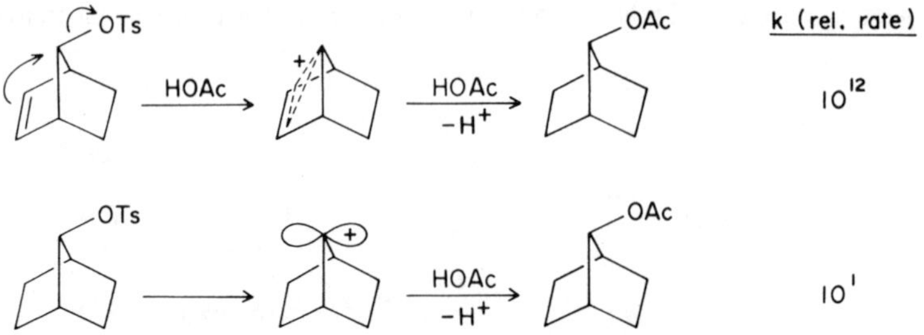

Figure 10.270 Anchimeric assistance and its effect on reaction rate.

10.280 S_E1 (Substitution Electrophilic Unimolecular)

A multistep electrophile interchange with bond-breaking (the rate-limiting step) proceding bond-making. The reaction is first order in substrate and zero order in attacking electrophile:

$$R{-}E \xrightarrow{\text{slow}} R^- + E^+ \xrightarrow{X^+} R{-}X$$

The intermediate (R^-) is a carbanion, which generally inverts more rapidly that it is attacked by X^+, leading to racemization (or epimerization) if the reaction site is a chiral center. The most common examples of the S_E1 mechanism involve removal of H^+ attached to carbon by reaction with strong base. S_E1 is the electrophile analog of the S_N1 mechanism.

Example. Figure 10.280.

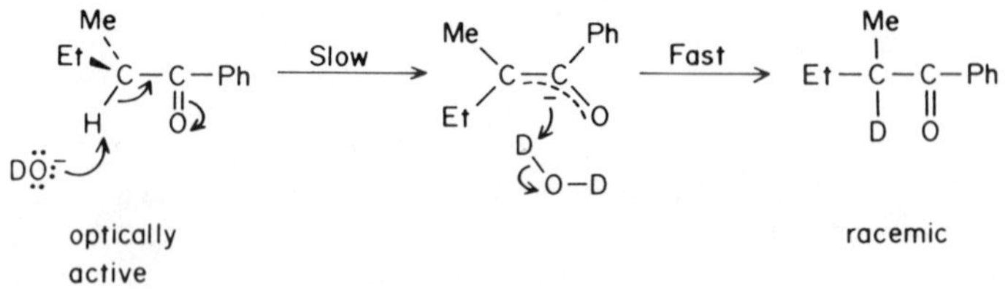

Figure 10.280 An S_E1 reaction (racemization).

10.290 S_E2 (Substitution Electrophilic Bimolecular)

The concerted displacement of one electrophile by another. This mechanism, which is relatively uncommon, involves stereospecific frontside approach by the attacking electrophile relative to the electrofuge, causing retention of configuration. The reaction is kinetically first order in both substrate and attacking electrophile and thus second order overall. S_E2 is the electrophile analog of the S_N2 mechanism.

Examples. Figure 10.290.

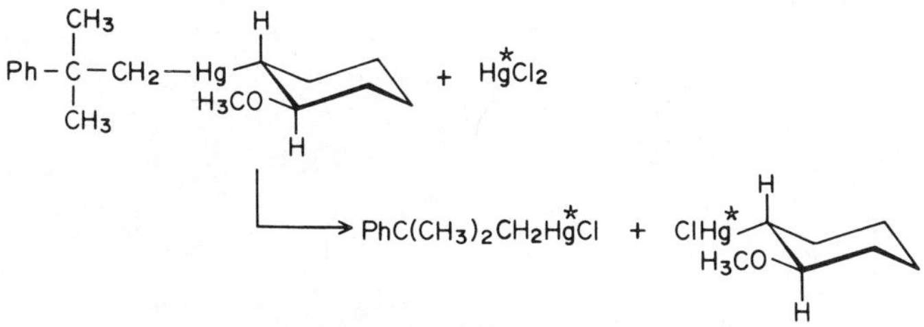

Figure 10.290 An S_E2 reaction (retention).

10.300 S_E2' (Substitution Electrophilic Bimolecular with Rearrangement)

A mechanism kinetically identical to the S_E2 process, but involving formation of a rearranged product.

Example. Figure 10.300.

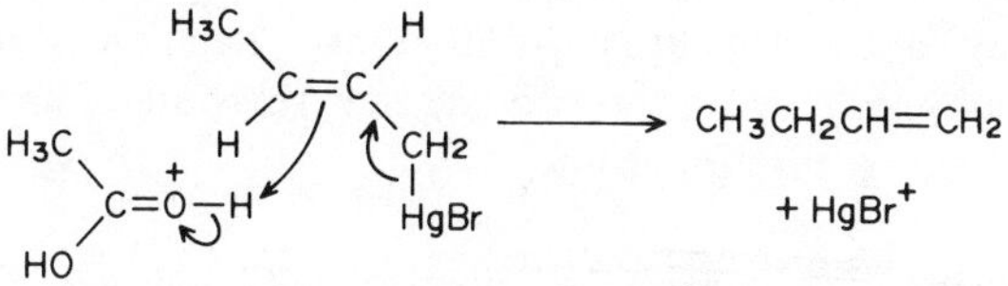

Figure 10.300 An S_E2' reaction (rearrangement).

10.310 S_Ei (Substitution Electrophilic Internal)

A concerted four-center electrophilic interchange, kinetically and stereochemically indistinguishable from the S_E2, except that at the transition state the nucleophile associated with the attacking electrophile becomes coordinated to the electrofuge in a cyclic transition state.

Examples. Figure 10.310.

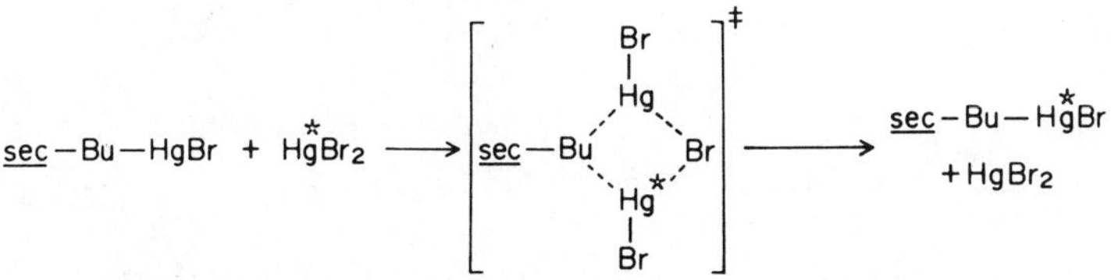

Figure 10.310 An S_Ei reaction (internal; no rearrangement).

10.320 S_Ei' (Substitution Electrophilic Internal with Rearrangement)

A concerted electrophile interchange, kinetically identical to the S_Ei, but involving a multicenter transition state and formation of rearranged product.

Example. Figure 10.320.

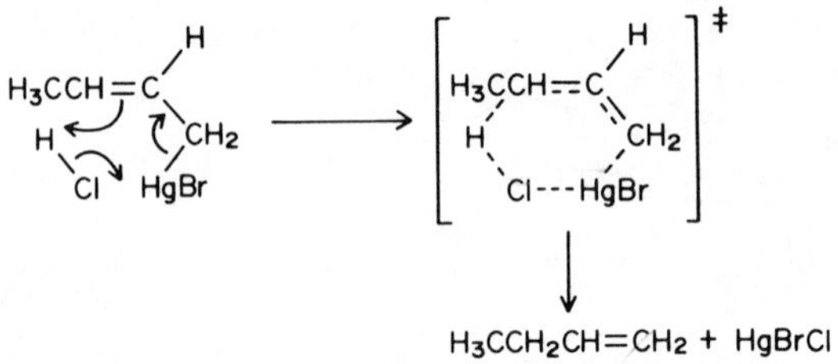

$$H_3CCH{=}C(H)CH_2HgBr + HCl \longrightarrow [H_3CCH{=}C(H)CH_2 \cdots H \cdots Cl \cdots HgBr]^{\ddagger} \longrightarrow H_3CCH_2CH{=}CH_2 + HgBrCl$$

Figure 10.320 An S_Ei' reaction (rearrangement).

10.330 Elimination

A reaction that involves a loss of two groups or atoms from the substrate, and therefore an increase in the degree of unsaturation of the substrate.

10.340 β (or 1,2) Elimination

An elimination reaction in which the two leaving groups (usually an electrofuge and a nucleofuge) are attached to adjacent atoms. Their departure causes formation of a new π bond between the two adjacent substrate atoms originally connected to the leaving groups (or atoms).

Example. Figure 10.340 shows a β elimination that involves a dehydrohalogenation.

$$Br{-}C(CH_3)_2{-}CH_2{-}H + NaOH \longrightarrow (H_3C)_2C{=}CH_2 + HOH + NaBr$$

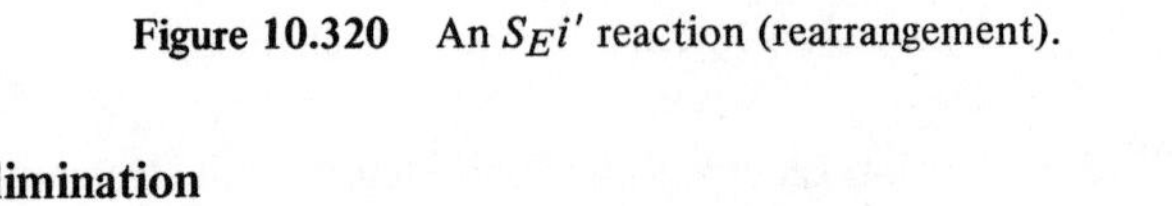

Figure 10.340 A β elimination to dehydrohalogenated product.

10.350 Saytzeff (Zaitsev) Elimination (A. M. Saytzeff, 1841-1910)

Elimination that results in the most substituted alkene product.

Example. Figure 10.350 shows a dehydrohalogenation that characteristically leads to predominant Saytzeff orientation.

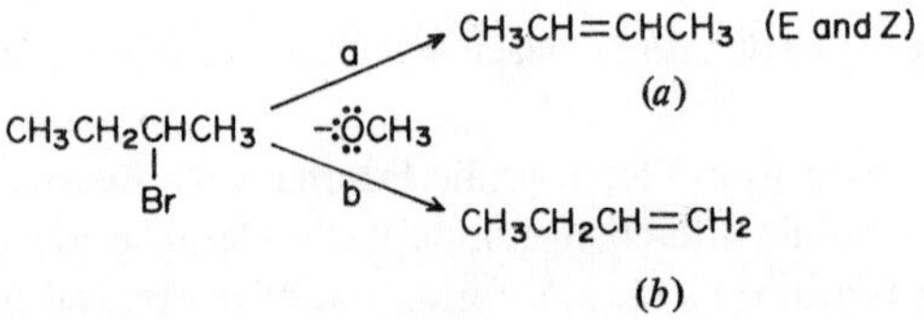

$$CH_3CH_2CH(Br)CH_3 \xrightarrow{^-:\ddot{O}CH_3} \begin{cases} \text{a: } CH_3CH{=}CHCH_3 \text{ (E and Z)} & (a) \\ \text{b: } CH_3CH_2CH{=}CH_2 & (b) \end{cases}$$

Figure 10.350 A β elimination: (*a*) to Saytzeff product; (*b*) to Hofmann product.

10.360 Hofmann Orientation (A. W. Hofmann, 1818-1892)

Elimination that results in the least substituted alkene product.

Example. The β elimination achieved by pyrolysis of onium hydroxides, Fig. 10.360*a* and *b*, characteristically leads to almost exclusive Hofmann orientation.

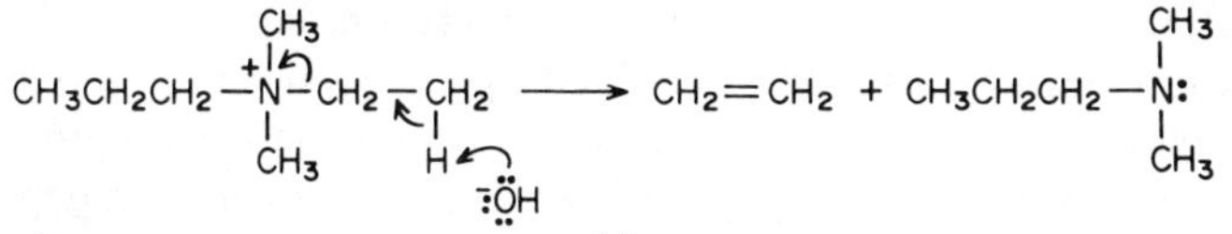

$$CH_3CH_2CH_2-\overset{+}{N}(CH_3)_2-CH_2-CH_2-H \;\; {}^{-}\!:\!OH \longrightarrow CH_2{=}CH_2 + CH_3CH_2CH_2-N(CH_3)_2\!:$$

(*a*)

$$CH_3CH(H)-CH_2-\overset{+}{N}(CH_3)_2-CH_2CH_3 \;\; HO\!:^{-} \longrightarrow CH_3CH{=}CH_2 + CH_3CH_2-N(CH_3)_2\!:$$

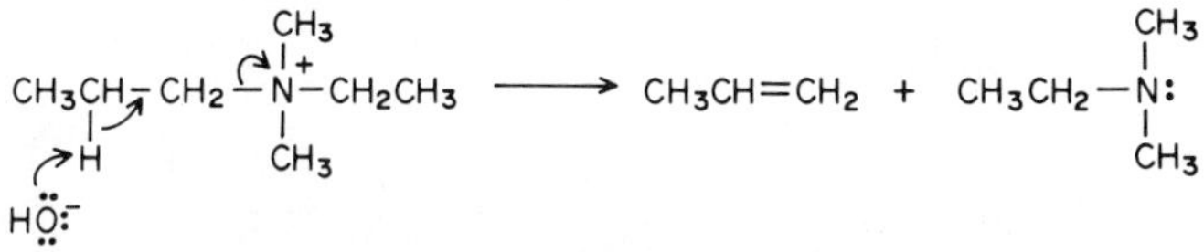

(*b*)

Figure 10.360 A Hofmann elimination: (*a*) the least substituted olefin (Hofmann orientation); (*b*) the most substituted olefin (Satyzeff orientation) not formed.

10.370 Bredt's Rule (J. Bredt, 1855-1937)

When β eliminations in bicyclic systems occur, the bridgehead atoms are not involved unless the ring bearing the incipient "trans" double bond has at least eight atoms, (i.e., when $n + m > 4$, Fig. 10.370).

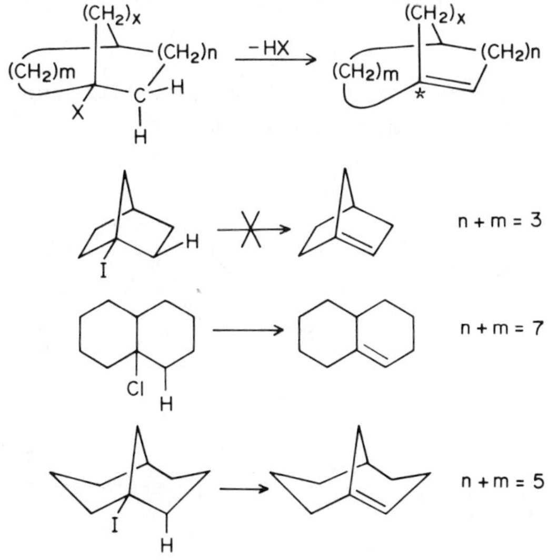

Figure 10.370 Application of Bredt's rule.

Examples. Figure 10.370. When $n + m > 4$, the bridgehead carbon can be involved, but bridgehead double bonds in molecules with four or fewer total carbon atoms between the bridgehead atoms ($n + m \leqslant 4$) are too strained and reactive to survive isolation.

10.380 *trans* (or *anti*) Elimination

A description of the stereochemistry of β elimination, in which the two leaving groups depart from opposite faces of the incipient double bond. Other than the stereochemical result, no other mechanistic information is necessarily implied. Note that *trans* elimination may give *cis* or *trans* (*Z* or *E*) product depending on the conformer (or stereoisomer) of the starting material.

Example. Figure 10.380.

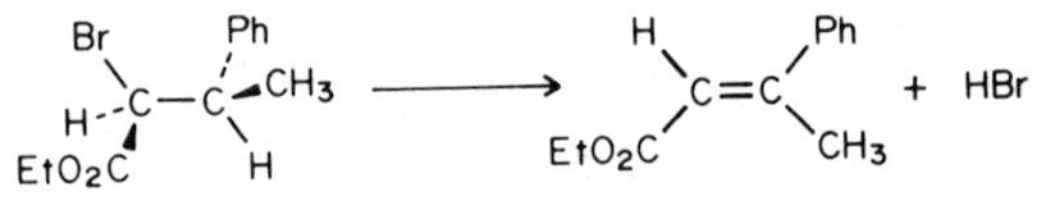

Figure 10.380 A *trans* elimination.

10.390 *cis* (or *syn*) Elimination

The stereochemistry of β elimination in which the two leaving groups depart from the same face of the incipient double bond. Note that *cis* elimination may give *cis* or *trans* (*Z* or *E*) product depending on the conformer (or stereoisomer) of the starting material.

Examples. Figure 10.390*a* and *b*; part *b* shows acetate pyrolysis leading to *Z* isomer but the *E* isomer is formed as well in this case.

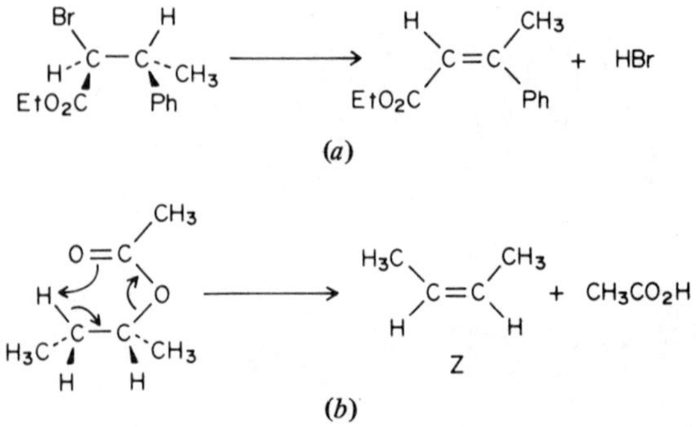

(*a*)

(*b*)

Figure 10.390 A *cis* elimination: (*a*) dehydrohalogenation; (*b*) acetate pyrolysis.

10.400 *E*1 (Elimination Unimolecular)

A multistep β elimination mechanism in which the nucleofuge is lost in the first and rate-limiting step and the electrofuge is lost in the second step. The intermediate is a carbocation, and this mechanism often competes with S_N1 processes. Saytzeff orientation is generally preferred. The stereochemistry of the elimination depends on the exact reaction conditions; often competitive *cis* and *trans* elimination is observed. Since the first step is rate-limiting, the elimination/substitution ratio ($E1/S_N1$) is expected to be independent of the nature of the leaving group.

Example. Figures 10.400*a*, *b* and *c* show three possible products from a common intermediate.

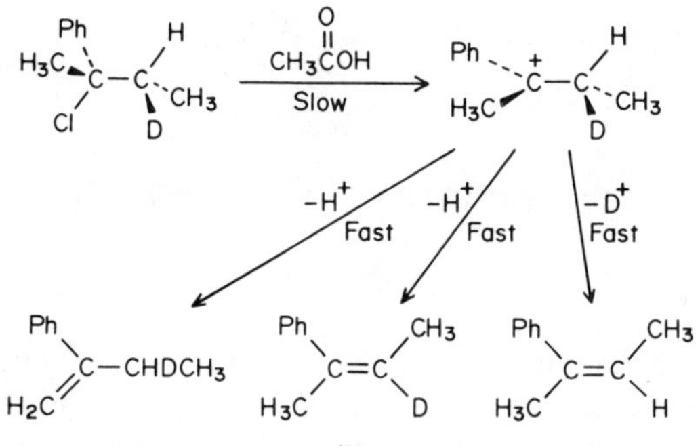

Figure 10.400 An *E*1 elimination: (*a*) Hofmann product; (*b*) Saytzeff elimination (*trans*); (*c*) Saytzeff elimination (*cis*).

10.410 *E*2 (Elimination Bimolecular)

A concerted elimination that is first order in both substrate and base (nucleophile). *trans* Elimination is generally favored. Because the attacking base is sensitive to the steric environment around the electrophile, Hofmann orientation is generally preferred. The *E*2 mechanism often competes with the S_N2 reactions.

Example. Figure 10.410.

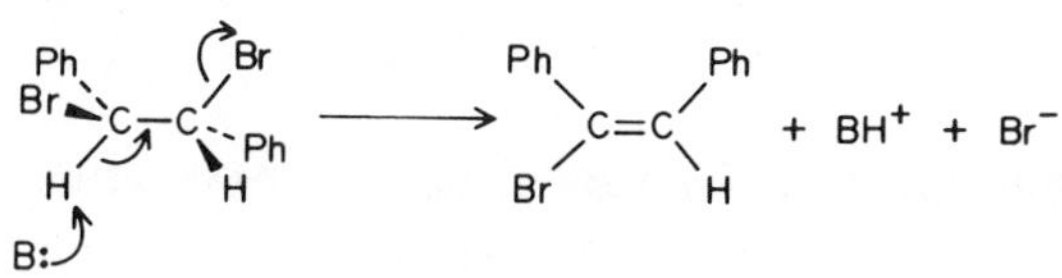

Figure 10.410 An *E*2 reaction.

10.420 E_1cb (Elimination Unimolecular Conjugate Base)

A multistep elimination mechanism in which loss of the electrofuge (usually H^+) occurs before loss of the nucleofuge. This mechanism involves a reversibly formed carbanion intermediate whose formation requires the presence of an electron withdrawing substituent on the carbanion carbon. Usually the reaction is first order in both substrate and base (nucleophile), hence second order overall (in contrast to the title). Hofmann orientation is generally preferred; the elimination stereochemistry may be *cis*, *trans*, or a combination.

Example. Figure 10.420.

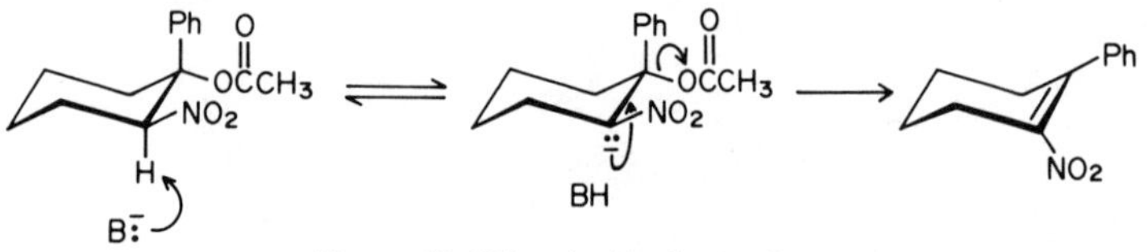

Ph O OCCH3 NO2 H B: ⇌ Ph O OCCH3 NO2 BH → Ph NO2

Figure 10.420 An E_1cb reaction.

10.430 γ (or 1,3) Elimination

An elimination reaction in which the two leaving groups are originally on atoms separated by another atom. The loss of the two groups causes formation of a new σ bond between the two atoms originally connected to the leaving groups, forming a three-membered ring. The E_1cb mechanism is generally required.

Example. Figure 10.430.

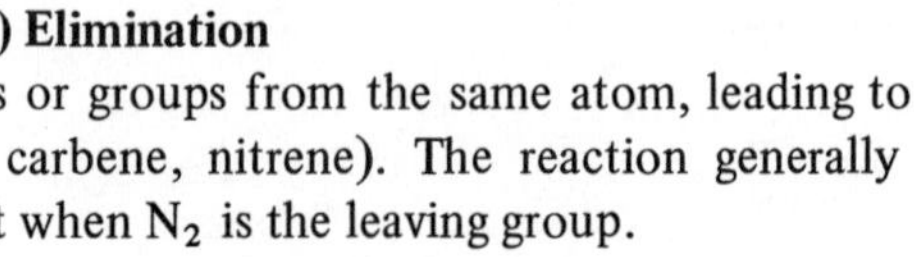

H O Cl $^-$:ÖCH3 ⇌ O Cl → O

Figure 10.430 A γ elimination.

10.440 α (or 1,1) Elimination

Loss of two atoms or groups from the same atom, leading to a **hypovalent** neutral species (e.g., carbene, nitrene). The reaction generally follows an E_1cb mechanism, except when N_2 is the leaving group.

Examples. Figure 10.440*a*. There are certain rare cases where loss of the nucleofuge precedes loss of the electrofuge, Fig. 10.440*b*.

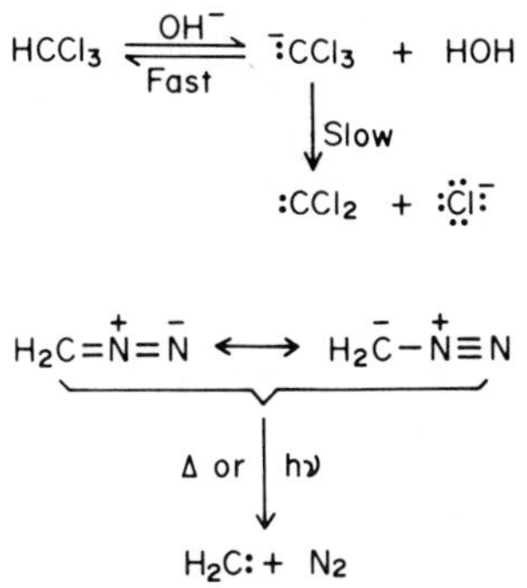

$$HCCl_3 \underset{Fast}{\overset{OH^-}{\rightleftharpoons}} \bar{:}CCl_3 + HOH \xrightarrow{Slow} :CCl_2 + :\ddot{Cl}:^-$$

$$H_2C{=}\overset{+}{N}{=}\overset{-}{N} \longleftrightarrow H_2\overset{-}{C}{-}\overset{+}{N}{\equiv}N \xrightarrow{\Delta \text{ or } h\nu} H_2C: + N_2$$

(*a*)

$$(CH_3S)_3CH \xrightarrow[BF_4^-]{Ph_3C^+} (CH_3S)_2\overset{+}{C}H\ BF_4^- + Ph_3C{-}SCH_3 \xrightarrow{-HBF_4} (CH_3S)_2C:$$

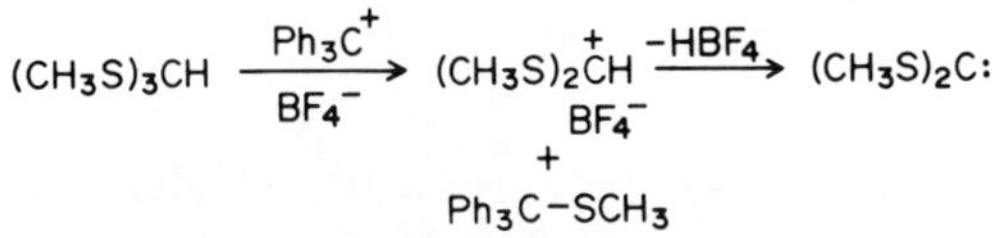

(*b*)

Figure 10.440 α Elimination to form carbenes: (*a*) via carbanion; (*b*) via carbenium ion.

10.450 Concerted Unimolecular Elimination

A one-step elimination, generally following a *cis* stereochemical course. These reactions are discussed more fully in the context of pericyclic cycloreversions (see Sect. 10.1100).

Examples. Xanthate ester pyrolysis, Fig. 10.450, and acetate ester pyrolysis, Fig. 10.390*b*.

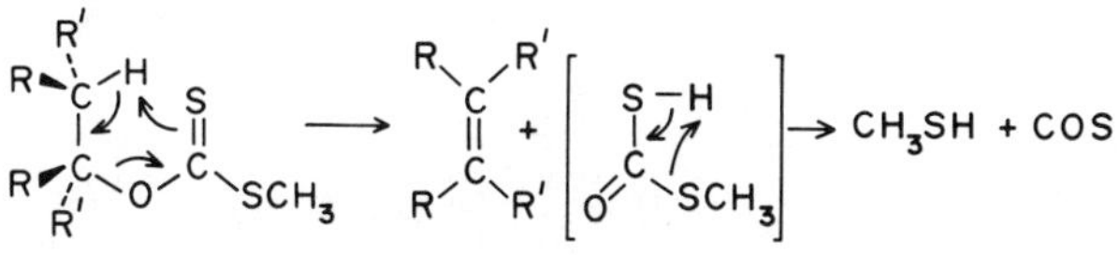

$$RR'C(H){-}C(R)(R'){-}O{-}C(=S){-}SCH_3 \longrightarrow RR'C{=}CRR' + [O{=}C(SH){-}SCH_3] \longrightarrow CH_3SH + COS$$

Figure 10.450 Xanthate ester pyrolysis.

10.460 Addition

A reaction that involves an increase in the number of groups attached to the substrate, and therefore a decrease in the degree of unsaturation of the substrate. Addition reactions are the reverse of elimination reactions. Most commonly, an addition involves the gain of two groups or atoms (one electrophile and one nucleophile) at each end of a π bond (1,2 addition) or ends of a π system (e.g., 1,4 or 1,6 addition). There are, however, examples of addition to certain highly reactive σ bonds (e.g., cyclopropane addition).

Examples. Figure 10.460.

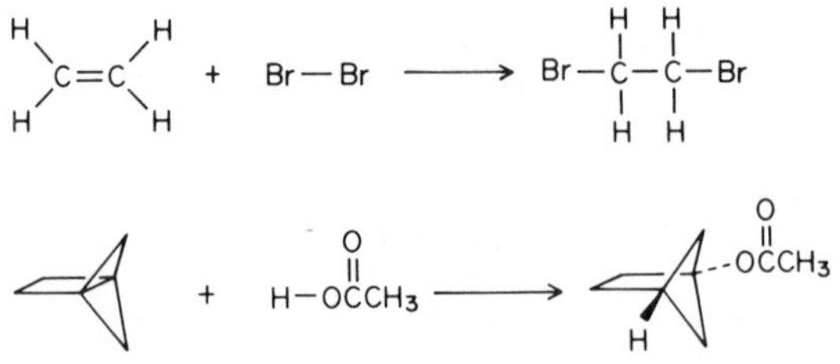

Figure 10.460 1,2 Additions.

10.470 Nucleophilic Addition to Carbon-Carbon Multiple Bonds

An addition reaction in which attachment of the nucleophile procedes attachment of the electrophile. Such reactions (usually occurring under basic conditions) involve carbanion intermediates, and thus are important only when the nucleophile is extremely reactive (e.g., is itself a carbanion) and/or the incipient carbanion is stabilized by electron-withdrawing substituents. Carbon-carbon triple bonds are somewhat more reactive toward nucleophilic addition than comparably substituted double bonds.

Examples. Figure 10.470*a* and *b*.

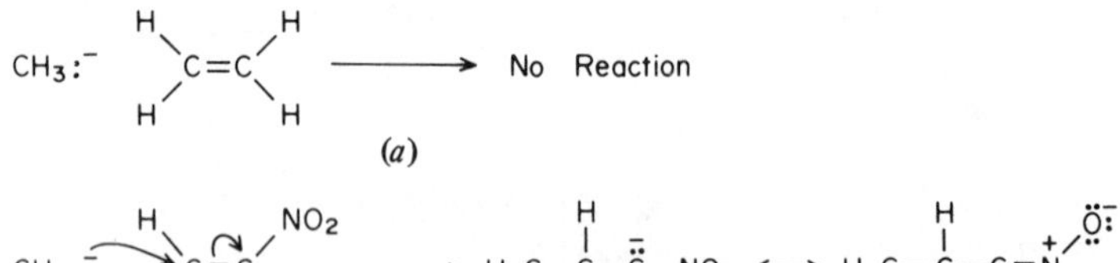

(*a*)

(*b*)

Figure 10.470 Nucleophilic addition to a double bond: (*a*) alkenes do not react; (*b*) reaction via stabilized anion.

10.480 Conjugate Nucleophilic (or Michael) Addition (A. Michael, 1853-1942)
Nucleophilic addition to a carbon-carbon π bond that is conjugated with an electron withdrawing group such as a carbonyl. The reaction involves first a 1,4 addition to the ends of the conjugated system, followed by a migration of the electrophile (usually H^+) to the "2" position (ketonization of the enol); the reaction results in net 1,2 addition to the olefin π bond.

Example. Figure 10.480.

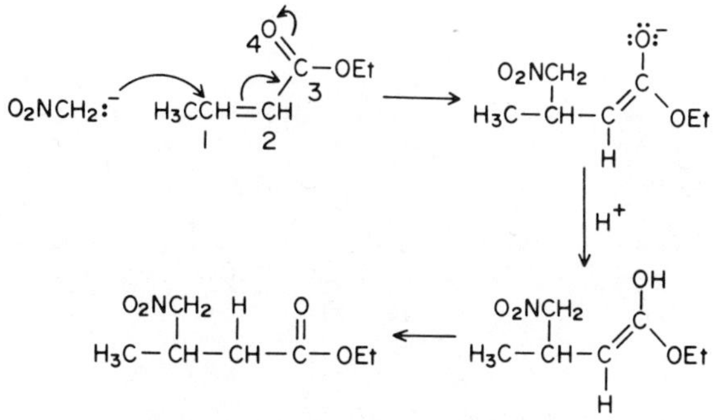

$O_2NCH_2:^-$ + $H_3CCH{=}CH{-}C({=}O)OEt$ (positions 1, 2, 3, 4) $\longrightarrow$ $H_3C{-}CH(CH_2NO_2){-}CH{=}C(O^-)OEt$ $\xrightarrow{H^+}$ $H_3C{-}CH(CH_2NO_2){-}CH{=}C(OH)OEt$ $\longrightarrow$ $H_3C{-}CH(CH_2NO_2){-}CH_2{-}C({=}O){-}OEt$

Figure 10.480 Michael addition.

10.490 Electrophilic Addition to Carbon-Carbon Multiple Bonds
An addition reaction in which electrophilic attachment precedes nucleophilic attachment, hence involves carbocation intermediates. Such reaction (occurring generally under acidic or neutral conditions) can involve cyclic as well as open cation intermediates. Olefins are generally more reactive than acetylenes in electrophilic additions.

Example. Figure 10.490 (see also Sect. 10.500).

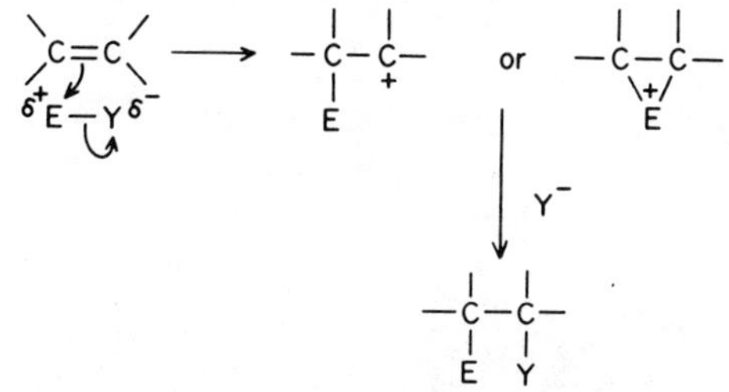

$>C{=}C<$ + $^{\delta+}E{-}Y^{\delta-}$ $\longrightarrow$ $-C(E){-}C^+-$ or $-C{-}C-$ bridged by E^+ (cyclic cation) $\xrightarrow{Y^-}$ $-C(E){-}C(Y)-$

Figure 10.490 Stepwise electrophilic addition.

10.500 Markovnikov's Rule (V. W. Markovnikov, 1838-1904)

As originally stated, the electrophilic addition of HX (X = halogen) to an unsymmetrically substituted alkene leads to attachment of the hydrogen to the less substituted carbon. More generally, the rule states that in the addition of $\overset{\delta^+}{A}$—$\overset{\delta^-}{B}$ to a carbon-carbon multiple bond, the more positive group A adds to the least substituted carbon. It is now recognized that this preference is a result of formation of the more stable carbenium ion intermediate (see also Sect. 10.730).

Examples. Figure 10.500*a* and *b*.

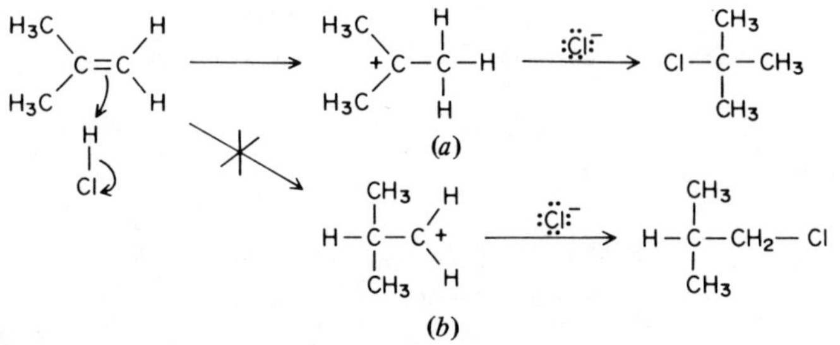

Figure 10.500 Markovnikov addition: (*a*) Markovnikov product via tertiary carbenium ion; (*b*) anti-Markovnikov product (minor product) via less stable primary carbenium ion.

10.510 π Complex

Most commonly, the species that results from interaction of π electrons with an electrophile; the electrophile is not localized on a particular atom. An empty orbital on the electrophile overlaps a filled π-type orbital of the donor. A π complex may be regarded as a **charge-transfer complex**. Formation of the π complex is often difficult to establish, but it is believed to be rapid and reversible.

Examples. Figure 10.510*a*, *b* and *c*.

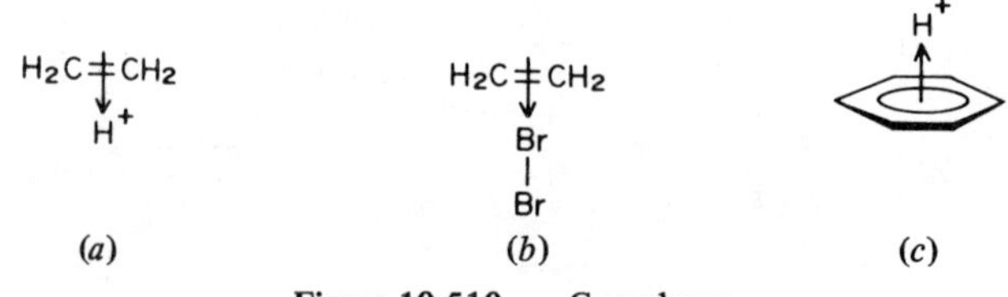

Figure 10.510 π Complexes.

10.520 σ Complex

The species that results from interaction of an electrophile with a σ bond. An empty orbital on the electrophile overlaps a filled σ orbital in the donor.

Examples. Figure 10.520 shows edge-protonated cyclopropane and protonated methane. The complexes shown in Fig. 10.520*b*, *c*, and *d*, resulting, respectively, from the electronic rearrangement of the π complexes shown in Fig. 10.510*a*, *b*, and *c*, are also called σ complexes to distinguish them from precursor π complexes.

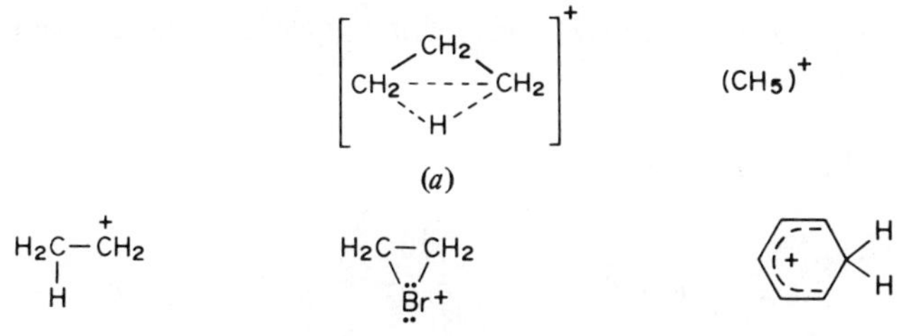

Figure 10.520 σ Complexes: (*a*) edge-protonated cyclopropane and protonated methane; (*b*) ethyl cation; (*c*) a bromonium ion; (*d*) cyclohexadienyl cation.

10.530 *cis* Addition

The stereochemical result of an addition reaction involving formal attachment of both groups (or atoms) to the same face of the π bond.

Example. See Fig. 10.540*a*.

10.540 *trans* Addition

The stereochemical result of an addition reaction involving formal attachment of both groups (or atoms) to opposite faces of the π bond.

Example. Figure 10.540*b*.

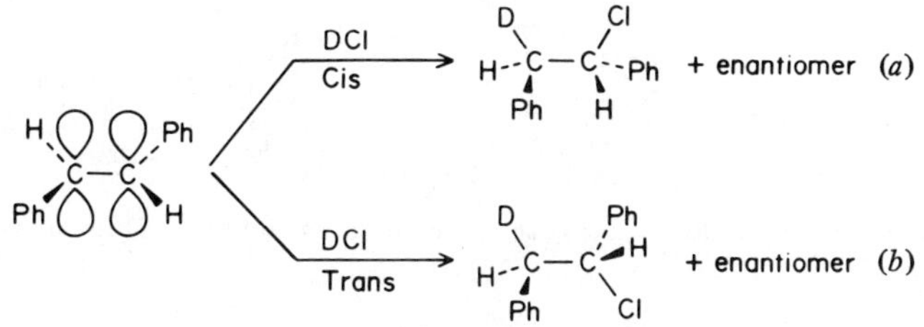

Figure 10.540 Addition reactions to an *E* (or *trans*) isomer: (*a*) *cis* addition giving *threo* product; (*b*) *trans* addition giving *erythro* product.

10.550 Ad_E2 (Addition Electrophilic Bimolecular)

A two-step addition in which attachment of the electrophile occurs first (the rate-determining step) and is followed by rapid nucleophilic attack on the cationic intermediate. The reaction is first order in substrate and first order in electrophile.

The stereochemistry of the addition depends on the structure of the intermediate and the degree of association of the nucleophile to the electrophile. Certain electrophiles such as Br^+ (from Br_2), HgX^+ (from HgX_2), and OH^+ (from RCO_3H) give cyclic intermediates (unless the open ion is sufficiently stabilized by electron donating substitutes), followed by nucleophilic attack which leads to *trans* addition.

Examples. Figure 10.550*a*, *b*, and *c*.

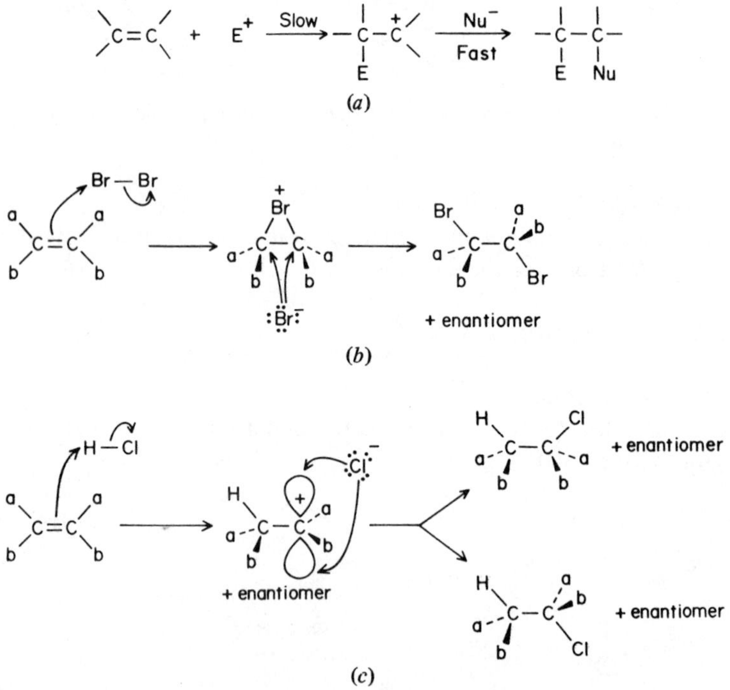

(*a*)

(*b*)

(*c*)

Figure 10.550 Ad_E2 reactions: (*a*) general reaction; (*b*) addition of bromine; (*c*) addition of HCl.

10.560 Cycloaddition

An addition of two connected atoms or groups to the ends of a π system. The two atoms remain connected and thus the reaction generates a product that has a new ring. The mechanisms of cycloaddition are often, but not always, concerted. (See **pericyclic reactions.**)

Examples. The Diels-Alder reaction, Fig. 10.560*a*, and a 1,3-dipolar addition, Fig. 10.560*b*.

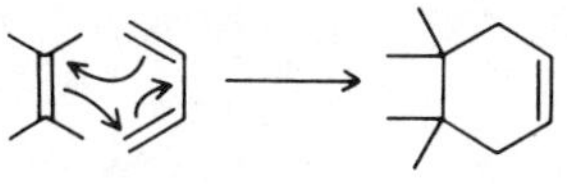

(*a*)

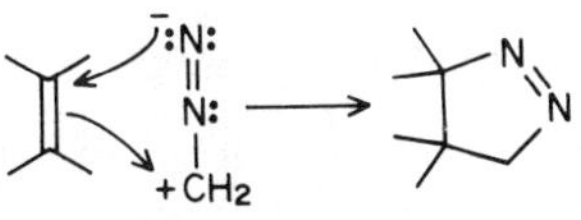

(*b*)

Figure 10.560 Cycloadditions: (*a*) the Diels-Alder reaction; (*b*) a 1,3-dipolar addition.

10.570 Concerted Addition

The one-step addition of two atoms or groups to a π (or σ) bond with *cis* stereochemistry via a cyclic transition state. Certain cycloadditions such as those shown in Fig. 10.560, as well as diimide reductions, are believed to be concerted.

Example. Figure 10.570, the reduction of an alkene by diimide.

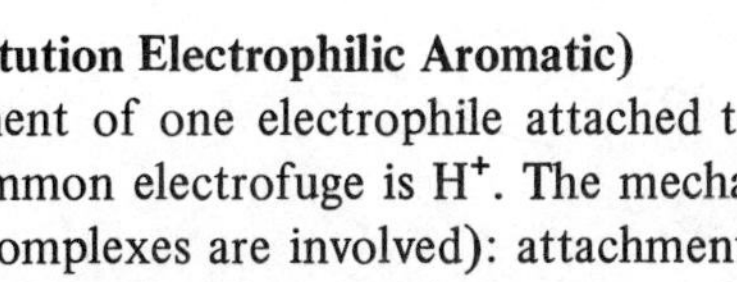

Figure 10.570 Concerted addition: alkene reduction by diimide.

10.580 S_EAr (Substitution Electrophilic Aromatic)

The stepwise replacement of one electrophile attached to an aromatic ring by another. The most common electrofuge is H^+. The mechanism involves at least two steps (more if π complexes are involved): attachment of the attacking electrophile to give a cyclohexadienyl cation (sometimes referred to as the σ complex), which then loses an electrofuge *from the same atom that was attacked* to reform the aromatic π system. Nucleophilic attachment to the intermediate (i.e., net addition to the aromatic π system) occurs only very rarely because of the cost of losing the resonance energy associated with an aromatic system.

Example. Figure 10.580. Electron-withdrawing substituents on the aromatic ring inhibit attack by the electrophile, electron-donating groups activate the ring toward electrophilic attack.

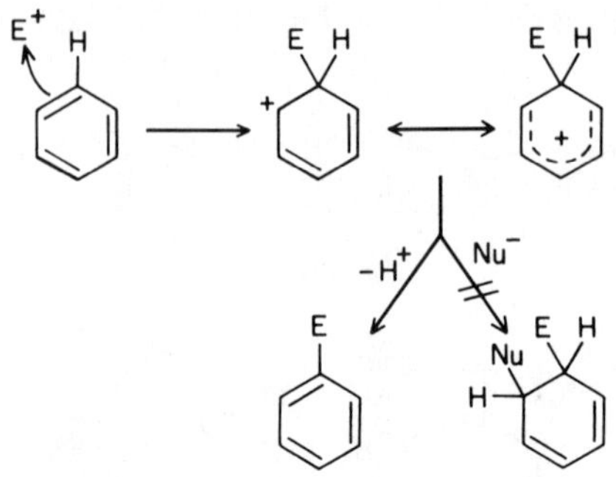

Figure 10.580 Aromatic electrophilic substitution, S_EAr.

10.590 *ortho, para* Director

A substituent on an aromatic ring that facilitates electrophilic substitution at the positions *ortho* and *para* to itself. Electron donating groups (alkyl, $-\ddot{O}R$, $-\ddot{N}R_2$, etc.) and halogens fall into this category; attachment of an electrophile *ortho* or *para* to them leads to a more stable σ complex because of resonance involving the substituent. On the other hand, *meta* attack does not permit such resonance stabilization.

Examples. See Fig. 9.1040*a* and *b*.

10.600 *meta* Director

A substituent on an aromatic ring that directs electrophilic substitution *meta* to itself. This occurs because the σ complex resulting from *meta* attack is less destabilized by like-charge repulsions than are σ complexes from *ortho* or *para* attack. Typically, *meta* directing groups are electron-withdrawing such as $-NO_2$, $-CN$, and $-\overset{\overset{\displaystyle O}{\|}}{C}-R$.

Example. Figure 10.600.

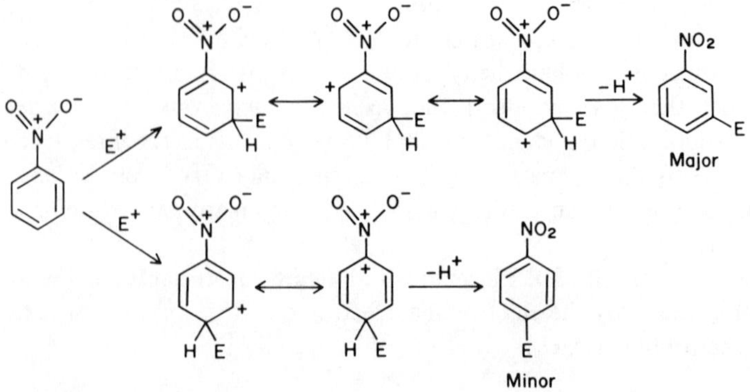

Figure 10.600 *meta* preferred substitution with an electron-withdrawing substituent.

10.610 S_NAr (Substitution Nucleophilic Aromatic)

Substitution of one nucleophile for another at an aromatic carbon. Although typical aromatic compounds are resistant toward nucleophilic attack, substitution by one or more strong electron-withdrawing groups can stabilize the anionic intermediate (a **Meisenheimer complex**; J. Meisenheimer, 1879-1934), which subsequently loses a nucleofuge to reform the aromatic π system.

Example. Figure 10.610. This type of reaction is frequently used for determining the *N*-terminal amino acid residue in a polypeptide; 2,4-dinitrofluorobenzene is attacked by the *N*-terminal amino group and labels it in the subsequent hydrolysis.

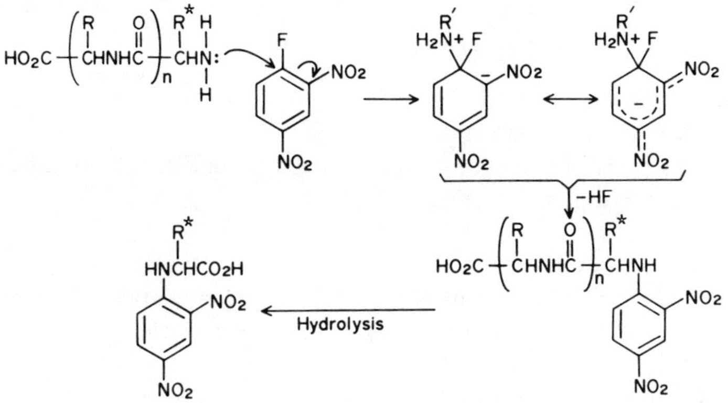

Figure 10.610 Nucleophilic aromatic substitution, S_NAr: *N*-terminal amino acid identification.

10.620 Benzyne (or Aryne) Mechanism

A nucleophilic substitution at an aromatic carbon atom under strongly basic conditions involving an elimination-addition mechanism with a benzyne intermediate.

Example. Figure 10.620.

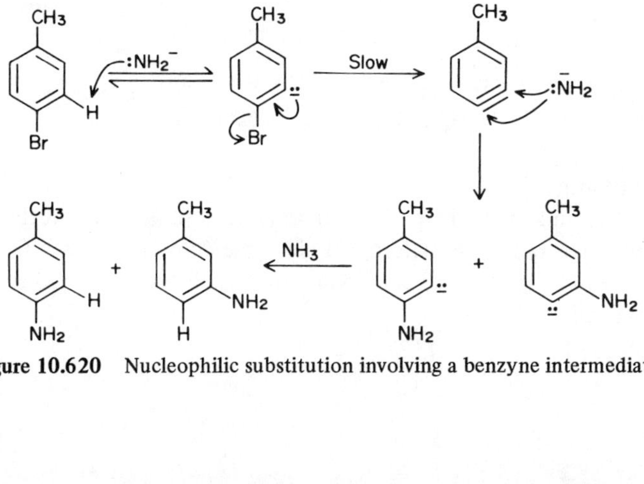

Figure 10.620 Nucleophilic substitution involving a benzyne intermediate.

10.630 Chain Reaction

A multistep mechanism, most commonly involving free radicals, in which an intermediate formed in one step brings about a second step that in turn generates an intermediate that brings about repetition of the first step.

Example. Figure 10.630, the chain reaction for the chlorination of alkanes.

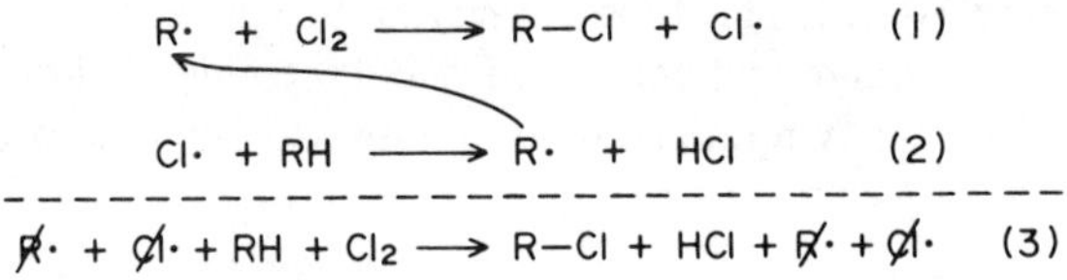

$$R\cdot + Cl_2 \longrightarrow R{-}Cl + Cl\cdot \qquad (1)$$

$$Cl\cdot + RH \longrightarrow R\cdot + HCl \qquad (2)$$

$$\cancel{R}\cdot + \cancel{Cl}\cdot + RH + Cl_2 \longrightarrow R{-}Cl + HCl + \cancel{R}\cdot + \cancel{Cl}\cdot \qquad (3)$$

Figure 10.630 A chain reaction.

10.640 Chain Propagating Steps

The repeating steps of a chain reaction [(1) and (2) in Fig. 10.630], the sum of which gives the overall net reaction [(3)].

10.650 Chain Carriers

The intermediates that are alternately formed then destroyed in a chain reaction. Both R· and Cl· in Fig. 10.630 can be regarded as chain carriers.

10.660 Chain Initiating Step

The nonrepeating preliminary step or steps in which the chain carrier is first formed.

Examples. Two possible chain initiating sequences are shown in Fig. 10.660 for the reaction in Fig. 10.630.

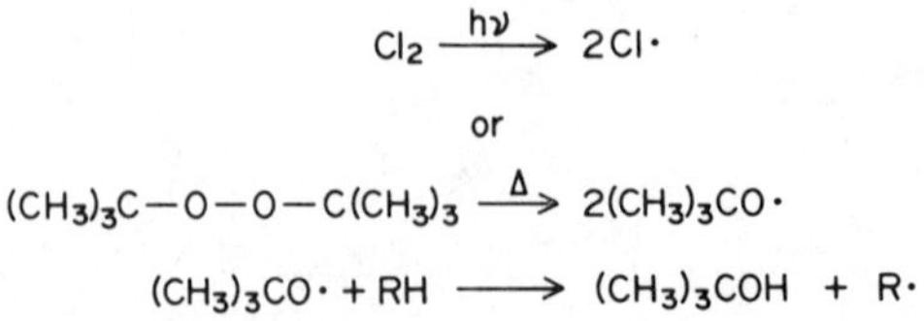

$$Cl_2 \xrightarrow{h\nu} 2Cl\cdot$$

or

$$(CH_3)_3C{-}O{-}O{-}C(CH_3)_3 \xrightarrow{\Delta} 2(CH_3)_3CO\cdot$$

peroxide

$$(CH_3)_3CO\cdot + RH \longrightarrow (CH_3)_3COH + R\cdot$$

Figure 10.660 Chain initiation.

10.670 Initiator

A reagent that undergoes thermolysis or photolysis to bring about the chain initiating step. Free radical initiators include those compounds such as peroxides (Fig. 10.660) and aliphatic azo compounds that when heated or irradiated, readily give free radicals.

Example. Figure 10.670, azobisisobutyronitrile (AIBN).

$$(CH_3)_2\underset{}{\overset{CN}{\overset{|}{C}}}-N=N-\overset{CN}{\overset{|}{C}}(CH_3)_2 \xrightarrow{\Delta} 2(CH_3)_2\dot{C}-CN + N_2$$

AIBN

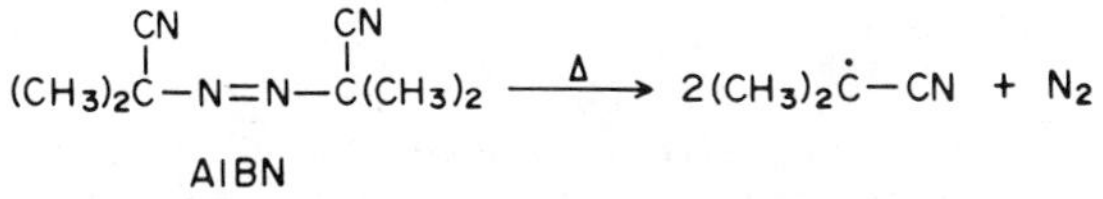

Figure 10.670 Azobisisobutyronitrile (AIBN), a common chain initiator.

10.680 Chain Terminating Step

A reaction in which two chain carriers react together, without regenerating new chain carriers. The chain reaction is therefore stopped.

Examples. Possible termination reactions, Fig. 10.680*a–d*, for Fig. 10.630; part *d* shows termination by **disproportionation.**

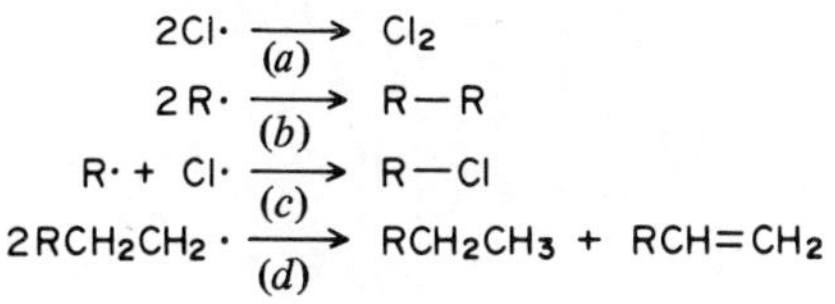

$$2Cl\cdot \xrightarrow[(a)]{} Cl_2$$
$$2R\cdot \xrightarrow[(b)]{} R-R$$
$$R\cdot + Cl\cdot \xrightarrow[(c)]{} R-Cl$$
$$2RCH_2CH_2\cdot \xrightarrow[(d)]{} RCH_2CH_3 + RCH=CH_2$$

Figure 10.680 Chain terminating reactions.

10.690 Inhibitor

A reagent that suppresses a chemical reaction. In free radical chain reactions certain compounds such as molecular oxygen intercept the chain carrier(s), diverting them from the chain reaction.

Example. Figure 10.690.

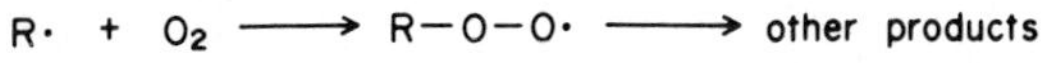

$$R\cdot + O_2 \longrightarrow R-O-O\cdot \longrightarrow \text{other products}$$

Figure 10.690 Molecular oxygen as an inhibitor.

10.700 S_H2 (Substitution Homolytic Bimolecular)

The attack of a free radical (or atom) on a terminal atom (an atom bonded to only one other atom, such as H) of another molecule. Such reactions are usually chain propagating steps in a free radical chain reaction mechanism.

Examples. Reactions (1) and (2) in Fig. 10.630 are both examples of S_H2 reactions.

10.710 $S_{RN}1$ (Substitution Radical Nucleophilic Unimolecular)

A nucleophilic substitution reaction that proceeds by a radical anion chain mechanism.

Example. The substitution of NH_2 for I in an aryl iodide can occur by the benzyne mechanism (Sect. 10.620) or the $S_{RN}1$ mechanism, Fig. 10.710.

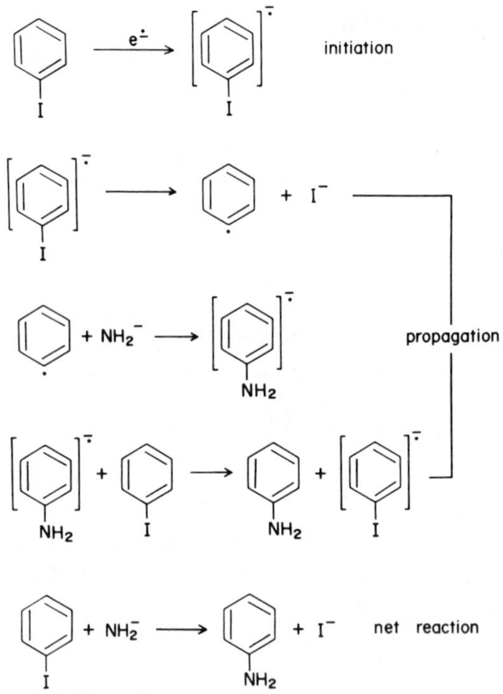

initiation

propagation

net reaction

Figure 10.710 Radical substitution, $S_{RN}1$.

10.720 Free Radical Addition

The addition of a free radical to an unsaturated center. The addition produces a new radical that has three reaction pathways open to it: (Fig. 10.720*a*) it may undergo elimination (hence net substitution); (Fig. 10.720*b*) it may abstract another atom (giving net addition via a chain reaction); or (Fig. 10.720*c*) it may add to another substrate molecule (leading to polymerization). Addition of a free radical to an aromatic π electron system usually follows the net substitution path to re-form the aromatic π system.

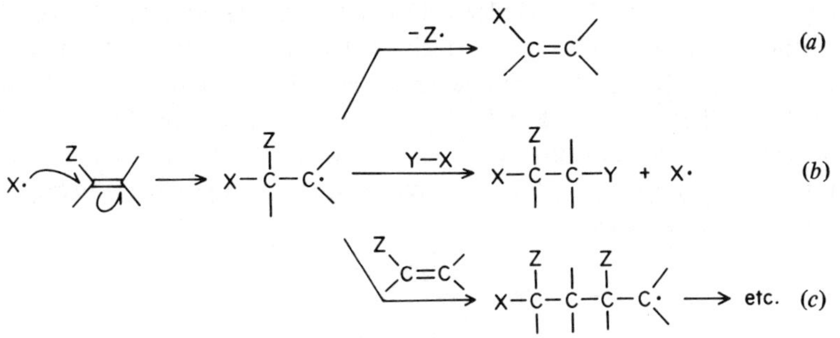

Figure 10.720 Free radical addition: (*a*) substitution reaction; (*b*) addition; (*c*) polymerization.

10.730 Anti-Markovnikov Addition

Addition of H—Z to a π bond, with Z becoming bonded to the less substituted atom and H to the most substituted atom. Z may be halogen, $—BR_2$, etc.

Examples. Figure 10.730*a*. Under free radical chain conditions addition of HBr to an unsymmetrical olefin proceeds via the more stable free radical intermediate (tertiary > secondary > primary) to give the opposite (anti-Markovnikov) product from that produced under ionic conditions. Boron hydrides also give the anti-Markovnikov adduct because in this case $—BR_2$ is the electrophile and H the nucleophile, Fig. 10.730*b*.

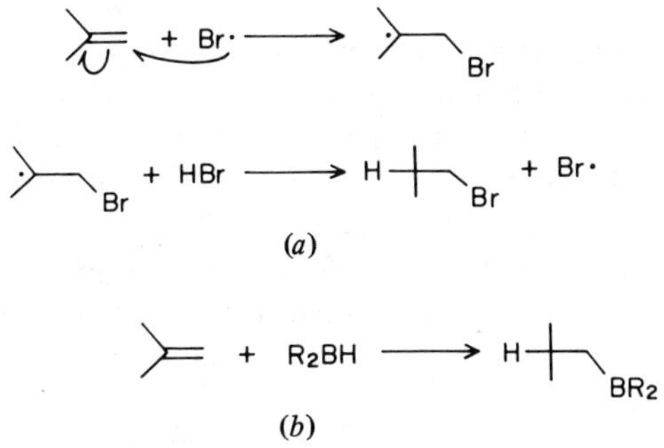

Figure 10.730 Anti-Markovnikov addition: (*a*) addition of HBr; (*b*) addition of boranes.

10.740 Addition to the Carbonyl

Addition to the carbonyl group occurs with attachment of the nucleophile to the carbonyl carbon and of the electrophile to the carbonyl oxygen.

Examples. Figure 10.740. Under acid conditions the electrophile (H^+) adds first, whereas under basic conditions the nucleophile adds first. Addition is irreversible if Nu = $H:^-$ or $R:^-$, but generally reversible (favoring the carbonyl) if Nu = ^-OR, $:NHR_2$, halide. Both C=N and C≡N multiple bonds behave similarly to the carbonyl function. Because the coordination number of the carbonyl carbon increases from three in the carbonyl to four in the product, the reaction is sensitive to steric factors.

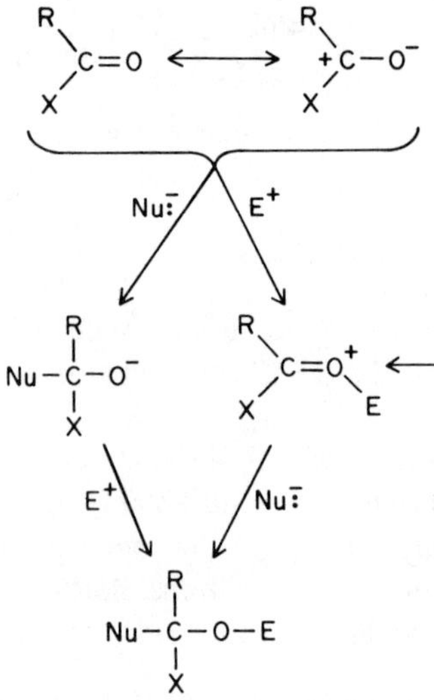

Figure 10.740 Carbonyl addition.

10.750 Cram's Rule of Asymmetric Induction (D. J. Cram, 1919-)

A rule for predicting the epimer that is preferentially formed by attack at a carbonyl group with a neighboring chiral center. The addition of $H:^-$ or $R:^-$ to the carbonyl group of such a substrate (the most reactive conformation of which is shown in Fig. 10.750, where L, M, and S are the large, medium, and small groups around the chiral center and R′M is the attacking reagent) occurs with R′ preferentially attacking over S to give the indicated stereoisomer (see Sect. 5.180).

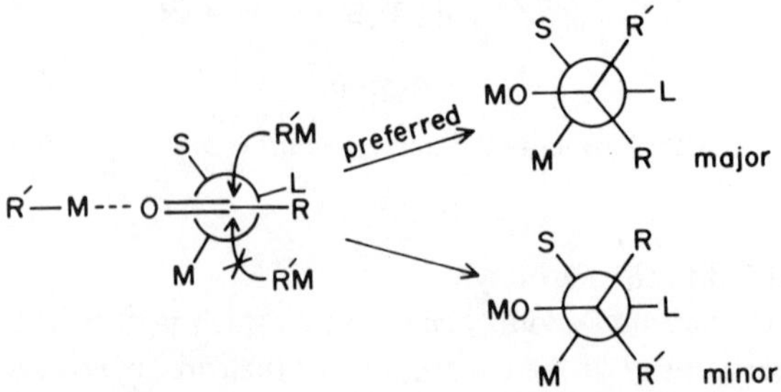

Figure 10.750 Application of Cram's rule.

Example. Figure 10.750. Since the stereoisomer that is preferentially formed may not necessarily be the more stable, this result is an example of kinetic control.

10.760 Ester Hydrolysis

The conversion of an ester into its acid and alcohol moieties according to the stoichiometry

$$R{-}\overset{\overset{\displaystyle O}{\|}}{C}{-}O{-}R' + H_2O \rightleftharpoons R{-}\overset{\overset{\displaystyle O}{\|}}{C}{-}O{-}H + H{-}O{-}R'$$

The hydrolysis may involve cleavage at either the acyl oxygen ($O{=}\overset{|}{C}\underset{\uparrow}{—}O{-}$) or alkyl oxygen ($-O\underset{\uparrow}{—}R'$) bond; the details of the mechanism are strongly dependent on the nature of R and R′, as well as the conditions of the reaction. Several such mechanisms are described below.

10.770 Saponification

Irreversible base-induced ester hydrolysis according to the equation

$$R\overset{\overset{\displaystyle O}{\|}}{C}{-}OR' \xrightarrow{OH^-} R\overset{\overset{\displaystyle O}{\|}}{C}O^- + HOR'$$

10.780 $A_{AC}2$ [Acid-Catalyzed (A) Bimolecular (2) Hydrolysis with Acyl-Oxygen (AC) Cleavage]

Nucleophilic attack by water on a protonated ester to give a tetrahedral intermediate that rapidly collapses to protonated acid and alcohol.

Example. Figure 10.780. This is the most common acid-catalyzed hydrolysis mechanism.

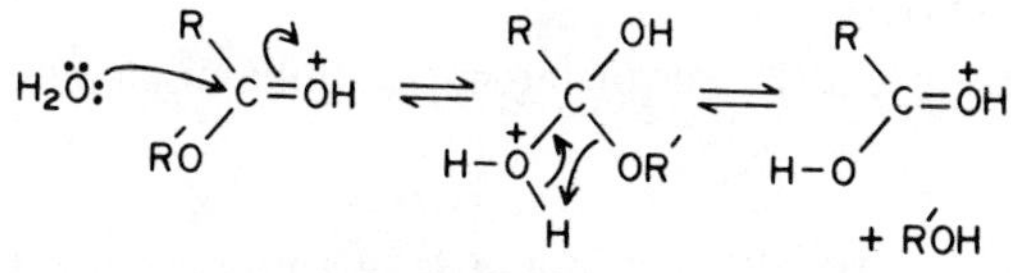

$$H_2\ddot{O}: + \underset{RO}{\overset{R}{>}}C{=}\overset{+}{O}H \rightleftharpoons H{-}\overset{+}{\underset{H}{O}}{-}\overset{R\quad OH}{C}{-}OR' \rightleftharpoons \underset{H-O}{\overset{R}{>}}C{=}\overset{+}{O}H + R'OH$$

Figure 10.780 Hydrolysis, $A_{AC}2$.

10.790 $B_{AC}2$ [Base-Catalyzed (B) Bimolecular (2) Hydrolysis with Acyl-Oxygen (AC) Cleavage]

Nucleophilic attack by base on the ester to give a tetrahedral ion, followed by rapid collapse to the alcohol and the conjugate base of the acid (with consumption of one mole equivalent of base).

Example. Figure 10.790. This is the most common mechanism for base-catalyzed ester hydrolysis. If R′ = phenyl, the phenol is also converted to its conjugate base, with overall consumption of two mole equivalents of base.

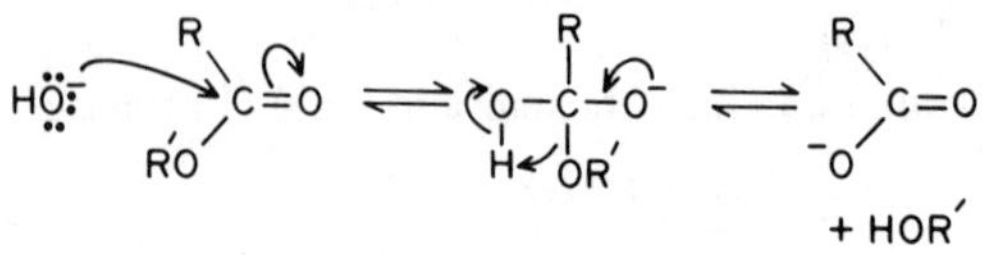

Figure 10.790 Hydrolysis, $B_{AC}2$.

10.800 $A_{AL}1$ [Acid-Catalyzed (A) Unimolecular (1) Hydrolysis with Alkyl-Oxygen (AL) Cleavage]

Fragmentation of a protonated ester to give the carboxylic acid and a carbenium ion, which is rapidly trapped by water.

Example. Figure 10.800. This mechanism occurs only when R'^+ is a relatively stable carbocation. When R'^+ is a stable carbocation and RCO_2^- is an exceptionally stable (nonnucleophilic) anion, the hydrolysis (e.g., of *tert*-butyl *p*-nitrobenzoate) can occur without acid catalyst.

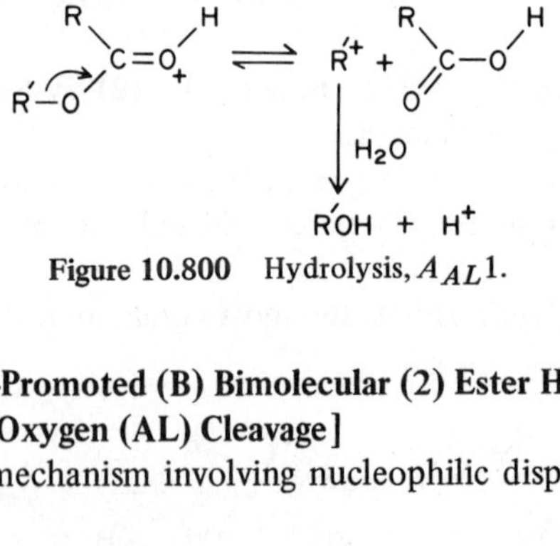

Figure 10.800 Hydrolysis, $A_{AL}1$.

10.810 $B_{AL}2$[Base-Promoted (B) Bimolecular (2) Ester Hydrolysis with Alkyl-Oxygen (AL) Cleavage]

An ester hydrolysis mechanism involving nucleophilic displacement at the alcohol carbon.

Example. Figure 10.810, although not strictly an hydrolysis is an ester cleavage involving a similar mechanism.

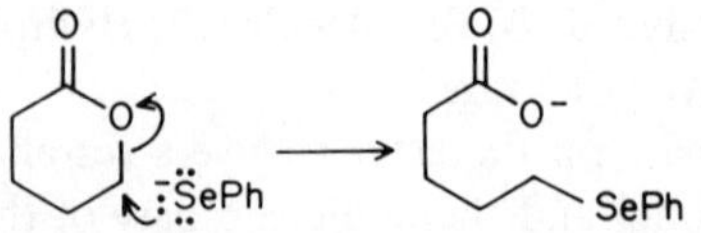

Figure 10.810 Ester cleavage, $B_{AL}2$.

10.820 Esterification

The reaction that creates an ester from its constituent acid and alcohol according to the stoichiometry

$$RCO_2H + HOR' \rightleftharpoons R\overset{\overset{O}{\|}}{C}{-}O{-}R' + H_2O$$

The mechanisms of esterification are the microscopic reverses of hydrolysis mechanisms (Sects. 10.780–10.810). Since each step is generally reversible, it is necessary to remove the water by-product (usually azeotropically) to force the reaction to completion. Esterification is an example of a **condensation reaction**; such reactions are characterized by either intra- or intermolecular loss of a small molecule such as H_2O, ROH, etc., between two reacting sites.

10.830 Oxidation Number

A number assigned to an atom in a molecule, representing its formal ownership of the valence electrons around it. All the electrons in a given bond are formally assigned to the more electronegative partner. Thus divalent oxygen is usually assigned an oxidation number of −2 (exceptions: OF_2 = +2; O_2 = 0; H_2O_2 = −1), hydrogen is usually +1 (exceptions: metal hydrides = −1; H_2 = 0), halogens are generally −1 (except in the elemental state). The algebraic sum of the oxidation numbers of all the atoms in a molecule must equal the charge on the molecule or ion. The oxidation number of carbon ranges from −4 to +4.

Examples. Figure 10.830.

−4	−3	−2	−1	0	+1	+2	+3	+4
$\underline{C}H_4$	$\cdot\underline{C}H_3$	$Cl\underline{C}H_3$	$R\underline{C}H_2Cl$	$\underline{C}H_2Cl_2$	$R_3\underline{C}OH$	$R_2\underline{C}{=}O$	$R\underline{C}O_2H$	$\underline{C}O_2$
	$H_3\underline{C}{-}R$	$H_3\underline{C}OH$		$\underline{C}$(solid)	$R\underline{C}HO$	$H\underline{C}O_2H$		$\underline{C}F_4$
				$\underline{C}H_2O$		$:\underline{C}{=}\ddot{O}:$		

Figure 10.830 Oxidation numbers for underlined carbon (R ≠ H).

10.840 Oxidation State

Synonymous with oxidation number.

10.850 Oxidation

The increase of oxidation number to a more positive value.

Examples. Figure 10.850*a* and *b* are oxidations, but *c* is not.

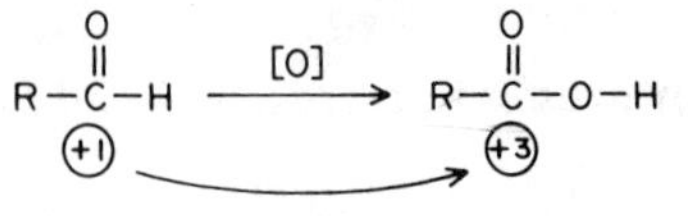

$$R-\overset{O}{\overset{\|}{C}}-H \xrightarrow{[O]} R-\overset{O}{\overset{\|}{C}}-O-H$$
(+1) → (+3)

(*a*)

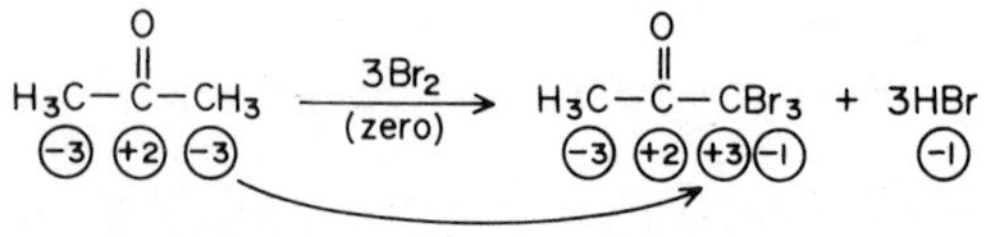

$$H_3C-\overset{O}{\overset{\|}{C}}-CH_3 \xrightarrow[(zero)]{3Br_2} H_3C-\overset{O}{\overset{\|}{C}}-CBr_3 + 3HBr$$
(−3) (+2) (−3) → (−3) (+2) (+3) (−1) (−1)

(*b*)

not oxidation

$$H_2CCl_2 \xrightarrow{2H_2O} [H_2C(OH)_2] \xrightarrow{-H_2O} H_2C=O$$
(0) (0) (0)

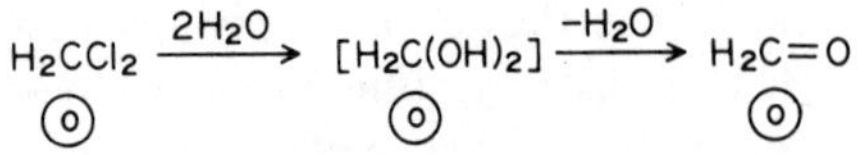

(*c*)

Figure 10.850 Oxidations of carbon: (*a*) two-electron oxidation; (*b*) six-electron oxidation. (*c*) An apparent oxidation (it is not).

10.860 Reduction

The decrease of oxidation number to a more negative value.

Examples. An oxidation must always be accompanied by a reduction somewhere in the system; i.e., the total net change in oxidation numbers must be zero. Stated alternatively, the loss of electrons (oxidation) must be matched by the gain of electrons (reduction). Thus in the example, Fig. 10.850*b*, the loss of six electrons from carbon is matched by the gain of six electrons by bromine. The reduction of carbonyl, Fig. 10.860*a*, and an intramolecular reduction-oxidation, Fig. 10.860*b* are other examples.

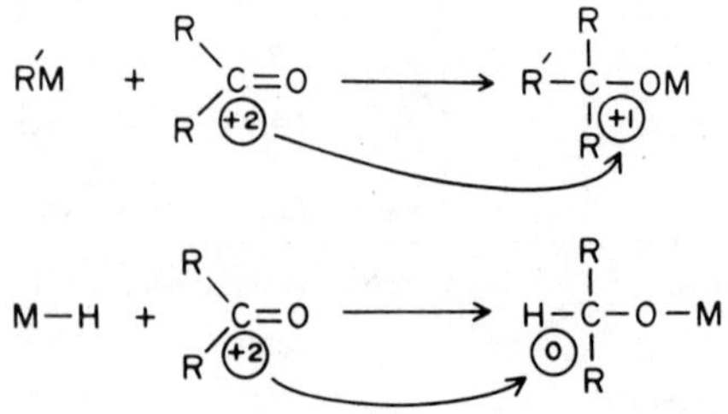

$$R'M + R_2C=O \longrightarrow R'-CR_2-OM$$
(+2) → (+1)

$$M-H + R_2C=O \longrightarrow H-CR_2-O-M$$
(+2) → (0)

(*a*)

$$Ph_2C(OH)-C(OH)Ph_2 \xrightarrow[-H_2O]{H^+} Ph-C(=O)-C(Ph)_3$$
(+1) (+1) → (+2) (0)

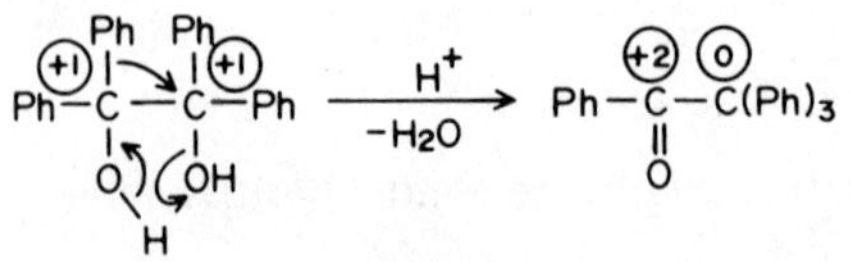

(*b*)

Figure 10.860 Reductions: (*a*) carbonyl reductions; (*b*) intramolecular oxidation-reduction.

10.870 Redox Reaction

A chemical reaction involving **red**uction and accompanying **ox**idation.

10.880 Intramolecular

Within a single molecule or species.

10.890 Intermolecular

Involving two or more molecules or species, either identical or different.

10.900 Isomerization

A reaction wherein a molecule or species is transformed into a different molecule or species with the same molecular formula (i.e., an isomer, see Sect. 5.020*a*).

10.910 Rearrangement

A reaction involving some change in the bonding sequence (which atoms are bonded) within a molecule. Most organic chemists use the term to denote reactions involving a change in the bonding sequence of the carbon skeleton of a molecule, although reactions such as olefin isomerizations are rearrangements as well.

10.920 Degenerate Rearrangement

A rearrangement that leads to a product which is indistinguishable from the reactant. All fluxional molecules (Sect. 5.090) undergo degenerate rearrangements.

Example. Figure 10.920.

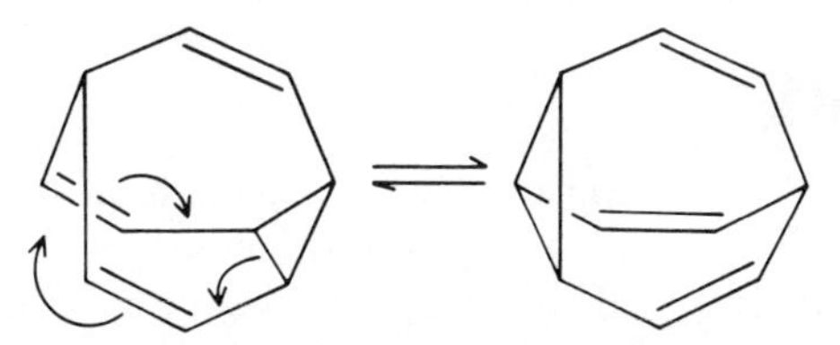

Figure 10.920 Degenerate rearrangement.

10.930 Pericyclic Reaction

Intra- or intermolecular processes involving *concerted* reorganization of electrons within a closed loop of interacting orbitals. Pericyclic reactions are subdivided into five classes: cycloaddition; electrocyclic; sigmatropic; chelatropic; and group transfer reactions.

10.940 Principle of Conservation of Orbital Symmetry

Overall orbital symmetry of a pericyclic reaction is said to be conserved (i.e., maintained) when each occupied orbital of the reactant transforms smoothly into an occupied orbital of the product with the same symmetry. Under these conditions there is no symmetry-imposed energy barrier to the reaction, although other factors (e.g., steric) generally give rise to a finite **activation energy.** By the **principle of microscopic reversibility**, if a reaction involves conservation of orbital symmetry in the forward direction, it also does in the reverse direction.

Examples. Figure 10.940. In the pericyclic conversion (by a conrotatory elec-

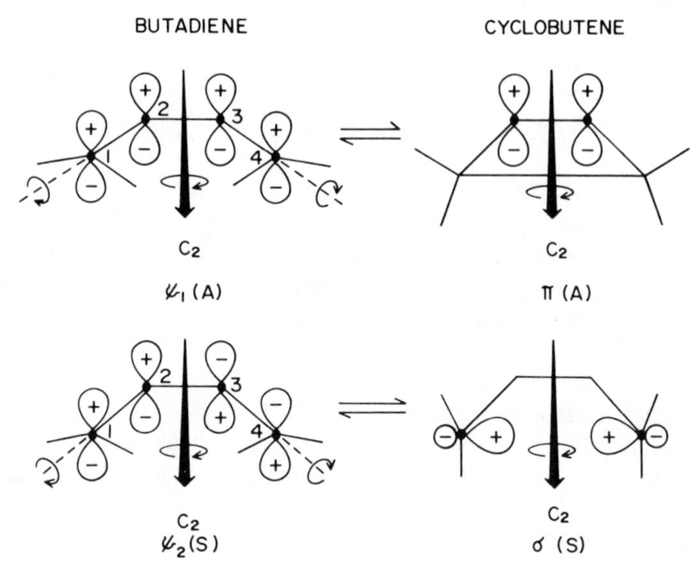

Figure 10.940 Symmetry correlations for conrotatory electrocyclization of butadiene.

trocyclization, *vide infra*) of butadiene to cyclobutene, the C_2 axis is preserved along the entire reaction coordinate. The occupied π orbitals of butadiene in the ground state are ψ_1 and ψ_2, and these are antisymmetric (A) and symmetric (S), respectively, with respect to the C_2 axis. The occupied orbitals of cyclobutene are the newly formed σ orbital between the C-1 and C-4 and the π orbital between C-2 and C-3; these orbitals are symmetric (S) and antisymmetric (A) with respect to the C_2 axis, respectively. Thus the occupied orbitals of the reactant tranform into occupied orbitals of the product with the same symmetry, and the overall symmetry has been conserved (see Fig. 10.1000*a*).

10.950 Allowed Reaction

A pericyclic reaction for which orbital symmetry is conserved. Such reactions must involve conversion of the ground (electronic) state of reactant into the ground state of product, or the first excited state of reactant into the first excited state of the product. The former type is frequently referred to as "thermally allowed," the latter "photochemically allowed." (See Fig. 10.1000*a* and *b*, respectively.)

10.960 Forbidden Reaction

A pericyclic reaction in which orbital symmetry is not conserved, i.e., in which occupied orbitals of the reactant do *not* transform into occupied orbitals of the product with the same symmetry. (See Fig. 10.1000*b*.) If a given reaction is forbidden in the concerted mode, it may still occur via some other *non*concerted pathway.

10.970 Electrocyclic Reaction

An intramolecular pericyclic reaction involving opening (or closing) of a ring by conversion of σ to π bonds (or the reverse).

Examples. Figure 10.940.

10.980 Conrotatory

During an electrocyclic reaction the termini (last carbon atoms and their substituents) of the open isomer must each rotate approximately 90° around the bond to which they are attached to form the σ bond of the product (or vice versa). If both termini rotate clockwise (or both counterclockwise), the ring closing (or opening) is said to be conrotatory.

Examples. Figures 10.940 and 10.990.

10.990 Disrotatory

When the termini in an electrocyclic reaction rotate in opposite directions, one clockwise and the other counterclockwise, the cyclization (or ring opening) is said to be disrotatory.

Examples. Figure 10.990. The electrocyclic ring opening of cyclobutene can occur by either of two conrotatory paths or by two disrotatory paths. (The two conrotatory paths give the same product.) It turns out (see Sect. 10.1000) that the ground state conrotatory reaction is allowed and disrotation is thermally forbidden. In the first excited state disrotation is allowed and conrotation is forbidden. This rule is generalized in Sect. 10.1140, Woodward-Hoffman rules.

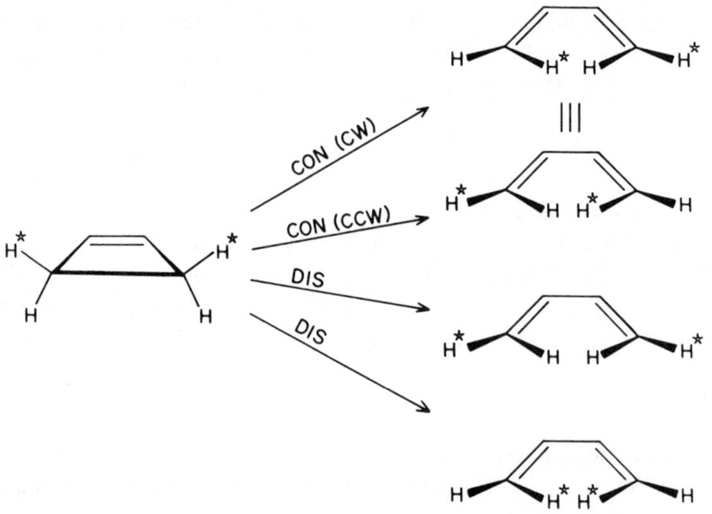

Figure 10.990 Ring opening of cyclobutene.

10.1000 Orbital Correlation Diagram

A diagram that shows the correspondence in energy and symmetry between relevant reactant and product orbitals in a pericyclic reaction. The relevant orbitals, frequently called the **basis set of orbitals**, are those that undergo change during the reaction. The diagram is constructed as follows: in separate columns the relevant orbitals of reactant and product are listed in their order of relative energies; each orbital is classified on the basis of the symmetry elements retained at all points along the reaction coordinate; lines are drawn between the reactant and product orbitals connecting the lowest energy orbitals of the same symmetry type.

Examples. In the ring opening of cyclobutene, Fig. 10.1000*a*, conrotation preserves the C_2 axis of symmetry. The four relevant orbitals, or basis set of orbitals, of this molecule are (lowest energy first) σ, π, π^*, and σ^*, which are respectively symmetric (S), antisymmetric (A), (S), and (A) with respect to the C_2 axis. Similarly, the product butadiene orbitals are $\psi_1(A)$, $\psi_2(S)$, $\psi_3(A)$, and $\psi_4(S)$. Note that the two lowest energy occupied orbitals (ground state) of cyclobutene correlate (have the same symmetry) with the two lowest energy orbitals (the ground state) of 1,3-butadiene. This orbital correlation diagram thus shows that the conrotatory conversion is thermally (ground state) allowed.

Next consider the disrotatory path on which a plane of symmetry is pre-

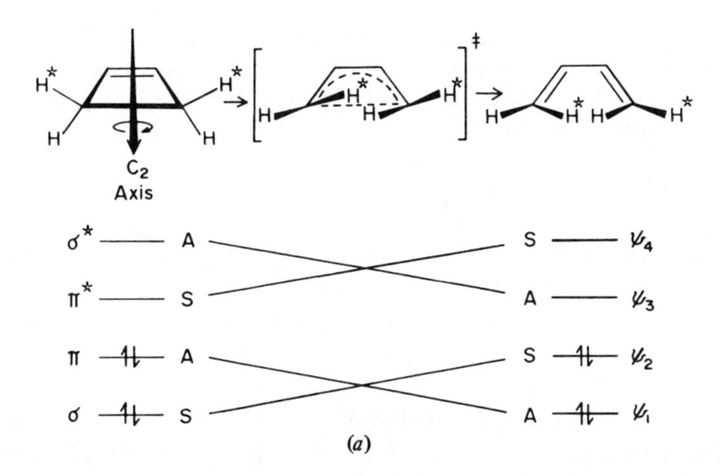

(a)

served, Fig. 10.1000*b*. The same basis set of orbitals is involved, but their symmetry is now determined with respect to the plane of symmetry. The two lowest energy occupied orbitals of cyclobutene do not correlate with the two lowest energy orbitals of 1,3-butadiene. The opening of ground state cyclobutene by the disrotatory pathway would require populating a high energy orbital (ψ_3) of butadiene, clearly an energetically unfavorable process. However, if the cyclobutene were excited so that one electron were in π and the other in π^*, the orbital correlation diagram would show that such a reaction is allowed and occurs with the conservation of orbital symmetry.

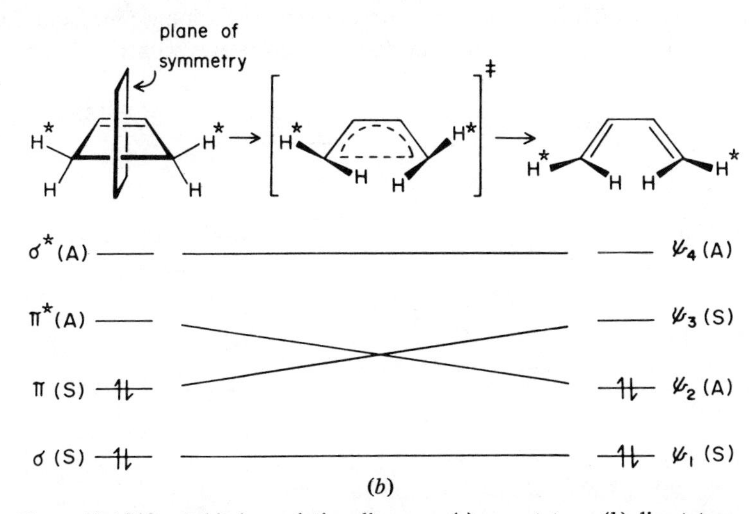

(b)

Figure 10.1000 Orbital correlation diagrams: (*a*) conrotatory; (*b*) disrotatory.

For the conversion of 1,3-cyclohexadiene to 1,3,5-hexatriene the result is exactly the opposite; now disrotation becomes thermally allowed and conrotation is photochemically allowed. For a summary of these results see Sect. 10.1140, Woodward-Hoffmann rules.

10.1010 State Correlation Diagram

A correlation diagram for a pericyclic reaction in which the symmetry of the electronic states (Sect. 2.580) are compared rather than symmetry of the orbitals of reactants and products. As with the orbitals, the states are listed in the increasing order of their energies, their symmetries determined, and correlations made analogously to orbital correlations.

Examples. The symmetry of an electronic state is determined from the product of the symmetries of the occupied orbitals making up the state. Suppose that one wants to develop the state correlation diagram for the disrotatory conversion of cyclobutene to butadiene, shown in Fig. 10.1000*b*. The ground state electronic configuration of cyclobutene is $\sigma^2\pi^2$ and, because the symmetries of these orbitals are S and S, respectively, the symmetry of the ground state is $S^2S^2 = S$. The electronic configuration of the first excited state of cyclobutene is $\sigma^2\pi\pi^*$, of symmetry $S^2SA = A$ (the product of $S \times S = A \times A = S$, but the product of $A \times S = A$). These states (and one other excited state of cyclobutene) are listed on the left side of Fig. 10.1010. The same principles are applied to the product, butadiene, using the orbitals shown in Fig. 10.1000*b* to develop the symmetry states shown on the right side of the state correlation diagram, Fig. 10.1010. To make the appropriate correlation we see from the orbital correlation diagram that σ of cyclobutene correlates with ψ_1 of butadiene

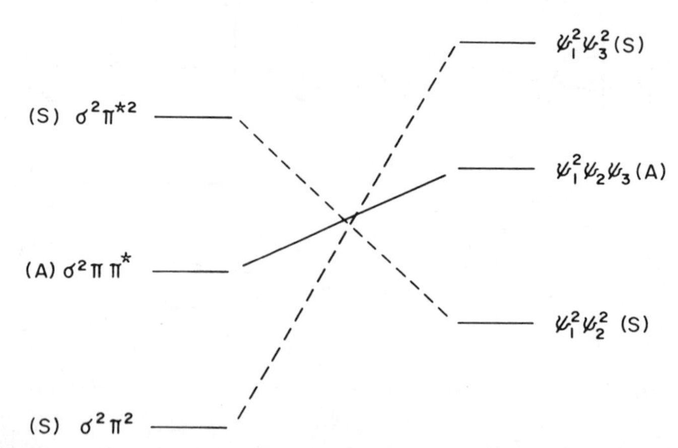

Figure 10.1010 State correlation diagram for disrotatory mode of cyclobutene⇌butadiene reaction.

but that π of cyclobutene correlates with ψ_3 of butadiene. Thus during disrotation the ground state of cyclobutene ($\sigma^2\pi^2$) correlates with $\psi_1^2\psi_3^2$ of butadiene, which is not the ground state but a doubly excited state of butadiene. Thus we see from the symmetry state correlation diagram that the ground state to ground state disrotatory ring opening of cyclobutene is a forbidden (high energy) process. Similar reasoning shows that the photochemical ring opening is an allowed process in the disrotatory mode.

10.1020 Suprafacial, *s*

A term describing the stereochemical relationship between the reacting termini of a molecule involved in a pericyclic reaction, where attachment occurs on the same face (side) of the molecule.

Examples. Figure 10.1030.

10.1030 Antarafacial, *a*

Opposite of suprafacial; attachment to the reacting termini on opposite sides of the molecule.

Examples. Figure 10.1030*a*–*d*. Note that a disrotatory ring closure is considered a suprafacial process, whereas conrotation is antarafacial (see Fig. 10.1030*a*). In the cycloaddition shown in Fig. 10.1030*d*, the suprafacial process proceeds with retention at centers 1 and 2 while the antarafacial process results in retention of configuration at center 1 and inversion at center 2.

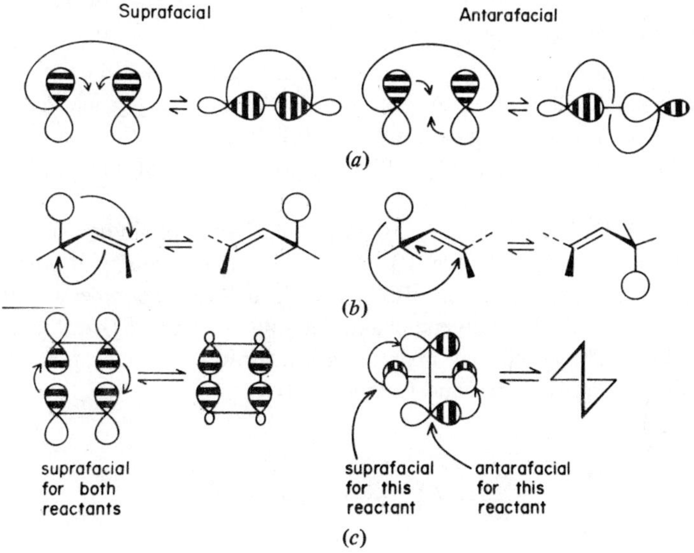

Suprafacial Antarafacial

(*a*)

(*b*)

suprafacial for both reactants

suprafacial for this reactant

antarafacial for this reactant

(*c*)

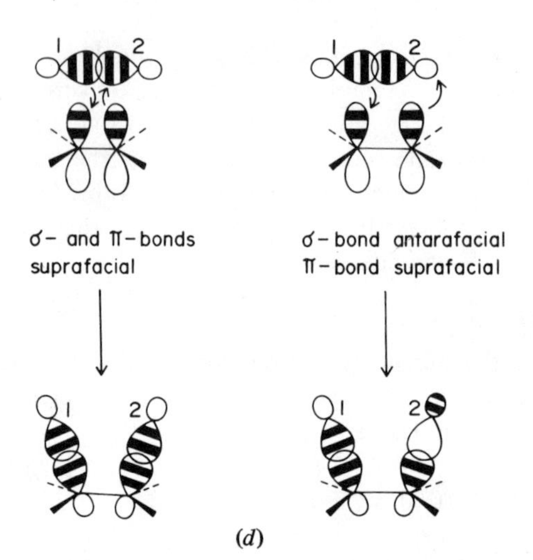

(*d*)

Figure 10.1030 Pericyclic reactions: (*a*) electrocyclization; (*b*) sigmatropic shift; (*c* and *d*) cycloadditions.

10.1040 Sigmatropic Shifts

Rearrangements that formally consist of the migration of a σ bond (actually the σ electrons) and the group attached to this bond from one position in a chain or ring to a new position in the chain or ring. The migrating σ bond is often at an allylic position, and during the rearrangement the σ electrons are accepted into a p orbital at the other terminus of the allylic system.

10.1050 [1, *n*]-Sigmatropic Shift

A sigmatropic shift in which there is no rearrangement within the migrating group. The number 1 is given to that atom in the migrating group that remains attached to the migrating sigma bond. The n indicates that the point of attachment to the chain has migrated from the first to the nth position along the chain.

Examples. A [1,2]-shift, Fig. 10.1050*a*; a [1,3]-shift or allylic rearrangement Fig. 10.1050*b*; a [1,5]-shift, Fig. 10.1050*c*. The stereochemistry at both the migrating group and the chain must also be described. These examples are all suprafacial with respect to the polyene chain. An antarafacial [1,5]-shift is shown in Fig. 10.1050*d*. The group may migrate with inversion or retention at the point of attachment, Fig. 10.1050*e* and *f*, respectively.

(*a*)

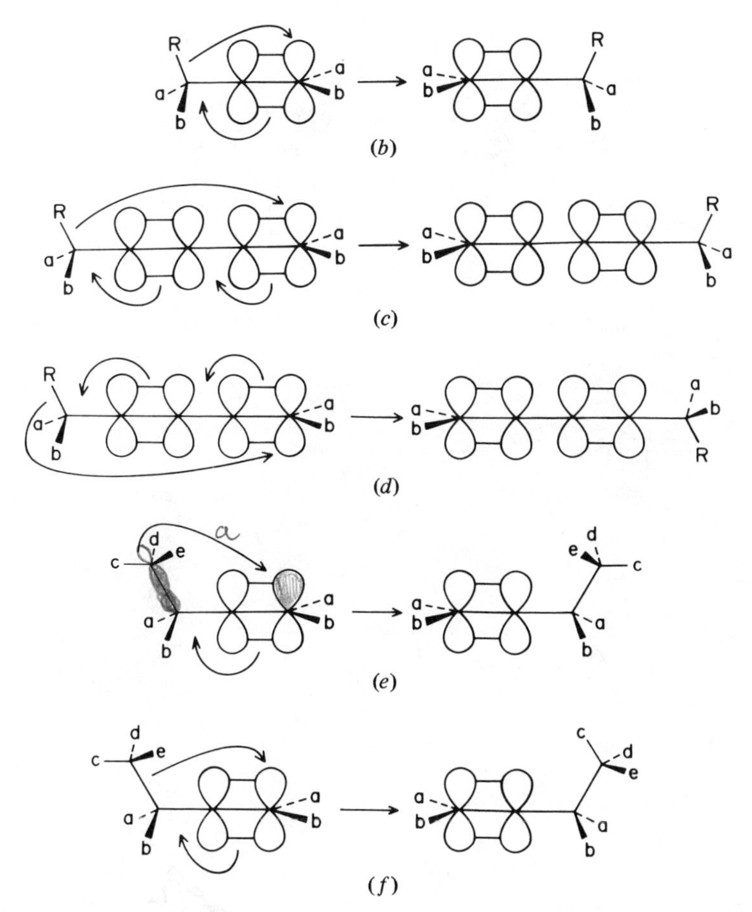

Figure 10.1050 [1,*n*]-Sigmatropic shifts: (*a*) [1,2]-shift; (*b*) [1,3]-shift; (*c*) [1,5]-shift; (*d*) antarafacial [1,5]-shift without inversion; (*e*) [1,3]-antarafacial shift with inversion; (*f*) [1,3]-suprafacial shift with retention.

10.1060 [*m*,*n*]-Sigmatropic Shift, $m \neq 1$

A sigmatropic shift in which the migrating group also undergoes rearrangement. The number *m* denotes the number of the atom on the migrating group that is attached to the migrating bond after the migration. The number *n*, as in [1,*n*]-shift, specifies the number of the atom in the chain that receives the migrating group. The numbering of atoms (*m*, *n*) on both the chain and migrating group starts at the atoms originally attached to the migrating bond.

Examples. The Cope rearrangement, Fig. 10.1060*a*; the Claisen rearrangement, Fig. 10.1060*b*; enolization, a [1,3]-shift, Fig. 10.1060*c*. An electron-pushing scheme for the Cope rearrangement is shown in Fig. 10.1060*d*.

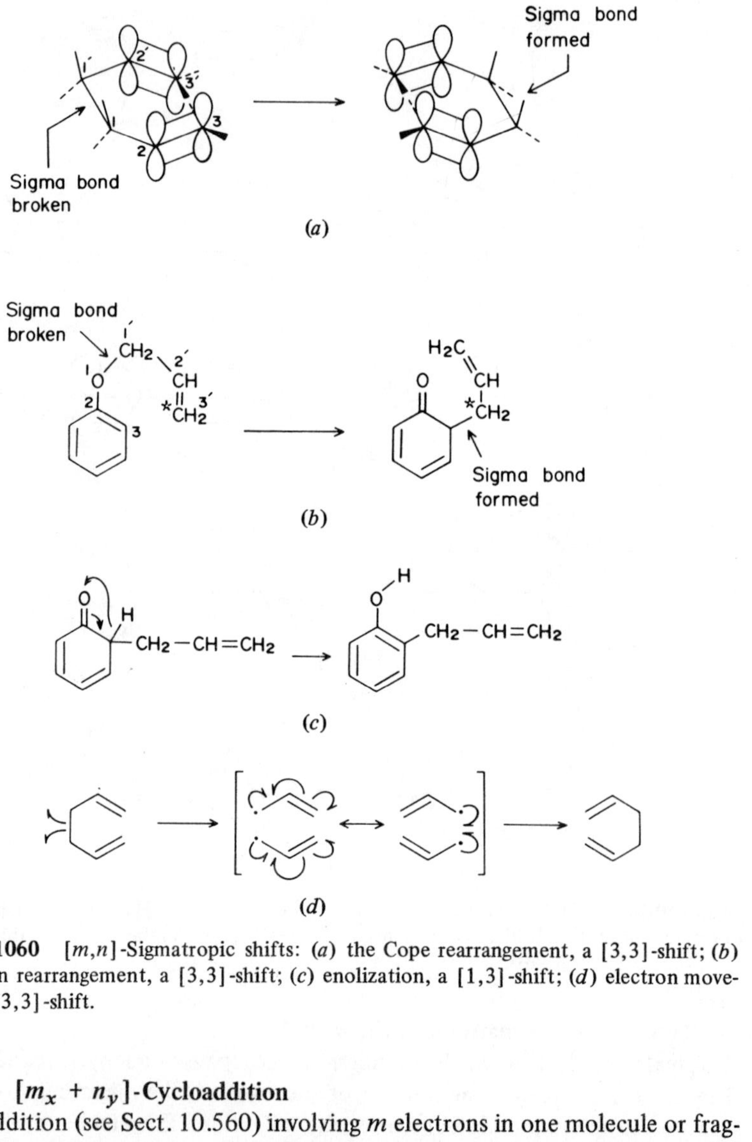

(a)

(b)

(c)

(d)

Figure 10.1060 [*m*,*n*]-Sigmatropic shifts: (*a*) the Cope rearrangement, a [3,3]-shift; (*b*) the Claissen rearrangement, a [3,3]-shift; (*c*) enolization, a [1,3]-shift; (*d*) electron movement in a [3,3]-shift.

10.1070 $[m_x + n_y]$-Cycloaddition

A cycloaddition (see Sect. 10.560) involving *m* electrons in one molecule or fragment and *n* electrons in the other. The subscripts *x* and *y* are either *s* or *a* depending on whether attachment to that molecule or fragment is suprafacial (*s*) or antarafacial (*a*). The type of electrons involved [σ, π, or ω (nonbonding)] is often noted by a preliminary subscript, e.g., ${}_{\pi}2_s + {}_{\pi}4_s$ or ${}_{\sigma}2_s + {}_{\sigma}2_a$.

Examples. Three different types of cycloadditions are shown in Fig. 10.1070*a–c*. Both ${}_{\pi}2_s + {}_{\pi}2_s$ and ${}_{\pi}2_s + {}_{\pi}2_a$ cycloadditions are shown in Fig. 10.1030*c*.

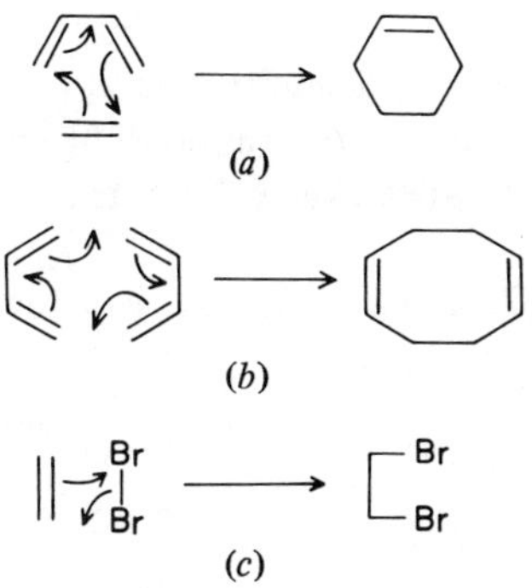

(a)

(b)

(c)

Figure 10.1070 Thermally allowed cycloadditions: (*a*) ${}_{\pi}4_s + {}_{\pi}2_s$; (*b*) ${}_{\pi}4_s + {}_{\pi}4_a$; (*c*) ${}_{\pi}2_s + {}_{\sigma}2_a$.

10.1080 1,3-Dipole

A three atom linkage, one resonance form of which has a negative charge at one end and a positive charge (hence an empty orbital) at the other. Such species are considered to have $4p\pi$ electrons.

Examples. Ozone, Fig. 10.1080*a*; a nitrile oxide, Fig. 10.1080*b*; a diazo compound, Fig. 10.1080*c*.

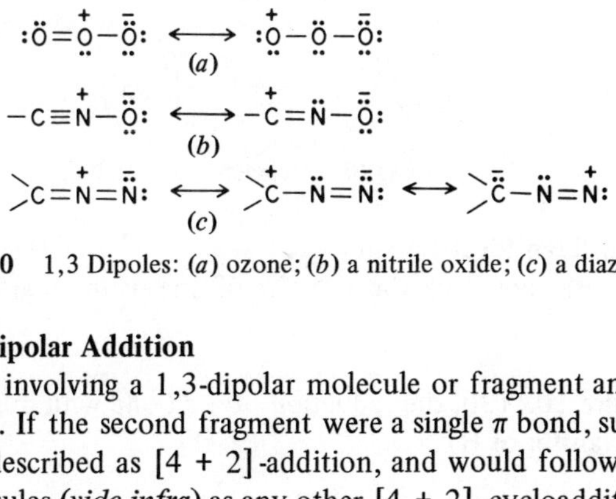

(a)

(b)

(c)

Figure 10.1080 1,3 Dipoles: (*a*) ozone; (*b*) a nitrile oxide; (*c*) a diazo compound.

10.1090 1,3-Dipolar Addition

A cycloaddition involving a 1,3-dipolar molecule or fragment and another (usually π) fragment. If the second fragment were a single π bond, such a cycloaddition would be described as [4 + 2]-addition, and would follow the same symmetry selection rules (*vide infra*) as any other [4 + 2]-cycloaddition.

Example. Figure 10.1090.

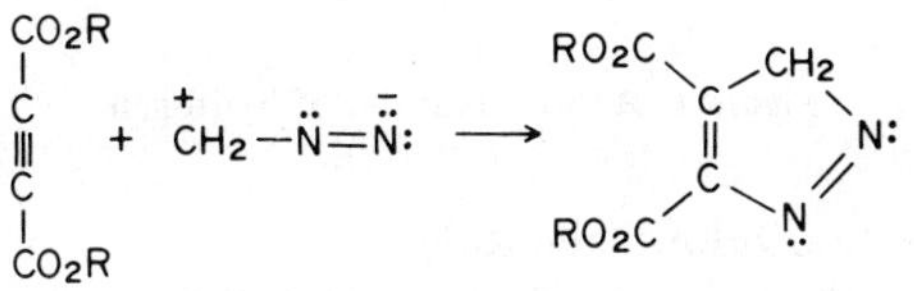

Figure 10.1090 1,3-Dipolar addition.

10.1100 Cycloreversion

The microscopic reverse of a cycloaddition reaction. Such a fragmentation follows the same symmetry selection rules as does the cycloaddition itself. A minus sign before the designation indicates a cycloreversion.

Example. Figure 10.1100 shows $-[_{\sigma}2_s + _{\sigma}2_s]$-cycloreversion.

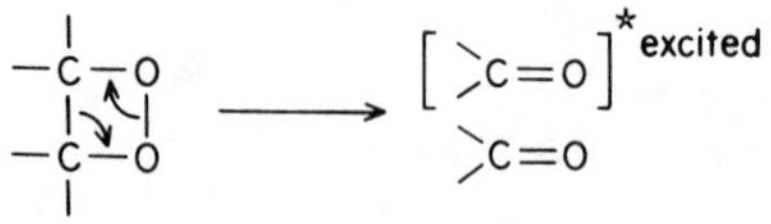

Figure 10.1100 A $-[_{\sigma}2_s + _{\sigma}2_s]$-cycloreversion.

10.1110 Retrograde Cycloaddition

Synonymous with cycloreversion.

10.1120 Cheletropic Reaction

A cycloaddition in which one of the reacting species or fragments acts through a single atom possessing both a filled and empty orbital.

Example. Figure 10.1120, the addition of a carbene to an olefin.

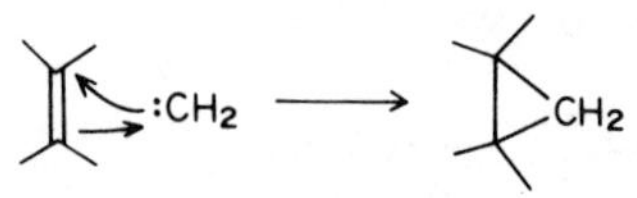

Figure 10.1120 A cheletropic reaction.

10.1130 Group Transfer Reaction

A pericyclic process involving the transfer of one or more groups or atoms to another molecule.

Example. Figure 10.1130, the reduction of a π bond with diimide (N_2H_2), involving formal transfer of H_2.

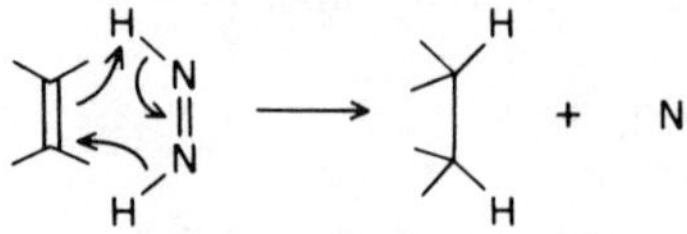

Figure 10.1130 Transfer of hydrogen.

10.1140 The Woodward-Hoffmann Rules

A series of generalized symmetry selection rules elaborated by R. B. Woodward (1917–1979) and R. Hoffmann (1937–), based on arguments such as correlation diagrams (Sect. 10.1000), which predict whether a given pericyclic reaction will be allowed under a given set of conditions. In its most general form the rules state that a pericyclic reaction is ground state allowed (excited state forbidden) if the total number of suprafacial $(4q + 2)$ and antarafacial $(4r)$ reacting components (electronic fragments) is odd (q and r are any integer including zero). Note: the antarafacial $(4q + 2)$ fragments and suprafacial $(4r)$ components are not counted. The rules can be restated more clearly for individual subclasses of pericyclic reactions:

(a) Electrocyclic processes. Here there is only one reacting component to be considered. If it contains $(4q + 2)$ π electrons (2, 6, 10, . . .), then the process is suprafacial (or disrotatory) in the ground state and antarafacial (conrotatory) in the excited state. Just the opposite occurs when the system comprises $4r$ electrons (4, 8, 12, . . .).

Examples. Figure 10.1140*a*.

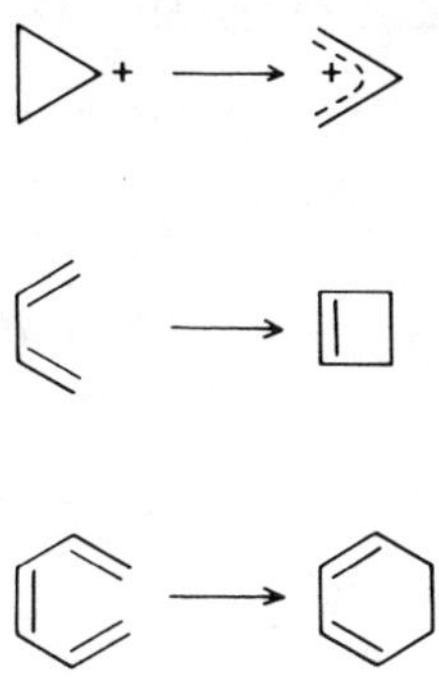

# electrons	ground state reaction
2	disrotatory
4	conrotatory
6	disrotatory

(*a*)

(b) Cycloadditions.

Examples. Figure 10.1140*b*.

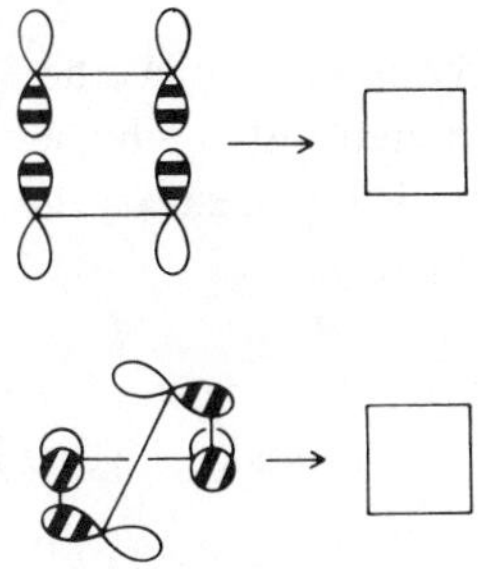

designation	components #$(4q+2)_s$	components #$(4r)_a$	total	ground state
2s + 2s	2	0	2	forbidden
2s + 2a	1	0	1	allowed

	designation				
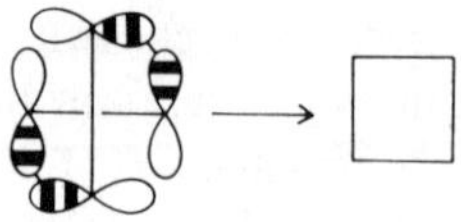	2a + 2a	0	0	0	forbidden
	4s + 2s	1	0	1	allowed

(*b*)

(c) Sigmatropic Shifts. Here the "migrating bond" is treated as though it were heterolytically cleaved, then the Woodward-Hoffmann rules are applied.

Examples. Figure 10.1140*c*.

	designation	components # (4q + 2)s	# (4r)a	total	ground state
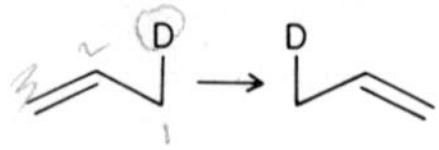 (D migration)	[1s,3s]				
treat as 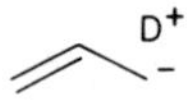 D^+ / $-$		0	0	0	forbidden
or 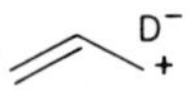 D^- / $+$		2	0	2	forbidden
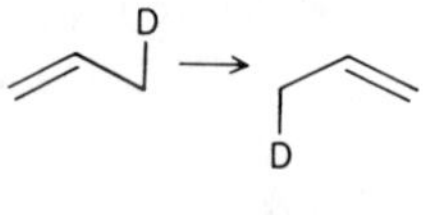 (D migration)	[1s,3a]	0 or 1	1 or 0	1	allowed
				1	allowed

(*c*)

Figure 10.1140 Woodward-Hoffmann rules: (*a*) electrocyclic processes; (*b*) cycloadditions; (*c*) sigmatropic shifts.

10.1150 Frontier-Orbital Approach

A methodology (developed by K. Fukui) for quickly predicting whether a given pericyclic reaction is allowed by examining the symmetry of the highest occupied molecular orbital (HOMO) and, if the reaction is bimolecular, the lowest unoccupied molecular orbital (LUMO) of the second partner.

Examples. The HOMO of ground state butadiene is ψ_2; in order for the σ bond to form between the termini it is predicted to close conrotatorially, Fig. 10.1150*a*.

The HOMO populated by the lowest energy excitation of butadiene is ψ_3; it closes disrotatorially, Fig. 10.1150*b*. Cycloadditions are handled by examining

ψ_2 , conrotatory (antarafacial)

(*a*)

ψ_3 , disrotatory (suprafacial)

(*b*)

the HOMO of one fragment (the donor) and the LUMO of the other (the acceptor). Although the more electronegative fragment is considered the acceptor, the results are unaffected by the roles of the fragments. The hydrogenation of ethylene, Fig. 10.1150*c*; the dimerization of ethylene, Fig. 10.1150*d*; the Diels-Alder reaction of ethylene and butadiene, Fig. 10.1150*e*.

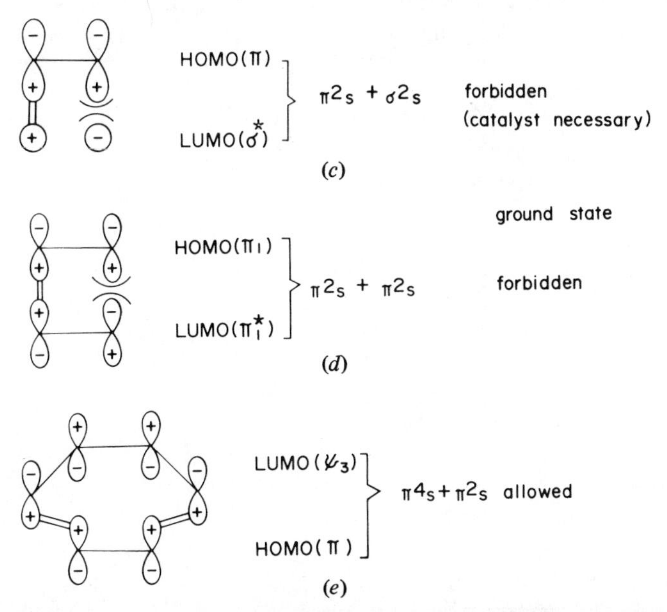

Figure 10.1150 Frontier-orbital approach using the HOMO-LUMO rules: (*a*) conrotatory cyclization; (*b*) disrotatory cyclization; (*c*) hydrogenation of ethylene; (*d*) dimerization of ethylene; (*e*) Diels-Alder reaction.

10.1160 Hückel-Möbius (H-M) Approach

Another method for quickly assessing whether a given pericyclic process is allowed (popularized by H. Zimmerman and M. J. S. Dewar) is to examine the cyclic array of orbitals at the transition state of the pericyclic reaction. If this array has no nodes or an even number of nodes between atoms (called a Hückel array), then it is stabilized (and the reaction allowed) if it contains $(4n + 2)$ electrons. If, on the other hand, it has an odd number of nodes between atoms [called a Möbius array (Sec. 3.660)], then it must possess $4n$ electrons to be stabilized. In assessing whether the array is Hückel or Möbius, nodes through atoms are not counted.

Examples

Process	Figure showing transition state	Number of nodes	Array	Number of e^-	Allowed?
Disrotatory Opening or closing	10.1160*a*	0	Hückel	4	No
Conrotatory Opening or closing	10.1160*b*	1	Möbius	4	Yes
$[4_s + 2_s]$-Cycloaddition	10.1160*c*	0	Hückel	6	Yes
$[2_s + 2_s]$-Cycloaddition	10.1160*d*	0	Hückel	4	No
$[1_a, 5_s]$-Sigmatropic shift	10.1160*e*	1	Möbius	6	No

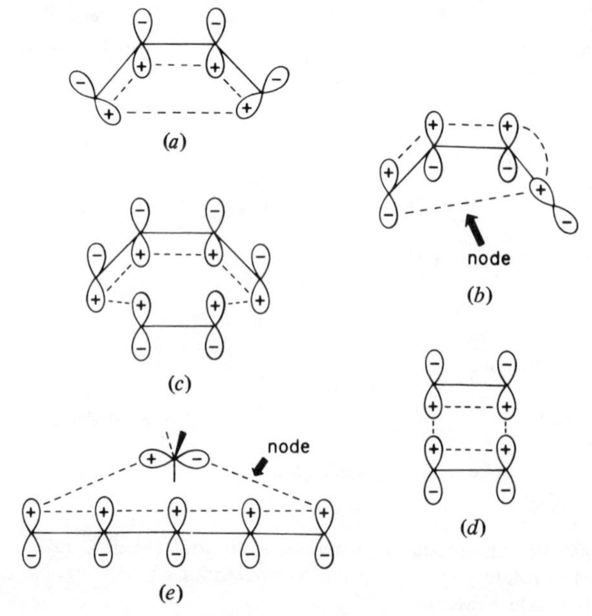

Figure 10.1160 Application of Hückel-Möbius rules: (*a*) forbidden; (*b*) allowed; (*c*) allowed; (*d*) forbidden; (*e*) forbidden.

11 Name Reactions, Type Reactions, and Their Mechanistic Pathways

CONTENTS

A large number of organic reactions used in synthesis are, as a matter of convenience, named after their discoverer(s) or after an investigator who either exploited or determined the scope of the reaction and in this way became identified with it. These reactions generally bear no particular relationship to each other and are called "name" reactions. Another large group of reactions are modeled on a reaction that was originally employed to synthesize a particular compound whose procedure was subsequently generalized to prepare related compounds. These "type" reactions are frequently named after the particular compound that was originally synthesized by that procedure.

Some of the name reactions described herein are now no longer recognized by their original names. Modern teaching, which generally tries to minimize memory work but instead emphasizes generalizations, is mostly responsible for this trend but a certain convenience remains in using the original names. In any case, the elaboration of the pathways for these reactions provides an opportunity to display the modern electron-pushing pictures (Sect. 10.010) for a wide variety of reactions, and perusal of this material will acquaint readers with this very important part of the organic chemistry vocabulary, thus providing the justification for this archival exercise. The pathways shown for these many reactions focus on the way electrons are moved to formally show bond-making and bond-breaking processes and should not be construed as "mechanisms" in the true meaning of the word. The reactions are not grouped together according to the type but are simply listed alphabetically.

11.010 Acetoacetic Ester Synthesis

The base-catalyzed alkylation or acylation of acetoacetic esters, Scheme 11.010*a* and *b*, respectively. The alkylation product is particularly useful for the synthesis of ketones via decarboxylation, Scheme 11.010*c*, or acids via deacetylation, Scheme 11.010*d*.

$H_3CC(=O)CH_2COEt$ (O O) + $EtO^- Na^+$ ⟶ $H_3CC(=O)\underline{C}HC(=O)OEt$ Na^+ + EtOH

+ $(H_3C)_2CH{-}Cl$ ⟶ $H_3CC(=O)CH(CH(CH_3)_2)C(=O)OEt$ + NaCl

(*a*)

$H_3CC(=O)\underline{C}HC(=O)OEt$ Na^+ + $Cl{-}C(=O){-}CH_2{-}C_6H_5$ ⟶ $H_3CC(=O)CH[C(Cl)(O^-)CH_2C_6H_5]C(=O)OEt$ Na^+ ⟶ $H_3CC(=O)CH[C(=O)CH_2C_6H_5]C(=O)OEt$ + NaCl

(*b*)

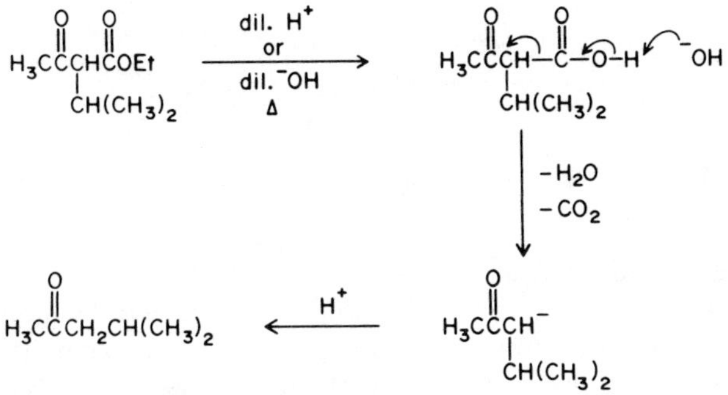

$$H_3CCCHCOEt \text{ (with } CH(CH_3)_2\text{)} \xrightarrow[\text{dil. }^-OH,\ \Delta]{\text{dil. } H^+ \text{ or}} H_3CCCH{-}C{-}O{-}H \ \ ^-OH \xrightarrow{-H_2O,\ -CO_2} H_3CCCH^- \xrightarrow{H^+} H_3CCCH_2CH(CH_3)_2$$

(*c*)

$$H_3CCCHCOEt \xrightarrow[^-OH]{\text{conc. alc.}} H_3CC{-}CHCOEt \xrightarrow{-H_3CCO_2H} {}^-CHCOEt \xrightarrow{RO{-}H} CH_2COEt \xrightarrow[\text{(saponification)}]{^-OH} CH_2C{-}O^- \xrightarrow{H^+} (H_3C)_2CHCH_2COH$$

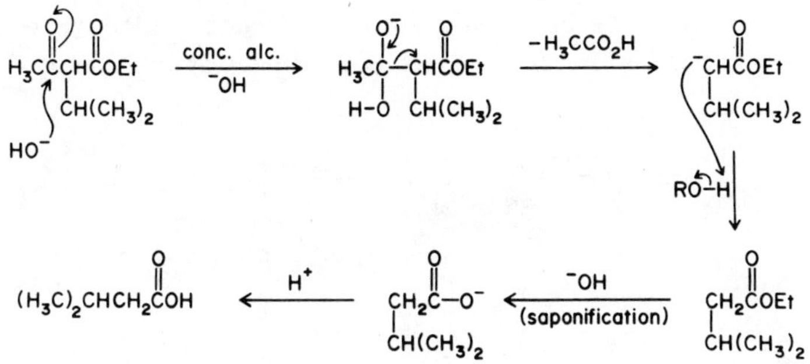

(*d*)

Scheme 11.010 Acetoacetic ester syntheses: (*a*) alkylation; (*b*) acylation; (*c*) ketone synthesis; (*d*) carboxylic acid synthesis.

11.020 Acyloin Condensation

The intermolecular, sodium-promoted condensation of two moles of ester, Scheme 11.020*a*, or the intramolecular condensation of a diester, Scheme 11.020*b*, to produce an α-hydroxyketone (acyloin). The latter, which leads to cyclic acyloins, is particularly valuable for the preparation of medium-size ring compounds.

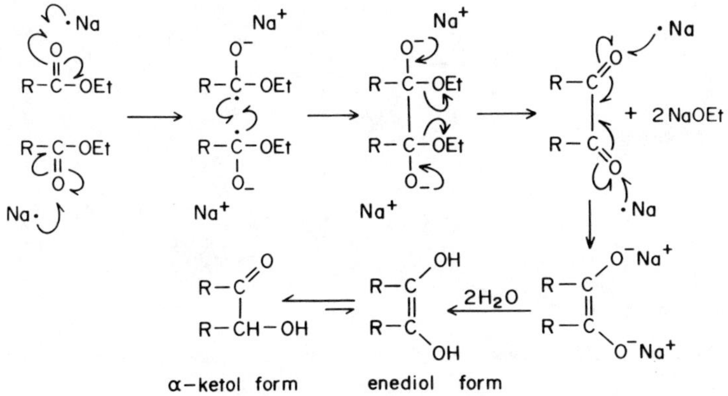

$$2\,R{-}C({=}O){-}OEt + 2\,Na\cdot \longrightarrow R{-}\dot{C}(O^-)(OEt){-}\dot{C}(O^-)(OEt){-}R \longrightarrow R{-}C(O^-)(OEt){-}C(O^-)(OEt){-}R \longrightarrow R{-}C({=}O){-}C({=}O){-}R + 2\,NaOEt$$

$$R{-}C(O^-Na^+){=}C(O^-Na^+){-}R \xrightarrow{2H_2O} R{-}C(OH){=}C(OH){-}R \rightleftharpoons R{-}C({=}O){-}CH(OH){-}R$$

α-ketol form   enediol form

(*a*)

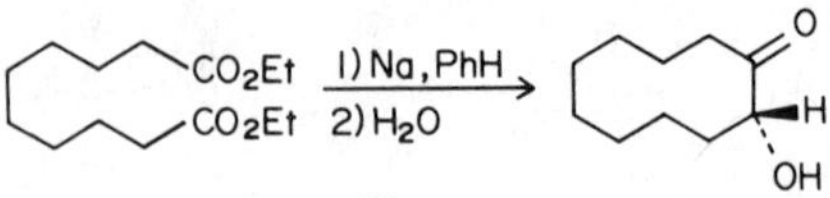

CO2Et CO2Et 1) Na, PhH 2) H2O O H OH

(*b*)

Scheme 11.020 Acyloin condensations: (*a*) intermolecular; (*b*) intramolecular.

11.030 Aldol Condensation

The condensation of two moles of aldehyde or two moles of ketone or of a dialdehyde or diketone to provide a β-hydroxycarbonyl compound. Two moles of acetaldehyde give the first compound in the series called aldol, $CH_3CHOHCH_2$-CHO. The aldol synthesis may be base- or acid-catalyzed, Scheme 11.030*a* and *b*, respectively. Both the base- and acid-catalyzed condensation are reversible. The reverse reaction is known as the **retro-aldol condensation.** Although ketones

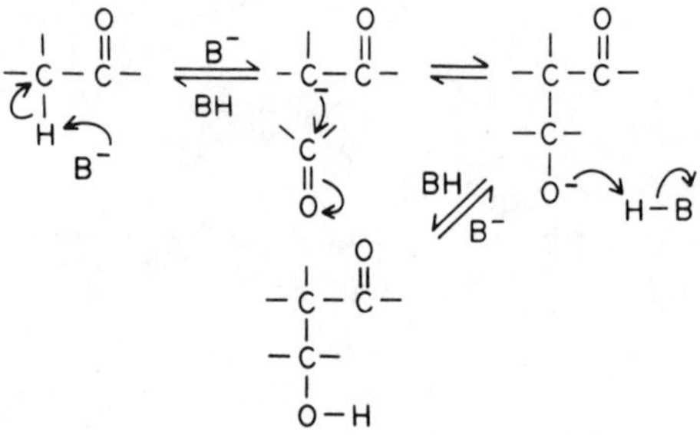

(*a*)

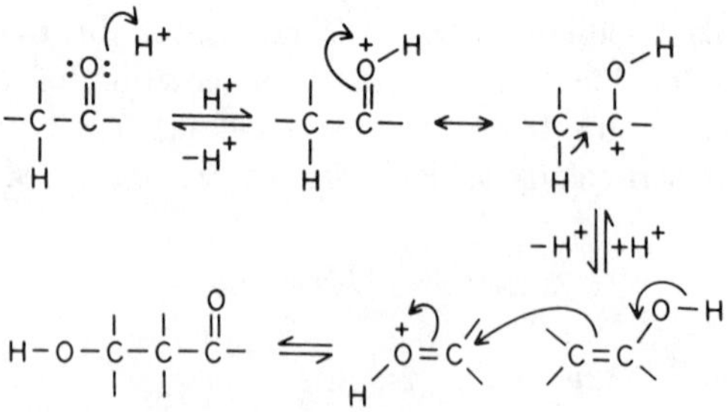

(*b*)

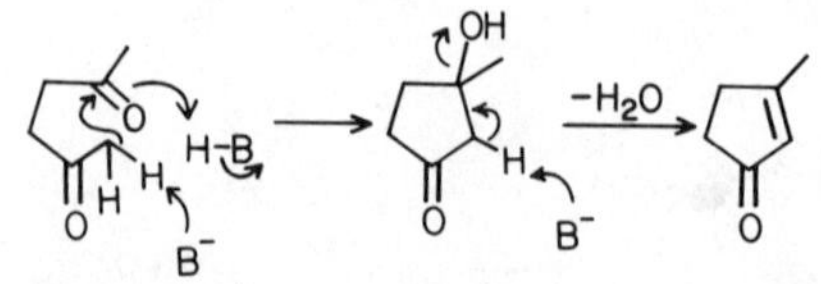

(*d*)

Scheme 11.030 Aldol condensation: (*a*) base-catalyzed; (*b*) acid-catalyzed; (*c*) cyclic aldol.

react more sluggishly than aldehydes, diketones react readily to give cyclic aldols, Scheme 11.030*c*; aldols readily lose water.

11.040 Arndt-Eistert Reaction

[L. Wolff, *Ann.*, **394**, 25 (1912); F. Arndt and B. Eistert, *Ber.*, 68, 200 (1935).] The conversion of a carboxylic acid, usually via its acyl chloride, to its next higher homolog on treatment with diazomethane. An intermediate in the synthesis is a diazoketone, which on treatment with colloidal silver undergoes a **Wolff rearrangement** (Sect. 11.1420) to a ketene. The rearrangement may be conducted in water, amines (other than tertiary), or alcohol to give, respectively, an acid, an amide, or an ester, Scheme 11.040*a*. As is apparent from this scheme, two moles of diazomethane are required to generate the diazoketone. The conversion of pivalic acid to its homolog is shown in Scheme 11.040*b*.

$R-C(=O)-Cl + {}^{-}CH_2-\overset{+}{N}\equiv N \longrightarrow R-C(-O^{-})(Cl)-CH_2-\overset{+}{N}\equiv N$

$\xrightarrow{-Cl^{-}}$

No excess CH_2N_2: $R-C(=O)-CH_2-\overset{+}{N_2}\ {}^{-}Cl \xrightarrow{-N_2} R-C(=O)-CH_2-Cl$

Excess CH_2N_2: $R-C(=O)-CH(H)-\overset{+}{N_2} + {}^{-}CH_2-\overset{+}{N}\equiv N \longrightarrow R-C(=O)-\overset{-}{C}H-\overset{+}{N_2} + H_3C-\overset{+}{N_2}\ Cl^{-} \xrightarrow{-N_2} H_3C-Cl$

$R-C(=O)-\overset{-}{C}H-\overset{+}{N_2} \xrightarrow{Ag,\ \text{or moist}\ Ag_2O} R-C(=O)-\ddot{C}H + N_2 \longrightarrow O{=}C{=}CHR$ (ketene)

$O{=}C{=}CHR \xrightarrow{H_2O} HOC(=O)CH_2R$

$O{=}C{=}CHR \xrightarrow{NH_3} H_2NC(=O)CH_2R$

$O{=}C{=}CHR \xrightarrow{R'OH} R'OC(=O)CH_2R$

(*a*)

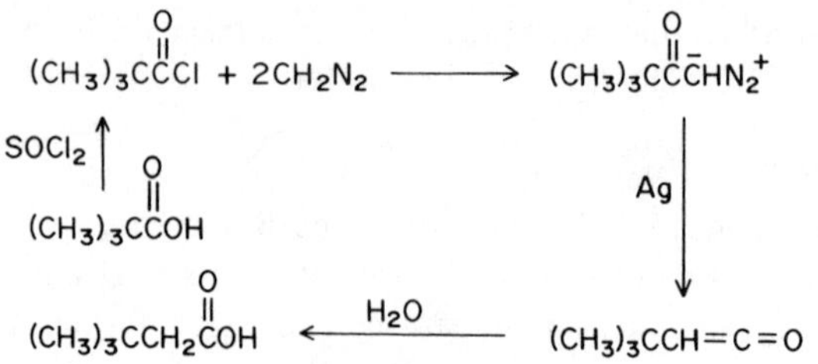

$$(CH_3)_3C\overset{O}{\overset{\|}{C}}Cl + 2CH_2N_2 \longrightarrow (CH_3)_3C\overset{O}{\overset{\|}{C}}\bar{C}HN_2^+$$

$$(CH_3)_3C\overset{O}{\overset{\|}{C}}OH \xrightarrow{SOCl_2} (CH_3)_3C\overset{O}{\overset{\|}{C}}Cl$$

$$(CH_3)_3C\overset{O}{\overset{\|}{C}}\bar{C}HN_2^+ \xrightarrow{Ag} (CH_3)_3CCH{=}C{=}O \xrightarrow{H_2O} (CH_3)_3CCH_2\overset{O}{\overset{\|}{C}}OH$$

(*b*)

Scheme 11.040 Arndt-Eistert synthesis: (*a*) general scheme; (*b*) homologation of pivalic acid.

11.050 Arbuzov Reaction (Michaelis-Arbuzov Reaction)

[A. Michaelis and R. Kaehne, *Ber.*, **31**, 1048 (1898); A. E. Arbuzov, *J. Russ. Phys. Chem. Soc.*, 38, 687 (1906).] The preparation of alkylphosphonate diesters from trialkylphosphites and alkyl halides, Scheme 11.050*a*. The preparation of diethyl benzylphosphonate is shown in Scheme 11.050*b*. The intermediate phosphonium salts are generally unstable and are not isolated.

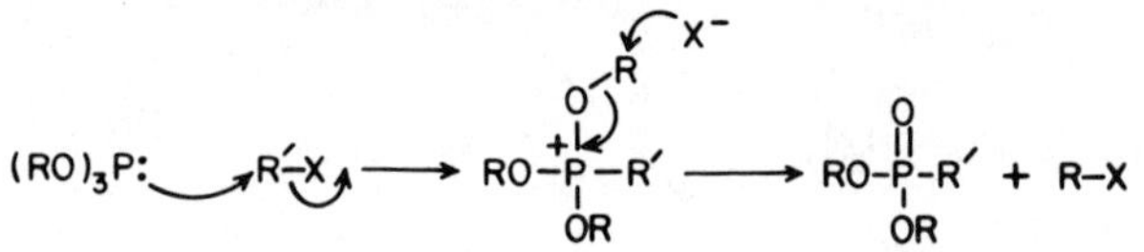

$$(RO)_3P: + R'{-}X \longrightarrow RO{-}\overset{+}{P}(OR)(OR){-}R'\;X^- \longrightarrow RO{-}\overset{O}{\overset{\|}{P}}(OR){-}R' + R{-}X$$

(*a*)

$$(EtO)_3P + PhCH_2Br \longrightarrow [(EtO)_3\overset{+}{P}{-}CH_2Ph]\overset{-}{Br} \longrightarrow (EtO)_2\overset{O}{\overset{\|}{P}}{-}CH_2Ph + EtBr$$

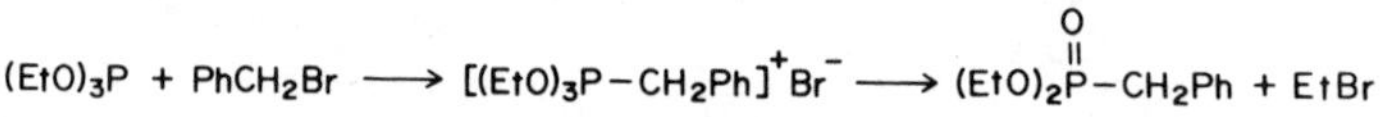

(*b*)

Scheme 11.050 Arbuzov reaction: (*a*) general scheme; (*b*) synthesis of diethyl benzylphosphonate.

11.060 Baeyer-Villiger Oxidation

[A. von Baeyer and V. Villiger, *Ber.*, 32, 3625 (1899).] The oxidation of a ketone to an ester by a peracid, Scheme 11.060*a*. Where there is more than one possibility for rearrangement, the migratory aptitude of the group controls which group will preferentially migrate. The conversion of isopropyl methyl ketone to isopropyl acetate is shown in Scheme 11.060*b*. Oxidation of a cyclic ketone, e.g., cyclopentanone, gives a lactone, Scheme 11.060*c*.

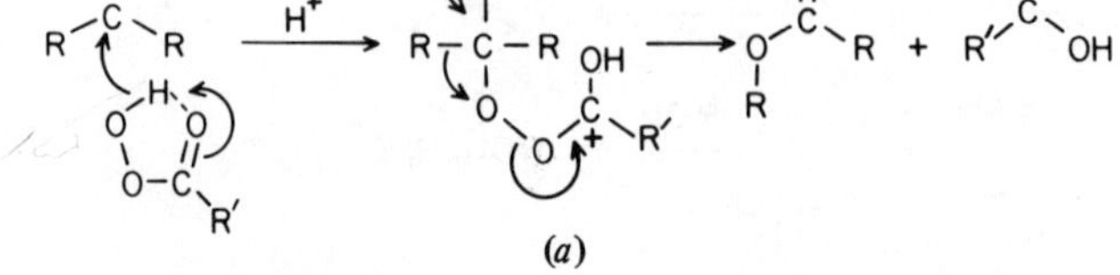

$$R{-}\overset{O}{\overset{\|}{C}}{-}R + R'CO_3H \xrightarrow{H^+} R{-}C(OH)(R){-}O{-}O{-}\overset{+}{C}(OH){-}R' \longrightarrow R{-}\overset{O}{\overset{\|}{C}}{-}OR + R'\overset{O}{\overset{\|}{C}}OH$$

(*a*)

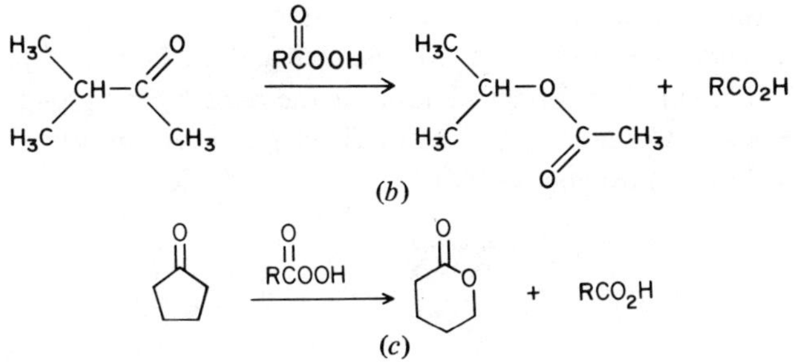

$$(CH_3)_2CH\text{-}C(=O)CH_3 \xrightarrow{RC(=O)OOH} (CH_3)_2CH\text{-}O\text{-}C(=O)CH_3 + RCO_2H$$

(*b*)

$$\text{cyclopentanone} \xrightarrow{RC(=O)OOH} \text{δ-valerolactone} + RCO_2H$$

(*c*)

Scheme 11.060 Baeyer-Villiger oxidation: (*a*) general scheme; (*b*) synthesis of isopropyl acetate; (*c*) conversion of a cyclic ketone to a lactone.

11.070 Bamford-Stevens Reaction

[W. R. Bamford and T. S. Stevens, *J. Chem. Soc.*, 4735 (1952).] The thermal decomposition of tosylhydrazones of aliphatic ketones under basic conditions to afford alkenes and cyclopropanes. In aprotic solvents the reaction proceeds through a carbene mechanism, and in protic solvents through a carbenium ion mechanism, Scheme 11.070.

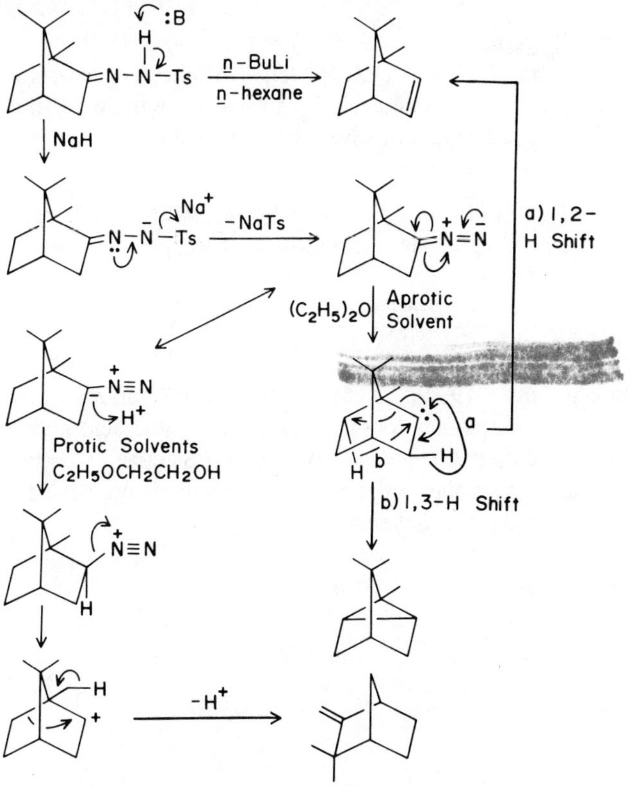

:B
H
=N−N−Ts
n-BuLi
n-hexane
NaH
Na⁺
=N−N−Ts
−NaTs
=N=N
a) 1,2-H Shift
$(C_2H_5)_2O$
Aprotic Solvent
N≡N
H⁺
Protic Solvents
$C_2H_5OCH_2CH_2OH$
a
H
b
H
b) 1,3-H Shift
N≡N
H
H
+
$-H^+$

Scheme 11.070 Bamford-Stevens reaction.

11.080 Barbier-Wieland Degradation

[H. Wieland, *Ber.*, **45**, 484 (1912); P. Barbier and R. Loquin, *Compt. Rend.*, **156**, 1445 (1913).] The formal removal of the α-methylene group of an ester to give the next lower homologous acid. The degradation consists of a four-step sequence as shown in Scheme 11.080.

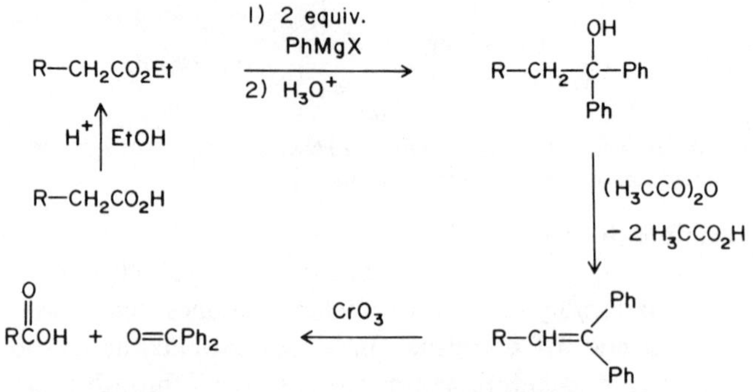

$R{-}CH_2CO_2H \xrightarrow{H^+,\ EtOH} R{-}CH_2CO_2Et \xrightarrow[2)\ H_3O^+]{1)\ 2\ \text{equiv. } PhMgX} R{-}CH_2{-}C(OH)Ph_2$

$R{-}CH_2{-}C(OH)Ph_2 \xrightarrow[-\,2\ H_3CCO_2H]{(H_3CCO)_2O} R{-}CH{=}CPh_2 \xrightarrow{CrO_3} RCOOH + O{=}CPh_2$

Scheme 11.080 Barbier-Wieland degradation.

11.090 Bart Reaction

[H. Bart, *Ger. Patent* 250264 (1910).] The conversion of an aromatic amine via its diazonium salt to an arsonic acid by treatment with an alkali arsenite in the presence of powdered copper or silver, Scheme 11.090.

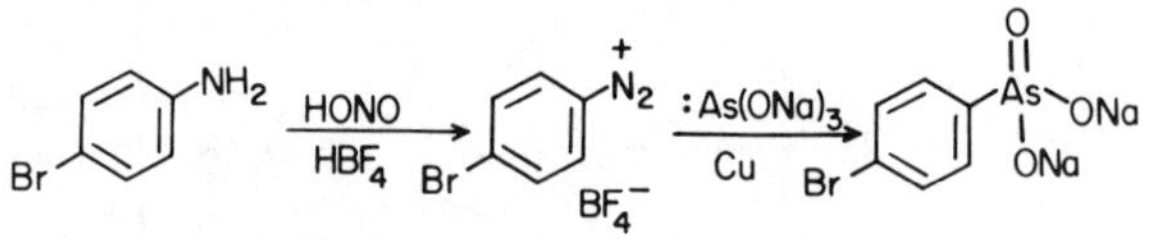

$Br{-}C_6H_4{-}NH_2 \xrightarrow[HBF_4]{HONO} Br{-}C_6H_4{-}N_2^+\ BF_4^- \xrightarrow[Cu]{:As(ONa)_3} Br{-}C_6H_4{-}As(=O)(ONa)_2$

Scheme 11.090 Bart reaction.

11.100 Beckmann Rearrangement

[E. Beckmann, *Ber.*, **19**, 988 (1886).] The acid-catalyzed (protic or Lewis) rearrangement of ketoximes to substituted carboxylic acid amides. The group *trans* to the nucleofuge (Sect. 10.090) migrates, Scheme 11.100*a*. The commercially important 6-caprolactam (a nylon precursor) is prepared from cyclohexanone oxime by this route, Scheme 11.100*b*.

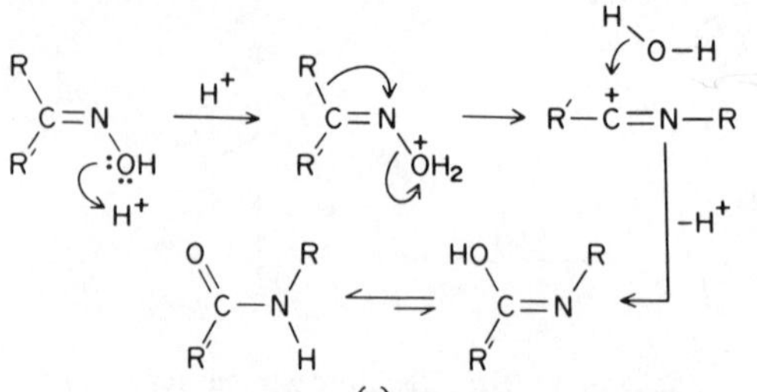

$RR'C{=}N{-}OH \xrightarrow{H^+} RR'C{=}N{-}OH_2^+ \longrightarrow R'{-}\overset{+}{C}{=}N{-}R \xrightarrow{H_2O,\ -H^+} R'(HO)C{=}N{-}R \rightleftharpoons R'C(=O){-}NHR$

(*a*)

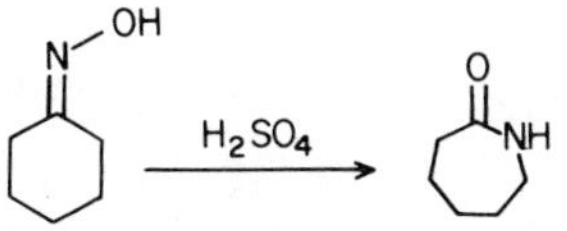

H₂SO₄

(*b*)

Scheme 11.100 Beckmann rearrangement: (*a*) general scheme; (*b*) 6-caprolactam synthesis.

11.110 Beckmann Rearrangement, Abnormal (Second Order)

The acid-catalyzed fragmentation of ketoximes to nitriles instead of the usual acid amide (Beckmann rearrangement). This reaction occurs when fragmentation leads to an unusually stable carbocation (see also Lossen Rearrangement, Sect. 11.800, and Ritter Reaction, Sect. 11.1090), Scheme 11.110*a* and *b*.

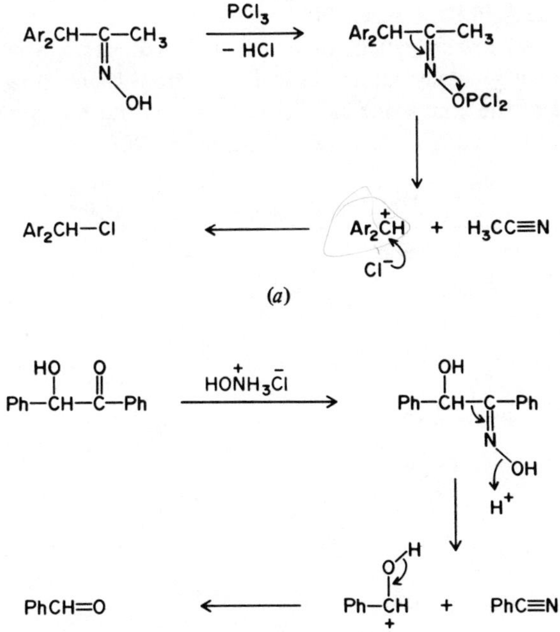

$$Ar_2CH{-}C(=N{-}OH){-}CH_3 \xrightarrow[-\,HCl]{PCl_3} Ar_2CH{-}C(=N{-}OPCl_2){-}CH_3 \longrightarrow Ar_2\overset{+}{C}H + H_3CC{\equiv}N;\quad Ar_2\overset{+}{C}H + Cl^- \longrightarrow Ar_2CH{-}Cl$$

(*a*)

$$Ph{-}CH(OH){-}C(=O){-}Ph \xrightarrow{HO\overset{+}{N}H_3\overset{-}{Cl}} Ph{-}CH(OH){-}C(=N{-}OH){-}Ph \xrightarrow{H^+} Ph{-}\overset{+}{C}H{-}OH + PhC{\equiv}N;\quad Ph{-}\overset{+}{C}H{-}OH \longrightarrow PhCH{=}O$$

(*b*)

Scheme 11.110 Abnormal Beckmann rearrangement: (*a*) a cleavage to acetonitrile; (*b*) a cleavage to benzonitrile.

11.120 Benzidine Rearrangement

The acid-catalyzed rearrangement of hydrazobenzenes to 4,4′-diaminobiphenyls, Scheme 11.120. If a substituent occupies the *para* position of the hydrazobenzene, a *p*-aminodiphenylamine, *p*-$NH_2{-}C_6H_4{-}NH{-}C_6H_4R$, results; this is called the **semidine rearrangement**.

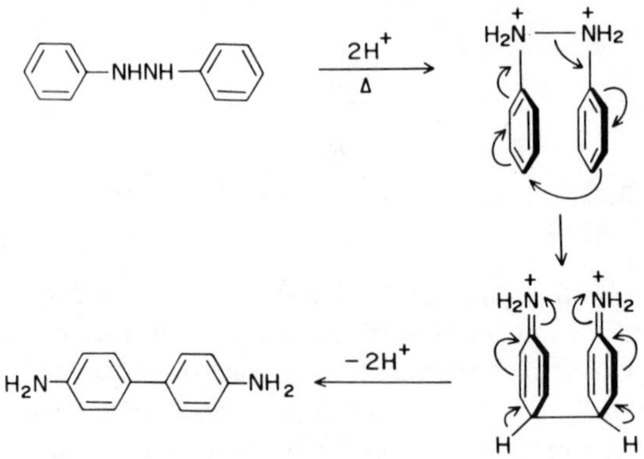

Scheme 11.120 Benzidine rearrangement.

11.130 Benzilic Acid Rearrangement

The base-catalyzed rearrangement of benzil to benzilic acid, Scheme 11.130*a*. The reaction is not generally applicable to 1,2-diketones containing α-methylene groups because of the competing aldol condensations, but the synthesis of citric acid from ketopinic acid is an exception, Scheme 11.130*b*.

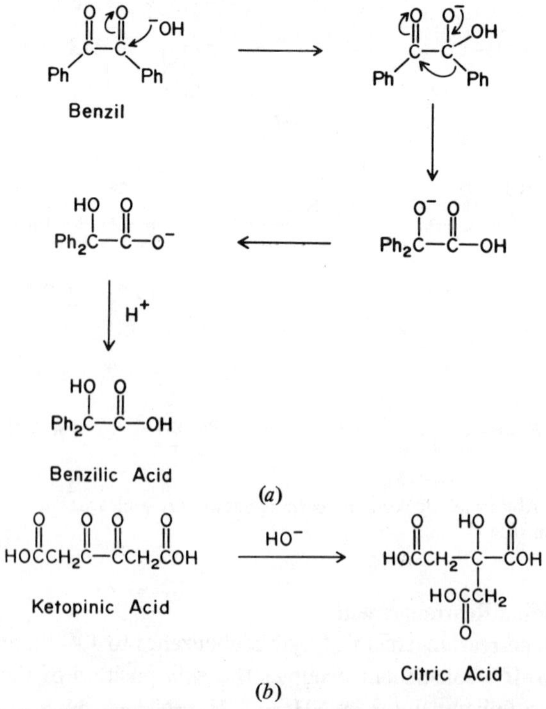

Benzil

Benzilic Acid

(*a*)

Ketopinic Acid

Citric Acid

(*b*)

Scheme 11.130 Benzilic acid rearrangement: (*a*) benzilic acid synthesis; (*b*) citric acid synthesis.

11.140 Benzoin Condensation

The cyanide ion–catalyzed, intermolecular condensation of an aromatic aldehyde to give an acyloin. The condensation of benzaldehyde gives benzoin, Scheme 11.140.

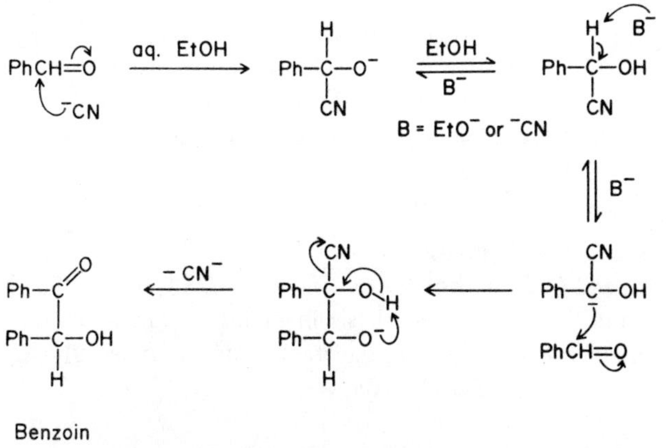

Scheme 11.140 Benzoin condensation.

11.150 Bischler-Napieralski Synthesis

[A. Bischler and B. Napieralski, *Ber.*, **26**, 1903 (1893).] The two-step conversion of *N*-acyl-β-phenylamines to 1-substituted isoquinolines, Scheme 11.150.

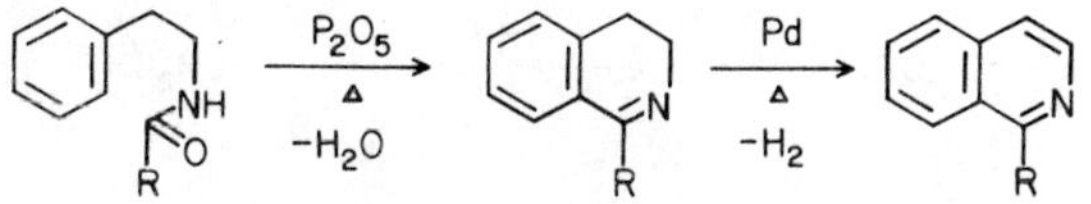

Scheme 11.150 Bischler-Napieralski synthesis.

11.160 Birch Reduction

[A. J. Birch, *J. Chem. Soc.*, 430 (1944).] A two-electron reduction of aromatic rings by means of a solution of an alkali metal in liquid ammonia, involving a radical anion intermediate, Scheme 11.160*a*. When a substituted aromatic compound is employed, a mixture of isomers may be obtained whose composition depends on the nature of the substituent(s), Scheme 11.160*b*.

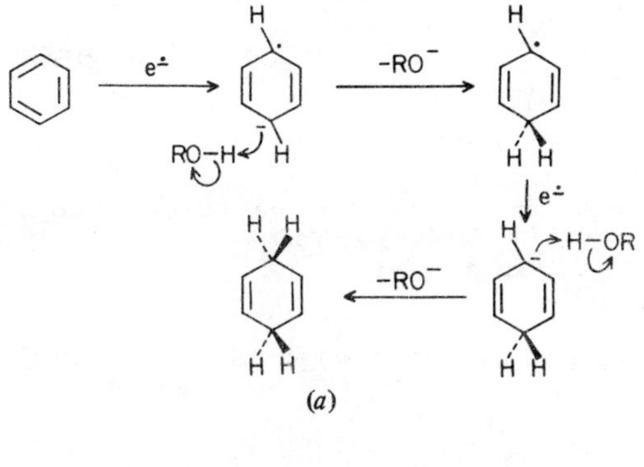

(*a*)

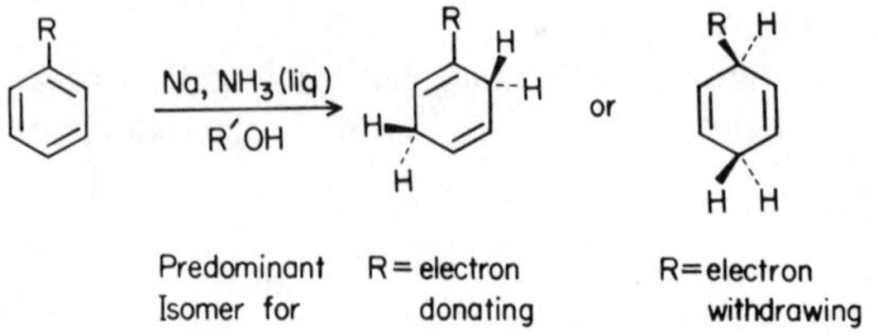

Na, NH_3 (liq), R′OH

or

Predominant Isomer for R = electron donating

R = electron withdrawing

(*b*)

Scheme 11.160 Birch reduction: (*a*) reduction of an unsubstituted aromatic; (*b*) reduction of a substituted aromatic.

11.170 Boord Olefin Synthesis

[L. C. Swallen and C. E. Boord, *J. Am. Chem. Soc.*, **52**, 651 (1930).] A versatile regiospecific, multistep olefin synthesis involving an aldehyde and a Grignard reagent, Scheme 11.170*a*. The use of acetaldehyde and *n*-pentyl magnesium bromide gives 1-heptene, Scheme 11.170*b*.

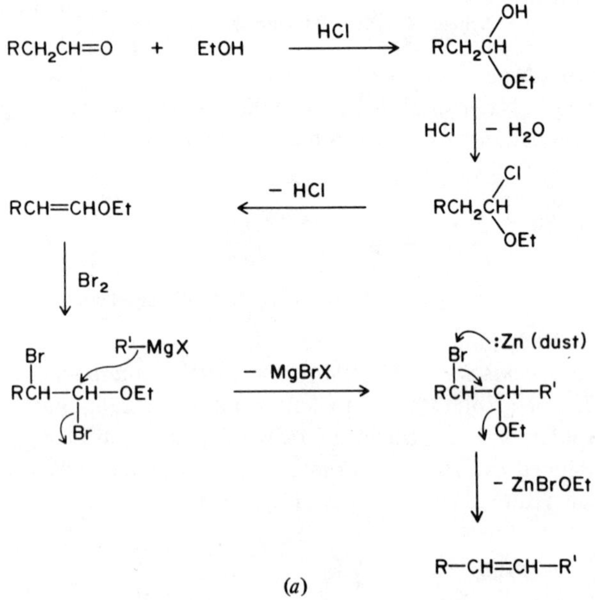

$$RCH_2CH{=}O + EtOH \xrightarrow{HCl} RCH_2CH(OH)(OEt)$$

$$RCH_2CH(OH)(OEt) \xrightarrow[-\,H_2O]{HCl} RCH_2CH(Cl)(OEt)$$

$$RCH_2CH(Cl)(OEt) \xrightarrow{-\,HCl} RCH{=}CHOEt$$

$$RCH{=}CHOEt \xrightarrow{Br_2} RCH(Br){-}CH(Br){-}OEt$$

$$RCH(Br){-}CH(Br){-}OEt + R'{-}MgX \xrightarrow{-\,MgBrX} RCH(Br){-}CH(OEt){-}R'$$

$$RCH(Br){-}CH(OEt){-}R' \xrightarrow[-\,ZnBrOEt]{:Zn\ (dust)} R{-}CH{=}CH{-}R'$$

(*a*)

$$CH_3CH{=}O \xrightarrow[\text{3) } CH_3(CH_2)_4{-}MgBr \text{; 4) Zn (dust)}]{\text{1) EtOH, HCl; 2) } Br_2} CH_2{=}CH(CH_2)_4CH_3$$

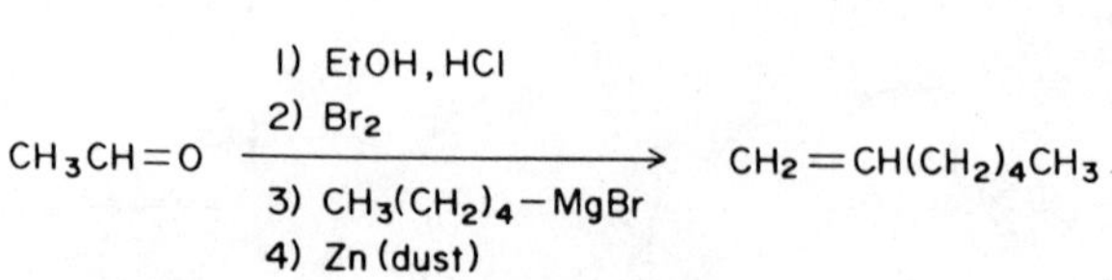

(*b*)

Scheme 11.170 Boord olefin synthesis: (*a*) general scheme; (*b*) synthesis of 1-heptene.

11.180 von Braun Degradation

[J. von Braun, *Ber.*, **33**, 1438, 2734 (1900).] The displacement of one of the R groups of a tertiary amine by the CN of cyanogen bromide to give an alkyl bromide and an *N,N*-disubstituted cyanamide, Scheme 11.180*a*. Cyclic amines undergo the reaction to give acyclic cyanamides, Scheme 11.180*b*. The reaction of an *N*-acyl cyclic amine with PX_5 is a modification that is useful for the synthesis of α, ω-dihaloalkanes, Scheme 11.180*c*.

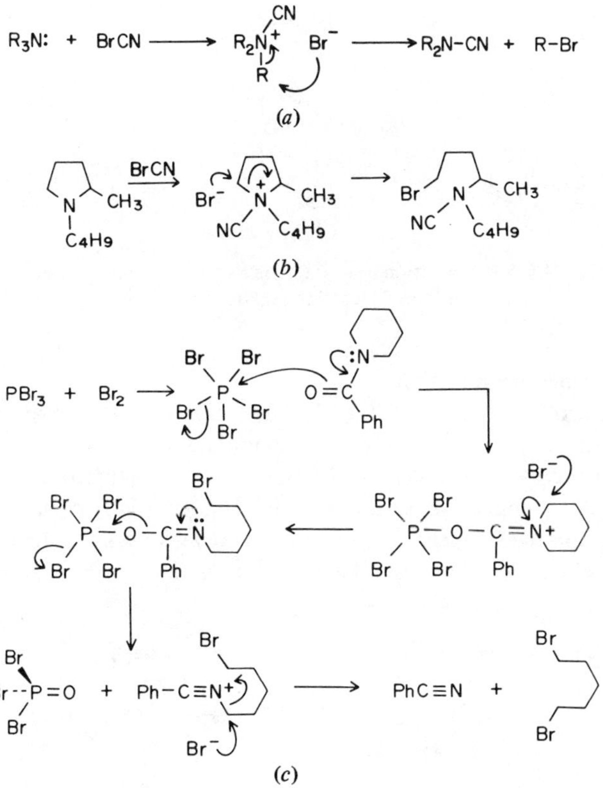

$$R_3N: + BrCN \longrightarrow R_2\overset{+}{N}(CN)R \;\; Br^- \longrightarrow R_2N{-}CN + R{-}Br$$

(*a*)

(*b*)

(*c*)

Scheme 11.180 von Braun degradation: (*a*) general scheme; (*b*) reaction of cyclic amines; (*c*) dihaloalkane synthesis.

11.190 Bucherer Reaction

[H. T. Bucherer, *J. Prakt. Chem.*, [2] **69**, 49 (1904).] The conversion of a naphthol to a naphthylamine or vice versa, Scheme 11.190*a*. 2-Naphthol can be converted to 2-aminonaphthalene by heating in aqueous ammonium sulfite in a sealed tube. The reverse reaction is obtained by heating 2-aminonaphthalene with sodium bisulfite, Scheme 11.190*b*.

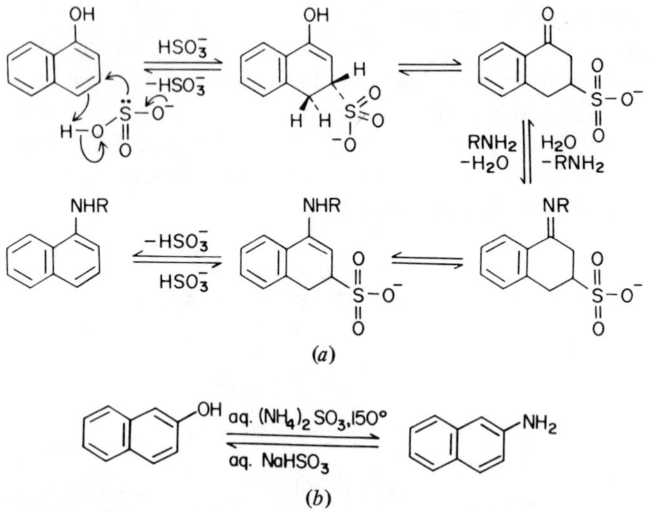

(*a*)

aq. $(NH_4)_2SO_3$, 150°

aq. $NaHSO_3$

(*b*)

Scheme 11.190 Bucherer reaction: (*a*) interconversion of naphthol and naphthylamine; (*b*) the interconversion under specific reaction conditions.

11.200 Cannizzaro Reaction

[S. Cannizzaro, *Ann.*, 88, 129 (1853).] The base-catalyzed disproportionation of two moles of aldehyde to give an alcohol and an acid (or its derivative) depending on the nature of the base, Scheme 11.200. If α-hydrogens are present in the aldehyde, competitive reactions (aldol condensation) occur, and thus the reaction is confined largely to aromatic aldehydes or alkyl aldehydes in which the carboxaldehyde group is attached to a tertiary carbon atom (see Sect. 11.1320).

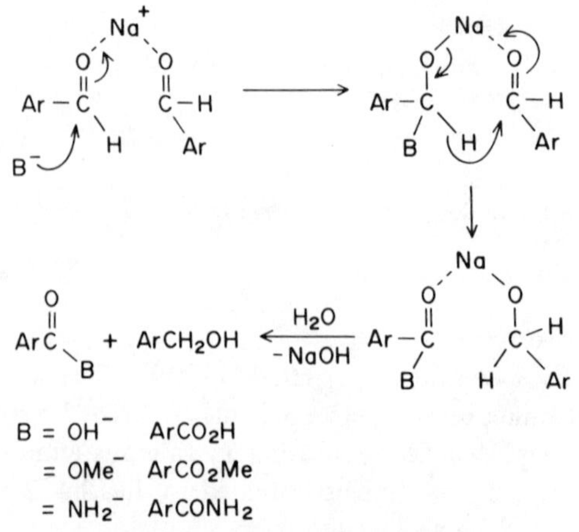

$B = OH^-$ $ArCO_2H$

$= OMe^-$ $ArCO_2Me$

$= NH_2^-$ $ArCONH_2$

Scheme 11.200 Cannizzaro reaction.

11.210 Chapman Rearrangement

[A. W. Chapman, *J. Chem. Soc.*, **127**, 1992 (1925).] The pyrolysis of *N,O*-diarylbenzimidates to give *N,N*-diarylbenzamides, Scheme 11.210*a* and *b*.

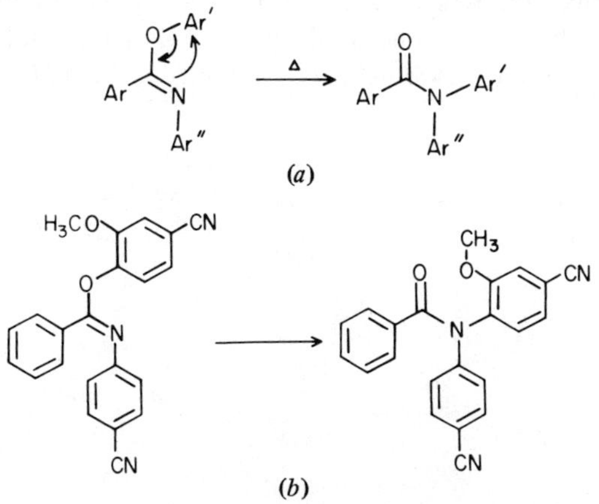

O—Ar′ Ar N Ar″ Δ → O Ar N—Ar′ Ar″

(*a*)

H_3CO CN O N CN → CH_3 O O CN N CN

(*b*)

Scheme 11.210 Chapman rearrangement: (*a*) general scheme; (*b*) reaction of a substituted aryl compound.

11.220 Chichibabin Reaction

[A. E. Chichibabin and O. A. Zeide, *J. Russ. Phys. Chem. Soc.*, **46**, 1216 (1914).] The amination of a pyridine ring by amide ion, Scheme 11.220. If the *ortho* positions are blocked, nucleophilic substitution occurs at the *para* position.

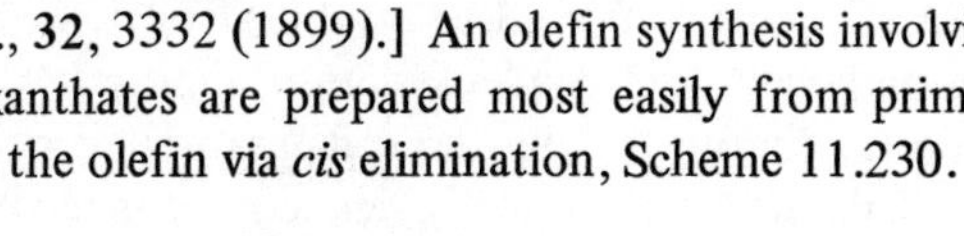

N $\bar{N}H_2$ —$NaNH_2$→ N NH_2 H —$-NaH$→ N NH_2

Scheme 11.220 Chichibabin reaction.

11.230 Chugaev Reaction

[L. Chugaev, *Ber.*, **32**, 3332 (1899).] An olefin synthesis involving pyrolysis of a xanthate. Such xanthates are prepared most easily from primary alcohols and the reaction gives the olefin via *cis* elimination, Scheme 11.230.

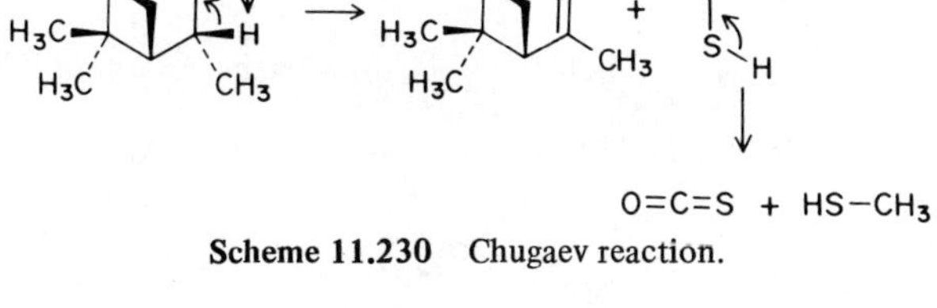

SCH_3 O S H_3C H_3C H CH_3 → H_3C H_3C CH_3 + O SCH_3 S H

O=C=S + HS—CH_3

Scheme 11.230 Chugaev reaction.

11.240 Claisen Condensation

[L. Claisen and O. Lowman, *Ber.*, **20**, 651 (1887).] The base-catalyzed condensation of two moles of an ester to give a β-keto ester. Thus ethyl acetate reacts to give ethyl acetoacetate, Scheme 11.240*a*. If two different esters are used, four products are possible, but if one of the two esters has no α-hydrogens, good yields of a single β-keto ester often may be obtained, Scheme 11.240*b*.

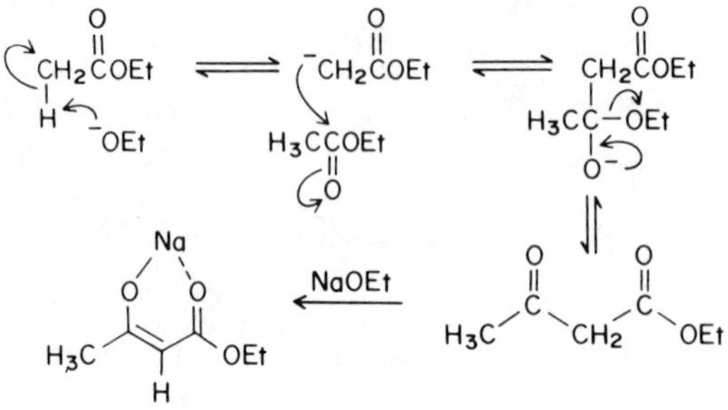

(*a*)

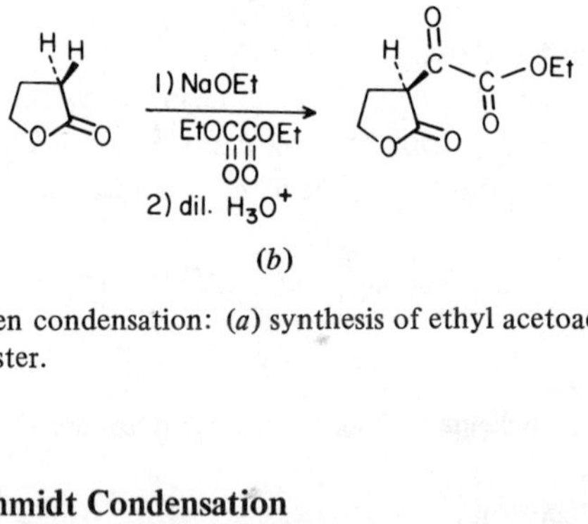

(*b*)

Scheme 11.240 Claisen condensation: (*a*) synthesis of ethyl acetoacetate; (*b*) reaction of a cyclic plus an acyclic ester.

11.250 Claisen-Schmidt Condensation

[L. Claisen and A. Claparède, *Ber.*, **14**, 2460 (1881); J. G. Schmidt, *Ber.*, **14**, 1459 (1881).] A base-catalyzed condensation of an aromatic aldehyde with an aliphatic aldehyde (or ketone) to give an α,β-unsaturated carbonyl compound, Scheme 11.250. This reaction is really an aldol condensation followed by dehydration.

$$PhC(=O)-CH_3 + PhCHO \xrightarrow[-H_2O]{NaOEt} PhC(=O)-CH=CHPh$$

Scheme 11.250 Claisen-Schmidt condensation.

11.260 Claisen Rearrangement

[L. Claisen, *Ber.*, **45**, 3157 (1912).] The thermally induced rearrangement of an allyl phenyl ether to an *o*-allylphenol, Scheme 11.260*a*. If both *ortho* positions are occupied, the rearrangement gives the *para* isomer via a subsequent Cope rearrangement, Scheme 11.260*b*. These are examples of [3,3]-sigmatropic shifts (Sect. 10.1040). The reaction also proceeds with allyl vinyl ether, yielding γ,δ-unsaturated carbonyl compounds.

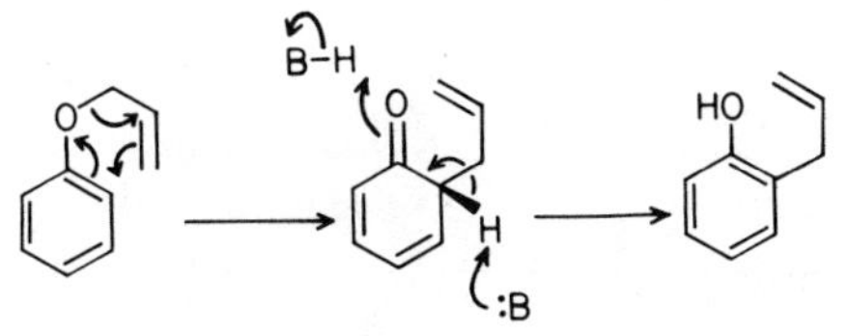

B–H O H :B HO

(*a*)

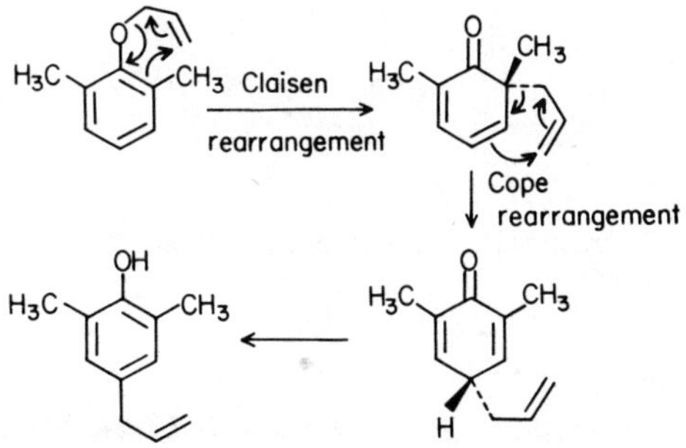

H_3C O CH_3 Claisen rearrangement H_3C O CH_3

Cope rearrangement

H_3C OH CH_3 H_3C O CH_3 H

(*b*)

Scheme 11.260 Claisen rearrangement: (*a*) reaction of allyl phenyl ether; (*b*) with the *ortho* positions blocked.

11.270 Clemmensen Reduction

[E. Clemmensen, *Ber.*, **46**, 1837 (1913).] The reduction of a carbonyl to a methylene group using activated zinc and hydrochloric acid, Scheme 11.270*a*. The zinc is activated by treating it briefly with a small quantity of aqueous $HgCl_2$. The reduction of acetophenone to ethylbenzene, Scheme 11.270*b*, is essentially quantitative.

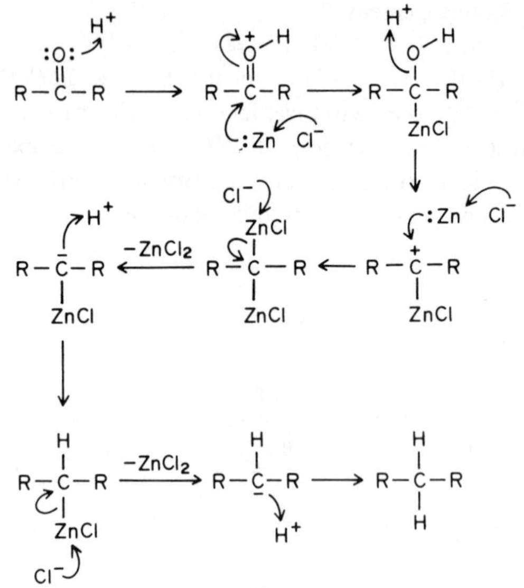

(*a*)

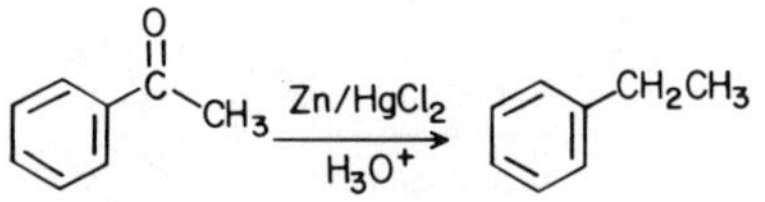

(*b*)

Scheme 11.270 Clemmensen reduction: (*a*) general scheme; (*b*) reduction of acetophenone.

11.280 Cope Elimination

[A. C. Cope, T. T. Foster, and P. H. Towle, *J. Am. Chem. Soc.*, **71**, 3929 (1949).] Pyrolysis of a trialkylamine oxide having a hydrogen β to the amine function to give an olefin and dialkylhydroxylamine by a *cis* elimination, Scheme 11.280.

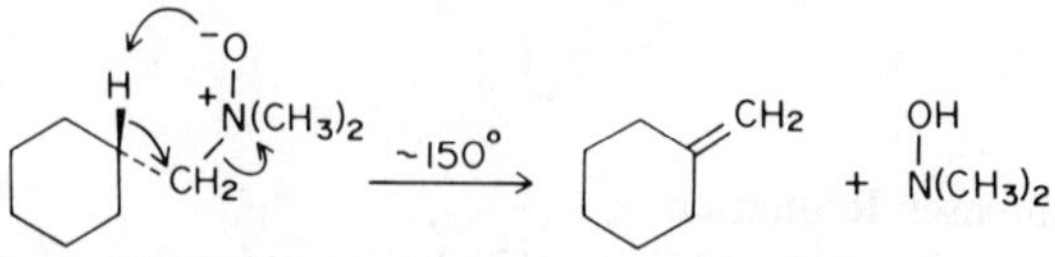

Scheme 11.280 Cope elimination.

11.290 Cope Rearrangement

[A. C. Cope and E. M. Hardy, *J. Am. Chem. Soc.*, **62**, 441 (1940).] The thermal [3,3]-sigmatropic shift in a 1,5-diene resulting in its isomerization, e.g., the conversion of *cis*-1,2-divinylcyclobutane to 1,5-cyclooctadiene, Scheme 11.290; analogous to the Claisen rearrangement (Sec. 11.260).

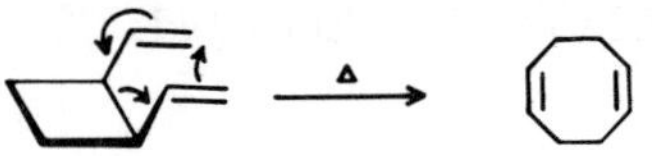

Scheme 11.290 Cope rearrangement.

11.300 Criegee Reaction

[R. Criegee, *Ber.*, **64**, 260 (1931).] Oxidative cleavage of a 1,2-glycol by lead tetraacetate, Scheme 11.300.

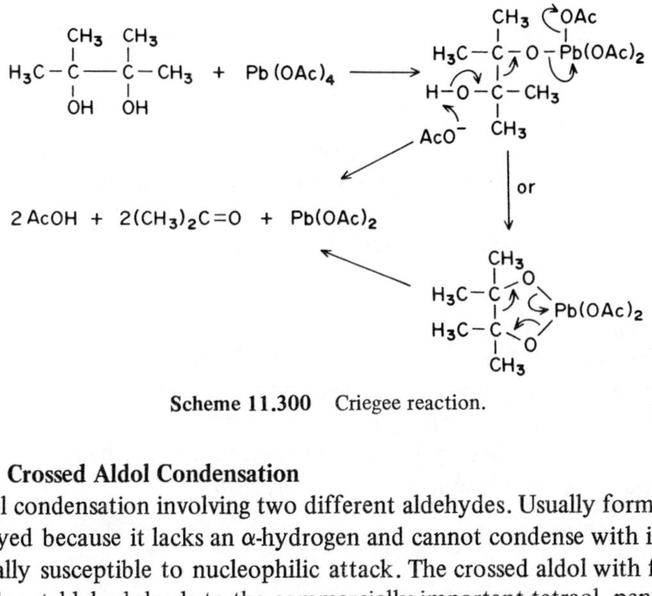

$$(CH_3)_2C(OH){-}C(OH)(CH_3)_2 + Pb(OAc)_4 \longrightarrow \text{intermediate} \longrightarrow 2\,AcOH + 2(CH_3)_2C{=}O + Pb(OAc)_2$$

Scheme 11.300 Criegee reaction.

11.310 Crossed Aldol Condensation

The aldol condensation involving two different aldehydes. Usually formaldehyde is employed because it lacks an α-hydrogen and cannot condense with itself, and is especially susceptible to nucleophilic attack. The crossed aldol with formaldehyde and acetaldehyde leads to the commercially important tetraol, pentaerythritol, Scheme 11.310. (See Sec. 11.200.)

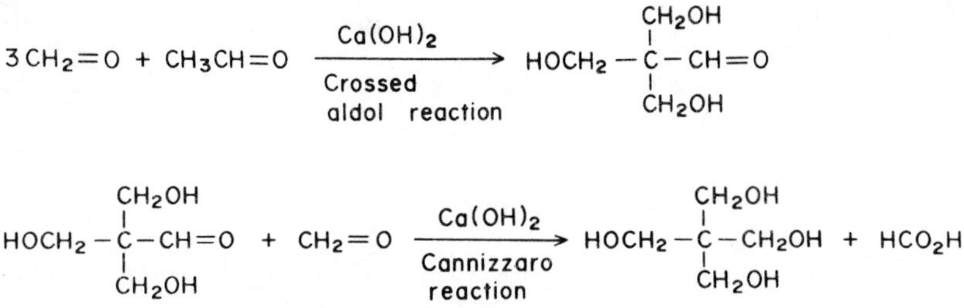

$$3\,CH_2{=}O + CH_3CH{=}O \xrightarrow[\text{Crossed aldol reaction}]{Ca(OH)_2} HOCH_2{-}C(CH_2OH)_2{-}CH{=}O$$

$$HOCH_2{-}C(CH_2OH)_2{-}CH{=}O + CH_2{=}O \xrightarrow[\text{Cannizzaro reaction}]{Ca(OH)_2} HOCH_2{-}C(CH_2OH)_2{-}CH_2OH + HCO_2H$$

Scheme 11.310 Crossed aldol condensation.

11.320 Curtius Degradation

[T. Curtius, *J. Prakt. Chem.*, [2] **50**, 275 (1894).] Conversion of a carboxylic acid through an acyl azide to an amine, RNH_2, with one less carbon, e.g., the conversion of benzoic acid to aniline, Scheme 11.320.

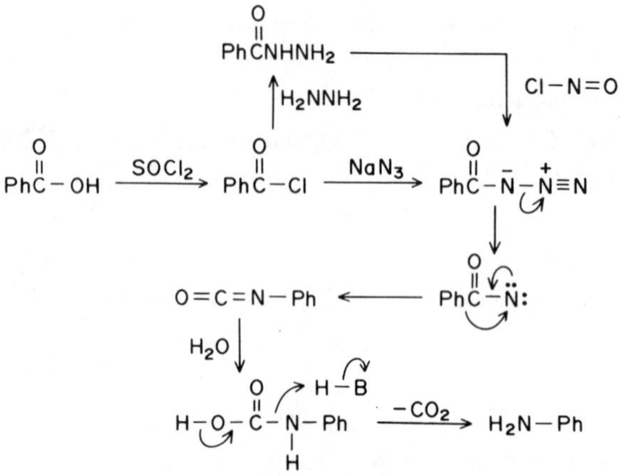

O
PhCNHNH2
H2NNH2
Cl−N=O
O O O
PhC−OH SOCl2 PhC−Cl NaN3 PhC−N−N≡N
O
O=C=N−Ph PhC−N:
H2O
O H−B
H−O−C−N−Ph −CO2 H2N−Ph
H

Scheme 11.320 Curtius degradation.

11.330 Darzens Glycidic Ester Synthesis

[G. Darzens, *Compt. Rend.*, **139**, 1214 (1904).] The reaction of an α-halo ester with an aldehyde (or ketone) in an aldol-type reaction to give a glycidic ester, Scheme 11.330*a*. The synthetic utility of the reaction hinges on two successive steps, hydrolysis, followed by decarboxylation, to give aldehydes that are difficult to prepare by other methods, Scheme 11.330*b*.

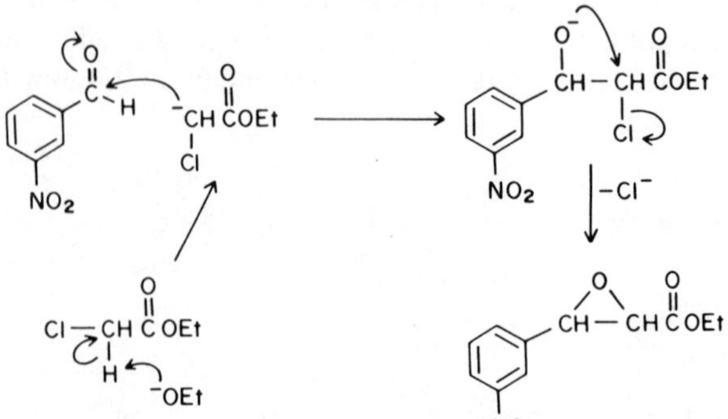

O O O− O
C−H −CH COEt CH−CH COEt
Cl Cl
NO2 NO2 −Cl−
O O O
Cl−CH C OEt CH−CH C OEt
H −OEt
NO2

(*a*)

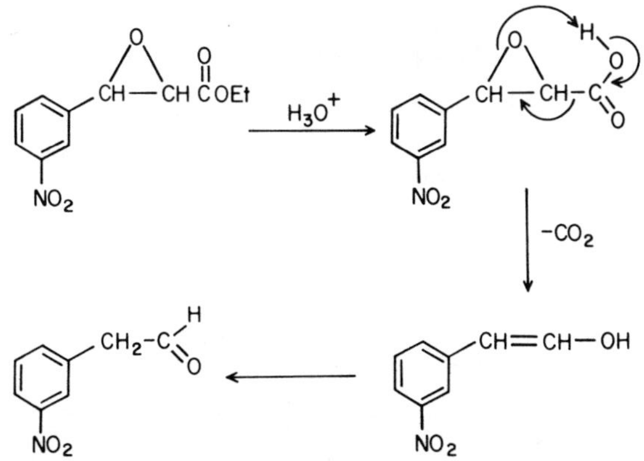

(b)

Scheme 11.330 Darzens glycidic ester synthesis: (a) synthesis of glycidic esters; (b) synthesis of aldehydes.

11.340 Demjanov Rearrangement

[N. J. Demjanov and M. Lushnikov, *J. Russ. Phys. Chem. Soc.*, **35**, 26 (1903).] The decomposition of an aliphatic diazonium compound to generate a carbenium ion that may then undergo rearrangement, Scheme 11.340*a*. This rearrangement is commonly employed in ring enlargement; **Tieffeneau-Demjanov** ring expansion (Scheme 11.340*b* and *c*), or ring contraction (Scheme 11.340*d*).

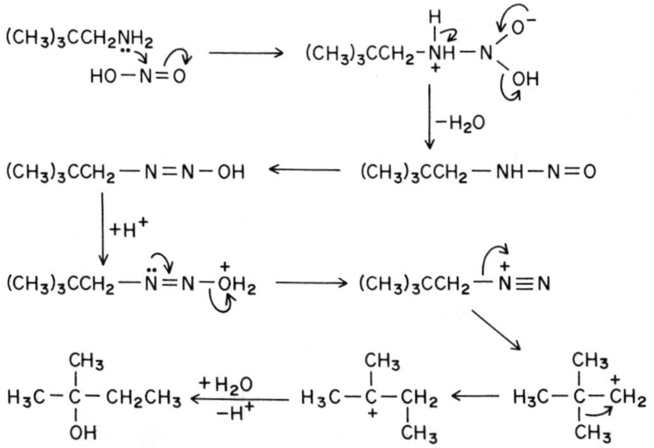

(a)

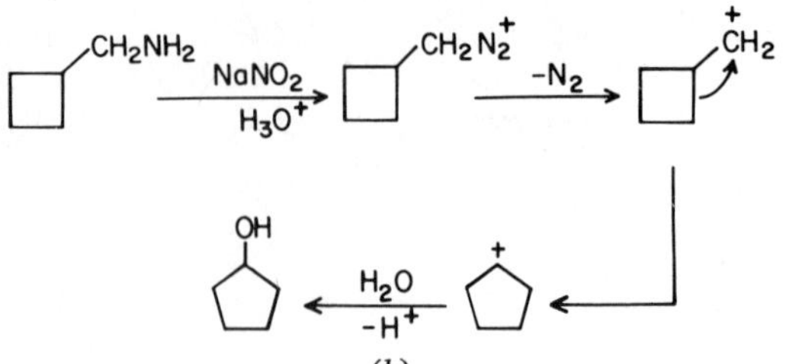

CH_2NH_2 — $NaNO_2$, H_3O^+ → $CH_2N_2^+$ — $-N_2$ → $\overset{+}{C}H_2$

OH ← H_2O, $-H^+$ ← +

(b)

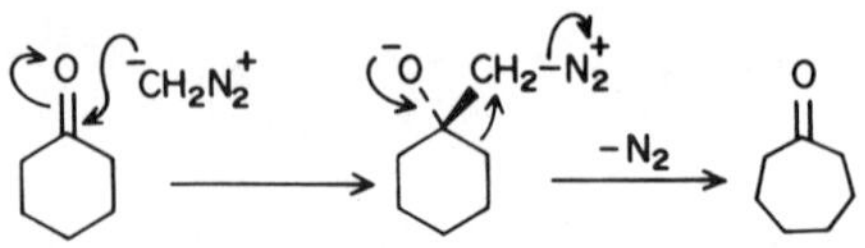

$CH_2N_2^+$ → $CH_2-N_2^+$ — $-N_2$ → O

(c)

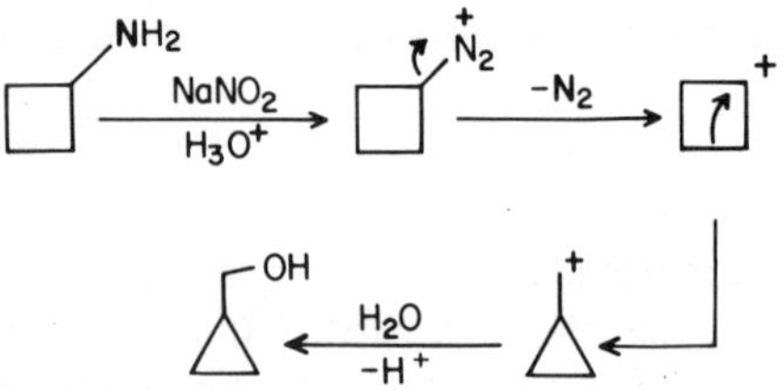

NH_2 — $NaNO_2$, H_3O^+ → $\overset{+}{N}_2$ — $-N_2$ → +

OH ← H_2O, $-H^+$ ← +

(d)

Scheme 11.340 Demjanov rearrangement: (*a*) reaction of an acyclic system; (*b*, *c*) ring expansions; (*d*) ring contraction.

11.350 Diazotization (or Griess) Reaction

[P. Griess, *Ann.*, **106**, 123 (1858).] The formation of diazonium salts by reaction of a primary amine with nitrous acid (see Scheme 11.340*a*). Although aliphatic diazonium compounds readily decompose, aromatic diazonium compounds are quite stable, Scheme 11.350. The value of the latter species resides in their use in the Sandmeyer reaction (Sect. 11.1130) or in the coupling reaction with aromatic phenols and/or amines to yield dyestuffs.

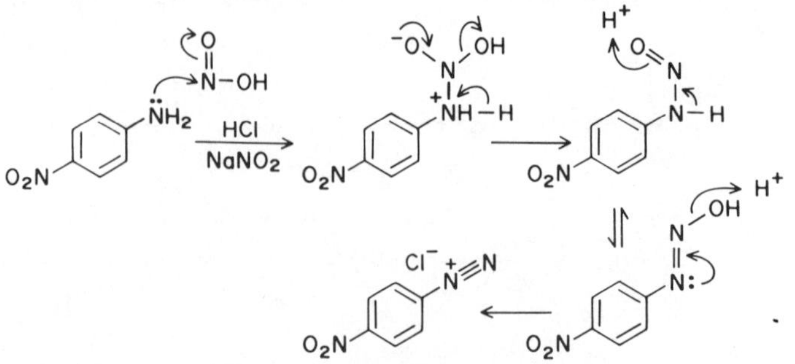

O_2N — $\ddot{N}H_2$, O, N–OH — HCl, $NaNO_2$ → O_2N — ^-O, N, OH, $\overset{+}{N}H-H$ → O_2N — H^+, O, N, N–H ⇌ O_2N — N, OH, H^+, N: → O_2N — Cl^-, $\overset{+}{N}\equiv N$

Scheme 11.350 Diazotization (or Griess) reaction.

11.360 Dieckmann Condensation

[W. Dieckmann, *Ber.*, 27, 102 (1894).] An intramolecular, base-catalyzed cyclization of a diester to give a β-ketoester that, after hydrolysis and decarboxylation, gives cyclopentanones or cyclohexanones, Scheme 11.360. This is an intramolecular Claisen-type condensation.

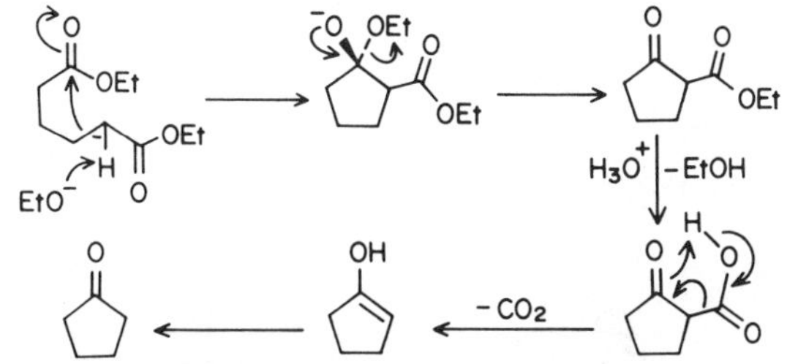

OEt OEt EtO⁻ H O OEt OEt OEt H_3O^+ – EtOH H O OH – CO_2

Scheme 11.360 Dieckmann condensation.

11.370 Diels-Alder Reaction

[O. Diels and K. Alder, *Ann.*, **460**, 98 (1928).] A synthesis of six-membered rings by cycloaddition of a 4π electron conjugated system to a 2π system, Scheme 11.370*a*. In the conventional reaction a diene reacts with an olefinic linkage (dienophile). There are many variations of the Diels-Alder reaction in which either the diene (Scheme 11.370*b*) or the dienophile (Scheme 11.370*c*) may contain hetero atoms. Singlet oxygen is characterized by its ability to act as a dienophile (Scheme 11.370*d*). The stereochemistry of the Diels-Alder reaction (Scheme 11.370*a*) is governed by the **Alder-Stein Rules** [K. Alder and G. Stein, *Angew. Chem.*, **50**, 510 (1937)]:

1. The addition to the dienophile as well as to the diene is *Z* (*cis*).
2. The reaction of a cyclic diene with a dienophile yields predominantly the *endo* product.

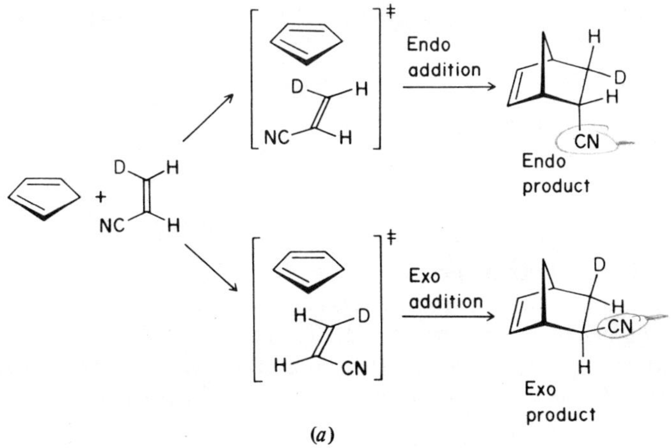

‡ Endo addition H D H CN Endo product D H NC H + ‡ Exo addition H D H CN D H CN H Exo product

(*a*)

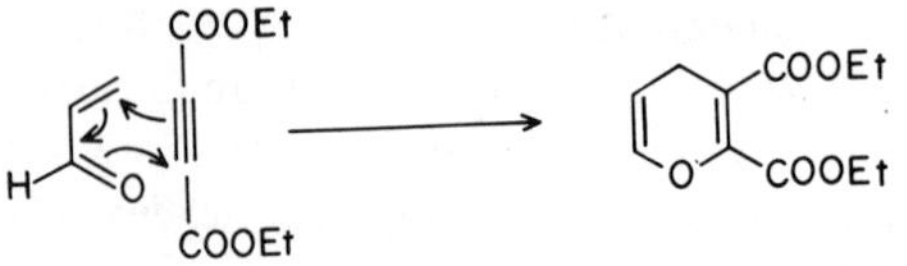

(*b*)

(*c*)

(*d*)

Scheme 11.370 Diels-Alder reaction with various dienophiles: (*a*) acrylonitrile; (*b*) diethyl acetylenedicarboxylate; (*c*) nitrosobenzene; (*d*) singlet oxygen.

11.380 Doebner–von Miller Reaction

[O. Doebner and W. von Miller, *Ber.*, **16**, 2464 (1883).] A synthesis of quinolines involving the reaction between aniline and two moles of an aldehyde (via the aldol condensation), Scheme 11.380.

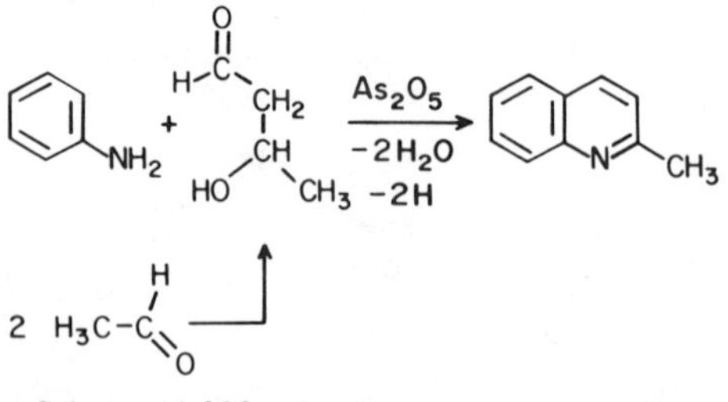

Scheme 11.380 Doebner–von Miller reaction.

11.390 Edman Degradation

[P. Edman, *Acta Chem. Scand.*, **4**, 277, 283 (1950).] A method for determining the *N*-terminal amino acid in a peptide. The reaction consists of treating the peptide with phenylisothiocyanate to form a phenylthiocarbamoyl derivative that on hydrolysis cleaves the *N*-terminal amino acid residue from the remainder of the peptide, Scheme 11.390.

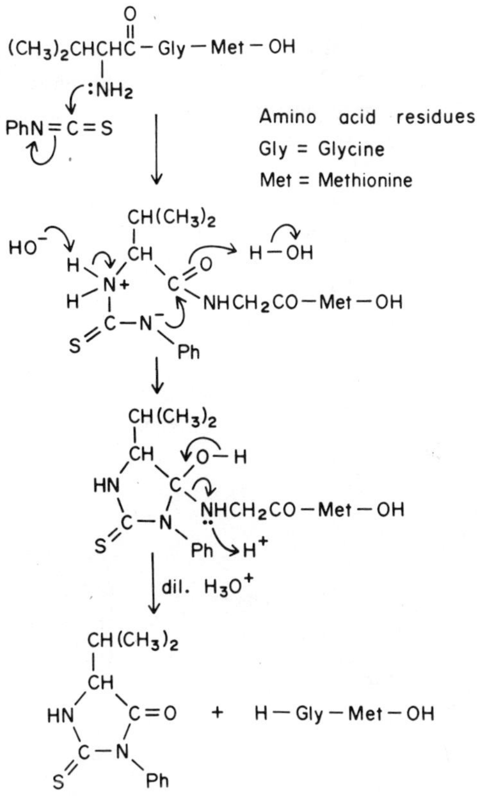

(CH3)2CHCHC(=O)–Gly–Met–OH
:NH2
PhN=C=S

Amino acid residues
Gly = Glycine
Met = Methionine

Scheme 11.390 Edman degradation.

11.400 Elbs Reaction

[K. Elbs and E. Larsen, *Ber.*, **17**, 2847 (1884).] The formation of anthracenes by thermal cyclodehydration of a diaryl ketone having a methyl (or methylene) group *ortho* to the carbonyl group, Scheme 11.400.

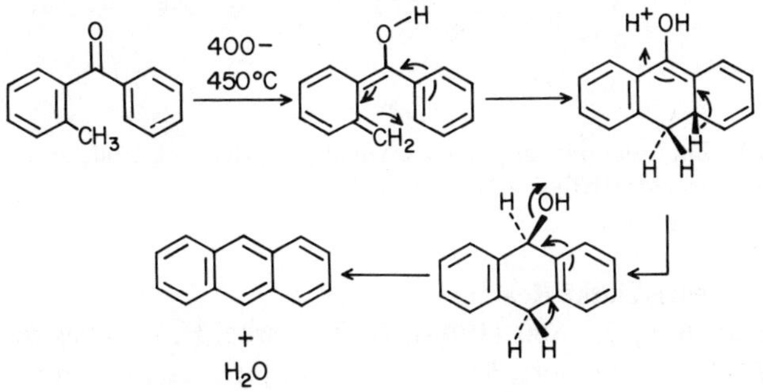

400–450°C

+ H_2O

Scheme 11.400 Elbs reaction.

11.410 Ene Reaction

A reaction of a double bond with an allylic system in which an allylic hydrogen and the olefinic terminus of the allylic system add across the double bond, Scheme 11.410*a*. This is a very general reaction that is insensitive to the substituents involved (Scheme 11.410*b–d*).

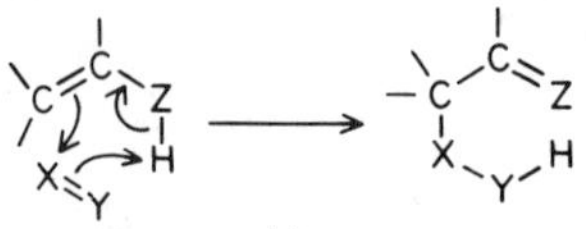

(*a*)

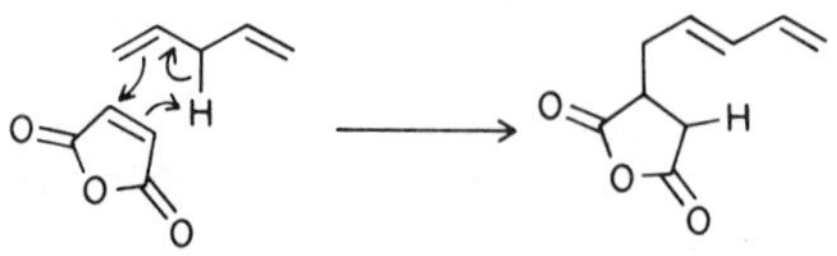

(*b*)

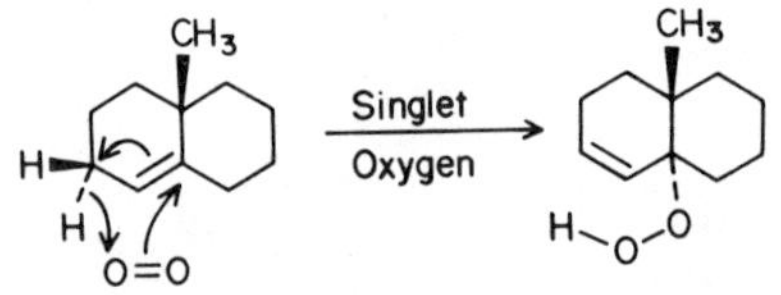

Singlet Oxygen

(*c*)

ene reaction

$-H_2O$

$+H_2O$
$-Se(OH)_2$

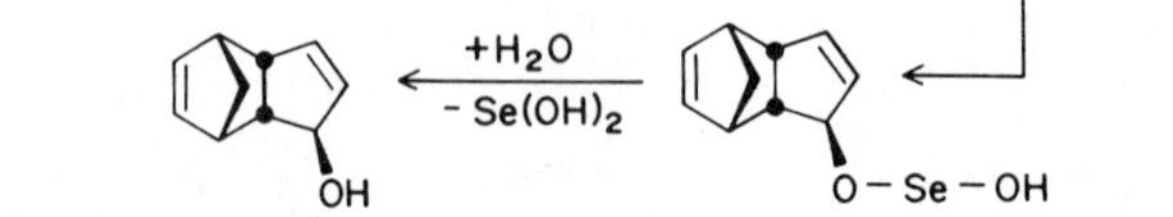

(*d*)

Scheme 11.410 Ene reaction: (*a*) general scheme; (*b*) maleic anhydride as an ene; (*c*) singlet oxygen as an ene; (*d*) $(HO)_2Se{=}O$ as an ene.

11.420 Eschweiler-Clarke Reaction

[W. Eschweiler, *Ber.*, 38, 880 (1905); H. T. Clarke, H. B. Gillespie, and S. Z. Weisshaus, *J. Am. Chem. Soc.*, **55**, 4571 (1933).] The conversion of primary or secondary amines to tertiary amines by reductive amination with a mixture of formaldehyde and formic acid, Scheme 11.420*a* and *b*. (See also Leuckart Reaction, Sect. 11.790.)

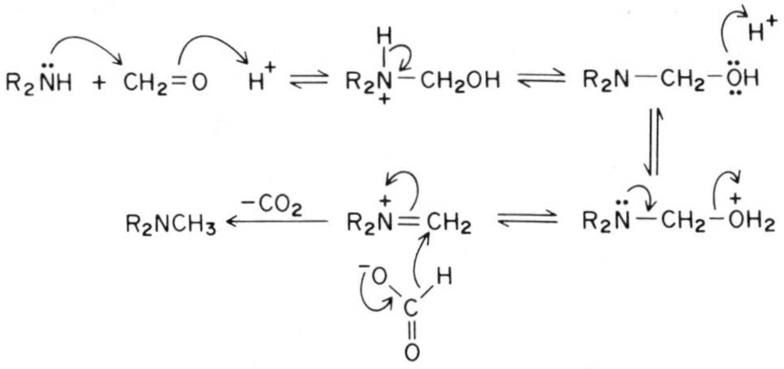

$$R_2\ddot{N}H + CH_2{=}O \quad H^+ \rightleftharpoons R_2\overset{+}{N}H{-}CH_2OH \rightleftharpoons R_2N{-}CH_2{-}\ddot{O}H \rightleftharpoons R_2\ddot{N}{-}CH_2{-}\overset{+}{O}H_2 \rightleftharpoons R_2\overset{+}{N}{=}CH_2 \xrightarrow[\text{HCO}_2^-]{-CO_2} R_2NCH_3$$

(*a*)

$$(H_3C)_3C{-}NH_2 \xrightarrow[HCO_2H]{CH_2O} (H_3C)_3C{-}N(CH_3)_2$$

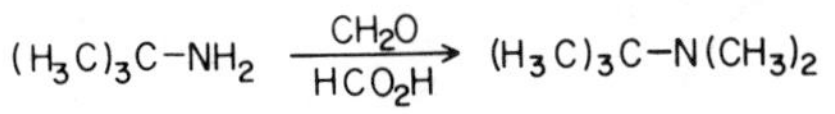

(*b*)

Scheme 11.420 Eschweiler-Clarke reaction: (*a*) with a secondary amine; (*b*) with a primary amine.

11.430 Étard Reaction

[A. L. Étard, *Compt. Rend.*, **90**, 534 (1880).] The oxidation of an arylmethyl group to an aldehyde by chromyl chloride, Scheme 11.430.

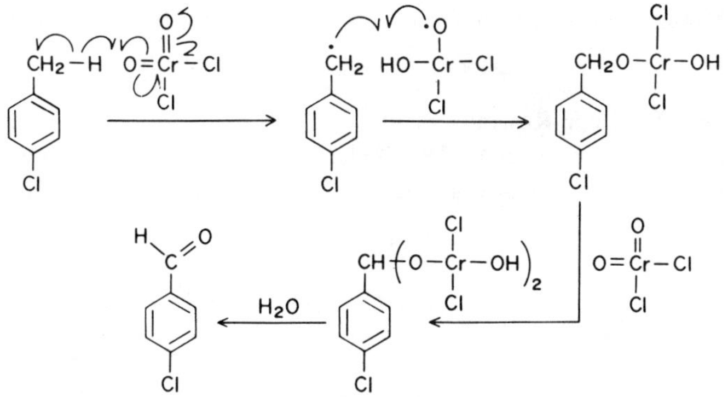

$$p\text{-}ClC_6H_4CH_2{-}H + O{=}CrO(Cl)_2 \longrightarrow p\text{-}ClC_6H_4\dot{C}H_2 \;\; HO{-}\dot{C}r(O)(Cl){-}Cl \longrightarrow p\text{-}ClC_6H_4CH_2O{-}Cr(Cl)_2{-}OH \xrightarrow{O=CrO(Cl)_2} p\text{-}ClC_6H_4CH{\left(O{-}Cr(Cl)_2{-}OH\right)}_2 \xrightarrow{H_2O} p\text{-}ClC_6H_4CHO$$

Scheme 11.430 Étard reaction.

11.440 Favorsky Rearrangement

[A. E. Favorsky, *J. Prakt. Chem.*, [2] **88**, 658 (1913).] The base-catalyzed conversion of α-haloketones via rearrangement to acids (or esters). The reaction may proceed by either of two mechanisms via the enolate when α-protons are available, Scheme 11.440*a* or via direct attack on the carbonyl group when α-protons are not available, Scheme 11.440*b*.

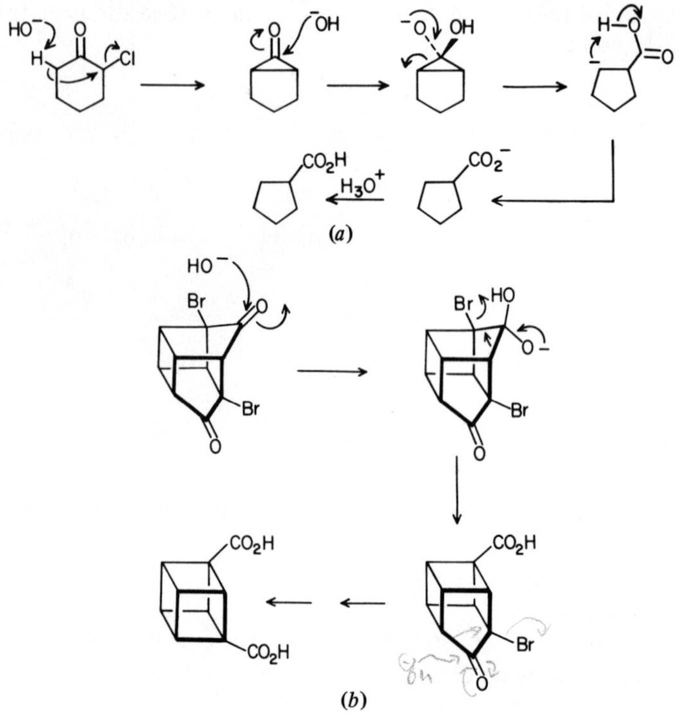

Scheme 11.440 Favorsky rearrangement: (*a*) with α-protons; (*b*) with unavailable α-protons.

11.450 Fischer-Hepp Rearrangement

[O. Fischer and E. Hepp, *Ber.*, **19**, 2991 (1886).] The rearrangement of the *N*-nitroso group of an aromatic amine to the *para* position of the ring, Scheme 11.450.

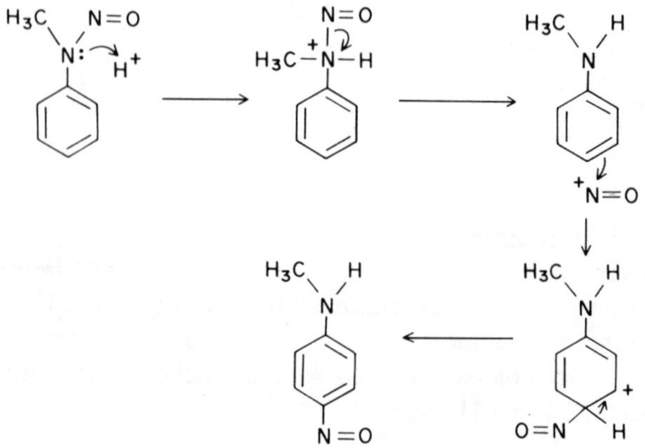

Scheme 11.450 Fischer-Hepp rearrangement.

11.460 Fischer Indole Synthesis

[E. Fischer and F. Jourdan, *Ber.*, **16**, 2241 (1883).] One of the most useful methods for preparation of indoles. Reaction occurs upon heating of phenylhydrazones with protic or Lewis acids, Scheme 11.460.

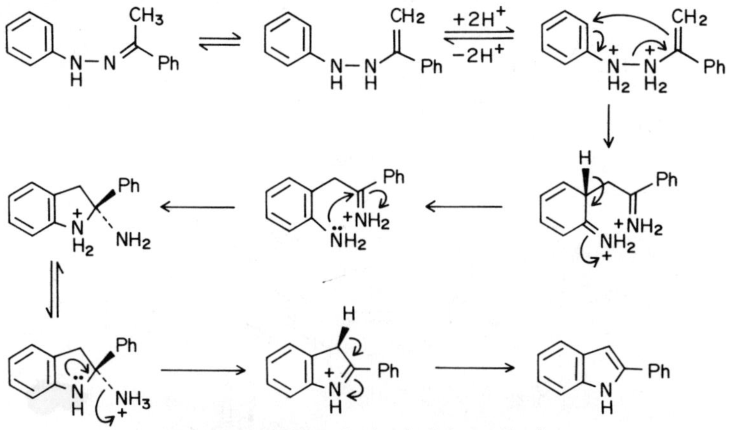

Scheme 11.460 Fischer indole synthesis.

11.470 Fischer Esterification

[E. Fischer and A. Speier, *Ber.*, **28**, 3252 (1895).] An esterification procedure using gaseous hydrochloric acid as the catalyst. Scheme 11.470.

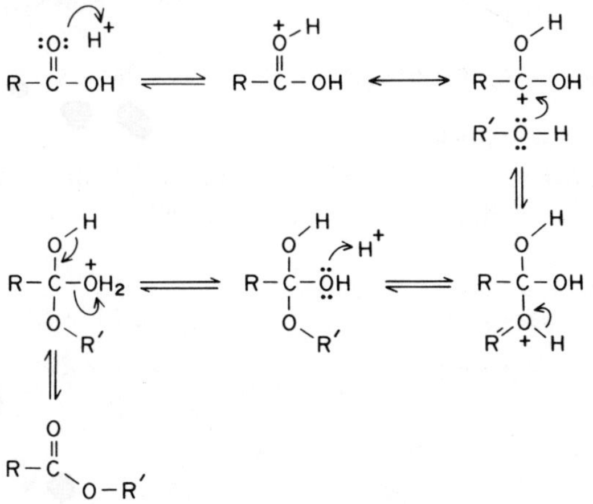

Scheme 11.470 Fischer esterification.

11.480 Friedel-Crafts Reaction

[C. Friedel and J. M. Crafts, *Compt. Rend.*, **84**, 1392, 1450 (1877).] An electrophilic substitution on an aromatic nucleus (catalyzed by a Lewis acid) in which either an alkyl group replaces a nuclear hydrogen atom (alkylation), Scheme

11.480*a*, or an acyl replaces a hydrogen atom (acylation), Scheme 11.480*b*. Alkylating agents may be alkyl halides, alcohols, or olefins, and acylating agents may be acyl halides or acid anhydrides. The most common catalyst is $AlCl_3$.

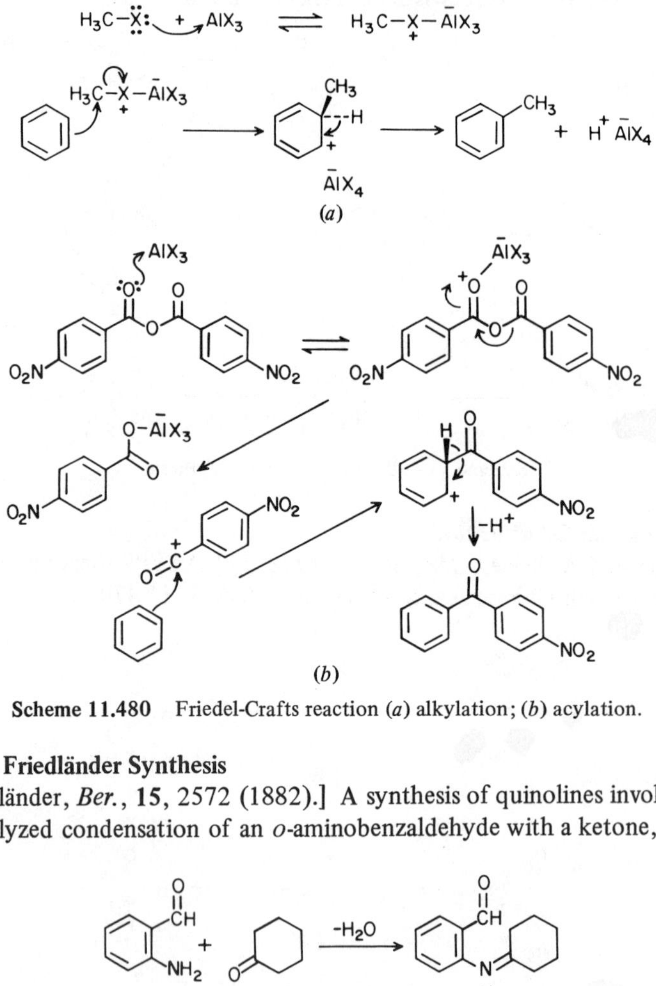

$$H_3C-\ddot{X}: + AlX_3 \rightleftharpoons H_3C-\overset{+}{X}-\overset{-}{Al}X_3$$

(*a*)

(*b*)

Scheme 11.480 Friedel-Crafts reaction (*a*) alkylation; (*b*) acylation.

11.490 Friedländer Synthesis

[P. Friedländer, *Ber.*, **15**, 2572 (1882).] A synthesis of quinolines involving the base-catalyzed condensation of an *o*-aminobenzaldehyde with a ketone, Scheme 11.490.

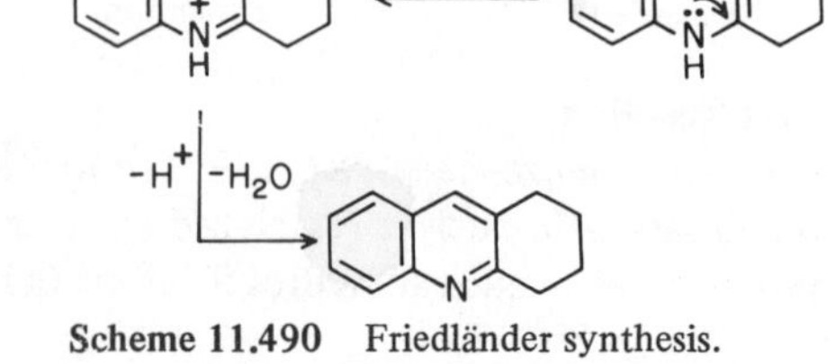

Scheme 11.490 Friedländer synthesis.

11.500 Fries Rearrangement

[A. Fries and G. Fink, *Ber.*, **41**, 4271 (1908).] The Lewis acid-catalyzed rearrangement of phenyl (or naphthyl) esters to *o*- or *p*-acylphenols, Scheme 11.500. (See also Photo-Fries Rearrangement, 11.1560.)

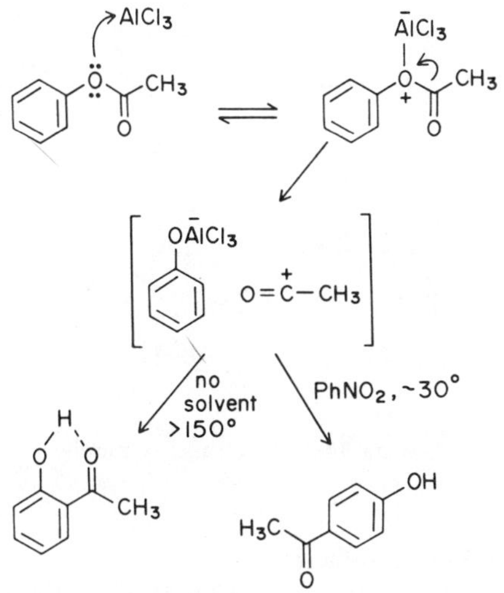

Scheme 11.500 Fries rearrangement.

11.510 Fujimoto-Belleau Reaction

[G. I. Fujimoto, *J. Am. Chem. Soc.*, **73**, 1856 (1951); B. Belleau, *J. Am. Chem. Soc.*, **73**, 5441 (1951).] The reaction of the Grignard reagent from a primary halide with an enol lactone to give a cyclic α-substituted, α,β-unsaturated ketone, Scheme 11.510.

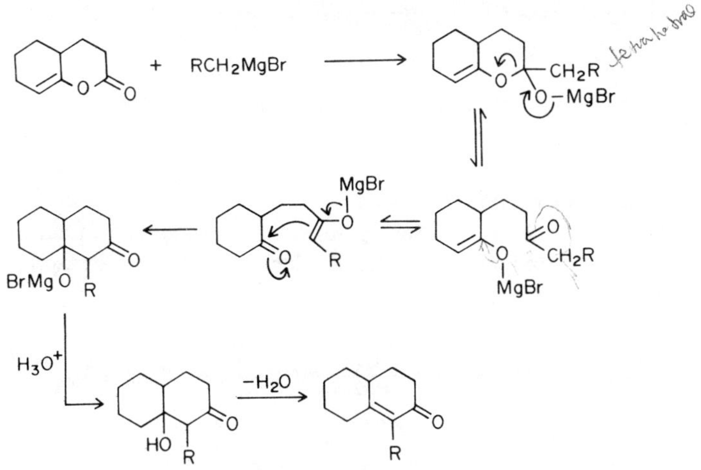

Scheme 11.510 Fujimoto-Belleau reaction.

11.520 Gabriel Synthesis

[S. Gabriel, *Ber.*, **20**, 2224 (1887).] A synthesis of primary alkyl amines by *N*-alkylation of phthalimide followed by treatment of the *N*-alkylphthalimide with hydrazine, Scheme 11.520.

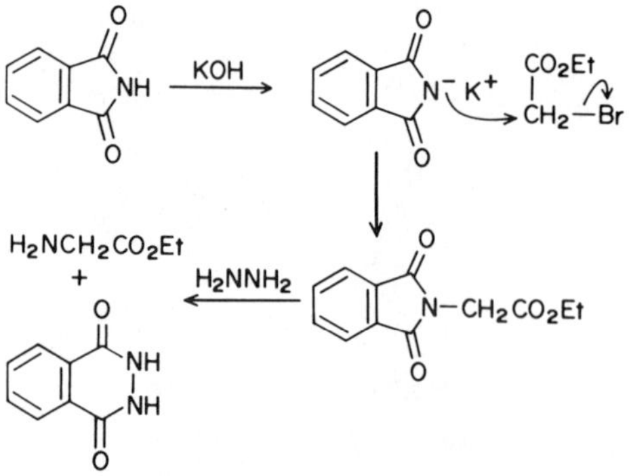

Scheme 11.520 Gabriel synthesis.

11.530 Gattermann-Koch Synthesis

[L. Gattermann and J. Koch, *Ber.*, **30**, 1622 (1897).] The direct substitution of the hydrogen of a substituted aromatic nucleus by a formyl group using a mixture of carbon monoxide and hydrochloric acid in the presence of cuprous chloride and aluminum trichloride. It is postulated that formyl chloride is an intermediate, Scheme 11.530.

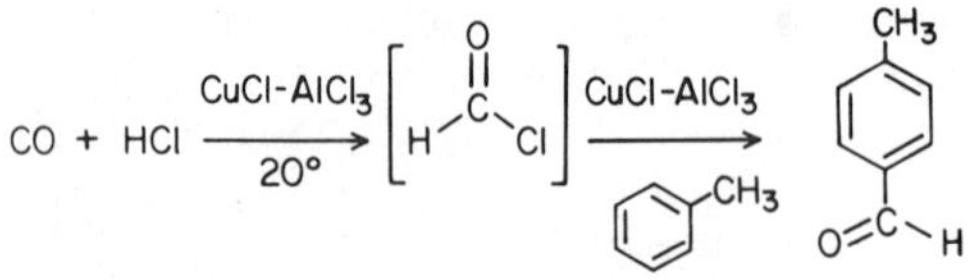

Scheme 11.530 Gattermann-Koch synthesis.

11.540 Glaser Reaction

[C. Glaser, *Ber.*, 2, 422 (1869).] An oxidative coupling of terminal alkynes to give diynes, catalyzed by cuprous ion in either water or pyridine, Scheme 11.540*a*, and *b*, respectively.

$$2\ CH_3CH_2C{\equiv}CH \xrightarrow[H_2O,\ O_2]{Cu_2Cl_2} CH_3CH_2C{\equiv}C{-}C{\equiv}CCH_2CH_3$$

(*a*)

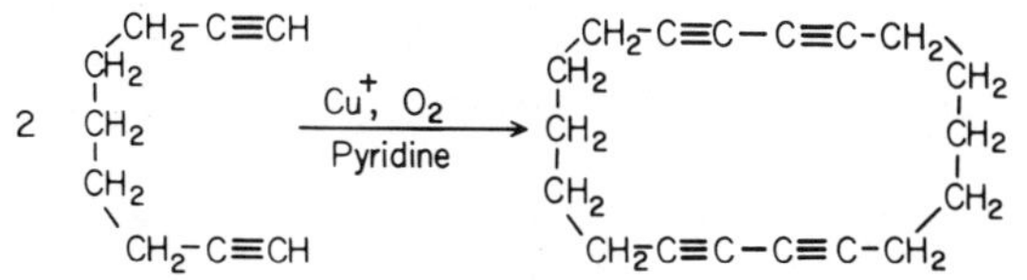

(b)

Scheme 11.540 Glaser reaction: (a) in water; (b) in pyridine.

11.550 Gomberg-Bachmann Synthesis

[M. Gomberg and W. E. Bachmann, *J. Am. Chem. Soc.*, **40**, 2339 (1924).] A synthesis of biaryls by a radical reaction involving the decomposition of diazonium salts in the presence of an aromatic solvent, Scheme 11.550.

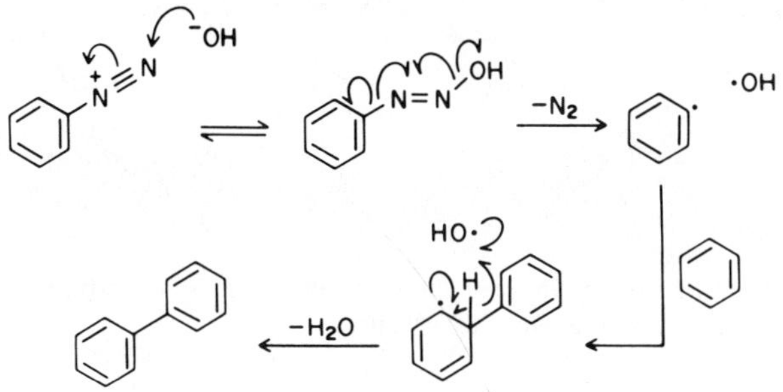

Scheme 11.550 Gomberg-Bachmann synthesis.

11.560 Grignard Reaction

[V. Grignard, *Compt. Rend.*, **130**, 1322 (1900).] Originally the use of the Grignard reagent RMgX to prepare tertiary alcohols from ketones or esters, Scheme 11.560*a* and *b*; primary alcohols from formaldehyde, Scheme 11.560*c*; secondary alcohols from aldehydes, Scheme 11.560*d*; or ketones from nitriles, Scheme 11.560*e*. The term is now used for any reaction involving the Grignard reagent.

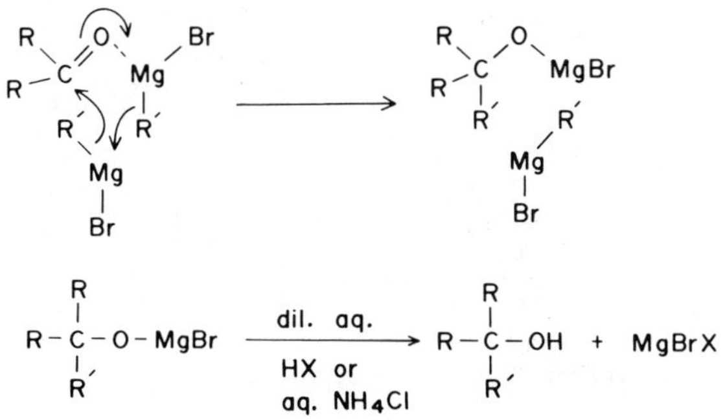

(a)

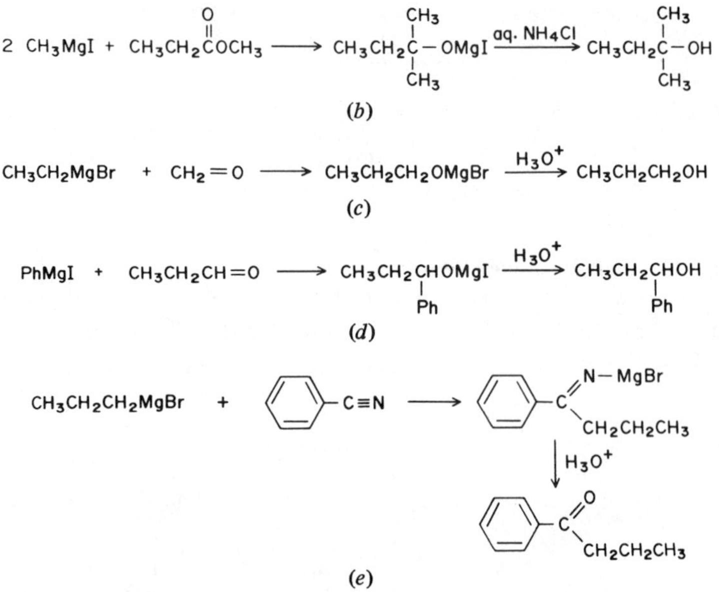

$$2\ CH_3MgI + CH_3CH_2\overset{O}{\overset{\|}{C}}OCH_3 \longrightarrow CH_3CH_2C(CH_3)_2{-}OMgI \xrightarrow{aq.\ NH_4Cl} CH_3CH_2C(CH_3)_2{-}OH$$

(*b*)

$$CH_3CH_2MgBr + CH_2{=}O \longrightarrow CH_3CH_2CH_2OMgBr \xrightarrow{H_3O^+} CH_3CH_2CH_2OH$$

(*c*)

$$PhMgI + CH_3CH_2CH{=}O \longrightarrow CH_3CH_2CH(Ph)OMgI \xrightarrow{H_3O^+} CH_3CH_2CH(Ph)OH$$

(*d*)

$$CH_3CH_2CH_2MgBr + C_6H_5{-}C{\equiv}N \longrightarrow C_6H_5{-}C({=}N{-}MgBr)CH_2CH_2CH_3 \xrightarrow{H_3O^+} C_6H_5{-}C({=}O)CH_2CH_2CH_3$$

(*e*)

Scheme 11.560 Grignard reactions for the preparation of: (*a*, *b*) tertiary alcohols; (*c*) primary alcohols; (*d*) secondary alcohols; (*e*) ketones.

11.570 Grob Fragmentation Reaction

[C. A. Grob and W. Baumann, *Helv. Chim. Acta*, 38, 594 (1955).] A method for carbon-carbon bond cleavage involving a four-atom chain. The reaction consists of cleavage at the 2,3-bond and involves neutralization of opposite charges at the 1 and 4 positions, Scheme 11.570*a*. An important application is the preparation of medium-size rings, Scheme 11.570*b*.

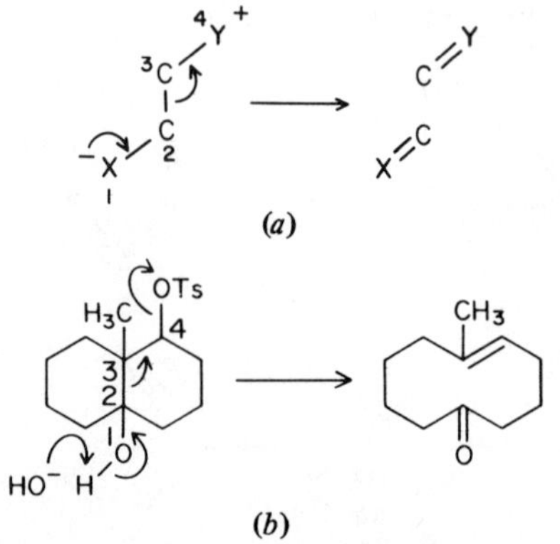

$$^{-}X^{1}{-}C^{2}{-}C^{3}{-}Y^{4\,+} \longrightarrow X{=}C \quad + \quad C{=}Y$$

(*a*)

(*b*)

Scheme 11.570 Grob fragmentation reactions: (*a*) general scheme; (*b*) synthesis of medium-sized rings.

11.580 Grundmann Aldehyde Synthesis

[C. Grundmann, *Ann.*, **524**, 31 (1936).] Overall, the conversion of an acyl or aroyl halide to the corresponding aldehyde. The reaction proceeds through an α-diazoketone to an α-acetoxyketone that is reduced to a 1,2-glycol and this is oxidized to the aldehyde, Scheme 11.580*a* and *b*.

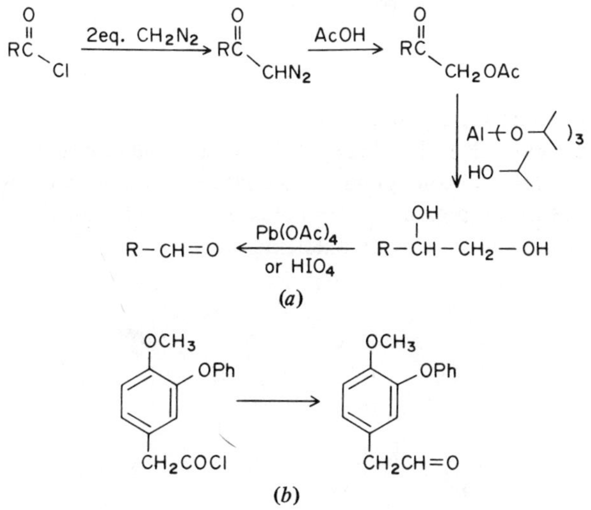

$$RCOCl \xrightarrow{2eq.\ CH_2N_2} RCOCHN_2 \xrightarrow{AcOH} RCOCH_2OAc$$

$$RCOCH_2OAc \xrightarrow[HO-CH(CH_3)_2]{Al(O-CH(CH_3)_2)_3} R-CH(OH)-CH_2-OH$$

$$R-CH(OH)-CH_2-OH \xrightarrow[\text{or } HIO_4]{Pb(OAc)_4} R-CH=O$$

(*a*)

$$\text{(3-OPh, 4-OCH}_3\text{)C}_6\text{H}_3\text{CH}_2\text{COCl} \longrightarrow \text{(3-OPh, 4-OCH}_3\text{)C}_6\text{H}_3\text{CH}_2\text{CH=O}$$

(*b*)

Scheme 11.580 Grundmann aldehyde synthesis: (*a*) general scheme; (*b*) conversion of a phenacyl chloride to a phenylacetaldehyde.

11.590 Haller-Bauer Reaction

[A. Haller and E. Bauer, *Compt. Rend.*, **148**, 70 (1909).] The conversion of a phenacyl compound, $R'CH_2COPh$, to a trisubstituted acetic acid, one group of which is the original R′ group. The intermediate is the amide of the trisubstituted acetic acid, Scheme 11.590*a*, which, if appropriately substituted, may be converted to a lactam, Scheme 11.590*b*.

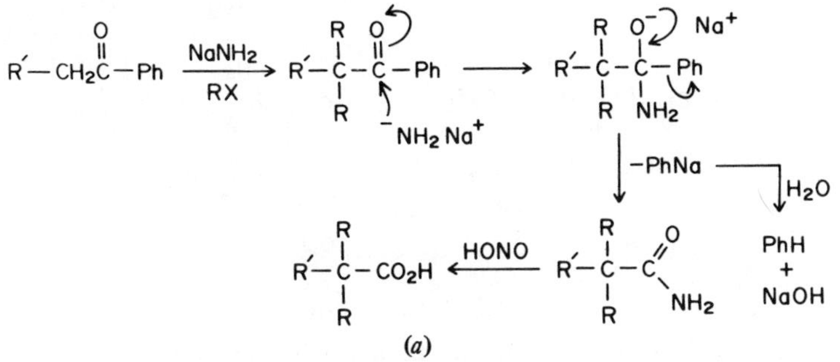

$$R'-CH_2COPh \xrightarrow[RX]{NaNH_2} R'-CR_2-CO-Ph \xrightarrow{^-NH_2\ Na^+} R'-CR_2-C(O^-)(NH_2)-Ph\ Na^+$$

$$\xrightarrow{-PhNa} R'-CR_2-CONH_2 \qquad (PhNa \xrightarrow{H_2O} PhH + NaOH)$$

$$R'-CR_2-CONH_2 \xrightarrow{HONO} R'-CR_2-CO_2H$$

(*a*)

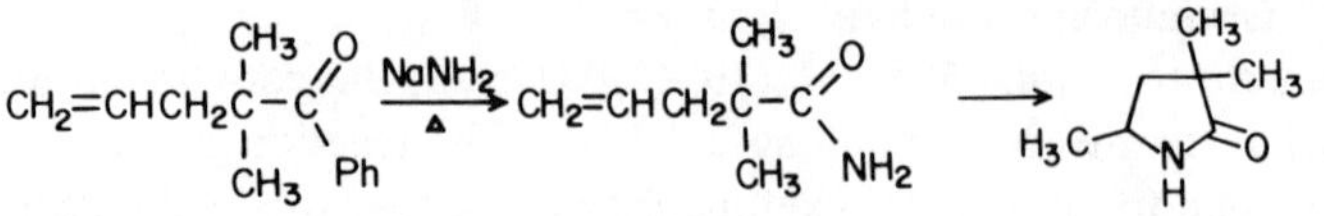

(*b*)

Scheme 11.590 Haller-Bauer reaction: (*a*) application to unsubstituted phenacyls; (*b*) application to substituted phenacyls.

11.600 Hantzsch Pyridine Synthesis

[A. Hantzsch, *Ann.*, **215**, 1 (1882).] A synthesis of alkylpyridines in which two moles of a β-dicarbonyl compound are involved in a condensation with one mole of an aldehyde in the presence of ammonia, Scheme 11.600. The resulting dihydropyridine is aromatized with an oxidizing agent.

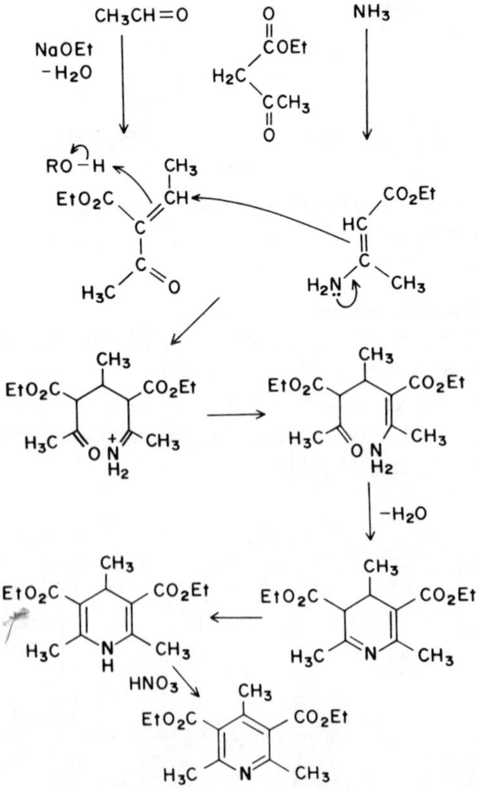

Scheme 11.600 Hantzsch pyridine synthesis.

11.610 Harries Ozonolysis Reaction

[C. Harries, *Ann.*, **343**, 311 (1905).] A classical method used for ascertaining the location of a double bond by treating the olefin with ozone and ultimately cleaving the molecule at the double bond, Scheme 11.610. Workup is accomplished by treating the mixture with a reducing agent to destroy all peroxides.

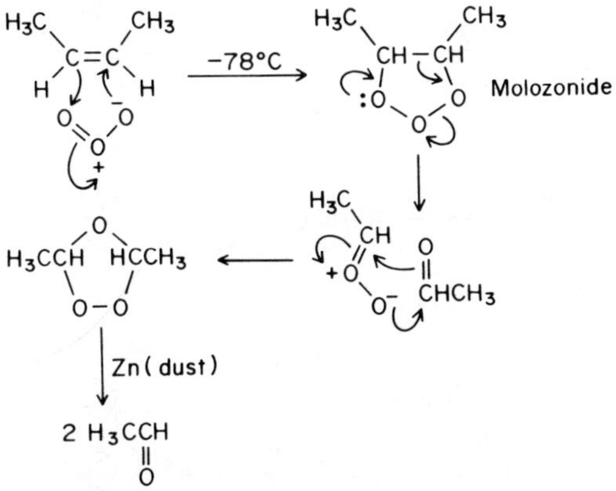

Scheme 11.610 Harries ozonolysis reaction.

11.620 Hell-Volhard-Zelinsky Reaction

[C. Hell, *Ber.*, **14**, 891 (1881); J. Volhard, *Ann.*, **242**, 141 (1887); N. Zelinsky, *Ber.*, **20**, 2026 (1887).] A method for preparing α-bromo esters via α-bromacyl bromides from carboxylic acids using phosphorus tribromide and bromine, Scheme 11.620*a* and *b*. This method is particularly useful because the ester cannot be brominated directly.

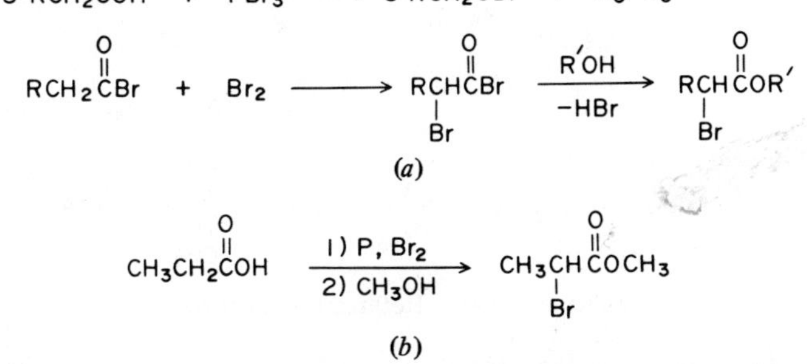

$$3\,RCH_2COOH + PBr_3 \longrightarrow 3\,RCH_2COBr + H_3PO_3$$

$$RCH_2COBr + Br_2 \longrightarrow RCH(Br)COBr \xrightarrow[-HBr]{R'OH} RCH(Br)COOR'$$

(*a*)

$$CH_3CH_2COOH \xrightarrow[2)\ CH_3OH]{1)\ P,\ Br_2} CH_3CH(Br)COOCH_3$$

(*b*)

Scheme 11.620 Hell-Volhard-Zelinsky reaction: (*a*) general scheme; (*b*) synthesis of methyl α-bromopropionate.

11.630 Hofmann Degradation (or Rearrangement)

[A. W. Hofmann, *Ber.*, **14**, 2725 (1881).] The conversion of an acid amide to an isocyanate and then to an amine by treating the amide with bromine in alkaline solution, Scheme 11.630*a* and *b*. Overall the reaction formally consists of the decarbonylation of an amide.

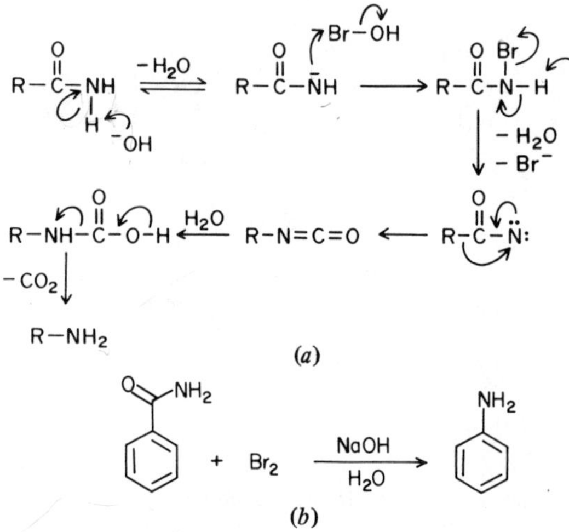

$$R{-}C(=O){-}NH{-}H \xrightarrow[\ ^-OH\ ]{-H_2O} R{-}C(=O){-}\bar{N}H \xrightarrow{Br{-}OH} R{-}C(=O){-}N(Br){-}H \ \ ^-OH \xrightarrow[-Br^-]{-H_2O} R{-}C(=O){-}\ddot{N}: \longrightarrow R{-}N{=}C{=}O \xrightarrow{H_2O} R{-}NH{-}C(=O){-}O{-}H \xrightarrow{-CO_2} R{-}NH_2$$

(*a*)

$$C_6H_5CONH_2 + Br_2 \xrightarrow[H_2O]{NaOH} C_6H_5NH_2$$

(*b*)

Scheme 11.630 Hofmann degradation: (*a*) general scheme; (*b*) conversion of benzamide to aniline.

11.640 Hofmann Elimination

[A. W. Hofmann, *Ber.*, **14**, 659 (1881).] The pyrolysis of a quaternary ammonium hydroxide to form an olefin. When there are alternatives, the least substituted olefin is formed, Scheme 11.640 (see also Sect. 10.360). The starting compound is an amine and this is converted to the quaternary salt using excess CH_3I as an alkylating agent in a process known as **exhaustive methylation.**

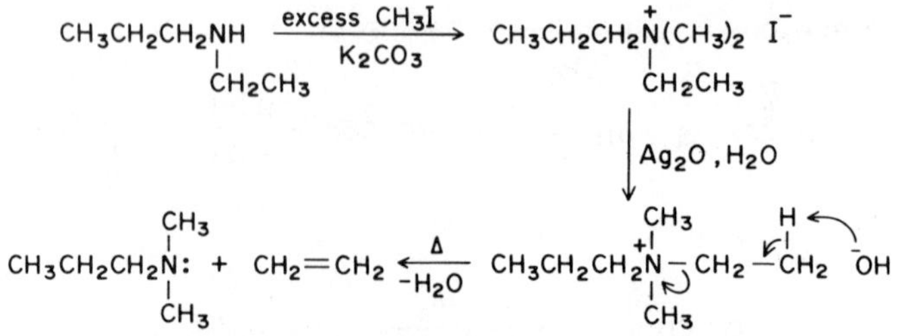

$$CH_3CH_2CH_2NH(CH_2CH_3) \xrightarrow[K_2CO_3]{\text{excess } CH_3I} CH_3CH_2CH_2\overset{+}{N}(CH_3)_2(CH_2CH_3)\ I^- \xrightarrow{Ag_2O,\ H_2O} CH_3CH_2CH_2\overset{+}{N}(CH_3)_2{-}CH_2{-}CH_2{-}H\ \ ^-OH \xrightarrow[-H_2O]{\Delta} CH_3CH_2CH_2\ddot{N}(CH_3)_2 + CH_2{=}CH_2$$

Scheme 11.640 Hofmann elimination.

11.650 Horner-Emmons Reaction

[L. Horner, H. Hoffman, and H. G. Wippel, *Ber.*, **91**, 61 (1958); W. S. Wadsworth and W. D. Emmons, *J. Am. Chem. Soc.*, **83**, 1733 (1961).] A method for converting a carbonyl group of an aldehyde or ketone to an olefinic group. The carbanion from a phosphonate attacks the carbonyl compound, Scheme 11.650*a*. The conversion of ethyl 4-oxopentanoate to diethyl 3-methyl-2-hexene-1,6-dioate is shown in Scheme 11.650*b*. These reactions are modifications of the Wittig reaction (Sect. 11.1400). If ketones are involved, the Horner-Emmons reaction generally gives higher yields and the phosphorus-containing by-products are water soluble.

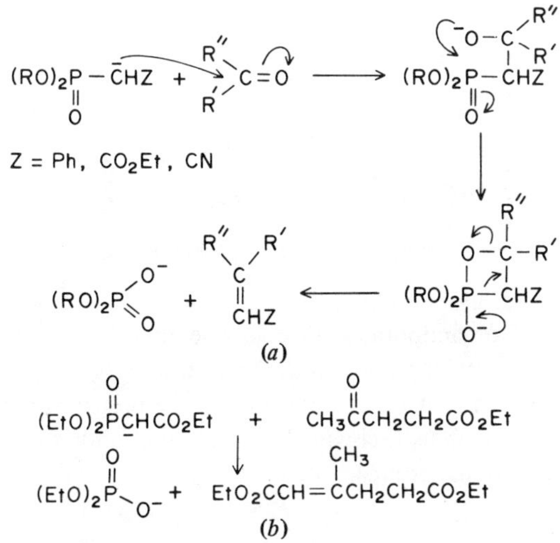

$(RO)_2P(=O)-\bar{C}HZ + R'R''C{=}O \longrightarrow (RO)_2P(=O)-CHZ-C(R')(R'')-O^- \longrightarrow$ cyclic intermediate $\longrightarrow (RO)_2PO_2^- + R''R'C{=}CHZ$

$Z = Ph,\ CO_2Et,\ CN$

(*a*)

$(EtO)_2P(=O)\underline{C}HCO_2Et + CH_3C(=O)CH_2CH_2CO_2Et \longrightarrow (EtO)_2P(=O)O^- + EtO_2CCH{=}C(CH_3)CH_2CH_2CO_2Et$

(*b*)

Scheme 11.650 Horner-Emmons reaction: (*a*) general scheme; (*b*) an α,β-unsaturated ester synthesis.

11.660 Houben-Hoesch Reaction

[K. Hoesch, *Ber.*, **48**, 1122 (1915); J. Houben, *Ber.*, **59**, 2878 (1926).] The condensation of aliphatic nitriles with phenols (or their methyl ethers) under acidic conditions to prepare the acylphenols (or their methyl ethers), Scheme 11.660.

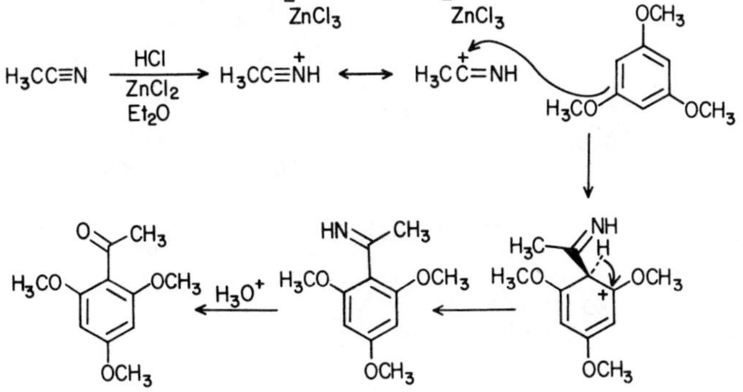

$H_3CC{\equiv}N \xrightarrow[ZnCl_2,\ Et_2O]{HCl} H_3CC{\equiv}\overset{+}{N}H\ \bar{Z}nCl_3 \longleftrightarrow H_3C\overset{+}{C}{=}NH\ \bar{Z}nCl_3$; + 1,3,5-trimethoxybenzene $\longrightarrow$ σ-complex $\longrightarrow$ 2,4,6-$(H_3CO)_3C_6H_2C(=NH)CH_3 \xrightarrow{H_3O^+}$ 2,4,6-$(H_3CO)_3C_6H_2C(=O)CH_3$

Scheme 11.660 Houben-Hoesch reaction.

11.670 Hunsdiecker Reaction

[C. Hunsdiecker, H. Hunsdiecker, and E. Vogt, U.S. Patent 2,176,181 (1939); *C. A.*, **34**, 1685 (1940).] The treatment of the silver (or mercury) salt of an aliphatic or aromatic acid with bromine in carbon tetrachloride to give an alkyl or aryl bromide, Scheme 11.670.

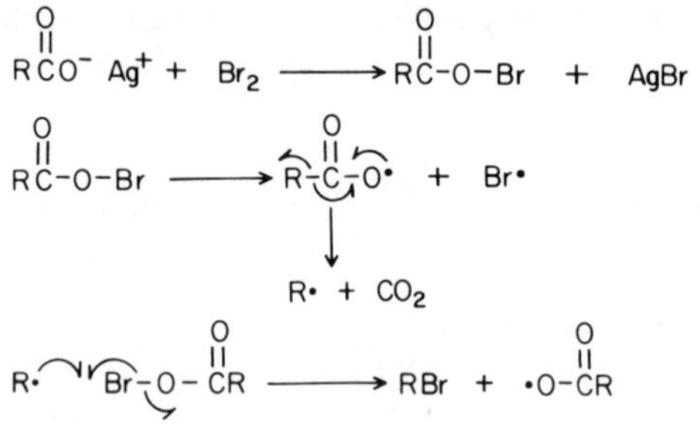

$$RC(=O)O^- Ag^+ + Br_2 \longrightarrow RC(=O)-O-Br + AgBr$$

$$RC(=O)-O-Br \longrightarrow R-C(=O)-O\cdot + Br\cdot$$

$$R-C(=O)-O\cdot \longrightarrow R\cdot + CO_2$$

$$R\cdot + Br-O-C(=O)R \longrightarrow RBr + \cdot O-C(=O)R$$

Scheme 11.670 Hunsdiecker reaction.

11.680 Hydroboration

The preparation of organoboranes through the concerted *cis* anti-Markovnikov addition of boron hydrides to olefins (Scheme 11.680*a*), allenes, and acetylenes. These organoboranes are of synthetic utility in the preparation of hydrocarbons (Scheme 11.680*b*), alcohols (Scheme 11.680*c*), and amines (Scheme 11.680*d*), as well as other functional groups.

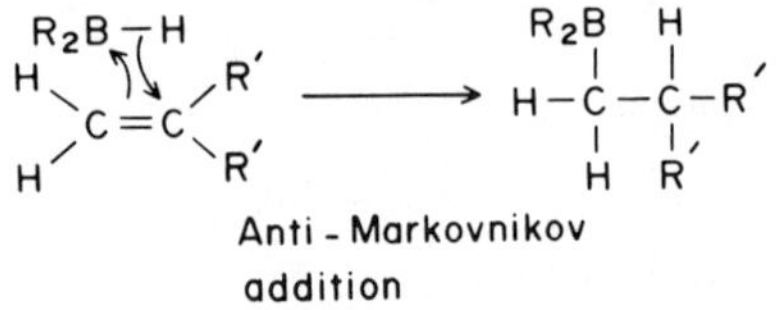

$$R_2B-H + H_2C=CR'_2 \longrightarrow H-C(H)(BR_2)-C(H)(R')-R'$$

Anti - Markovnikov addition

(*a*)

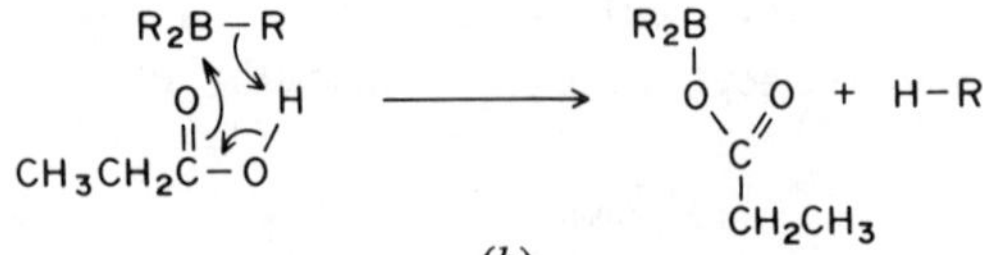

$$R_2B-R + CH_3CH_2C(=O)-OH \longrightarrow R_2B-O-C(=O)CH_2CH_3 + H-R$$

(*b*)

$$R_2B-R + {}^-O-OH \longrightarrow R_2\bar{B}(R)-O-OH \xrightarrow{-HO^-} R_2B-OR$$

repeat two times

$$(RO)_3B + {}^-OH \longrightarrow (RO)_2\bar{B}(OR)-O-H \longrightarrow ROH + (RO)_2B-O^-$$

$$\xrightarrow[\text{repeat two times}]{-2ROH} BO_3^{\equiv}$$

borate

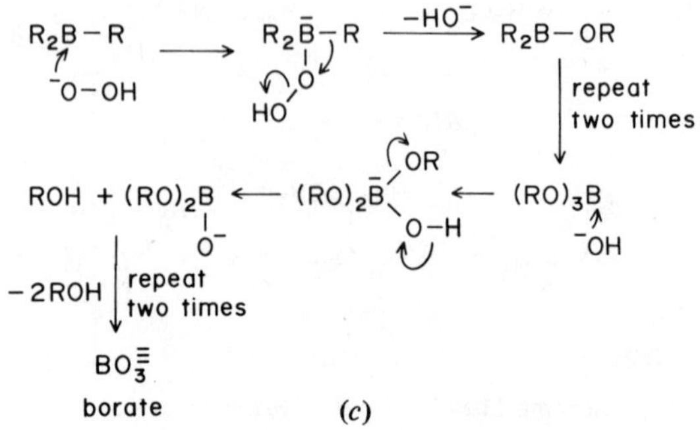

(*c*)

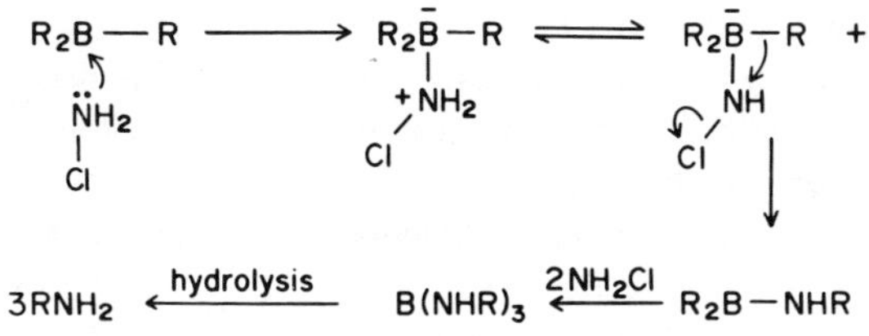

$R_2B{-}R$ + $\ddot{N}H_2Cl$ ⟶ $R_2\bar{B}{-}R$ ($^+NH_2$, Cl) ⇌ $R_2\bar{B}{-}R$ (NH, Cl) + H^+ ⟶ $R_2B{-}NHR$ $\xrightarrow{2NH_2Cl}$ $B(NHR)_3$ $\xrightarrow{\text{hydrolysis}}$ $3RNH_2$

(*d*)

Scheme 11.680 Hydroboration: (*a*) addition to olefins followed by conversion to (*b*) alkanes, (*c*) alcohols, and (*d*) amines.

11.690 Jacobsen Reaction

[O. Jacobsen, *Ber.*, **19**, 1209 (1886).] The treatment of a polymethylated benzene with concentrated sulfuric acid to give a rearranged polymethylbenzenesulfonic acid, Scheme 11.690*a*. In a subsequent reaction the sulfonic acid groups can be removed. Halogenated polymethylbenzenes can be made to undergo intermolecular halogen transfer under the same conditions, Scheme 11.690*b*.

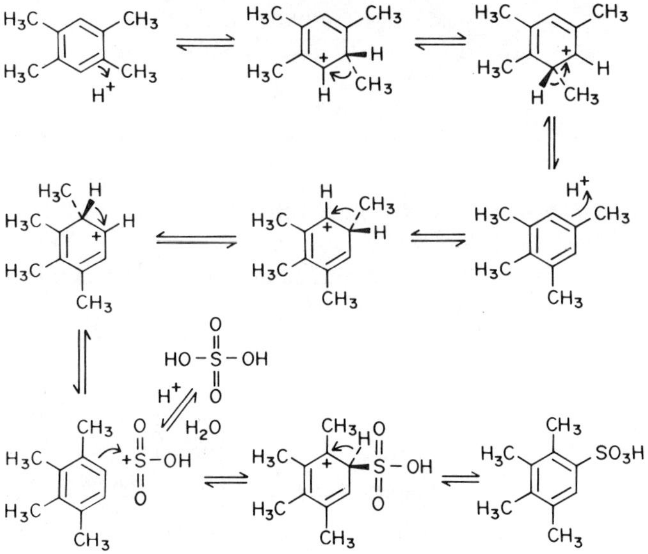

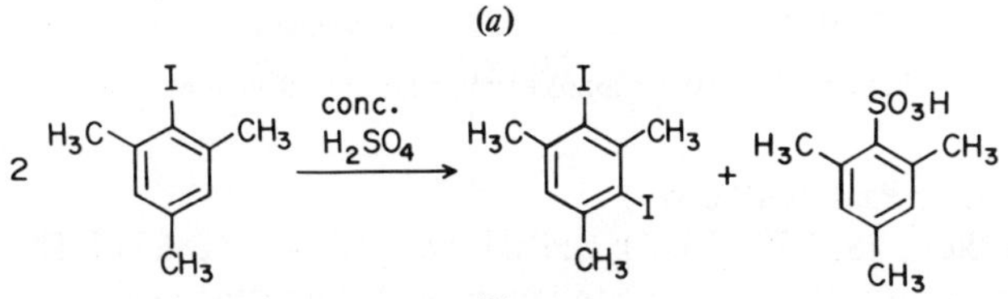

Scheme 11.690 Jacobsen reaction: (*a*) reaction of an aromatic hydrocarbon; (*b*) reaction of aryl iodide.

11.700 Japp-Klingemann Reaction

[F. R. Japp and F. Klingemann, *Ann.*, 247, 190 (1888).] The reaction of a diazonium ion and a stabilized carbanion derived by the removal of an activated hydrogen atom leading to the formation of a hydrazone, Scheme 11.700*a*. The reaction can be made to proceed intramolecularly as in Scheme 11.700*b*.

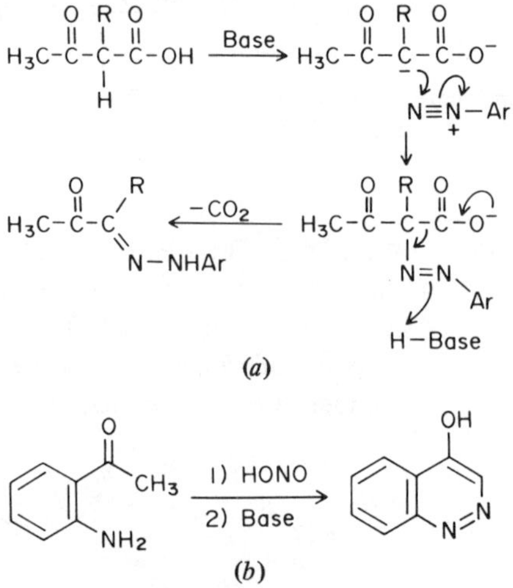

Scheme 11.700 Japp-Klingemann reaction: (*a*) intermolecular; (*b*) intramolecular.

11.710 Kiliani-Fischer Cyanohydrin Synthesis

[H. Kiliani, *Ber.*, 18, 3066 (1885); E. Fischer, *Ber.*, 22, 2204 (1889).] The conversion of an aldose to the homologous aldose wherein the additional carbon is provided by cyanide ion, Scheme 11.710.

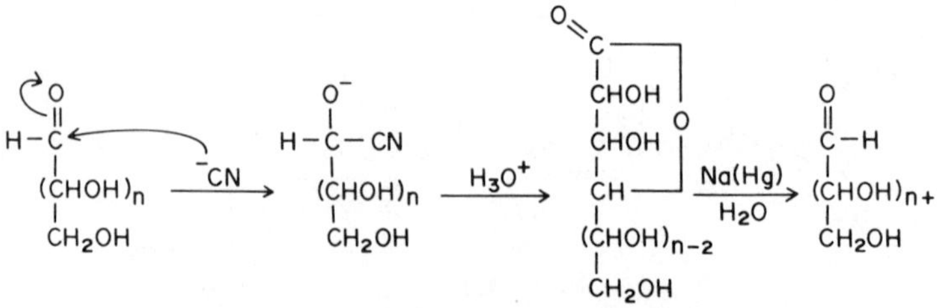

Scheme 11.710 Kiliani-Fischer cyanohydrin synthesis.

11.720 Knorr-Paal Synthesis

[L. Knorr, *Ber.*, 18, 299 (1885); C. Paal, *Ber.*, 18, 367 (1885).] The reaction of 1,4-diketones with the appropriate inorganic dehydrating reagent to give furans, pyrroles, or thiophenes, Scheme 11.720.

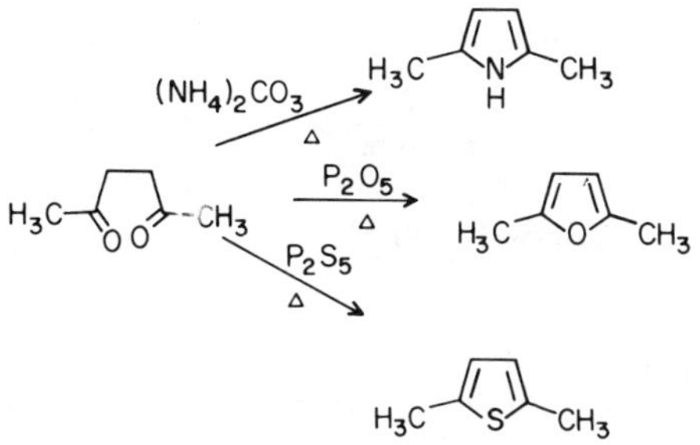

Scheme 11.720 Knorr-Paal synthesis.

11.730 Knoevenagel Reaction

[E. Knoevenagel, *Ber.*, **31**, 2596 (1898).] The synthesis of an α,β-unsaturated acid by the amine-catalyzed reaction between malonic acid or its esters and an aldehyde or ketone, Scheme 11.730. The malonate can be replaced by other compounds having activated hydrogen atoms.

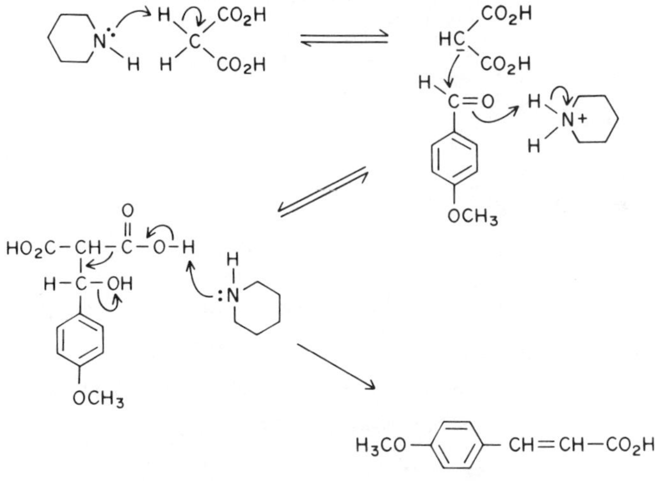

Scheme 11.730 Knoevenagel reaction.

11.740 Koch-Haaf Reaction

[Von H. Koch and W. Haaf, *Ann.*, **618**, 251 (1958).] A synthesis of branched chain carboxylic acids achieved by dissolving olefins in concentrated sulfuric acid and adding carbon monoxide, followed by quenching with water, Scheme 11.740.

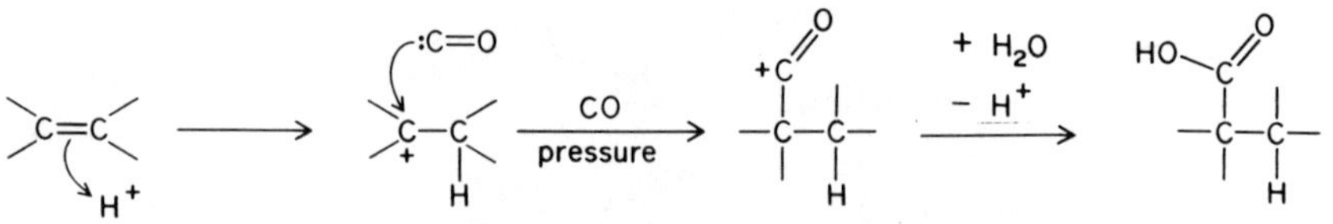

$$\mathrm{>C{=}C<} \xrightarrow{H^+} \mathrm{>\overset{+}{C}{-}C(H)<} \xrightarrow[\text{pressure}]{CO} \mathrm{-C(\overset{+}{C}{=}O){-}C(H)-} \xrightarrow[-H^+]{+H_2O} \mathrm{-C(COOH){-}C(H)-}$$

Scheme 11.740 Koch-Haaf reaction.

11.750 Kochi Reaction

[J. K. Kochi, *J. Am. Chem. Soc.*, **87**, 2500 (1965).] The replacement of a carboxy group with a chloro group by treatment of the acid with lead tetraacetate and lithium chloride, Scheme 11.750.

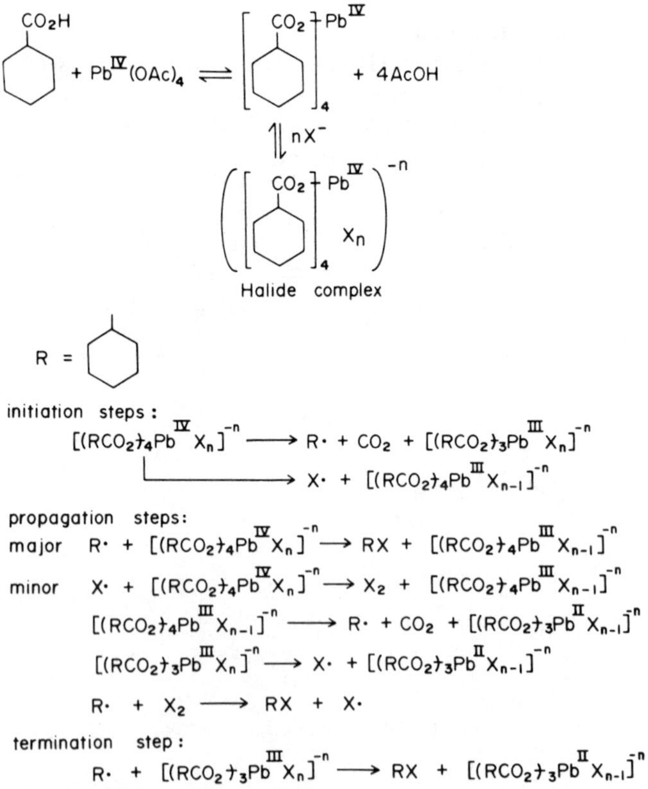

$$C_6H_{11}CO_2H + Pb^{IV}(OAc)_4 \rightleftharpoons [(C_6H_{11}CO_2)_4Pb^{IV}] + 4\,AcOH$$

$$[(C_6H_{11}CO_2)_4Pb^{IV}] \underset{}{\overset{nX^-}{\rightleftharpoons}} \left([(C_6H_{11}CO_2)_4Pb^{IV}]X_n\right)^{-n}$$

Halide complex

R = cyclohexyl

initiation steps:

$$[(RCO_2)_4Pb^{IV}X_n]^{-n} \longrightarrow R\cdot + CO_2 + [(RCO_2)_3Pb^{III}X_n]^{-n}$$

$$[(RCO_2)_4Pb^{IV}X_n]^{-n} \longrightarrow X\cdot + [(RCO_2)_4Pb^{III}X_{n-1}]^{-n}$$

propagation steps:

major $R\cdot + [(RCO_2)_4Pb^{IV}X_n]^{-n} \longrightarrow RX + [(RCO_2)_4Pb^{III}X_{n-1}]^{-n}$

minor $X\cdot + [(RCO_2)_4Pb^{IV}X_n]^{-n} \longrightarrow X_2 + [(RCO_2)_4Pb^{III}X_{n-1}]^{-n}$

$$[(RCO_2)_4Pb^{III}X_{n-1}]^{-n} \longrightarrow R\cdot + CO_2 + [(RCO_2)_3Pb^{II}X_{n-1}]^{-n}$$

$$[(RCO_2)_3Pb^{III}X_n]^{-n} \longrightarrow X\cdot + [(RCO_2)_3Pb^{II}X_{n-1}]^{-n}$$

$$R\cdot + X_2 \longrightarrow RX + X\cdot$$

termination step:

$$R\cdot + [(RCO_2)_3Pb^{III}X_n]^{-n} \longrightarrow RX + [(RCO_2)_3Pb^{II}X_{n-1}]^{-n}$$

Scheme 11.750 Kochi reaction: details of the radical chain mechanism (n = 1, 2).

11.760 Kolbe Hydrocarbon Synthesis

[H. Kolbe, *Ann.*, **69**, 257 (1849).] Electrolysis of the salt of a carboxylic acid, RCO_2H, to give the coupled product, R—R, Scheme 11.760*a*. Butane can be made according to Scheme 11.760*b*.

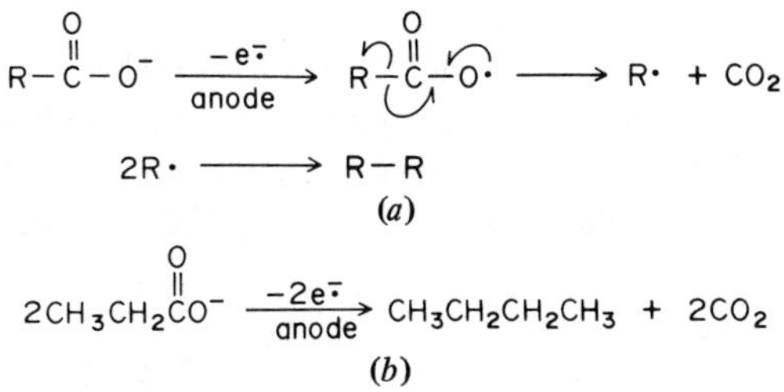

$$R{-}\overset{O}{\overset{\|}{C}}{-}O^- \xrightarrow[\text{anode}]{-e^-} R{-}\overset{O}{\overset{\|}{C}}{-}O\cdot \longrightarrow R\cdot + CO_2$$

$$2R\cdot \longrightarrow R{-}R$$

(*a*)

$$2CH_3CH_2\overset{O}{\overset{\|}{C}}O^- \xrightarrow[\text{anode}]{-2e^-} CH_3CH_2CH_2CH_3 + 2CO_2$$

(*b*)

Scheme 11.760 Kolbe hydrocarbon synthesis: (*a*) general scheme; (*b*) synthesis of butane.

11.770 Kostanecki Reaction

[S. von Kostanecki and A. Rozycki, *Ber.*, **34**, 102 (1901).] A synthesis of chromones and coumarins from *o*-acylphenols by *O*-acetylation followed by a base-catalyzed cyclization, Scheme 11.770.

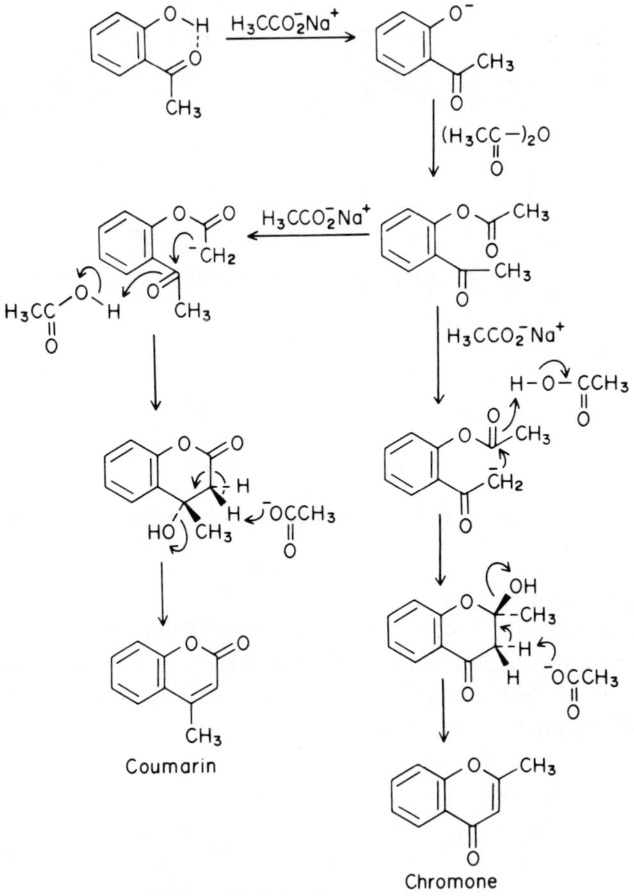

$H_3CCO_2^-Na^+$ $(H_3CC(O){-})_2O$ $H_3CCO_2^-Na^+$ $H_3CCO_2^-Na^+$

Coumarin

Chromone

Scheme 11.770 Kostanecki reaction.

11.780 Kröhnke Aldehyde Synthesis

[F. Kröhnke and E. Börner, *Ber.*, **69**, 2006 (1936).] The conversion of a benzyl halide to a benzaldehyde by treatment of the former with pyridine and conversion of the pyridinium salt to a nitrone by reaction with *p*-nitrosodimethylaniline. The aldehyde is obtained by hydrolysis of the nitrone, Scheme 11.780.

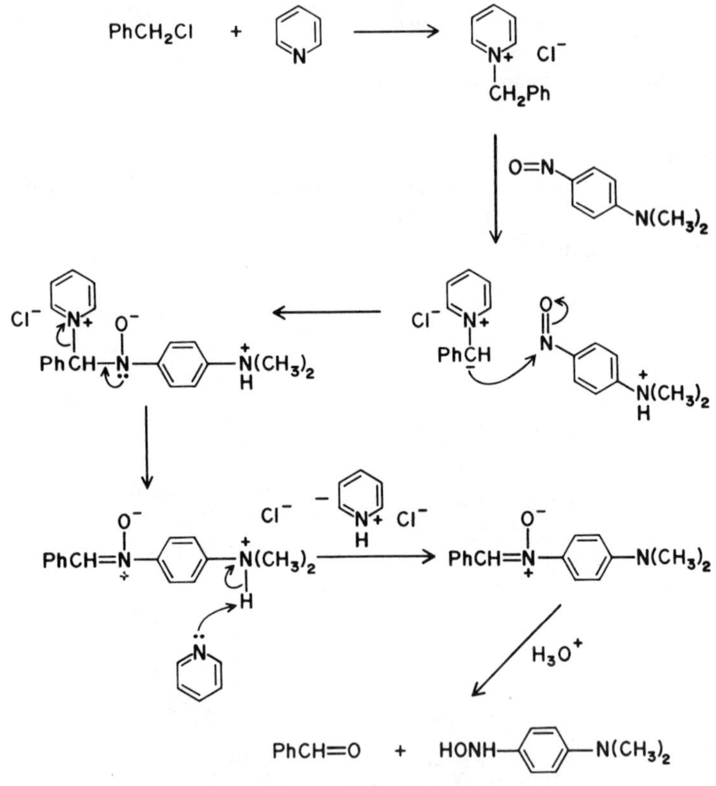

$PhCH_2Cl$ + pyridine → N-benzylpyridinium Cl^- (CH_2Ph)

O=N–C₆H₄–N(CH₃)₂

$PhCH{=}O$ + HONH–C₆H₄–$N(CH_3)_2$

Scheme 11.780 Kröhnke aldehyde synthesis.

11.790 Leuckart Reaction

[R. Leuckart, *Ber.*, **18**, 2341 (1885).] The preparation of amines by reaction of a ketone with ammonium formate, Scheme 11.790*a*, or with dimethylformamide, Scheme 11.790*b*. Methylamines may also be prepared via the **Eschweiler-Clarke modification** (see Sect. 11.420).

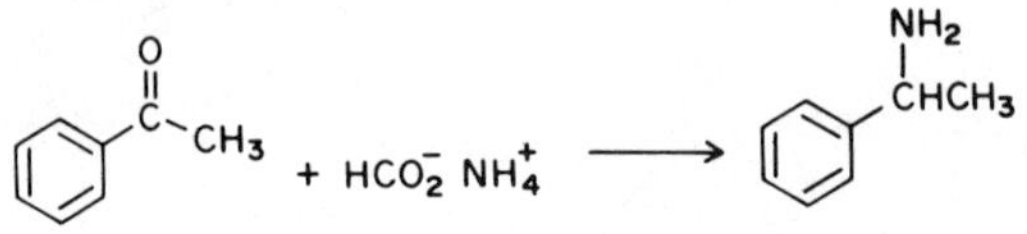

$C_6H_5COCH_3$ + $HCO_2^-\ NH_4^+$ → $C_6H_5CH(NH_2)CH_3$

(*a*)

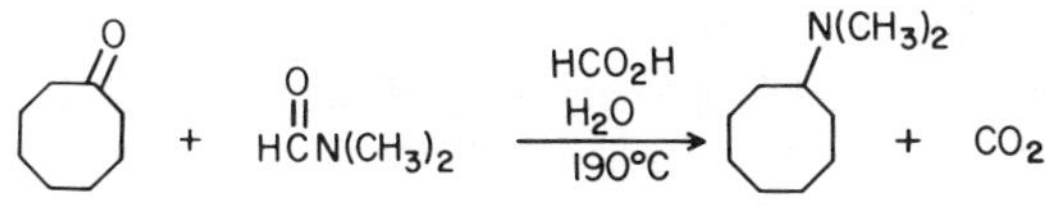

(*b*)

Scheme 11.790 Leuckart reaction: (*a*) with ammonium formate; (*b*) with dimethylformamide.

11.800 Lossen Rearrangement

[W. Lossen, *Ann.*, **161**, 347 (1872).] The dehydration of a hydroxamic acid, which results in a Hofmann-type rearrangement. The product is an amine containing one carbon less than the original hydroxamic acid, Scheme 11.800.

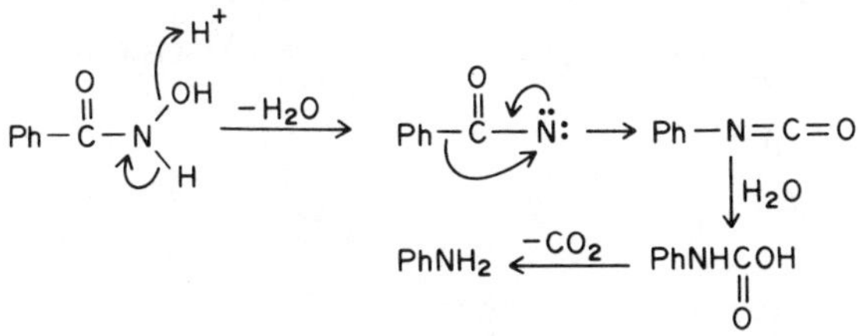

Scheme 11.800 Lossen rearrangement.

11.810 Madelung Indole Synthesis

[W. Madelung, *Ber.*, **45**, 1128 (1912).] A base-catalyzed thermal cyclization of *N*-acyl-*o*-toluidines to indoles, Scheme 11.810.

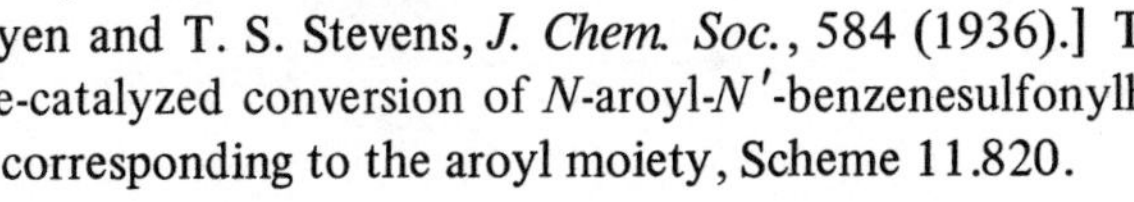

Scheme 11.810 Madelung indole synthesis.

11.820 McFadyen-Stevens Reaction

[J. S. McFadyen and T. S. Stevens, *J. Chem. Soc.*, 584 (1936).] The high temperature, base-catalyzed conversion of *N*-aroyl-*N*′-benzenesulfonylhydrazines to the aldehyde corresponding to the aroyl moiety, Scheme 11.820.

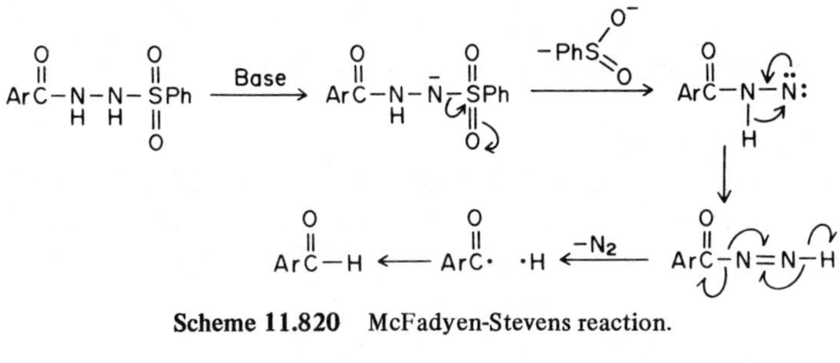

Scheme 11.820 McFadyen-Stevens reaction.

11.830 Malonic Ester Synthesis

The alkylation, Scheme 11.830*a*, or acylation, Scheme 11.830*b*, of malonic ester to provide substituted acetic acids.

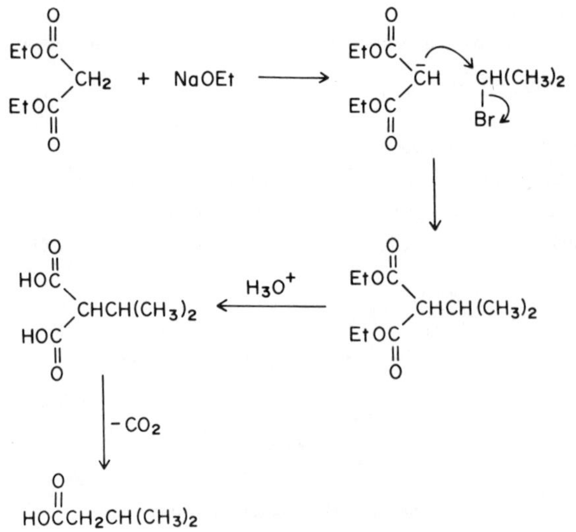

$$(EtOC(O))_2CH_2 + NaOEt \longrightarrow (EtOC(O))_2\bar{C}H + CH(CH_3)_2Br$$

$$\longrightarrow (EtOC(O))_2CHCH(CH_3)_2 \xrightarrow{H_3O^+} (HOC(O))_2CHCH(CH_3)_2 \xrightarrow{-CO_2} HOC(O)CH_2CH(CH_3)_2$$

(*a*)

$$(EtOC(O))_2\bar{C}H + ClC(O)Ph \longrightarrow (EtOC(O))_2CH{-}C(O^-)(Cl){-}Ph$$

$$\longrightarrow (EtOC(O))_2CH{-}C(O){-}Ph \xrightarrow{1)\ H_3O^+;\ 2)\ -CO_2} HOC(O)CH_2C(O)Ph$$

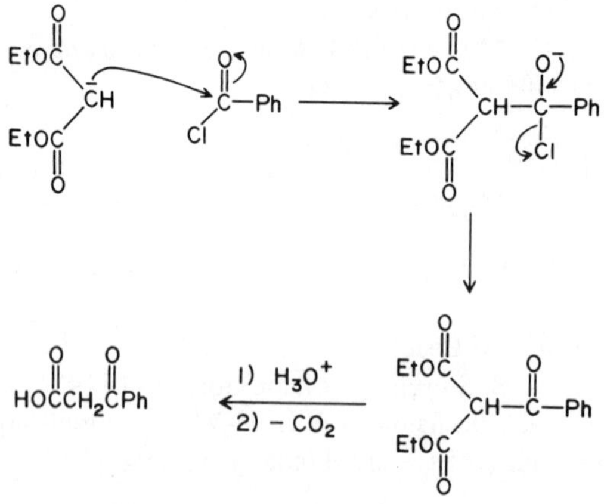

(*b*)

Scheme 11.830 Malonic ester synthesis: (*a*) alkylation; (*b*) acylation.

11.840 Mannich Reaction

[C. Mannich and W. Krosche, *Arch. Pharm.*, **250**, 647 (1912).] The substitution of an α-hydrogen atom of a ketone by the $-CR'_2-NR_2$ group, Scheme 11.840*a*. Since the reaction proceeds through the attack of an iminium ion, $>C=\overset{+}{N}< \longleftrightarrow$

$\gt\overset{+}{C}-\ddot{N}\lt$, on the enol form of the ketone, the reactions of iminium ions with electron-rich double bonds such as those present in phenols, aryl ethers, pyrroles, and indoles are also considered Mannich reactions, Scheme 11.840*b*.

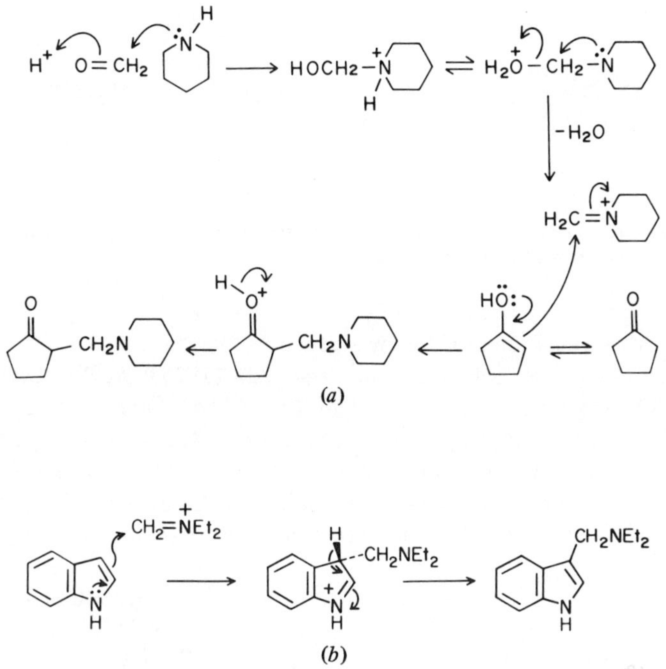

H^+ $O{=}CH_2$ (piperidine, N–H) ⟶ $HOCH_2{-}\overset{+}{N}H$ (piperidinium) ⇌ $H_2\overset{+}{O}{-}CH_2{-}N$ (piperidine)

$-H_2O$

$H_2C{=}\overset{+}{N}$ (piperidinium)

HÖ: (cyclopentenol) ⇌ O (cyclopentanone)

O (2-(piperidinomethyl)cyclopentanone), CH_2N ⟵ $H{-}\overset{+}{O}$, CH_2N ⟵ enol

(*a*)

$CH_2{=}\overset{+}{N}Et_2$; indole (N–H) ⟶ H, CH_2NEt_2, $\overset{+}{N}$–H ⟶ CH_2NEt_2 (3-substituted indole, N–H)

(*b*)

Scheme 11.840 Mannich reaction: (*a*) with cyclopentanone; (*b*) with indole.

11.850 Meerwein Arylation Reaction

[H. Meerwein, E. Büchner, and K. van Emster, *J. Prakt. Chem.*, [2] **152**, 237 (1939).] The replacement of an activated vinyl hydrogen atom by an aryl group achieved by the reaction of a diazonium salt in the presence of Cu^{2+} salts, Scheme 11.850*a*, *b*, and *c*.

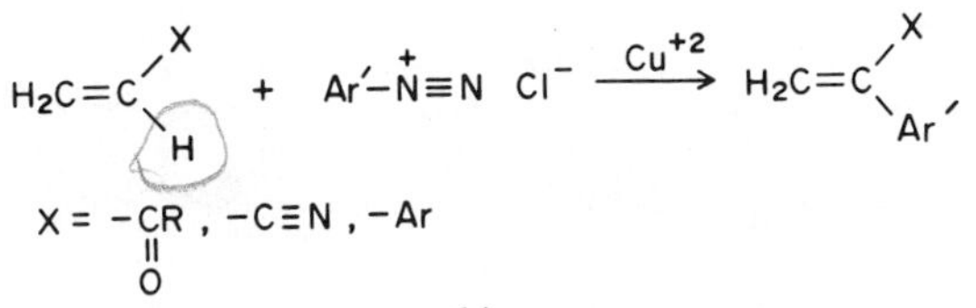

$$H_2C{=}C(X)H + Ar'{-}\overset{+}{N}{\equiv}N\ Cl^- \xrightarrow{Cu^{+2}} H_2C{=}C(X)Ar'$$

$X = -\underset{\underset{O}{\|}}{C}R,\ -C{\equiv}N,\ -Ar$

(*a*)

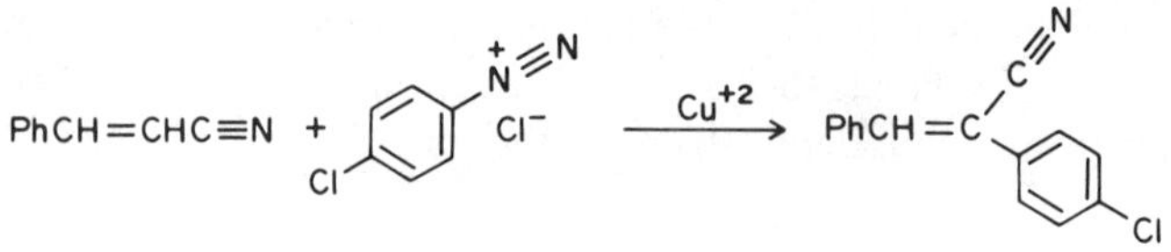

$PhCH{=}CHC{\equiv}N$ + p-$ClC_6H_4N_2^+ Cl^-$ $\xrightarrow{Cu^{+2}}$ $PhCH{=}C(C{\equiv}N)(C_6H_4Cl$-$p)$

(*b*)

2-hydroxy-1,4-naphthoquinone + $Ar{-}\overset{+}{N}{\equiv}N\ Cl^-$ $\xrightarrow{Cu^{+2}}$ 2-hydroxy-3-aryl-1,4-naphthoquinone

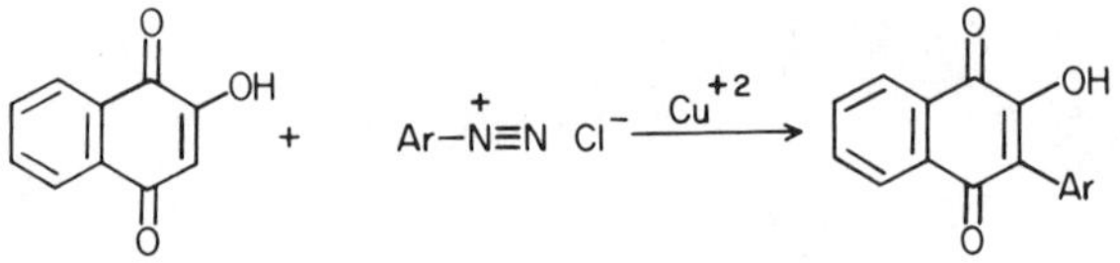

(*c*)

Scheme 11.850 Meerwein arylation reaction: (*a*) general scheme; (*b*) with substituted styrenes; (*c*) with substituted naphthoquinones.

11.860 Meerwein-Ponndorf-Verley Reduction

[H. Meerwein and R. Schmidt, *Ann.*, **444**, 221 (1925); W. Ponndorf, *Angew.* Chem., **39**, 138 (1926); A. Verley, *Bull. Soc. Chim.*, *France*, [4] **37**, 537,871 (1925).] The aluminum isopropoxide reduction of the carbonyl group of a ketone, Scheme 11.860*a*, or of an aldehyde, Scheme 11.860*b*. The reverse reaction, oxidation of a primary or secondary alcohol to the carbonyl function in the presence of an aluminum alkoxide, is the **Oppenauer oxidation** (*vide infra*).

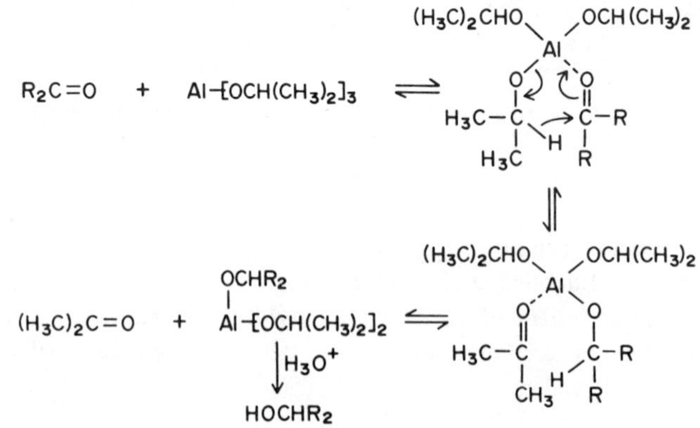

$R_2C{=}O$ + $Al{-}[OCH(CH_3)_2]_3$ ⇌ [cyclic complex: $(H_3C)_2CHO$, $OCH(CH_3)_2$ on Al; $H_3C{-}C(H_3C)(H){-}O{\cdots}Al{\cdots}O{=}C{-}R(R)$]

⇅

$(H_3C)_2C{=}O$ + $OCHR_2{-}Al{-}[OCH(CH_3)_2]_2$ ⇌ [cyclic complex: $(H_3C)_2CHO$, $OCH(CH_3)_2$ on Al; $H_3C{-}C(CH_3){=}O{\cdots}Al{-}O{-}C(H){-}R(R)$]

$\xrightarrow{H_3O^+}$ $HOCHR_2$

(*a*)

o-nitrobenzaldehyde ($H{-}C{=}O$, NO_2) $\xrightarrow[2)\ H_3O^+]{1)\ Al[OCH(CH_3)_2]_3,\ HOCH(CH_3)_2}$ o-nitrobenzyl alcohol (CH_2OH, NO_2)

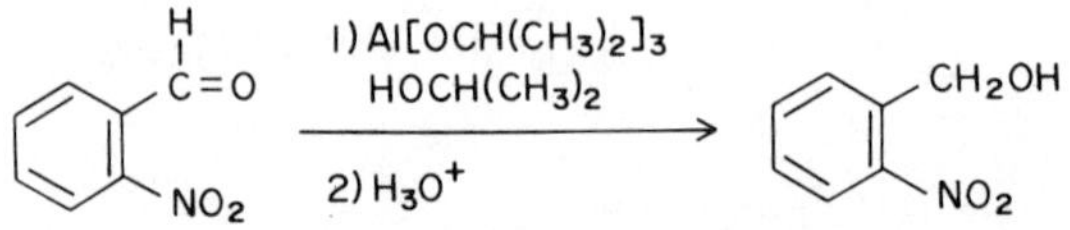

(*b*)

Scheme 11.860 Meerwein-Ponndorf-Verley reduction: (*a*) of a ketone; (*b*) of an aldehyde.

11.870 Meyer-Schuster Rearrangement

[K. H. Meyer and K. Schuster, *Ber.*, **55**, 819 (1922).] The acid-catalyzed rearangement of substituted propargylic alcohols to α,β-unsaturated ketones (or aldehydes if an alkyn-1-ol is used) via allenic intermediates, Scheme 11.870.

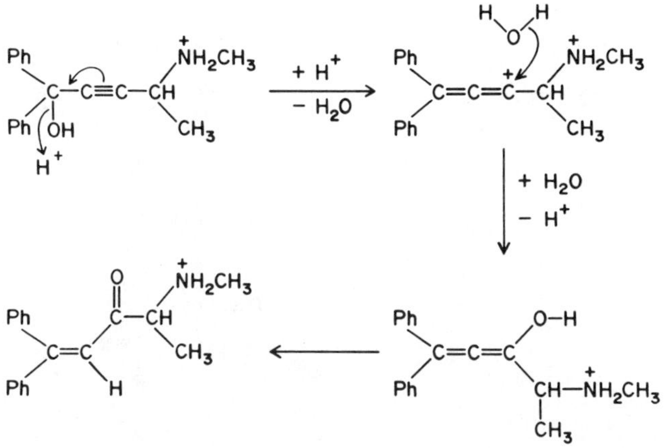

Scheme 11.870 Meyer-Schuster rearrangement.

11.880 Meyers Aldehyde Synthesis

[A. I. Meyers, A. Nabeya, H. Adickes, and I. Politzer, *J. Am. Chem. Soc.*, **91**, 763 (1969).] The conversion of an alkyl halide, R′X, to the aldehyde, $R'CH_2CHO$, using a 2-lithiomethyl-tetrahydro-3-oxazine in a coupling reaction followed by reduction and hydrolysis, Scheme 11.880.

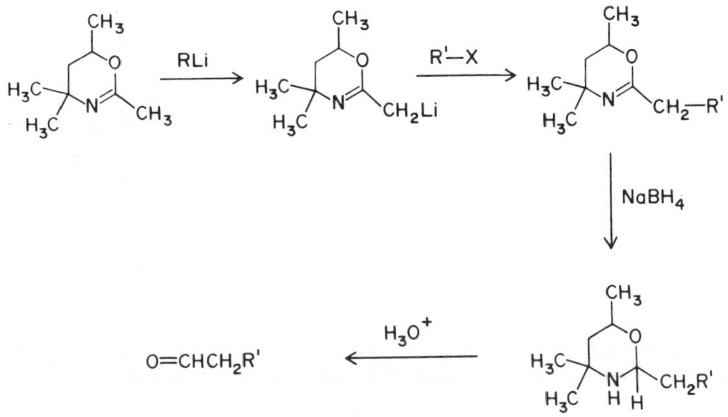

Scheme 11.880 Meyers Aldehyde synthesis.

11.890 Michael Addition

[A. Michael, *J. Prakt. Chem.*, [2] **35**, 349 (1887).] The addition of carbanions to electron deficient carbon-carbon double bonds, Scheme 11.890*a*. The olefinic linkage generally is part of a conjugated system. A wide variety of reaction partners undergo this fundamental reaction, Scheme 11.890*b*, *c*, and *d*.

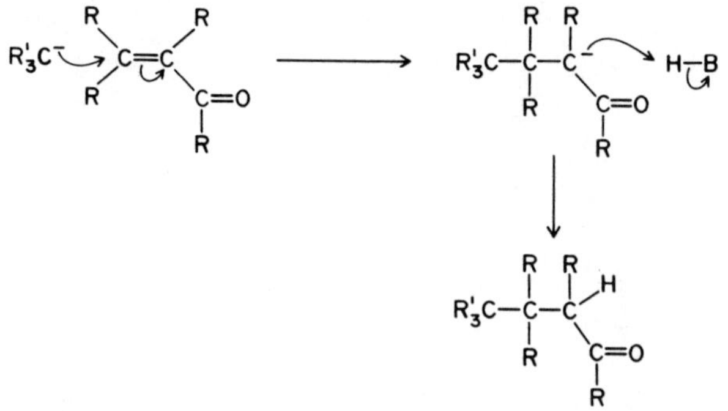

R'_3C^- + $R_2C{=}C(R){-}C({=}O)R$ → $R'_3C{-}CR_2{-}C^-(R){-}C({=}O)R$ + H–B → $R'_3C{-}CR_2{-}CH(R){-}C({=}O)R$

(*a*)

$(H_3C)_2C{=}CHNO_2$ —1) KCN, 2) H^+→ $(H_3C)_2C(CN){-}CH_2NO_2$

(*b*)

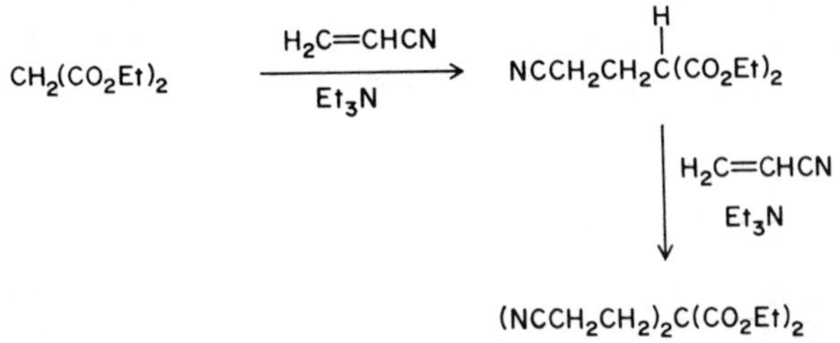

$CH_2(CO_2Et)_2$ —$H_2C{=}CHCN$, Et_3N→ $NCCH_2CH_2CH(CO_2Et)_2$ —$H_2C{=}CHCN$, Et_3N→ $(NCCH_2CH_2)_2C(CO_2Et)_2$

(*c*)

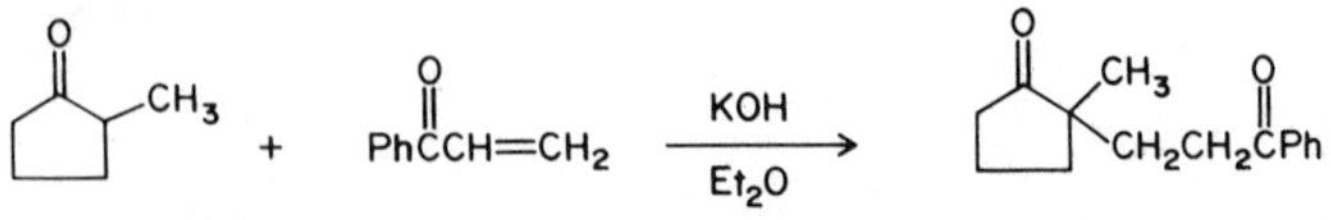

2-methylcyclopentanone + $PhCCH{=}CH_2$ (with C=O) —KOH, Et_2O→ 2-methyl-2-(CH_2CH_2CPh, with C=O)cyclopentanone

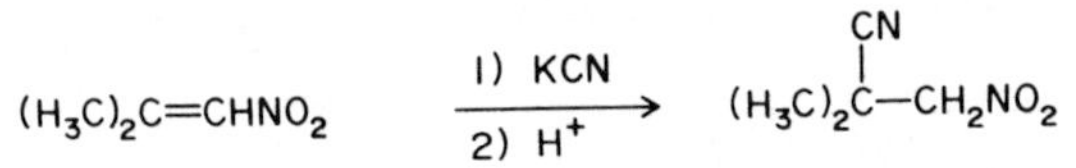

(*d*)

Scheme 11.890 Michael addition: (*a*) general scheme; (*b*) with an α,β-unsaturated nitro compound; (*c*) with an α,β-unsaturated nitrile; (*d*) with an α,β-unsaturated carbonyl compound.

11.900 Nenitzescu Indole Synthesis

[C. D. Nenitzescu, *Bull. Soc. Chim. Romania*, **11**, 37 (1929).] A synthesis of substituted 5-hydroxyindoles via the condensation between *p*-benzoquinones and *β*-aminoacrylic esters, Scheme 11.900*a* and *b*.

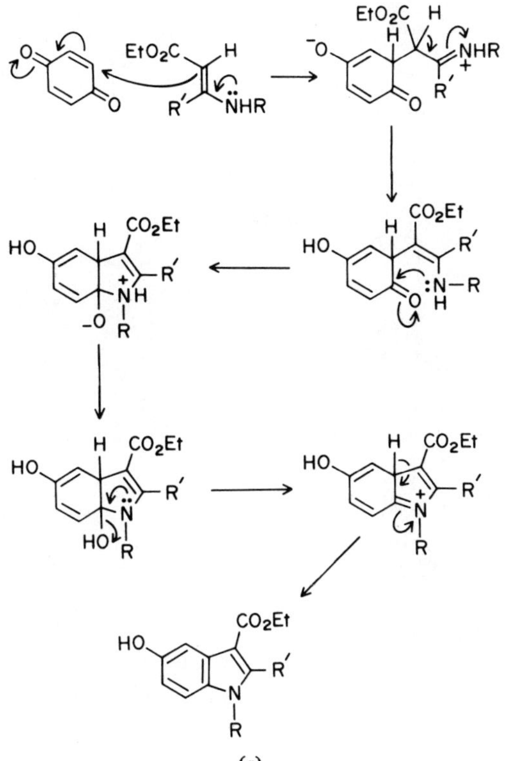

(*a*)

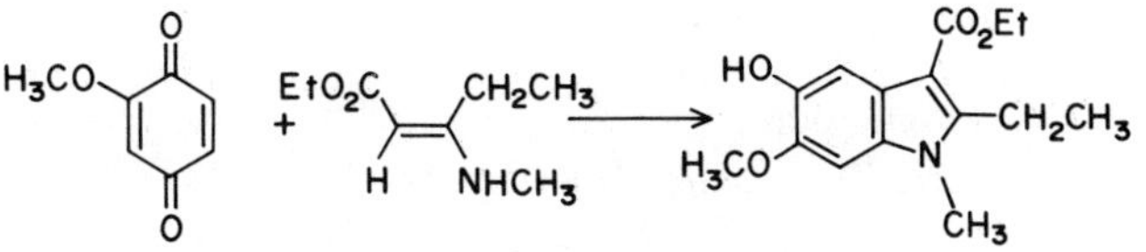

(*b*)

Scheme 11.900 Nenitzescu indole synthesis: (*a*) using *p*-benzoquinone; (*b*) using a substituted *p*-benzoquinone.

11.910 NIH Shift

[G. Guroff, J. W. Daly, D. M. Jerina, J. Renson, B. Witkop, and S. Undenfriend, *Science*, **157**, 1524 (1967).] The monooxygenase-catalyzed formation of arene oxides from alkylaromatics, followed by ring-opening and rearrangement, Scheme 11.910. The reaction occurs widely in biological systems and is named after the National Institute of Health (NIH), where it was discovered.

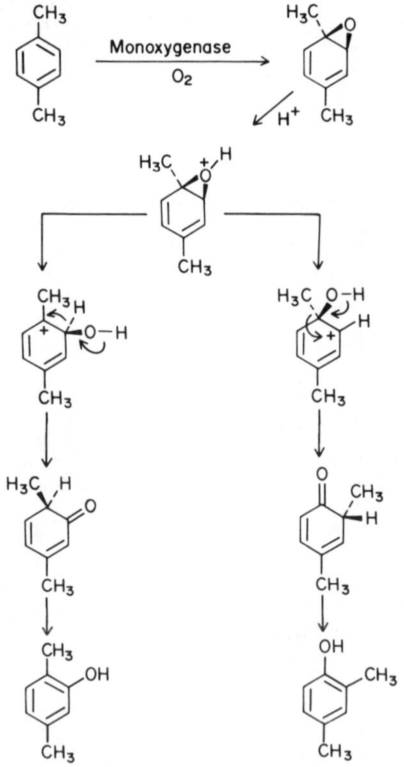

Scheme 11.910 NIH shift.

11.920 Oppenauer Oxidation

[R. V. Oppenauer, *Rec. Trav. Chim. Pays-Bas*, **56**, 137 (1937).] The aluminum alkoxide-catalyzed oxidation of secondary alcohols to the corresponding ketones, Scheme 11.920*a* and *b*. This reaction is the reverse of the Meerwein-Ponndorf-Verley reduction. This is a good method to oxidize allyl alcohols to α, β-unsaturated ketones.

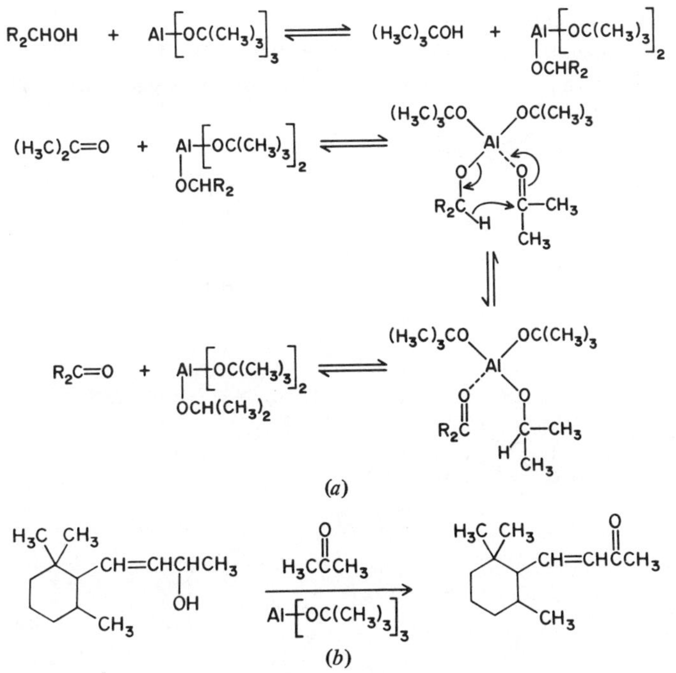

$R_2CHOH + Al[OC(CH_3)_3]_3 \rightleftharpoons (H_3C)_3COH + (R_2CHO)Al[OC(CH_3)_3]_2$

$(H_3C)_2C{=}O + (R_2CHO)Al[OC(CH_3)_3]_2 \rightleftharpoons$ [cyclic complex: $(H_3C)_3CO$, $OC(CH_3)_3$, Al, O, O, R_2C, H, $C{-}CH_3$, CH_3]

$R_2C{=}O + ((CH_3)_2CHO)Al[OC(CH_3)_3]_2 \rightleftharpoons$ [complex: $(H_3C)_3CO$, $OC(CH_3)_3$, Al, O, O, R_2C, H, $C{-}CH_3$, CH_3]

(*a*)

H_3C CH_3 $CH{=}CHCHCH_3$ OH CH_3 $\xrightarrow[Al[OC(CH_3)_3]_3]{H_3CCCH_3 (O)}$ H_3C CH_3 $CH{=}CHCCH_3$ (O) CH_3

(*b*)

Scheme 11.920 Oppenauer oxidation: (*a*) general scheme; (*b*) allyl alcohol oxidation.

11.930 Passerini Reaction

[M. Passerini, *Gazz. Chim. Ital.*, **51**, 126 (1921).] The synthesis of *N*-substituted α-acyloxycarboxylic acid amides through the reaction of an isonitrile with a carbonyl compound and a carboxylic acid, Scheme 11.930.

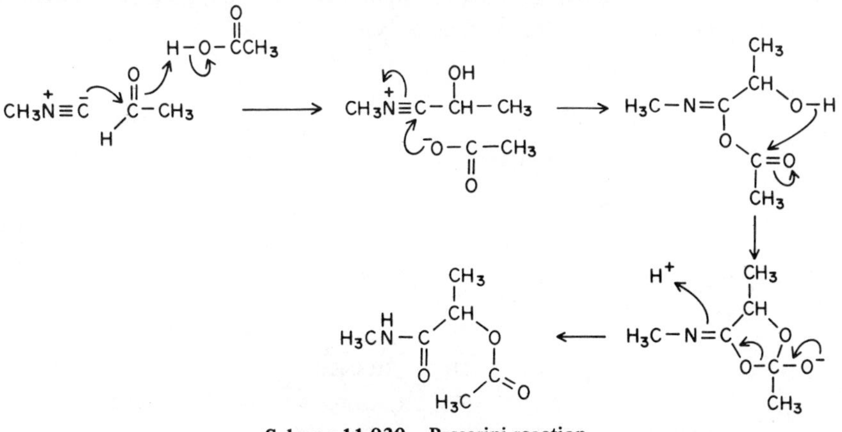

$CH_3\overset{+}{N}{\equiv}\overset{-}{C}$ + $H{-}C(=O){-}CH_3$ + $H{-}O{-}C(=O)CH_3$ $\longrightarrow$ $CH_3\overset{+}{N}{\equiv}C{-}CH(OH){-}CH_3$, $^{-}O{-}C(=O){-}CH_3$ $\longrightarrow$ $H_3C{-}N{=}C(O{-}C(=O)CH_3){-}CH(CH_3){-}O{-}H$

$\downarrow$

$H_3CNH{-}C(=O){-}CH(CH_3){-}O{-}C(=O){-}CH_3$ $\longleftarrow$ H^+ $H_3C{-}N{=}C$ [cyclic: CH, CH_3, O, $O{-}C{-}O^-$, CH_3]

Scheme 11.930 Passerini reaction.

11.940 Pechmann Synthesis

[H. von Pechmann and C. Duisberg, *Ber.*, **16**, 2119 (1883).] The synthesis of coumarins by the acid-catalyzed condensation of phenols with β-ketoesters, Scheme 11.940.

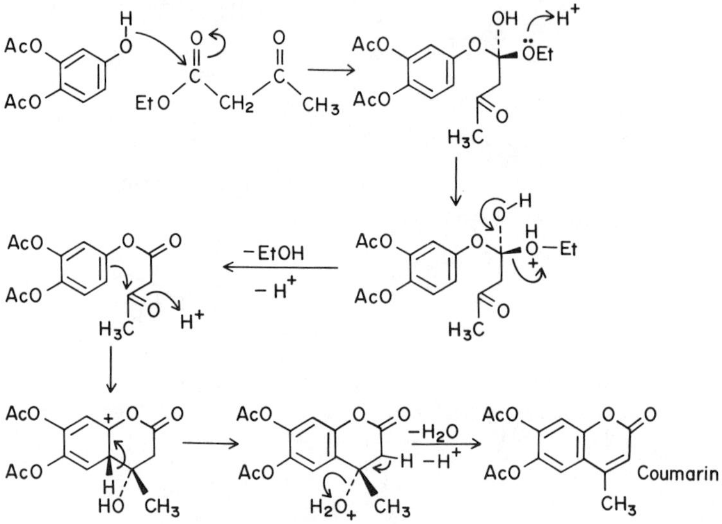

AcO AcO H O O C EtO CH2 O C CH3 → AcO AcO O OH ÖEt H+ H3C O

−EtOH −H+ H O O−Et H Coumarin CH3 −H2O H −H+ H2O+ HO CH3

Scheme 11.940 Pechmann synthesis.

11.950 Perkin-Markovnikov-Krestovnikov-Freund Synthesis

[W. Markovnikov and A. Krestovnikov, *Ann.*, **208**, 333 (1881); A. Freund, *Monatsh. Chem.*, **3**, 625 (1882); W. H. Perkin, *Ber.*, **16**, 208 (1883).] The condensation of α, ω-dihaloalkanes with compounds containing an active methylene group to produce alicyclic compounds, Scheme 11.950.

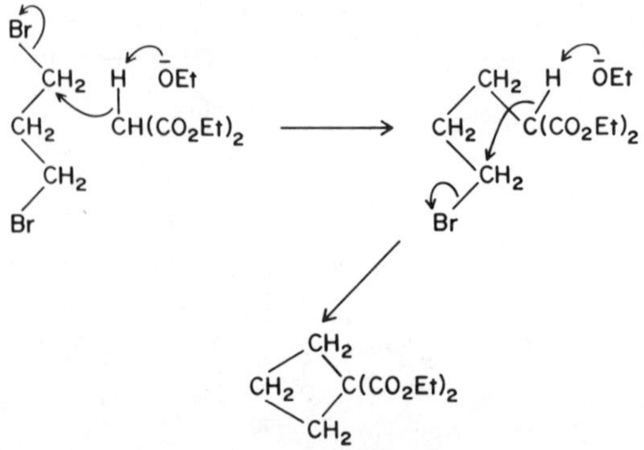

Br CH2 H OEt CH2 CH(CO2Et)2 CH2 Br → CH2 H OEt CH2 C(CO2Et)2 CH2 Br → CH2 CH2 C(CO2Et)2 CH2

Scheme 11.950 Perkin-Markovnikov-Krestovnikov-Freund synthesis.

11.960 Perkin Condensation

[W. H. Perkin, *J. Chem. Soc.*, **21**, 53 (1868).] An aldol-type condensation between acetic anhydride and aryl aldehydes in the presence of sodium acetate to produce cinnamic acid-type compounds, Scheme 11.960.

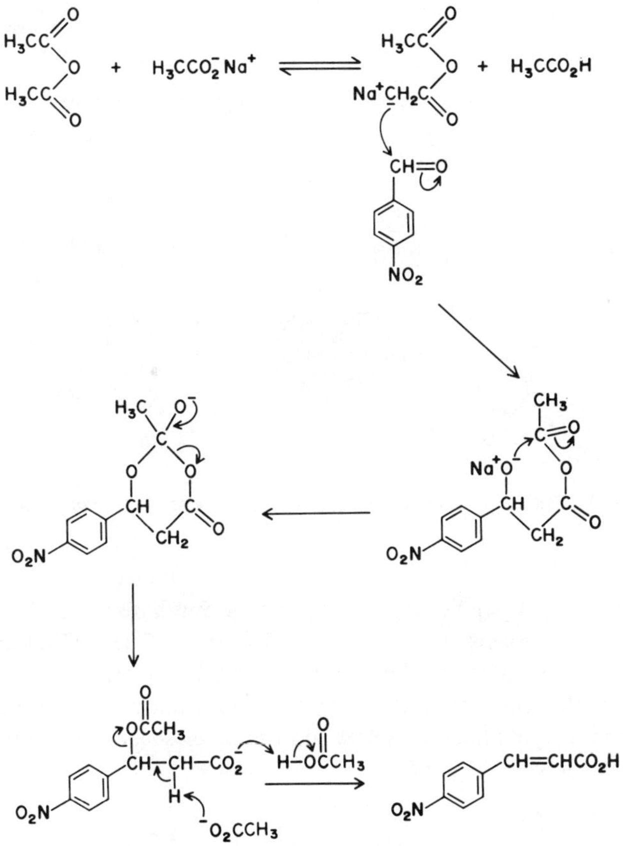

$H_3CC(=O)OC(=O)CH_3$ + $H_3CCO_2^- Na^+$ ⇌ $Na^+ \bar{C}H_2C(=O)OC(=O)CH_3$ + H_3CCO_2H

$CH{=}O$

NO_2

CH_3

$C{=}O$

Na^+O^-

O

CH

C

O

CH_2

O_2N

H_3C

O^-

C

O

O

CH

C

O

CH_2

O_2N

O

$OCCH_3$

$CH{-}CH{-}CO_2^-$

H

$^-O_2CCH_3$

O

$H{-}OCCH_3$

O_2N

$CH{=}CHCO_2H$

O_2N

Scheme 11.960 Perkin condensation.

11.970 Pfitzner-Moffatt Oxidation

[K. E. Pfitzner and J. G. Moffatt, *J. Am. Chem. Soc.*, **85**, 3027 (1963).] The oxidation of alcohols to carbonyls with dimethyl sulfoxide and dicyclohexylcarbodiimide, Scheme 11.970*a*. This reaction is most useful in the oxidation of primary alcohols to aldehydes without further oxidation to the carboxylic acid, Scheme 11.970*b*.

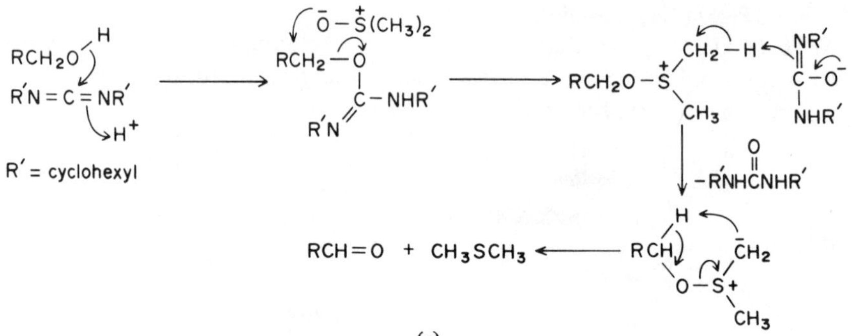

RCH₂O H
R'N=C=NR' H⁺
R' = cyclohexyl
O⁻–S⁺(CH₃)₂
RCH₂–O
C–NHR'
R'N
RCH₂O–S⁺ CH₂–H NR' C–O⁻ NHR'
CH₃
–R'NHCONHR'
RCH=O + CH₃SCH₃
RCH H CH₂ O–S+ CH₃

(a)

OH CH₃ N O N HOH₂C O OAc
$(CH_3)_2\overset{+}{S}-\overset{-}{O}$
$C_6H_{11}N=C=NC_6H_{11}$
H_3PO_4
OH CH₃ N O N O=CH O OAc

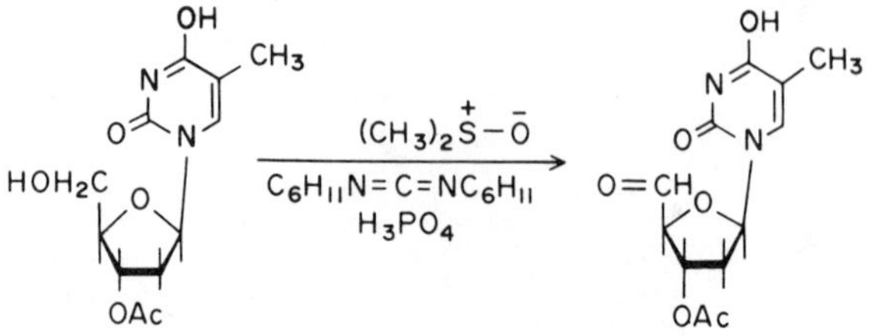

(b)

Scheme 11.970 Pfitzner-Moffatt oxidation: (*a*) general scheme; (*b*) application to a thymidine derivative.

11.980 Pictet-Spengler Tetrahydroisoquinoline Synthesis

[A. Pictet and T. Spengler, *Ber.*, **44**, 2030 (1911).] The cyclization of Schiff bases prepared from phenethylamines and aldehydes to tetrahydroisoquinolines, Scheme 11.980*a*. An interesting extension of this reaction involves the use of an aldehyde and a tryptamine derivative to form indoles related to naturally occurring alkaloids, Scheme 11.980*b*.

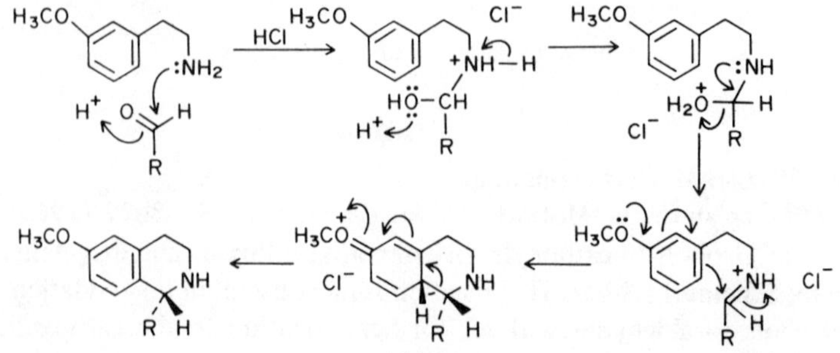

H₃CO :NH₂ H⁺ O H R
HCl
H₃CO Cl⁻ ⁺NH–H HÖ–CH H⁺ R
H₃CO :NH H₂Ö⁺ H Cl⁻ R
H₃CÖ ⁺NH Cl⁻ H R
H₃CO⁺ Cl⁻ NH H R H
H₃CO NH R H

(a)

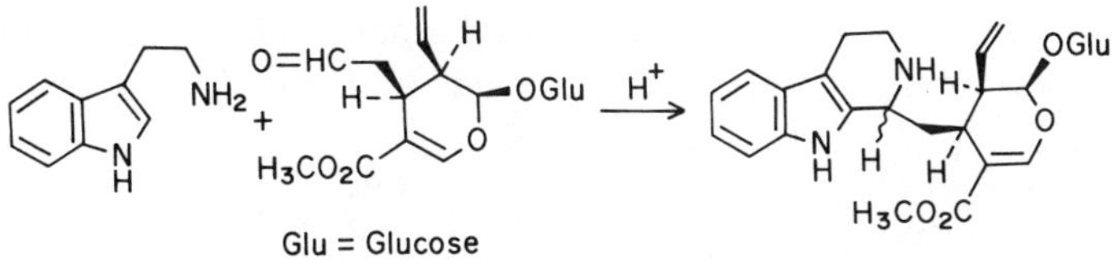

(*b*)

Scheme 11.980 Pictet-Spengler tetrahydroisoquinoline synthesis: (*a*) general scheme; (*b*) synthesis of an alkaloid.

11.990 Pinacol Rearrangement

Treatment of a 1,2-gylcol with strong acid to give an aldehyde or ketone via the Wagner-Meerwein-Whitmore rearrangement, Scheme 11.990*a*. *trans*-Cyclohexanediols give cyclopentyl carbonyls, Scheme 11.990*b*.

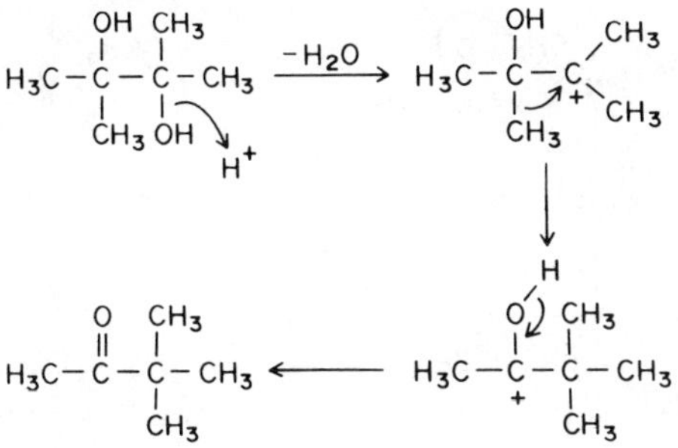

(*a*)

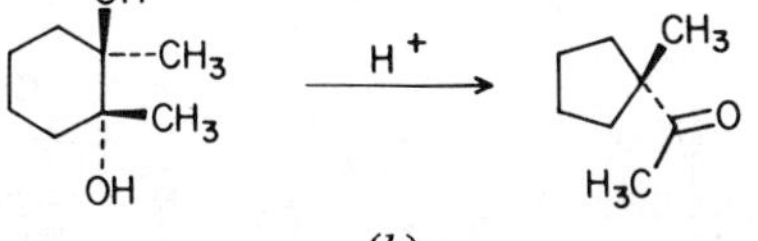

(*b*)

Scheme 11.990 Pinacol rearrangement: (*a*) formation of pinacol; (*b*) formation of a ketone with change in ring size.

11.1000 Polonovski Reaction

[M. Polonovski and M. Polonovski, *Bull. Soc. Chim. France*, **41**, 1190 (1927).] Reduction of *tert*-amine *N*-oxides during reaction with acetic anhydride. Substituted acid amides are the products of alkyl group oxidation and *O*-acetylated aminophenols are the products of aromatic ring oxidation, Scheme 11.1000.

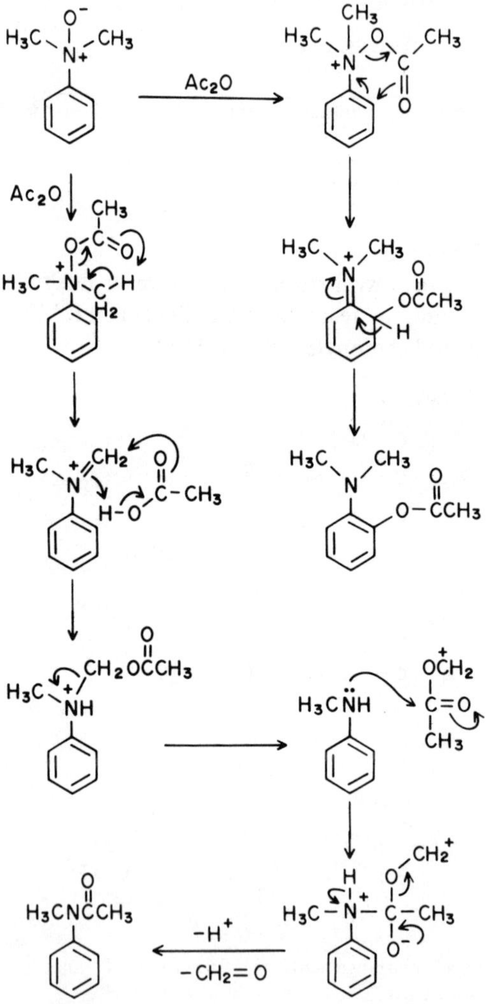

O⁻
H₃C N+ CH₃
Ac₂O
CH₃
H₃C O CH₃
+N C
O
Ac₂O CH₃
O C O
+
H₃C—N C H
H₂
H₃C + CH₃
N
O
OCCH₃
H
H₃C + CH₂ O
N C—CH₃
H—O
H₃C CH₃
N O
O—CCH₃
O
CH₂OCCH₃
H₃C +
NH
+
OCH₂
H₃CNH C=O
CH₃
CH₂⁺
H O
+
H₃C—N— C—CH₃
O⁻
O
H₃CNCCH₃
—H⁺
—CH₂=O

Scheme 11.1000 Polonovski reaction.

11.1010 Pomeranz-Fritsch Reaction

[C. Pomeranz, *Monatsh. Chem.*, **14**, 116 (1893); P. Fritsch, *Ber.*, **26**, 419 (1893).] The synthesis of isoquinolines by the cyclization of the Schiff bases prepared from benzaldehydes and β-aminoacetaldehyde acetal, Scheme 11.1010*a*. In a modification [Schlittler and Mueller, *Helv. Chim. Acta*, **31**, 914 (1948)] the aldehyde and amine moieties are interchanged, Scheme 11.1010*b*.

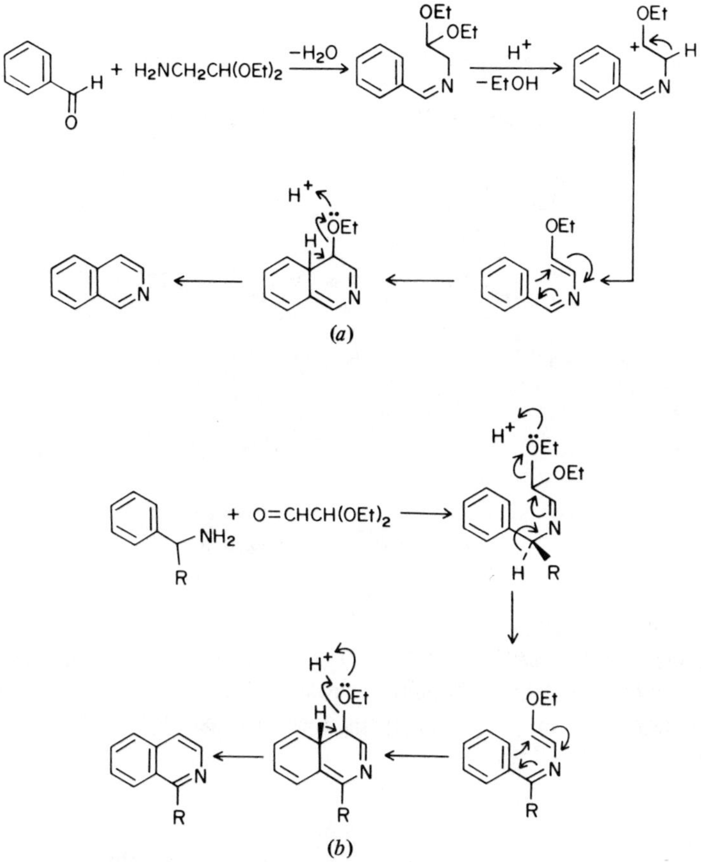

(*a*)

(*b*)

Scheme 11.1010 Pomeranz-Fritsch reaction: (*a*) using an aromatic aldehyde and an aliphatic amine; (*b*) using an aliphatic aldehyde and an aralkylamine.

11.1020 Prévost Reaction

[C. Prévost, *Compt. Rend.*, **196**, 1129 (1933).] The preparation of *trans*-1,2-glycols by benzoylation of cyclic olefins with iodine and silver benzoate in an aprotic solvent, followed by hydrolysis, Scheme 11.1020.

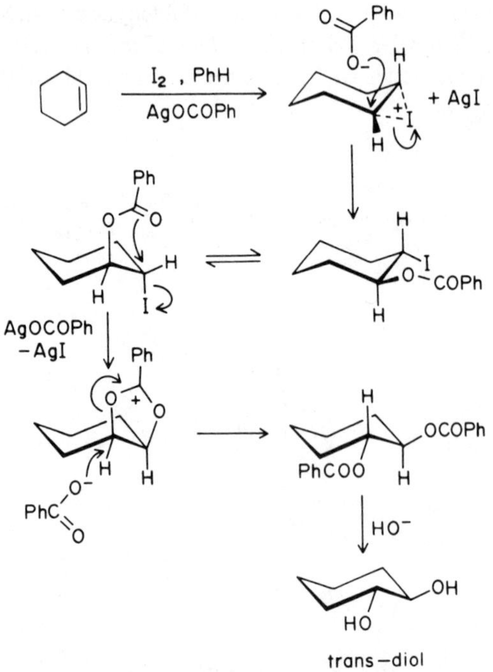

Scheme 11.1020 Prévost reaction.

11.1030 Prins Reaction

[H. J. Prins, *Chem. Weekbl.*, 16, 64 (1919).] The acid-catalyzed condensation of an olefin with an aldehyde (usually formaldehyde). The hydroxymethyl intermediate may react to give a 1,3-glycol, a substituted allyl alcohol, or a 1,3-dioxane, Scheme 11.1030.

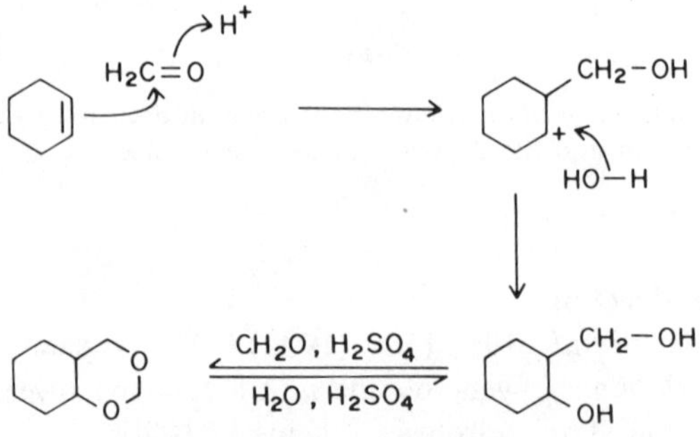

Scheme 11.1030 Prins reaction.

11.1040 Pschorr Synthesis

[R. Pschorr, *Ber.*, **29**, 496 (1896).] A synthesis of phenanthrenes starting with an *o*-nitrobenzaldehyde and a phenylacetic acid. The resulting α-phenyl-*o*-nitrocinnamic acid is reduced, diazotized, and cyclized, Scheme 11.1040.

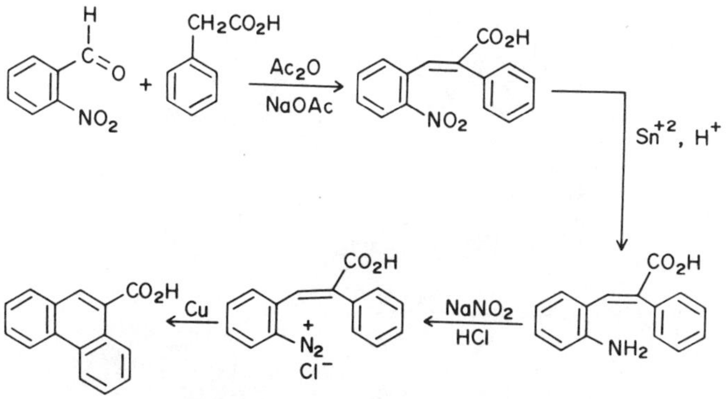

H
C=O
NO_2
+
CH_2CO_2H
Ac_2O
NaOAc
CO_2H
NO_2
Sn^{+2}, H^+
CO_2H
NH_2
$NaNO_2$
HCl
CO_2H
$\overset{+}{N}_2$
Cl^-
Cu
CO_2H

Scheme 11.1040 Pschorr synthesis.

11.1050 Ramberg-Backlund Reaction

[L. Ramberg and B. Backlund, *Arkiv. Kemi, Mineral Geol.*, **13A**[27], 1 (1940).] Reaction of α-halosulfones with strong base to yield olefins through the 1,3-elimination of the hydrogen halide, followed by the extrusion of sulfur dioxide, Scheme 11.1050*a* and *b*.

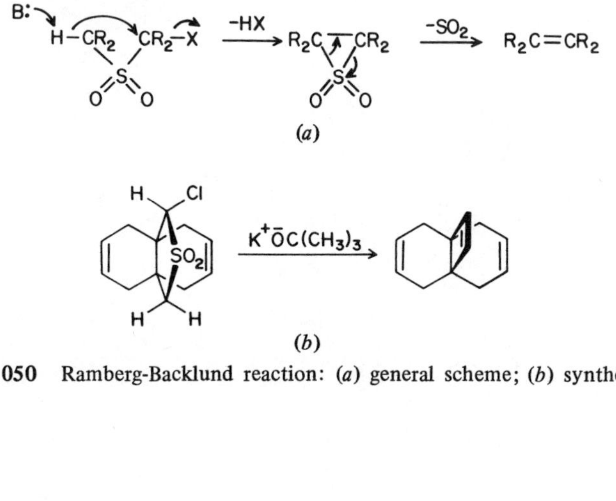

B:
H–CR_2 CR_2–X
S
O O
–HX
R_2C CR_2
S
O O
$-SO_2$
$R_2C{=}CR_2$

(*a*)

H Cl
SO_2
H H
$K^+\bar{O}C(CH_3)_3$

(*b*)

Scheme 11.1050 Ramberg-Backlund reaction: (*a*) general scheme; (*b*) synthesis of a tricyclotriene.

11.1060 Reformatsky Reaction

[S. Reformatsky, *Ber.*, **20**, 1210 (1887).] The metalation of an α-bromoester with zinc, followed by condensation with an aldehyde to give a β-hydroxyester or the α,β-unsaturated ester resulting from dehydration, Scheme 11.1060.

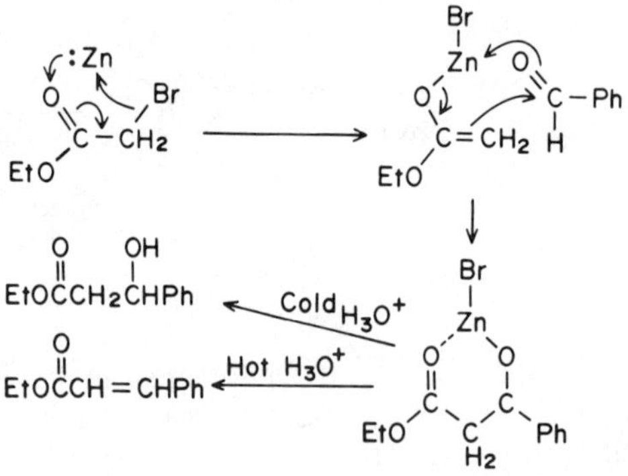

Scheme 11.1060 Reformatsky reaction.

11.1070 Reimer-Tiemann Reaction

[K. Reimer, *Ber.*, **9**, 423 (1876); K. Reimer and F. Tiemann, *Ber.*, **9**, 824 (1876).] The base-catalyzed condensation of chloroform with phenol. The reaction proceeds via electrophilic attack of dichlorocarbene on a phenoxide to ultimately give an *o*-hydroxybenzaldehyde, Scheme 11.1070.

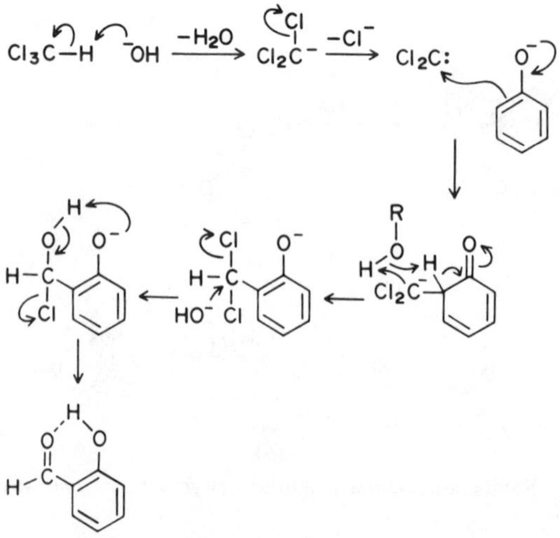

Scheme 11.1070 Reimer-Tiemann reaction.

11.1080 Reissert Reaction

[A. Reissert, *Ber.*, **38**, 1603 (1905).] The *N*-acylation of quinoline and derivatives with an acid chloride, followed by attack by cyanide to give a **Reissert compound** that can be hydrolyzed to give an aldehyde and 2-quinolinecarboxylic acid, Scheme 11.1080. The reaction may also be extended to isoquinolines.

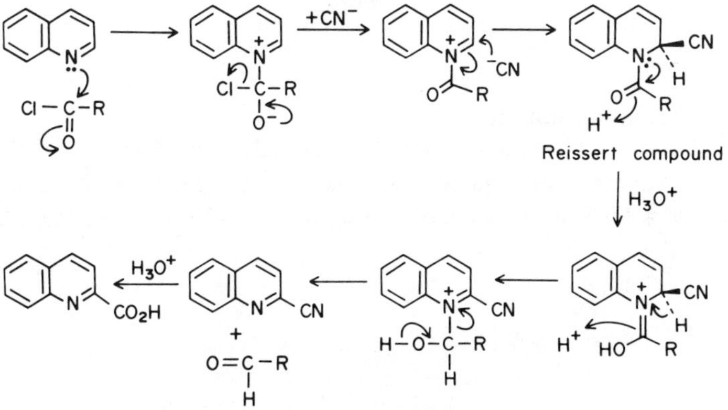

Cl—C—R
O
Cl—C—R
O⁻
+CN⁻
CN
R
CN
H
R
H^+
Reissert compound
H_3O^+
H_3O^+
CO_2H
CN
+
O=C—R
H
CN
H—O—C—R
H
CN
H
H^+
HO
R

Scheme 11.1080 Reissert reaction.

11.1090 Ritter Reaction

[J. J. Ritter and P. P. Minieri, *J. Am. Chem. Soc.*, **70**, 4045 (1948).] The reaction of concentrated sulfuric acid with carbenium ion precursors in the presence of HCN or RCN to give *N*-substituted formamides or amides, Scheme 11.1090.

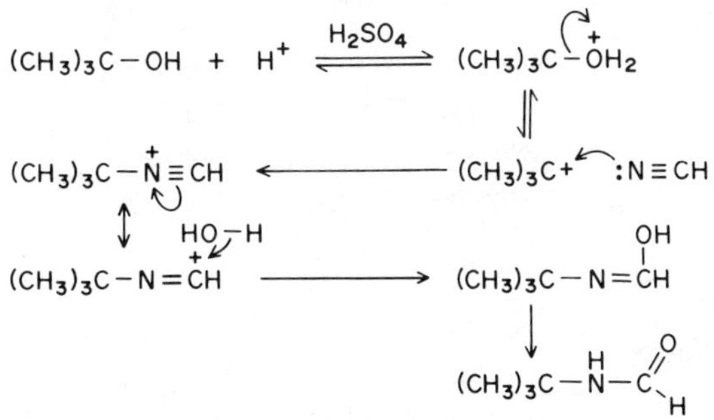

$$(CH_3)_3C-OH + H^+ \overset{H_2SO_4}{\rightleftharpoons} (CH_3)_3C-\overset{+}{O}H_2$$

$$(CH_3)_3C-\overset{+}{O}H_2 \rightleftharpoons (CH_3)_3C^+ + :N\equiv CH$$

$$(CH_3)_3C^+ + :N\equiv CH \longrightarrow (CH_3)_3C-\overset{+}{N}\equiv CH \leftrightarrow (CH_3)_3C-N=\overset{+}{C}H$$

$$(CH_3)_3C-N=\overset{+}{C}H + HO-H \longrightarrow (CH_3)_3C-N=CH(OH) \longrightarrow (CH_3)_3C-NH-CHO$$

Scheme 11.1090 Ritter reaction.

11.1100 Rosenmund Reduction

[K. W. Rosenmund, *Ber.*, **51**, 585 (1918).] The reduction of acid chlorides to aldehydes by poisoned palladium on barium sulfate catalyst, Scheme 11.1100.

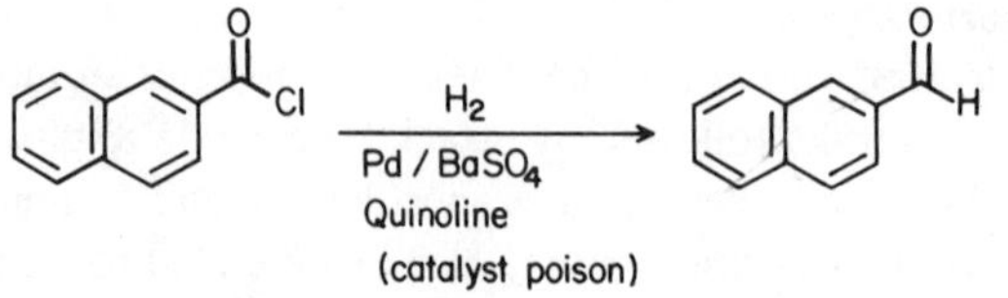

H_2

Pd / $BaSO_4$

Quinoline

(catalyst poison)

Scheme 11.1100 Rosenmund reduction.

11.1110 Robinson Annelation Reaction

[E. C. du Feu, F. J. McQuillin, and R. Robinson, *J. Chem. Soc.*, 53 (1937).] The fusion of a cyclohexenone ring onto an existing cyclohexanone ring (annelation) achieved by a Michael addition followed by an aldol condensation, Scheme 11.1110.

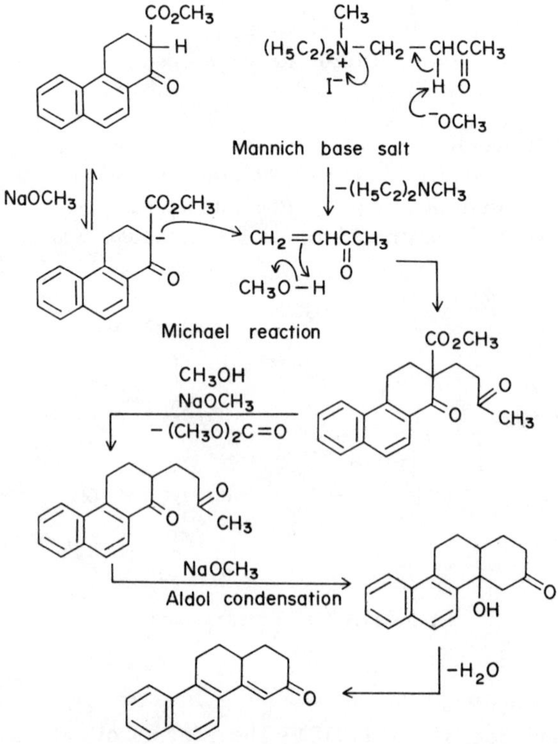

CO_2CH_3 H O

$(H_5C_2)_2\overset{+}{N}(CH_3)-CH_2-CH(H)CCH_3$ (C=O) I^- $^-OCH_3$

Mannich base salt

$NaOCH_3$

$-(H_5C_2)_2NCH_3$

$CH_2=CHCCH_3$ (C=O)

CH_3O-H

Michael reaction

CH_3OH

$NaOCH_3$

$-(CH_3O)_2C=O$

CO_2CH_3 O CH_3

$NaOCH_3$

Aldol condensation

OH O

$-H_2O$

Scheme 11.1110 Robinson annelation reaction.

11.1120 Rupe Rearrangement

[H. Rupe and E. Kambli, *Helv. Chim. Acta*, **9**, 672 (1926).] The acid-catalyzed rearrangement of a *tert*-propargylic alcohol to give an α,β-unsaturated ketone, Scheme 11.1120.

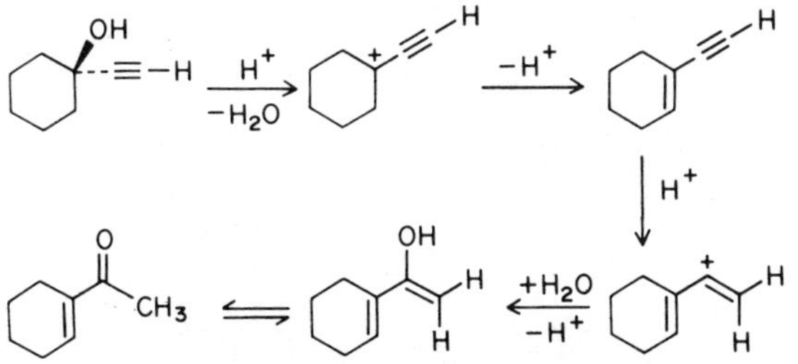

OH
---≡—H $\xrightarrow[-H_2O]{H^+}$ + ≡—H $\xrightarrow{-H^+}$ ≡—H

↓ H^+

O CH3 ⇌ OH H H $\xleftarrow[-H^+]{+H_2O}$ + H H

Scheme 11.1120 Rupe rearrangement.

11.1130 Sandmeyer Reaction

[T. Sandmeyer, *Ber.*, **17**, 1633 (1884.] The treatment of an aromatic diazonium salt with either cuprous cyanide or cuprous halide. The reaction is used to replace an aromatic amino group by either a cyano or halo group, Scheme 11.1130.

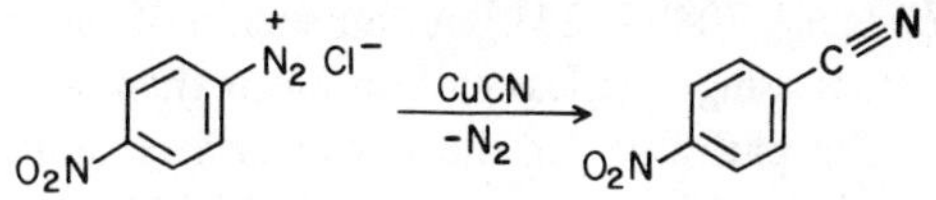

O_2N—C6H4—$\overset{+}{N_2}$ Cl^- $\xrightarrow[-N_2]{CuCN}$ O_2N—C6H4—C≡N

Scheme 11.1130 Sandmeyer reaction.

11.1140 Sanger *N*-Terminal Amino Acid Residue Identification

[F. Sanger, *Biochem. J.*, **39**, 507 (1945).] The use of 2,4-dinitrofluorobenzene to react with and subsequently identify the *N*-terminal amino acid residue of a polypeptide, Scheme 11.1140.

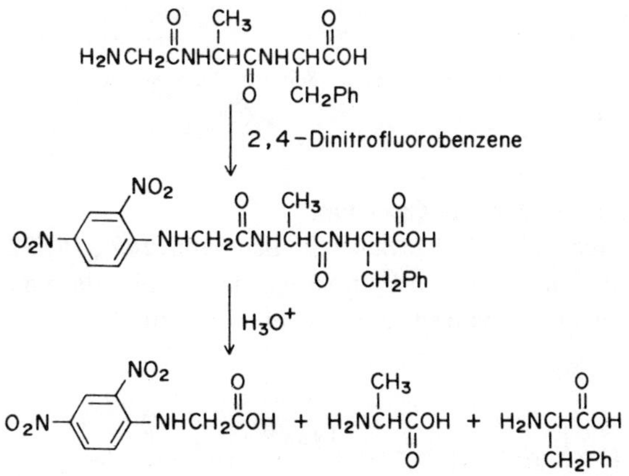

$H_2NCH_2C(=O)NHCH(CH_3)C(=O)NHCH(CH_2Ph)C(=O)OH$

↓ 2,4-Dinitrofluorobenzene

O_2N—C6H3(NO_2)—$NHCH_2C(=O)NHCH(CH_3)C(=O)NHCH(CH_2Ph)C(=O)OH$

↓ H_3O^+

O_2N—C6H3(NO_2)—$NHCH_2C(=O)OH$ + $H_2NCH(CH_3)C(=O)OH$ + $H_2NCH(CH_2Ph)C(=O)OH$

Scheme 11.1140 Sanger *N*-terminal amino acid residue identification.

11.1150 Schiemann Reaction

[G. Balz and G. Schiemann, *Ber.*, **60**, 1186 (1927).] The treatment of an aromatic diazonium salt with KBF_4. The reaction is used to replace an aromatic amino group by a fluorine atom, Scheme 11.1150.

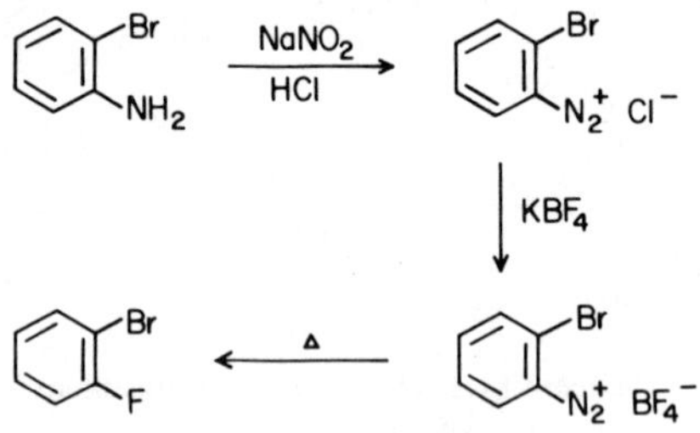

Scheme 11.1150 Schiemann reaction.

11.1160 Schmidt Degradation

[K. F. Schmidt, *Ber.*, **57**, 704 (1924).] A conversion of carboxylic acids via their azides to amines possessing one less carbon atom than the original compound, Scheme 11.1160. The pathway for the conversion is similar to the Hofmann, Lossen, and Curtius degradations (Sects. 11.630, 11.800, and 11.320).

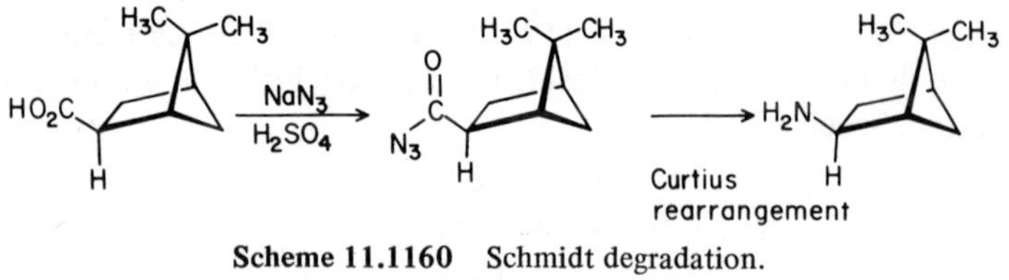

Scheme 11.1160 Schmidt degradation.

11.1170 Schotten-Baumann Procedure

[C. Schotten, *Ber.*, **17**, 2544 (1884); E. Baumann, *Ber.*, **19**, 3218 (1886).] A procedure for preparing an ester by treating an acid chloride and an alcohol with aqueous alkali in a two-phase system, Scheme 11.1170.

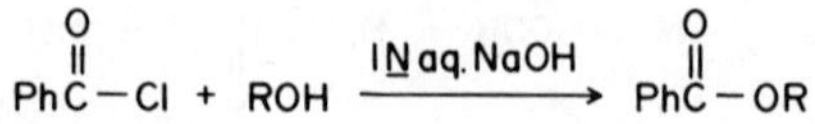

$$PhC(=O)-Cl + ROH \xrightarrow{1\underline{N}\ aq.\ NaOH} PhC(=O)-OR$$

Scheme 11.1170 Schotten-Baumann procedure.

11.1180 Skraup Synthesis

[Z. H. Skraup, *Ber.*, **13**, 2086 (1880).] A synthesis of the quinoline ring system from aromatic amines and glycerol in the presence of sulfuric acid and an oxidizing agent such as O_2. *o*-Diamines yield the phenanthroline ring system, Scheme 11.1180.

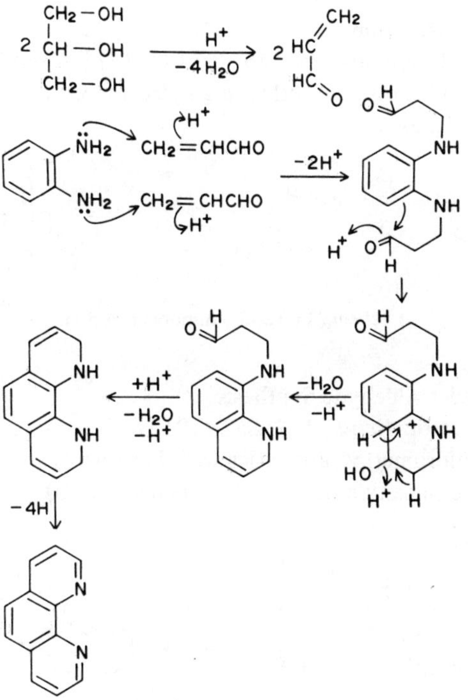

Scheme 11.1180 Skraup synthesis.

11.1190 Simmons-Smith Reaction

[H. E. Simmons and R. D. Smith, *J. Am. Chem. Soc.*, **80**, 5323 (1958).] The reaction of methylene iodide in the presence of a Zn-Cu couple with olefins to give cyclopropanes. The reaction proceeds through a zinc species with carbenoid character, Scheme 11.1190*a* and *b*.

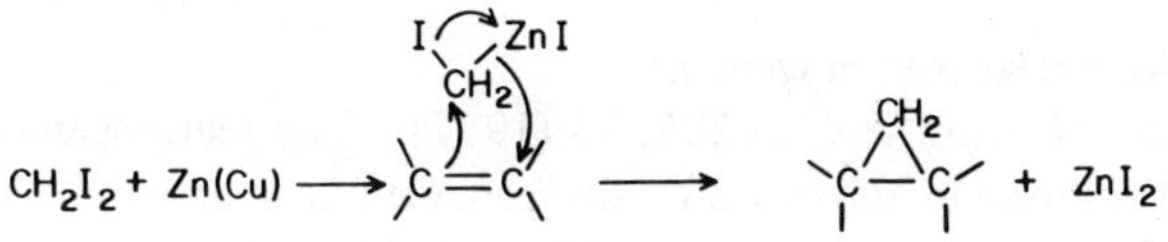

(*a*)

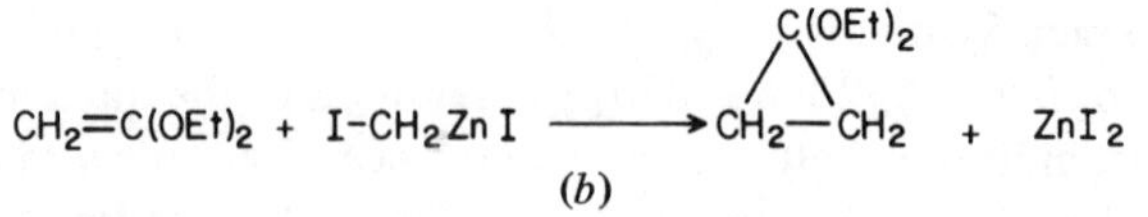

$CH_2{=}C(OEt)_2 + I{-}CH_2ZnI \longrightarrow$ cyclo-$C(OEt)_2CH_2CH_2 + ZnI_2$

(*b*)

Scheme 11.1190 Simmons-Smith reaction: (*a*) general scheme; (*b*) preparation of 1,1-diethoxycyclopropane.

11.1200 Simonis Reaction

[E. Petschek and H. Simonis, *Ber.*, **46**, 2014 (1913).] A synthesis of chromones from substituted phenols and ethyl α-alkylacetoacetates in the presence of P_2O_5, Scheme 11.1200.

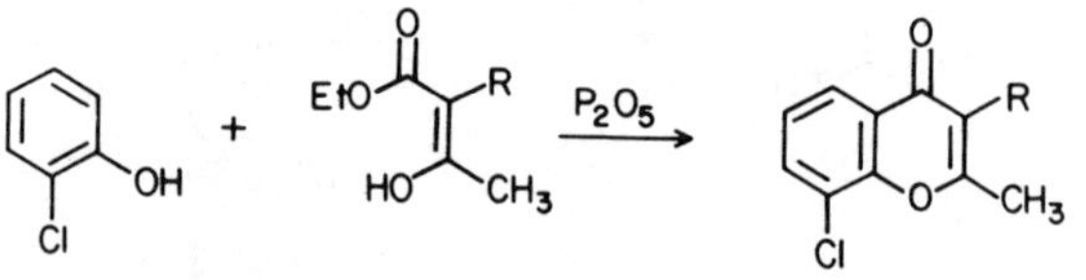

Scheme 11.1200 Simonis reaction.

11.1210 Sommelet Aldehyde Synthesis

[M. Sommelet, *Compt. Rend.*, **157**, 852 (1913).] The preparation of aldehydes by reaction of chloromethyl aromatics with hexamethylenetetramine followed by hydrolytic decomposition of the salt, Scheme 11.1210.

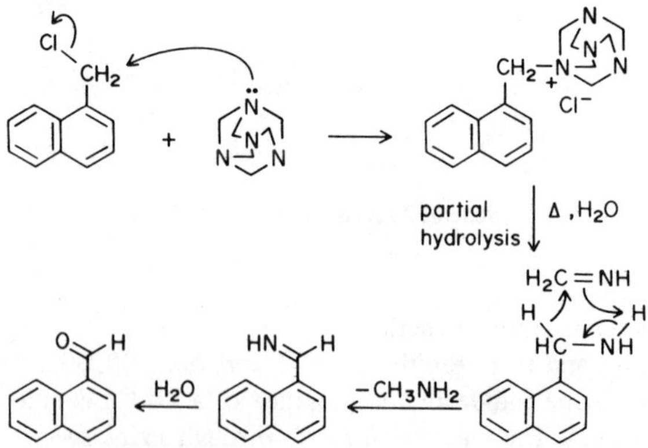

Scheme 11.1210 Sommelet aldehyde synthesis.

11.1220 Sommelet Rearrangement

[M. Sommelet, *Compt. Rend.*, **205**, 56 (1937).] The rearrangement of quaternary benzylammonium salts to *o*-methylbenzylamines, Scheme 11.1220; a special case of the Stevens rearrangement (see Sect. 11.1250).

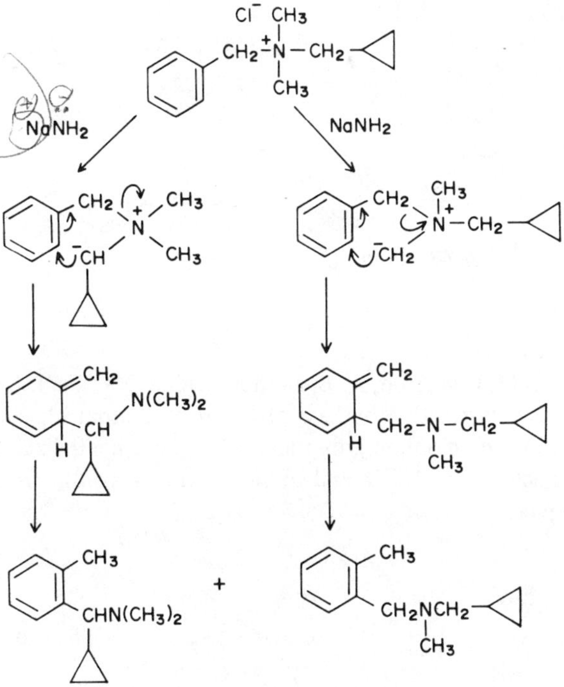

Scheme 11.1220 Sommelet rearrangement.

11.1230 Sonn-Müller Aldehyde Synthesis

[A. Sonn and E. Müller, *Ber.*, **52**, 1927 (1919).] Conversion of an aroyl halide to the corresponding aldehyde via reduction of an imino chloride and hydrolysis of the resulting Schiff base, Scheme 11.1230.

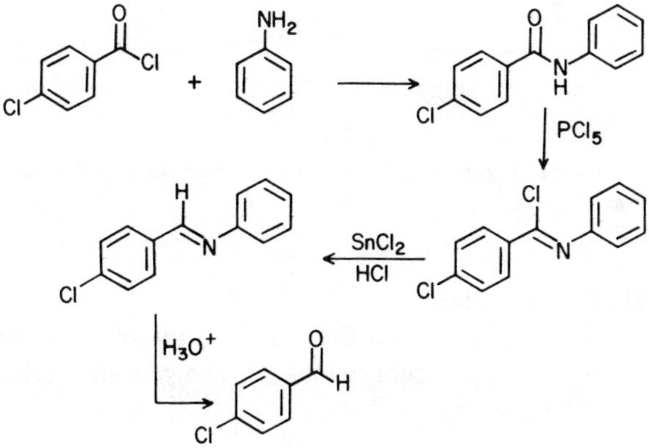

Scheme 11.1230 Sonn-Müller aldehyde synthesis.

11.1240 Stephen Aldehyde Synthesis

[H. Stephen, *J. Chem. Soc.*, **127**, 1874 (1925).] The reduction of an aromatic nitrile to the aldehyde, Scheme 11.1240.

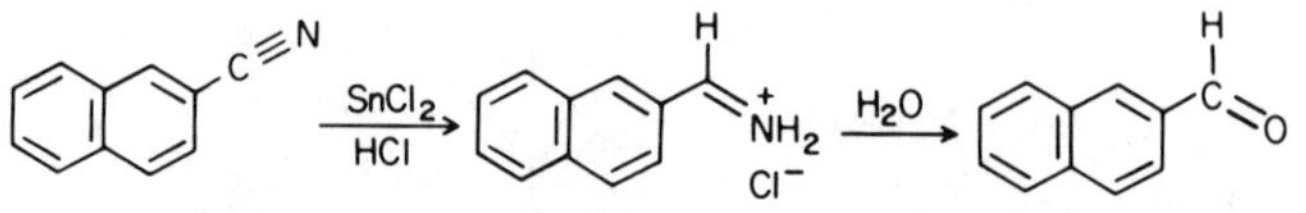

Scheme 11.1240 Stephen aldehyde synthesis.

11.1250 Stevens Rearrangement

[T. S. Stevens, E. M. Creighton, A. B. Gordon, and M. MacNicol, *J. Chem. Soc.*, 3193 (1928).] The abstraction by base of a hydrogen atom adjacent to a sulfonium or quaternary ammonium group and subsequent neutralization of charge by a 1,2,-shift, Scheme 11.1250*a* and *b*; the 1,2-shift may also proceed through a caged radical pair.

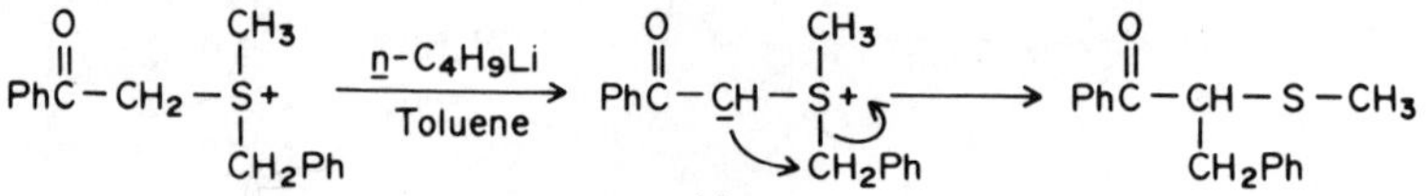

(*a*)

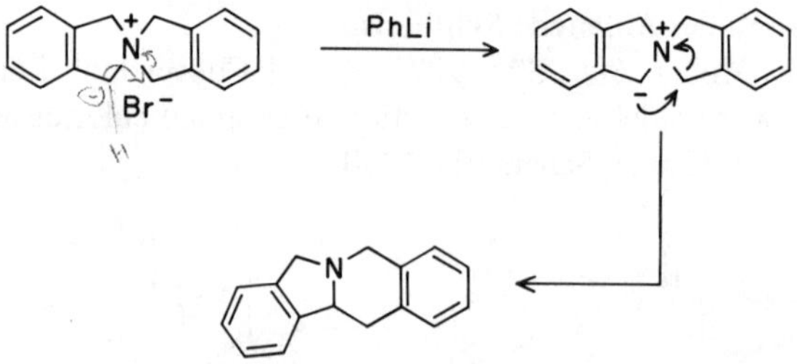

(*b*)

Scheme 11.1250 Stevens rearrangement: (*a*) with a sulfonium compound; (*b*) with an ammonium compound.

11.1260 Stobbe Condensation

[H. Stobbe, *Ber.*, **26**, 2312 (1893).] The condensation of a succinic ester with an aldehyde or ketone in an aldol-type reaction to give an alkylidene succinic acid ester, Scheme 11.1260.

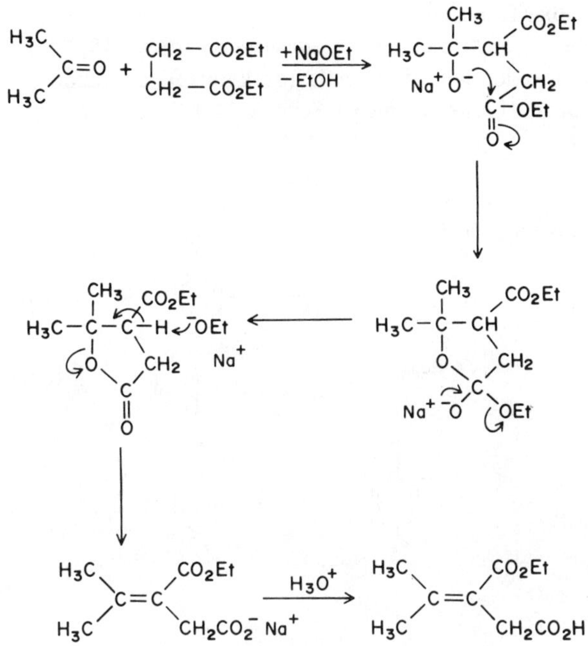

$(H_3C)_2C=O + CH_2-CO_2Et / CH_2-CO_2Et$ $\xrightarrow[-EtOH]{+NaOEt}$ $H_3C-C(CH_3)(O^- Na^+)-CH(CO_2Et)-CH_2-C(=O)-OEt$

$\downarrow$

$H_3C-C(CH_3)-CH(CO_2Et)-CH_2-C(O^- Na^+)(OEt)-O$ (ring)

$\leftarrow$

$H_3C-C(CH_3)-C(CO_2Et)-H \; ^-OEt \; Na^+$, O, CH_2, $C=O$ (ring)

$\downarrow$

$(H_3C)_2C=C(CO_2Et)CH_2CO_2^- Na^+$ $\xrightarrow{H_3O^+}$ $(H_3C)_2C=C(CO_2Et)CH_2CO_2H$

Scheme 11.1260 Stobbe condensation.

11.1270 Strecker Amino Acid Synthesis

[A. Strecker, *Ann.*, 75, 27 (1850).] The conversion of an aldehyde to an α-amino acid by reaction with cyanide ion and ammonia, Scheme 11.1270.

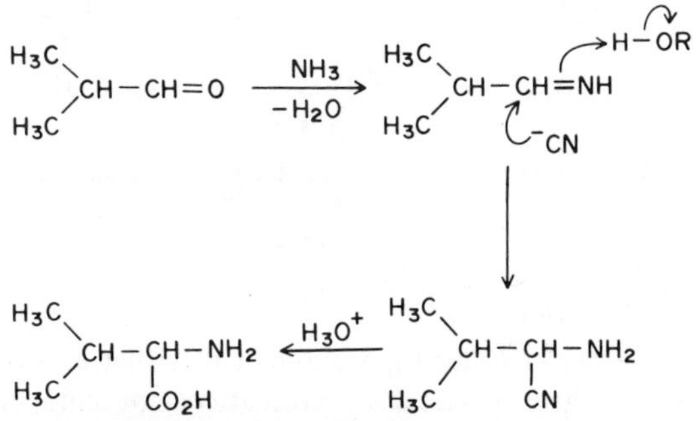

$(H_3C)_2CH-CH=O$ $\xrightarrow[-H_2O]{NH_3}$ $(H_3C)_2CH-CH=NH$ ($H-OR$, ^-CN)

$\downarrow$

$(H_3C)_2CH-CH(NH_2)-CO_2H$ $\xleftarrow{H_3O^+}$ $(H_3C)_2CH-CH(NH_2)-CN$

Scheme 11.1270 Strecker amino acid synthesis.

11.1280 Teuber Reaction

[H. J. Teuber and G. Jellinek, *Naturwissenschaften*, **38**, 259 (1951).] The radical-induced oxidation of phenols (or amines) to quinones by potassium nitrosodisulfonate **(Frémy's radical)**, Scheme 11.1280*a* and *b*.

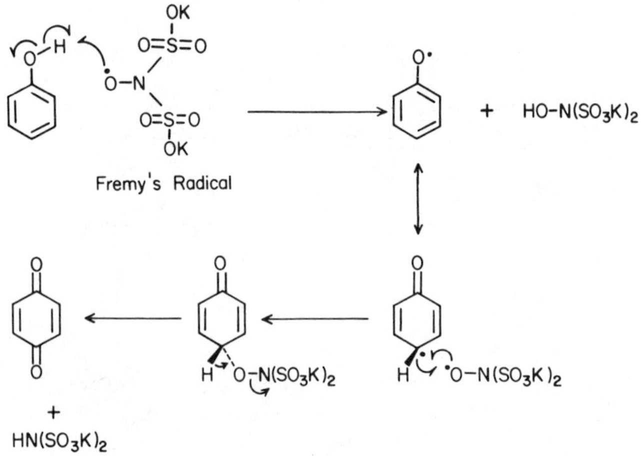

(*a*)

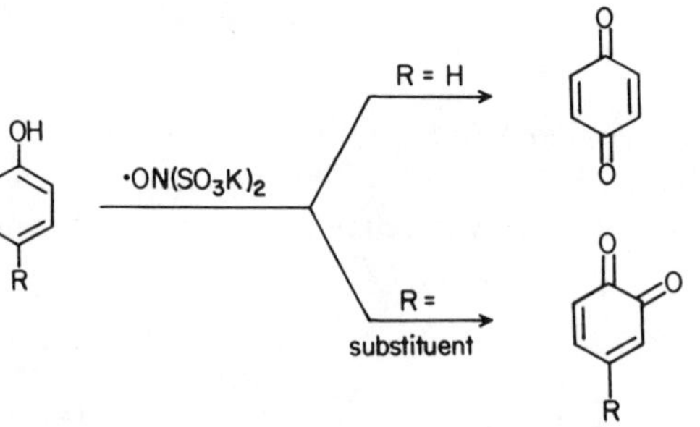

(*b*)

Scheme 11.1280 Teuber reaction: (*a*) using phenol; (*b*) using a substituted phenol.

11.1290 Thiele Reaction

[J. Thiele, *Ber.*, **31**, 1247 (1898).] Treatment of either a *p*- or *o*-quinone with acetic anhydride to give a triacetoxy aromatic compound, Scheme 11.1290*a* and *b*.

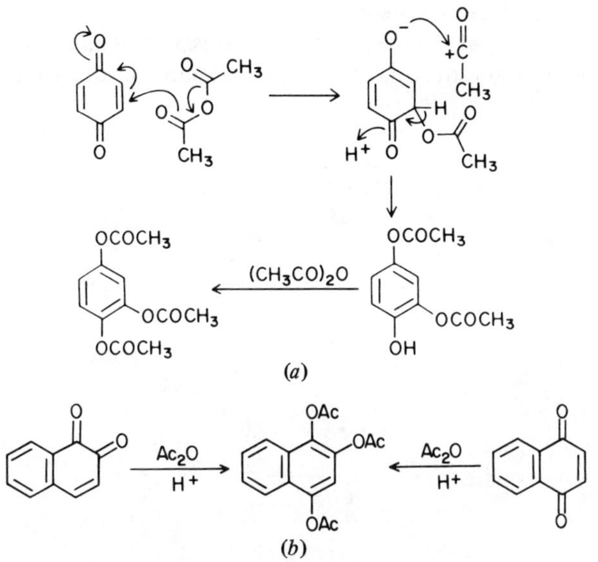

$$(CH_3CO)_2O$$

(a)

Ac_2O, H^+ — OAc, OAc, OAc — Ac_2O, H^+

(b)

Scheme 11.1290 Thiele reaction: (*a*) using *p*-benzoquinone; (*b*) using naphthoquinones.

11.1300 Thorpe-Ziegler Cyclization

[H. Baron, F. G. P. Remfry and J. F. Thorpe, *J. Chem. Soc.*, **85**, 1726 (1904); K. Ziegler, H. Eberle, and H. Ohlinger, *Ann.*, **504**, 94 (1933).] A synthesis of medium-size cyclic ketones from a long chain α, ω-dinitrile using the ether-soluble base lithium ethylphenylamide, Scheme 11.1300.

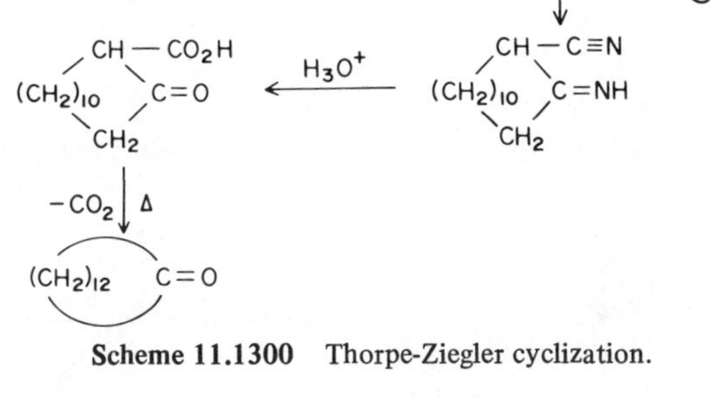

$$\begin{array}{l} H-CH-C\equiv N \\ (CH_2)_{10} \\ CH_2-C\equiv N \end{array} \xrightarrow[\ ]{Li^+\ \ Ph\bar{N}CH_2CH_3} \begin{array}{l} \bar{C}H-C\equiv N \\ (CH_2)_{10} \\ CH_2-C\equiv N \end{array} \longrightarrow \begin{array}{l} CH-C\equiv N \\ (CH_2)_{10}\ \ C=NH \\ CH_2 \end{array} \xrightarrow{H_3O^+} \begin{array}{l} CH-CO_2H \\ (CH_2)_{10}\ \ C=O \\ CH_2 \end{array} \xrightarrow[\Delta]{-CO_2} (CH_2)_{12}\ \ C=O$$

Scheme 11.1300 Thorpe-Ziegler cyclization.

11.1310 Tiemann Rearrangement

[F. Tiemann, *Ber.*, **24**, 4162 (1891).] The condensation of an amino-substituted oxime with benzenesulfonyl chloride to give a monosubstituted urea, Scheme 11.1310.

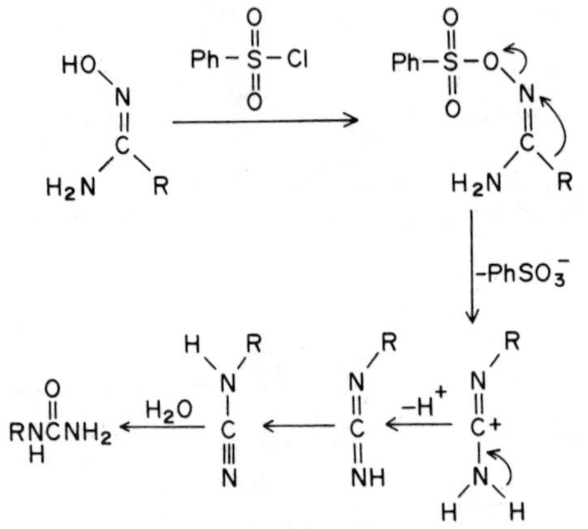

Scheme 11.1310 Tiemann rearrangement.

11.1320 Tishenko Reaction (Tishenko-Claisen Reaction)

[V. Tishenko, *J. Russ. Phys. Chem. Soc.*, **38**, 355 (1906); L. Claisen, *Ber.*, **20**, 646 (1887).] The alkoxide-catalyzed disproportionation of aldehydes to produce esters, Scheme 11.320*a*. If the alkoxide is an aluminum alkoxide, e.g., aluminium triisopropoxide, the isopropyl ester is obtained, Scheme 11.320*b*. The reaction is also catalyzed by Al_2O_3 (see Sect. 11.200).

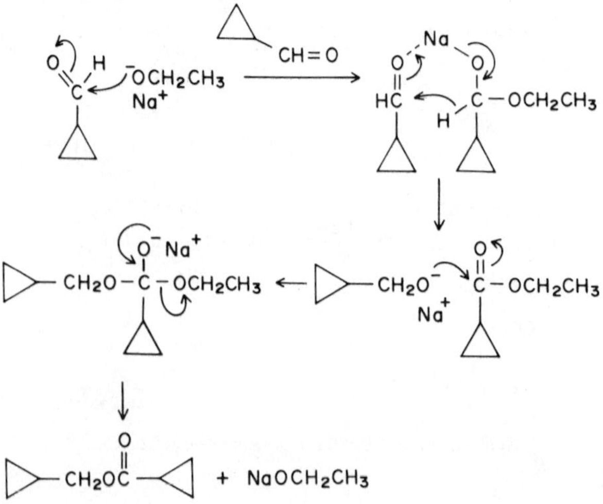

(*a*)

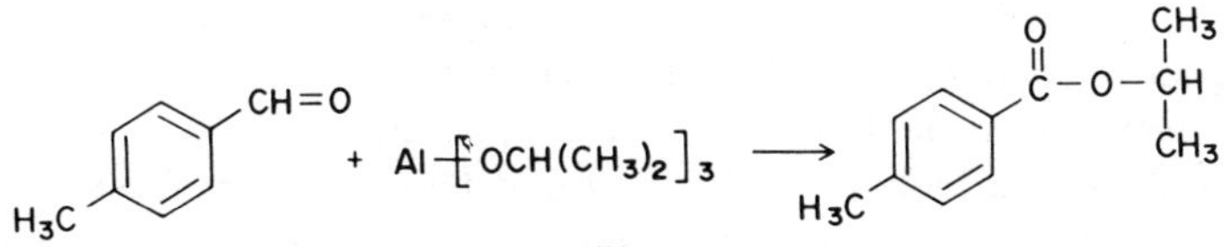

(*b*)

Scheme 11.1320 Tishenko reaction: (*a*) using sodium alkoxide; (*b*) using aluminum triisopropoxide.

11.1330 Ullmann Reaction

[F. Ullmann, *Ann.*, 332, 38 (1904).] A synthesis of biaryls through the reaction of two moles of aryl halide in the presence of copper, Scheme 11.1330.

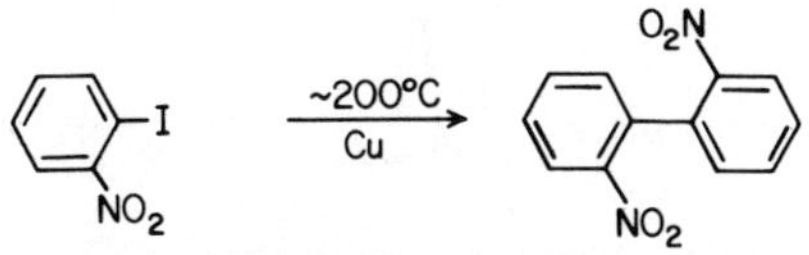

Scheme 11.1330 Ullmann reaction.

11.1340 Vilsmeier Reaction

[A. Vilsmeier and A. Haack, *Ber.*, **60**, 119 (1927).] An aldehyde synthesis using dimethylformamide for introducing the aldehyde functional group into an activated aromatic nucleus, Scheme 11.1340.

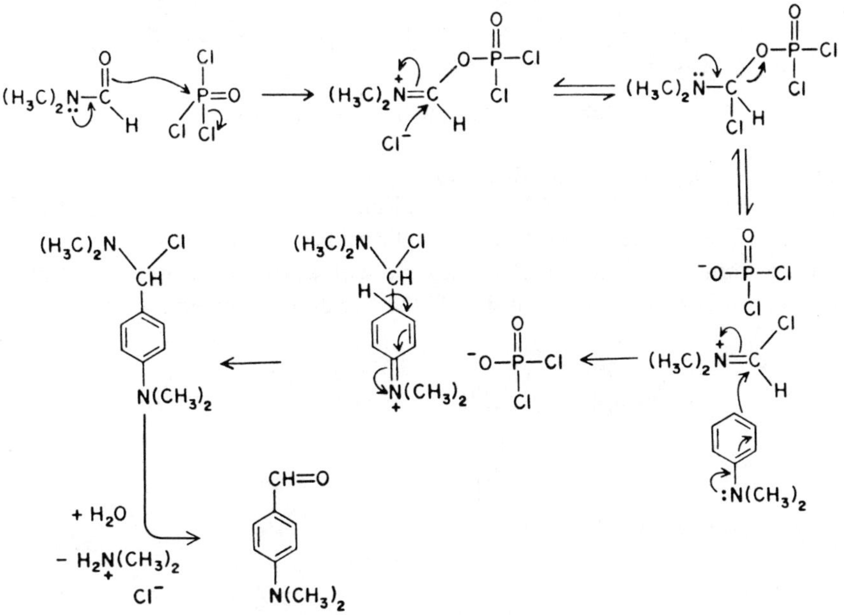

Scheme 11.1340 Vilsmeier reaction.

11.1350 Wagner-Meerwein-Whitmore Rearrangement

[G. Wagner, *J. Russ. Phys. Chem.*, **31**, 690 (1899); H. Meerwein, *Ann.*, **405**, 129 (1914); F. Whitmore and H. S. Rothrock, *J. Am. Chem. Soc.*, **55**, 1106 (1933).] The migration of a hydride or R^- to a carbenium ion center from an adjacent carbon atom, hence a 1,2-shift; especially important in the elucidation of some of the rearrangements of naturally occurring compounds, e.g., the dehydration of borneol to camphene, Scheme 11.1350.

Borneol $\xrightarrow[-H_2O]{H^+}$ $\xrightarrow{1,2-\text{shift}}$ ... Camphene

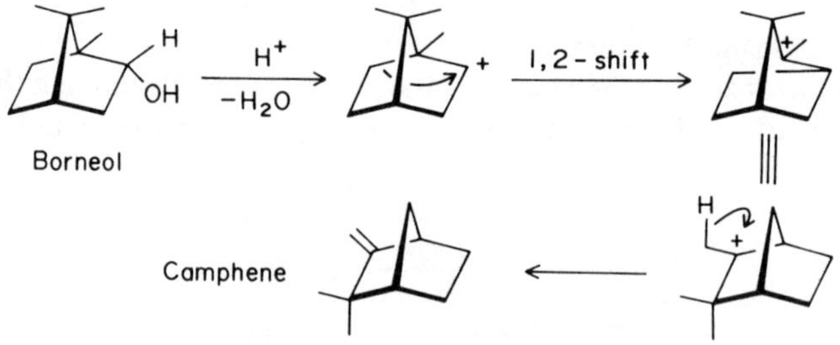

Scheme 11.1350 Wagner-Meerwein-Whitmore rearrangement.

11.1360 Wallach Rearrangement

[O. Wallach and L. Belli, *Ber.*, **13**, 525 (1880).] The acid-catalyzed rearrangement of azoxybenzenes to *p*- and sometimes *o*-hydroxyazobenzenes, Scheme 11.1360.

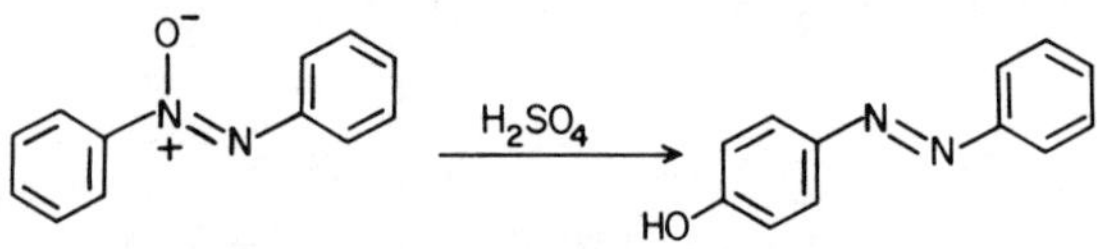

$\xrightarrow{H_2SO_4}$

Scheme 11.1360 Wallach rearrangement.

11.1370 Willgerodt-Kindler Reaction

[C. Willgerodt, *Ber.*, **20**, 2467 (1887); K. Kindler, *Ann.*, **431**, 193 (1923).] Conversion of an aralkyl ketone to a carboxylic acid containing the same number of carbon atoms by treatment of the ketone with sulfur in the presence of morpholine, followed by hydrolysis of the resulting thiomorpholide, Scheme 11.1370.

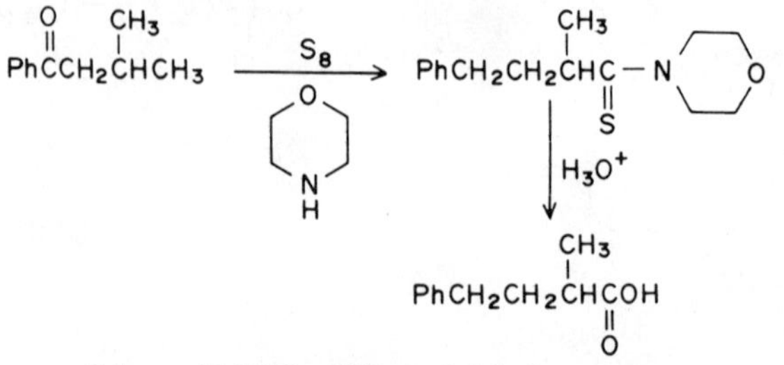

$PhC(=O)CH_2CH(CH_3)CH_3 \xrightarrow{S_8,\ \text{morpholine}} PhCH_2CH_2CH(CH_3)C(=S)-N(\text{morpholino}) \xrightarrow{H_3O^+} PhCH_2CH_2CH(CH_3)C(=O)OH$

Scheme 11.1370 Willgerodt-Kindler reaction.

11.1380 Williamson Ether Synthesis

[A. W. Williamson, *J. Chem. Soc.*, 229 (1852).] An ether synthesis involving the reaction between an alkyl halide and an alkoxide, Scheme 11.1380.

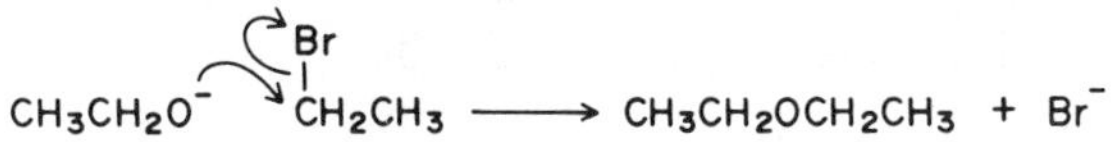

$$CH_3CH_2O^- + BrCH_2CH_3 \longrightarrow CH_3CH_2OCH_2CH_3 + Br^-$$

Scheme 11.1380 Williamson ether synthesis.

11.1390 Wittig Rearrangement

[G. Wittig and L. Löhmann, *Ann.*, **550**, 260 (1942).] The base-induced rearrangement of an ether to an alcohol via a 1,2-shift, Scheme 11.1390.

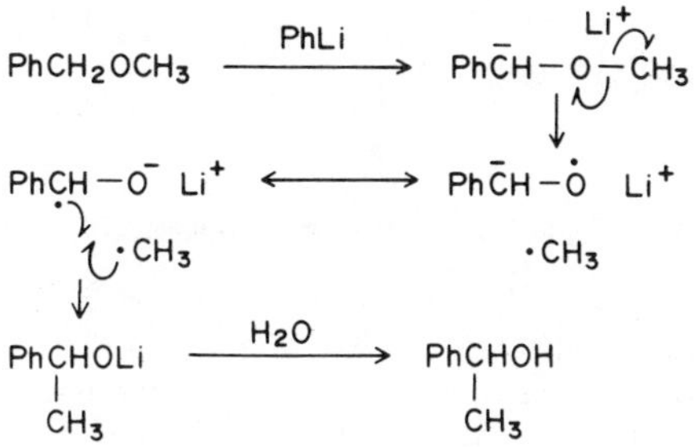

$$PhCH_2OCH_3 \xrightarrow{PhLi} Ph\bar{C}H{-}O{-}CH_3\ Li^+$$

$$PhCH{-}O^-\ Li^+\ \cdot CH_3 \longleftrightarrow Ph\bar{C}H{-}\dot{O}\ Li^+\ \cdot CH_3$$

$$PhCH(CH_3)OLi \xrightarrow{H_2O} PhCH(CH_3)OH$$

Scheme 11.1390 Wittig rearrangement.

11.1400 Wittig Reaction

[G. Wittig and G. Geissler, *Ann.*, **580**, 44 (1953).] Conversion of the $>C{=}O$ bond of an aldehyde or ketone to a $>C{=}C<$ bond. Reaction of an alkyl halide with Ph_3P forms a phosphonium halide that on treatment with base yields a phosphorus-carbon ylide. The ylide then reacts with the carbonyl component to form an olefin with simultaneous elimination of $Ph_3P{=}O$, Scheme 11.1400.

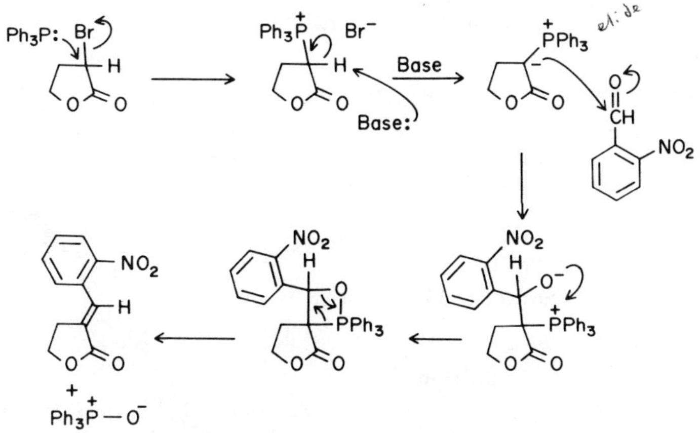

Ph₃P: + α-bromo-γ-butyrolactone → phosphonium bromide ($Ph_3\overset{+}{P}$, Br^-) $\xrightarrow{Base}$ ylide ($\overset{+}{P}Ph_3$) + o-nitrobenzaldehyde (O=CH, NO_2) → betaine (O^-, $\overset{+}{P}Ph_3$) → oxaphosphetane (O–PPh₃) → α-(o-nitrobenzylidene)-γ-butyrolactone + $Ph_3\overset{+}{P}{-}O^-$

Scheme 11.1400 Wittig reaction.

11.1410 Wohl-Ziegler Bromination

[A. Wohl, *Ber.*, 52, 51 (1919); K. Ziegler, A. Späth, E. Schaaf, W. Schumann, and E. Winkelmann, *Ann.*, **551**, 80 (1942).] Bromination of a benzylic (Scheme 11.1410*a*) or allylic position using *N*-bromosuccinimide, (NBS). The reaction is prone to allylic rearrangement, Scheme 11.1410*b*.

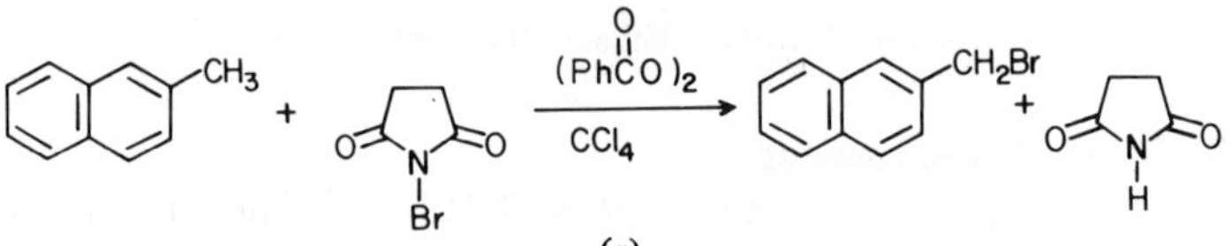

$$C_{10}H_7{-}CH_3 + \text{NBS} \xrightarrow[CCl_4]{(PhC(O)O)_2} C_{10}H_7{-}CH_2Br + \text{succinimide (N–H)}$$

(*a*)

$$H_3C(CH_2)_4CH_2CH{=}CH_2 + \text{NBS} \xrightarrow[CCl_4]{(PhC(O){-}O)_2} H_3C(CH_2)_4CH(Br)CH{=}CH_2 + H_3C(CH_2)_4CH{=}CHCH_2Br$$

cis and trans

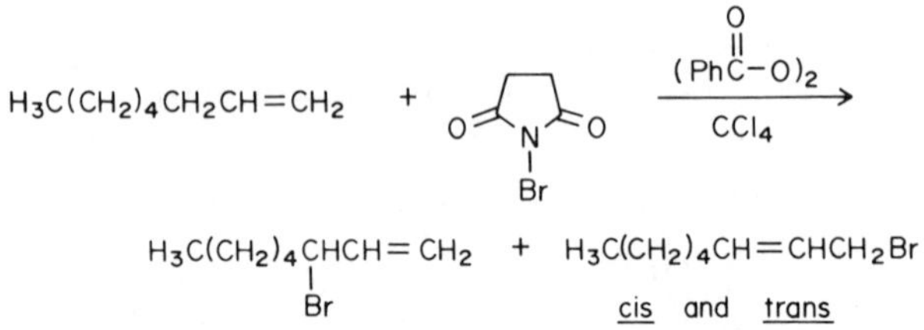

(*b*)

Scheme 11.1410 Wohl-Ziegler bromination: (*a*) at the benzylic position; (*b*) at the allylic position with rearrangement.

11.1420 Wolff Rearrangement

[L. Wolff, *Ann.*, **394**, 25 (1912).] The rearrangement of a diazoketone to a ketene achieved either by thermolysis (uncatalyzed or catalyzed by heavy metals) or photolysis. This reaction is a step in the **Arndt-Eistert synthesis** (see Sect. 11.040), Scheme 11.1420.

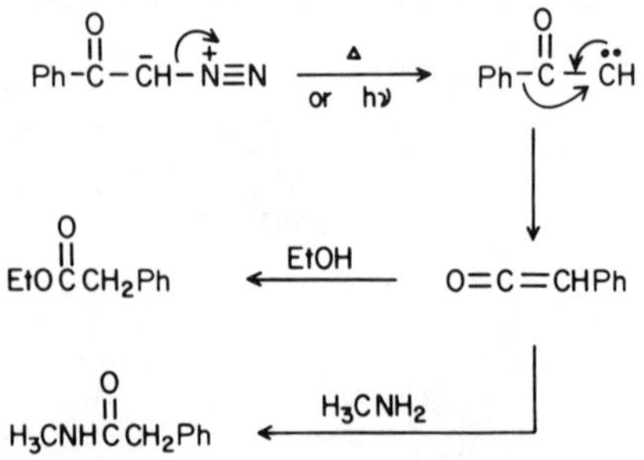

$$Ph{-}C(O){-}\bar{C}H{-}\overset{+}{N}{\equiv}N \xrightarrow[\text{or } h\nu]{\Delta} Ph{-}C(O){-}\ddot{C}H \longrightarrow O{=}C{=}CHPh$$

$$O{=}C{=}CHPh \xrightarrow{EtOH} EtOC(O)CH_2Ph$$

$$O{=}C{=}CHPh \xrightarrow{H_3CNH_2} H_3CNHC(O)CH_2Ph$$

Scheme 11.1420 Wolff rearrangement.

11.1430 Wolff-Kishner Reduction

[N. Kishner, *J. Russ. Phys. Chem. Soc.*, **43**, 582 (1911); L. Wolff, *Ann.*, **394**, 86 (1911).] The reduction of an aldehydic or ketonic carbonyl group to a methylene group by treatment of the corresponding hydrazone with aqueous alkali in a sealed tube or in a high boiling solvent **(Huang-Minlon modification)**, Scheme 11.1430.

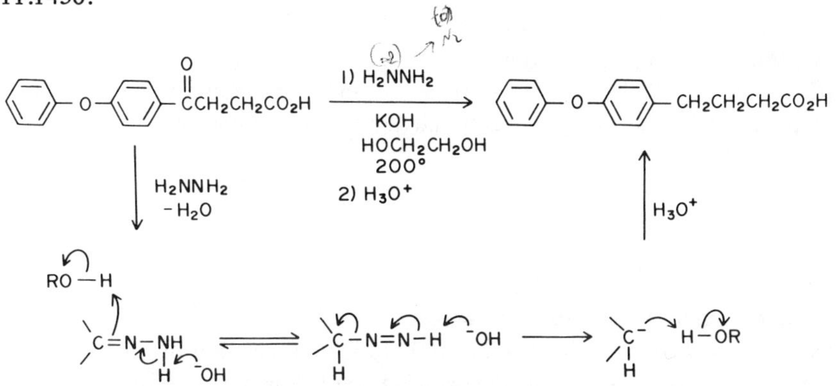

Scheme 11.1430 Wolff-Kishner reduction.

11.1440 Woodward Hydroxylation

[R. B. Woodward, U.S. Patent 2,687,435 (1954).] The *cis* hydroxylation of an olefin using iodine and silver acetate, Scheme 11.1440.

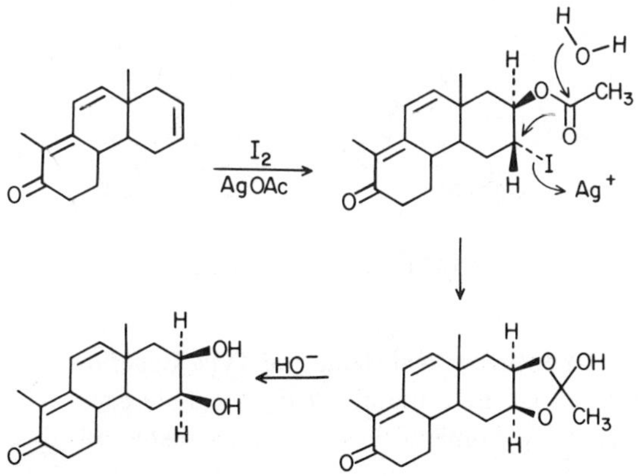

Scheme 11.1440 Woodward hydroxylation.

11.1450 Wurtz Coupling Reaction

[A. Wurtz, *Ann. Chim. Phys.*, [3] **44**, 275 (1855).] A synthesis of alkanes involving the coupling of alkyl groups through the reaction of two moles of alkyl halides, generally with sodium, Scheme 11.1450.

$$2\ CH_3CH_2CH_2CH_2{-}Cl \xrightarrow{Na} CH_3(CH_2)_6CH_3 + 2\ NaCl$$

Scheme 11.1450 Wurtz coupling reaction.

11.1460 Wurtz-Fittig Reaction

[B. Tollens and R. Fittig, *Ann.*, **131**, 303 (1864).] A modification of the Wurtz coupling in which an aryl group is coupled to an alkyl group, Scheme 11.1460.

$$Ph{-}Br + CH_3CH_2CH_2CH_2{-}Br \xrightarrow{Na} PhCH_2CH_2CH_2CH_3$$

Scheme 11.1460 Wurtz-Fittig reaction.

11.1470 Zinin Reduction of Nitroarenes

[N. Zinin, *J. Prakt. Chem.*, [1] **27**, 140 (1842).] Reduction of the nitro group to an amino group using S^{2-} as the reducing agent, Scheme 11.1470.

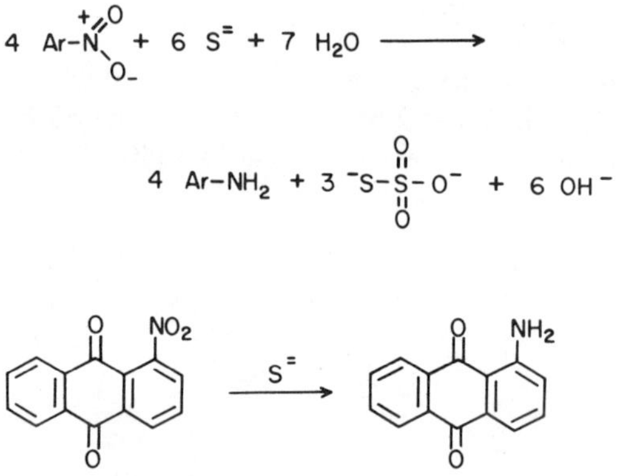

$$4\ Ar{-}\overset{+}{N}(=O){-}O^- + 6\ S^{=} + 7\ H_2O \longrightarrow$$

$$4\ Ar{-}NH_2 + 3\ {}^{-}S{-}S(=O)_2{-}O^- + 6\ OH^-$$

$$\text{1-nitroanthraquinone} \xrightarrow{S^{=}} \text{1-aminoanthraquinone}$$

Scheme 11.1470 Zinin reduction of nitroarenes.

Photochemical Name and Type Reactions

Special organic reactions are initiated by light and are particularly useful in synthetic organic chemistry. Some of these reactions are described below.

11.1480 Barton Reaction

[D. H. R. Barton, J. M. Beaton, L. E. Geller, and M. M. Pechet, *J. Am. Chem. Soc.*, **82**, 2640 (1960); **83**, 4076 (1961).] The transformation of nitrite esters to γ-oximino alcohols via homolytic cleavage to form nitrous oxide (NO) and an alkoxy radical. The alkoxy radical may then abstract remote and unactivated hydrogen atoms, Scheme 11.1480*a*. This reaction has found significant application in the synthesis of steroids, Scheme 11.1480*b*.

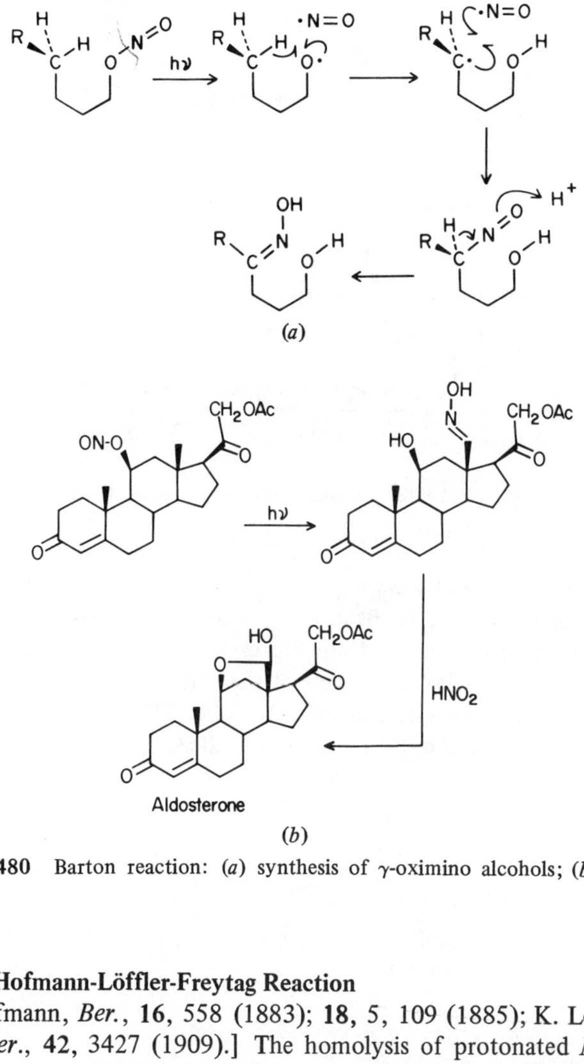

(*a*)

Aldosterone

(*b*)

Scheme 11.1480 Barton reaction: (*a*) synthesis of γ-oximino alcohols; (*b*) synthesis of aldosterone.

11.1490 Hofmann-Löffler-Freytag Reaction

[A. W. Hofmann, *Ber.*, **16**, 558 (1883); **18**, 5, 109 (1885); K. Löffler and C. Freytag, *Ber.*, **42**, 3427 (1909).] The homolysis of protonated *N*-haloamines either thermally or photochemically to form amine salts with halogenated alkyl substituents. Neutralization of these amine salts leads to formation of cyclic amines, Scheme 11.1490*a*. This reaction is particularly useful in preparing five-membered nitrogen-containing rings, Scheme 11.1490*b*.

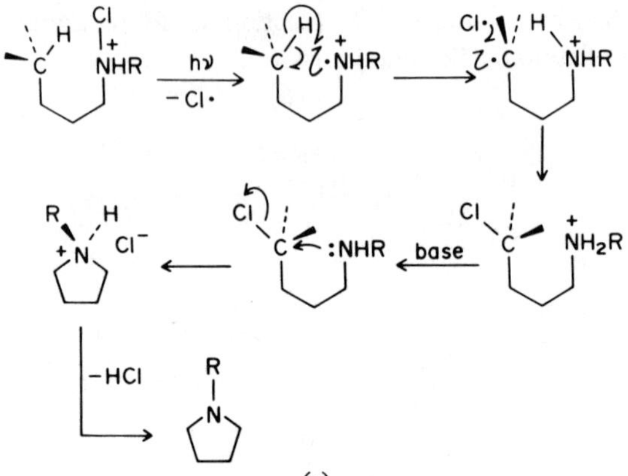

(*a*)

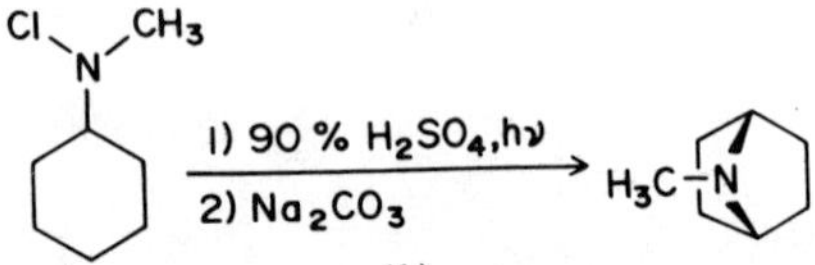

(*b*)

Scheme 11.1490 Hofmann-Löffler-Freytag reaction: (*a*) general scheme; (*b*) preparation of nitrogen-containing rings.

11.1500 Norrish Type I Cleavage

[R. G. W. Norrish and C. H. Bamford, *Nature*, **138**, 1016 (1936); **140**, 195 (1937).] A reaction originating from the carbonyl n, π^* state and proceeding through the homolytic cleavage of one of the carbonyl alkyl groups as the primary photochemical process, Scheme 11.1500*a*. The Norrish Type I process finds synthetic utility in the ring cleavage of cyclic ketones, Scheme 11.1500*b*.

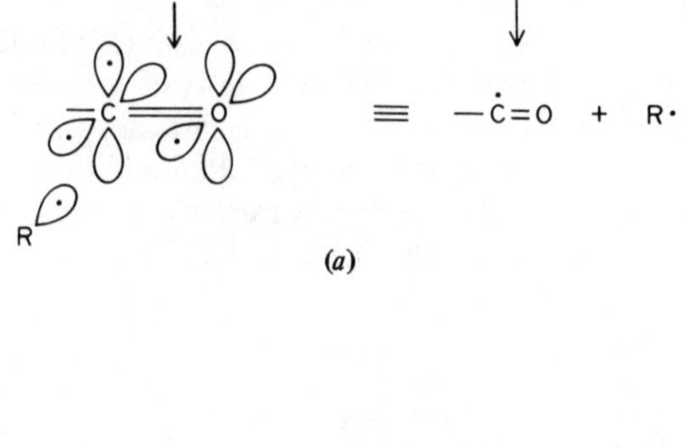

(*a*)

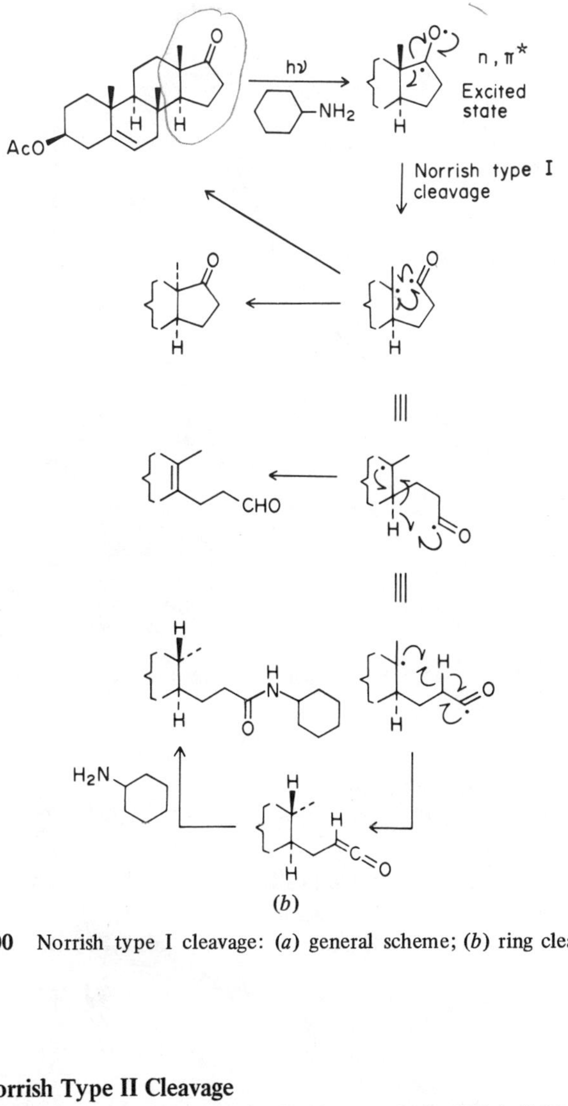

Scheme 11.1500 Norrish type I cleavage: (*a*) general scheme; (*b*) ring cleavage of cyclic ketones.

11.1510 Norrish Type II Cleavage

[R. G. W. Norrish and C. H. Bamford, *Nature*, **138**, 1016 (1936); **140**, 195 (1937).] A reaction originating from the carbonyl n, π^* state and proceeding through a six-centered fragmentation initiated by γ-hydrogen abstraction. The products of the fragmentation are an olefin and an enol which tautomerizes to the carbonyl compound, Scheme 11.510*a* and *b*. A similar reaction occurs in the mass spectral fragmentation of carbonyl compounds, also shown in Scheme 11.510*a*, where it is identified as the **McLafferty cleavage**.

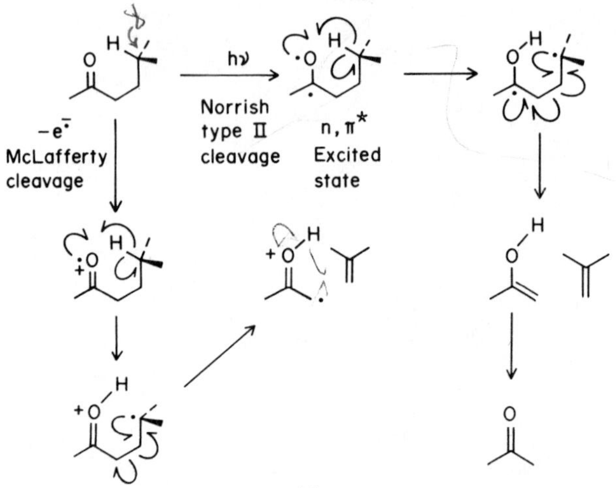

hν
Norrish type II cleavage
n, π* Excited state
$-e^{\cdot}$ McLafferty cleavage

(a)

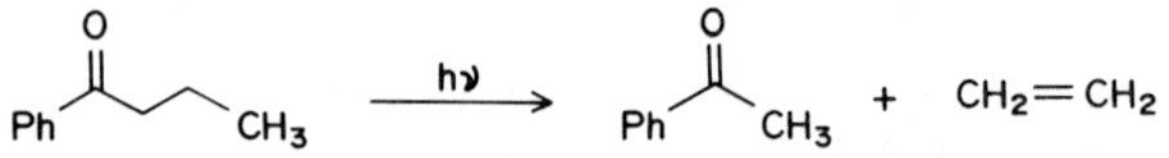

hν
Ph CH$_3$ → Ph CH$_3$ + $CH_2{=}CH_2$

(b)

Scheme 11.1510 Norrish type II cleavage; (*a*) general scheme; (*b*) cleavage of phenyl *n*-propyl ketone.

11.1520 Paterno-Büchi Reaction

[E. Paterno and G. Chieffi, *Gazz. Chim. Ital.*, **39**, 341 (1909); G. Büchi, C. G. Inman, and E. S. Lipinsky, *J. Am. Chem. Soc.*, **76**, 4327 (1954).] The cycloaddition reaction between an electronically excited carbonyl group and a ground state olefin to yield an oxetane, Scheme 11.1520*a*. This is the only good method for synthesizing four-membered ether rings, Scheme 11.1520*b*.

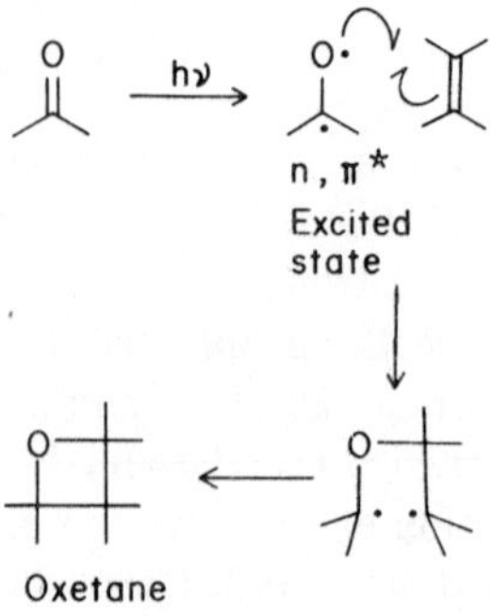

hν
n, π* Excited state
Oxetane

(*a*)

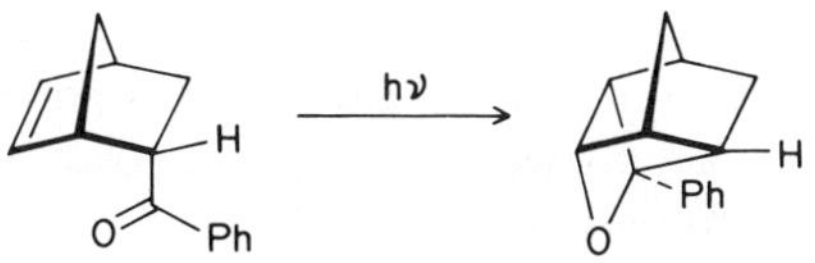

(*b*)

Scheme 11.1520 Paterno-Büchi reaction: (*a*) general scheme; (*b*) synthesis of tetracyclo-oxetane.

11.1530 Photoaddition

A diverse class of reactions in which polar molecules add to multiple bonds under the influence of light, Scheme 11.1530.

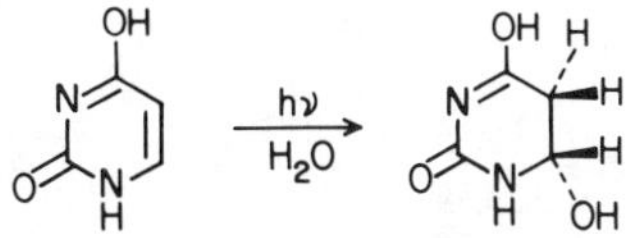

Scheme 11.1530 Photoaddition.

11.1540 Photocycloaddition

A diverse class of light-initiated reactions in which the termini of two multiple bond systems become joined through σ bonds. The most common example of this type of reaction is the formation of a cyclobutane ring from two ethylene moieties, Scheme 11.1540; see also Scheme 11.1520.

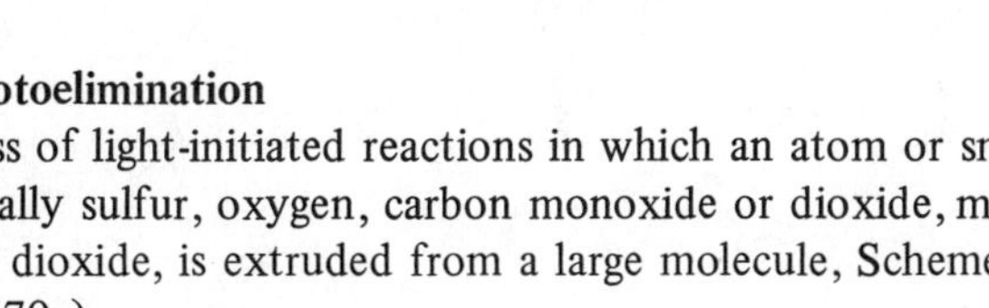

Scheme 11.1540 Photocycloaddition.

11.1550 Photoelimination

A diverse class of light-initiated reactions in which an atom or small molecular fragment, usually sulfur, oxygen, carbon monoxide or dioxide, molecular nitrogen, or sulfur dioxide, is extruded from a large molecule, Scheme 11.1550 (see Scheme 11.1570*a*).

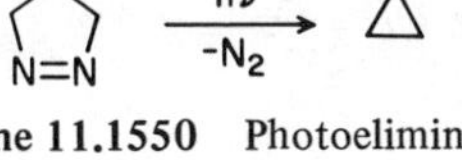

Scheme 11.1550 Photoelimination.

11.1560 Photo-Fries Reaction

The formation of a mixture of *o*- and *p*-acylphenols upon photolysis of phenol esters in solution. In the gas phase, where solvent cage effects are not operative, many other products are formed, Scheme 11.1560.

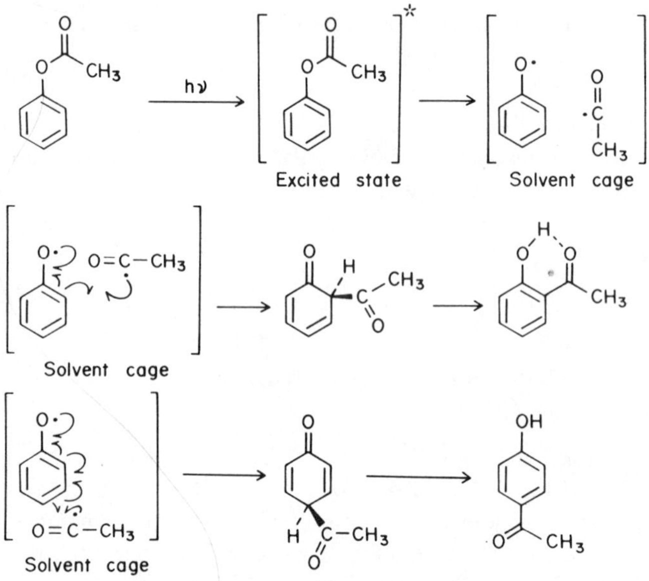

Scheme 11.1560 Photo-Fries reaction.

11.1570 Photooxygenation

The light-initiated reaction of triplet oxygen with radicals (**Type I photooxygenation**, Scheme 11.1570*a*), or of singlet oxygen with olefinic moieties (**Type II photooxygenation**, Schemes 11.410*c*, 11.370*d*, and 11.1570*b*).

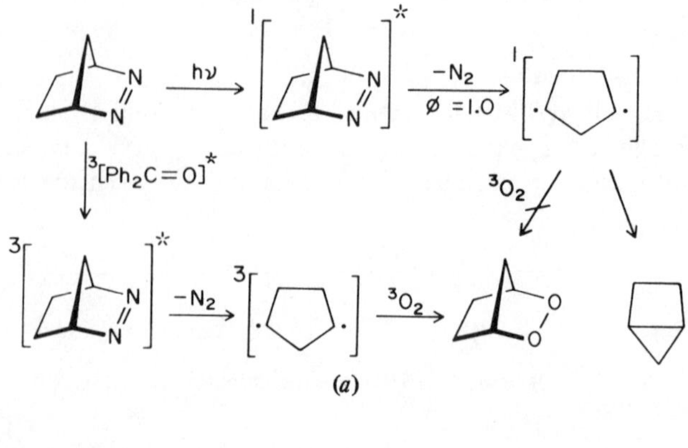

(*a*)

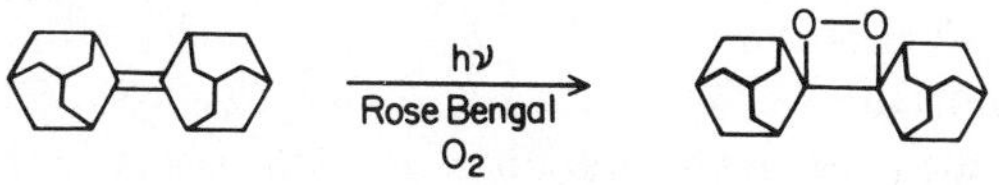

$$\text{(adamantylideneadamantane)} \xrightarrow[\text{Rose Bengal, } O_2]{h\nu} \text{(1,2-dioxetane)}$$

(*b*)

Scheme 11.1570 Photooxygenation: (*a*) Type I; (*b*) Type II.

11.1580 Photoreduction

The light-initiated reduction of a variety of functional groups in the presence of an electron donor (Scheme 11.1580*a*) or a hydrogen atom donor (Scheme 11.1580*b*).

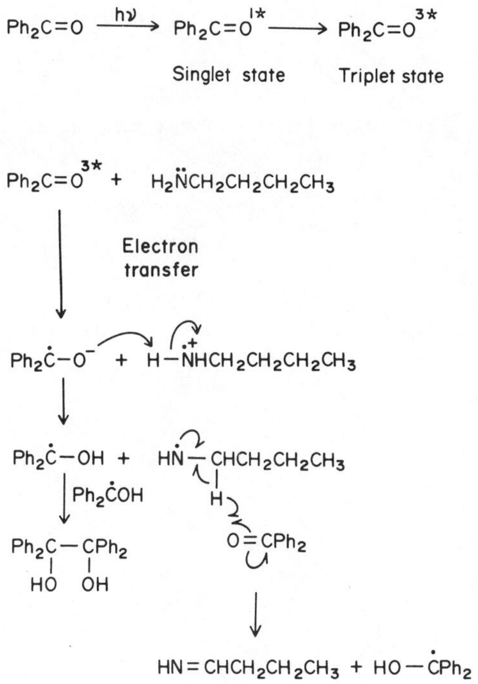

$$Ph_2C{=}O \xrightarrow{h\nu} Ph_2C{=}O^{1*} \longrightarrow Ph_2C{=}O^{3*}$$

Singlet state Triplet state

$$Ph_2C{=}O^{3*} + H_2\ddot{N}CH_2CH_2CH_2CH_3$$

Electron transfer

$$Ph_2\dot{C}{-}O^- + H{-}\overset{\cdot +}{N}HCH_2CH_2CH_2CH_3$$

$$Ph_2\dot{C}{-}OH + H\dot{N}{-}CHCH_2CH_2CH_3 \text{ (H abstracted by } O{=}CPh_2)$$

$$Ph_2\dot{C}{-}OH \xrightarrow{Ph_2\dot{C}OH} Ph_2C(OH){-}C(OH)Ph_2$$

$$HN{=}CHCH_2CH_2CH_3 + HO{-}\dot{C}Ph_2$$

(*a*)

$$Ph_2C{=}O^{3*} + H{-}C(OH)Ph_2 \longrightarrow Ph_2\dot{C}(OH) \;\; \dot{C}(OH)Ph_2 \longrightarrow Ph_2C(OH){-}C(OH)Ph_2$$

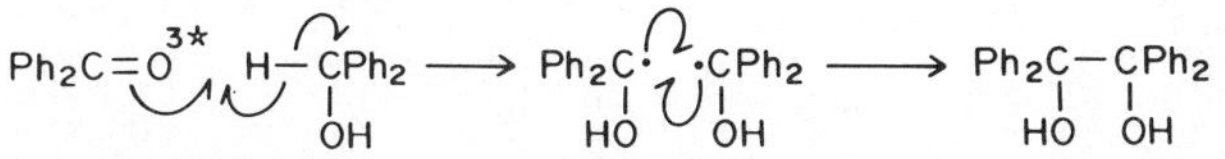

(*b*)

Scheme 11.1580 Photoreduction: (*a*) electron donor; (*b*) hydrogen atom donor.

11.1590 Yang Cyclization

[N. C. Yang and D-D. H. Yang, *J. Am. Chem. Soc.*, **80**, 2913 (1958).] The coupling rather than the fragmentation of the Norrish Type II biradical to form cyclobutanols in low to moderate yields, Scheme 11.1590.

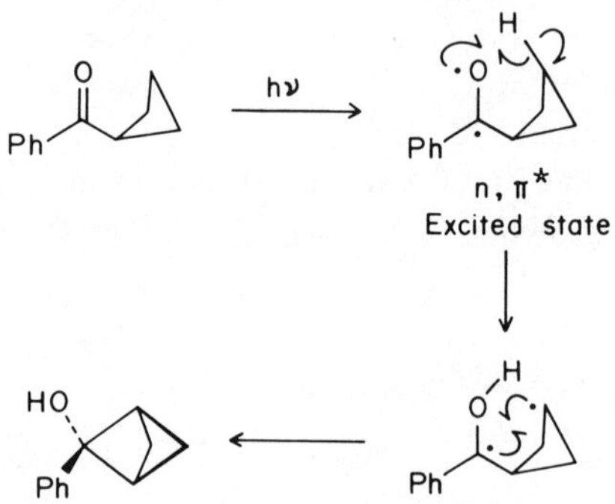

Ph O hν H O Ph n, π* Excited state H O Ph HO Ph

Scheme 11.1590 Yang cyclization.

12 Organometal Compounds

CONTENTS

12.010 Metals

Those elements that possess the following group characteristics: high electrical conductivity, decreasing with increasing temperature; high thermal conductivity; high ductility (easily stretched; not brittle); and malleability (can be hammered and formed without breaking).

12.020 The Periodic Classification

Figure 12.020 shows a version of the **periodic table**. Those elements that are considered metals are shaded either lightly (Group A) or more darkly (Group B). Those that are nonmetals are not shaded and those that have intermediate properties are crosshatched. The members of this last group are sometimes called

Period	I A	II A	III B	IV B	V B	VI B	VII B	VIII (VIII B)			I B	II B	III A	IV A	V A	VI A	VII A	0 (VIII A)
1	1 H 1.0080																1 H 1.0080	2 He 4.00260
2	3 Li 6.941	4 Be 9.01218											5 B 10.81	6 C 12.0111	7 N 14.0067	8 O 15.9994	9 F 18.9984	10 Ne 20.179
3	11 Na 22.9898	12 Mg 24.305											13 Al 26.9815	14 Si 28.086	15 P 30.9738	16 S 32.06	17 Cl 35.453	18 Ar 39.948
4	19 K 39.102	20 Ca 40.08	21 Sc 44.9559	22 Ti 47.90	23 V 50.9414	24 Cr 51.996	25 Mn 54.9380	26 Fe 55.847	27 Co 58.9332	28 Ni 58.71	29 Cu 63.546	30 Zn 65.37	31 Ga 69.72	32 Ge 72.59	33 As 74.9216	34 Se 78.96	35 Br 79.904	36 Kr 83.80
5	37 Rb 85.4678	38 Sr 87.62	39 Y 88.9059	40 Zr 91.22	41 Nb 92.9064	42 Mo 95.94	43 Tc 98.9062	44 Ru 101.07	45 Rh 102.9055	46 Pd 106.4	47 Ag 107.868	48 Cd 112.40	49 In 114.82	50 Sn 118.69	51 Sb 121.75	52 Te 127.60	53 I 126.9045	54 Xe 131.30
6	55 Cs 132.9055	56 Ba 137.34	*	72 Hf 178.49	73 Ta 180.9479	74 W 183.85	75 Re 186.2	76 Os 190.2	77 Ir 192.22	78 Pt 195.09	79 Au 196.9665	80 Hg 200.59	81 Tl 204.37	82 Pb 207.2	83 Bi 208.9806	84 Po (210)	85 At (210)	86 Rn (222)
7	87 Fr (223)	88 Ra 226.0254	†	104 Rf (260)	105 Ha	106												

Representative elements: I A, II A; III A – 0 (VIII A). Transition elements: III B – II B. Inner transition elements:

*Lanthanides	57 La 138.9055	58 Ce 140.12	59 Pr 140.9077	60 Nd 144.24	61 Pm (145)	62 Sm 150.4	63 Eu 151.96	64 Gd 157.25	65 Tb 158.9254	66 Dy 162.50	67 Ho 164.9303	68 Er 167.26	69 Tm 168.9342	70 Yb 173.04	71 Lu 174.97
†Actinides	89 Ac (227)	90 Th 232.0381	91 Pa 231.0359	92 U 238.029	93 Np 237.0482	94 Pu (242)	95 Am (243)	96 Cm (247)	97 Bk (247)	98 Cf (251)	99 Es (254)	100 Fm (257)	101 Md (256)	102 No (256)	103 Lr (257)

Figure 12.020 The periodic table. Note: Atomic weights based on $^{12}_{6}C = 12$. Numbers in parentheses are mass numbers of most stable isotopes.

metalloids or semi-metals; among these are boron, silicon, germanium, arsenic, antimony, and tellurium. There is no clear distinction between metals and metalloids and no universal agreement as to classification. The metals are characterized by low ionization energies; i.e., they tend to give up electrons readily. The properties of elements can be correlated with their position in a particular row (also called a period) in the periodic chart. Hydrogen and helium constitute the first row or first period elements. The second row begins with lithium, includes carbon, and ends with neon. There are seven such rows or periods. The elements are also classified according to the column or group in which they appear. There are nine such groups, 0 through VIII (Roman numerals are more commonly used). The valence electron configuration of all the elements in a column (also called a family) are identical. For example, the Group I elements all have one electron in their outermost orbital, an *s* orbital. The Group 0 elements all possess the full shell electronic configuration and hence are not prone to share, accept, or donate electrons; they occur in the monoatomic form and are called "noble" gases. (Some authors prefer to designate them as Group VIIIA because they possess the full complement of valence electrons, 2 for helium and 8 for the remaining ones). All the Groups except 0 and VIII are divided into A and B subgroups or families. However, authors who designate the noble gases as Group VIIIA usually call the Group VIII transition metals, Group VIIIB. The electronic configuration of atoms (Sect. 2.560) is designated by the principal quantum number, n, followed by the azimuthmal orbital designation, $s, p, d \cdots$ each bearing a superscript indicating the number of electrons in the orbital. Thus for the nitrogen atom (atomic no. 7) in the ground state the electronic configuration is $1s^2 2s^2 2p^3$. In the Bohr atomic theory the value of n corresponds to the number of the shell $n = 1, 2, 3, 4, 5 \cdots$, which is also designated by a corresponding capital letter K, L, M, N, O, respectively; these shells can accommodate a maximum of 2, 8, 18, 32, and 32 electrons respectively. The elements in the A families have one to seven electrons in their outer or valence shell and they have fixed valence states. The group number corresponds to the number of valence electrons, e.g., Ca(IIA), Al(IIIA), C(IVA), etc. The elements in the A families are frequently called the **representative elements** or the **main group elements** and the B families are the transition elements.

12.030 Alkali Metals

The elements in the first column (Group IA) of the periodic table. The correct position of hydrogen is still being debated. Its single valence electron entitles it to a place in Group IA with the alkali metals. But it simultaneously is one electron short of the first rare gas (helium) and this entitles it to a place in Group VIIA as a cousin or congener of the halogens.

12.040 Alkaline Earth Metals

The elements in the second column (Group IIA) of the periodic chart.

12.050 Transition Metals

Those elements that appear in the middle of the periodic table; they were thought to breach the transition from the alkali and alkaline earth metals on the left side to the nonmetals on the right side of the periodic table. There are several more specific definitions: (1) all the elements except those in the Group A families; (2) those elements that have partially filled *d* or *f* shells; (3) those elements that have partially filled *d* or *f* shells either as elements or in any of their *common* oxidation states. Most periodic charts use a classification based on definition (1). According to this definition Zn (as well as the other Group IIB metals) is a transition metal even though its electronic configuration is [argon core] $3d^{10}4s^2$ and in all its compounds it displays a 2+ valence state, which corresponds to a filled d ($3d^{10}$) configuration. In building up the electronic configuration of the fourth row elements, the first transition metal series begins (Sc) with electron occupation of the 3*d* level. Because there are five *d* orbitals which can accomodate 10 electrons, the 10 elements starting with Sc and ending with Zn can be thought of as being generated by successive addition of *d* electrons, providing the justification for including Zn as a transition metal. However, such successive addition does not correspond to the facts. Cr has the configuration [Ar] $3d^5 4s^1$; the next electron then goes into a 4*s* orbital rather than a *d* orbital to give, for Mn, the configuration [Ar] $3d^5 4s^2$. The preferred electronic configuration for both Cr and Mn is characterized by 4*s* orbital occupation in order to provide single electron occupation of the degenerate five *d* orbitals. This arises because such a configuration minimizes electron-electron interactions leading to the special stability of the "half filled" shell. Definition (2) excludes Group IB and IIB metals, since they have filled *d* (or *d* and *f*) orbitals; elements in these two groups are frequently called **post-transition metals**. Definition (3) includes the IB metals, Cu, Ag, and Au, because, e.g., Cu^{2+} has a d^9 configuration even though Cu^+ as well as the element have a d^{10} configuration. Also, since the chemical behavior of the IB metals is quite similar to that of the transition metals, organometal chemists generally prefer definition (3). Finally, the nine Group VIII metals are placed together even though some of them, e.g., Fe, Co, and Ni, possess different *d* configurations. Elements in each of the Group VIII columns, e.g., Fe, Ru, and Os, possess similar properties.

12.060 First, Second, and Third Transition Metal Series

The transition metals in the fourth, fifth, and sixth periods, respectively. Collectively these elements are called the main transition metals or ***d*-block elements**.

12.070 Inner Transition Metals

The first 15 transition metals in the sixth row of the periodic table constitute the first inner transition metal series, characterized by 4*f* and/or $5d^1$ orbital occupation. This series is also called the **lanthanides**. It starts with lanthanum (atomic no. 57) and runs through leutetium (atomic no. 71). The second inner transition metal series consists of the first 15 transition metals in the seventh period. They are characterized by possessing $5f$ and/or $6d^1$ electrons. The series starts with actinium (atomic no. 89) and ends with lawrencium (atomic no. 103). This series is also called the **actinide series.** Some chemists restrict the lanthanide and actinide series to the 14 elements following the first member but because of chemical similarities, the 15 in each series are generally grouped together.

12.080 Transuranium elements

All the elements following uranium (atomic no. 92) in the periodic classification; none of these are naturally occurring.

12.090 Coinage Metals

The Group IB metals, Cu, Ag, and Au, historically used to mint coins.

12.100 Organometal Compounds

In its most restrictive sense, compounds possessing a metal atom directly bonded to one or more carbon atoms. Such a definition would exclude compounds containing a metalloid bonded directly to carbon, such as $(CH_3)_4Si$. The definition would also exclude compounds that have no direct metal-carbon bond, although they may have organic groups bonded through oxygen, nitrogen, or phosphorus. Thus aluminum triisopropoxide, $Al[OCH(CH_3)_2]_3$; tris(ethylenediamine)cobalt(III) chloride, $[Co(en)_3]Cl_3$; or tetrakis[triphenylphosphineplatinum], $[Pt(PPh_3)_4]$; would not be considered as organometals. On the other hand, a compound such as Prussian blue, $Fe_4[Fe(CN)_6]_3$, ferric ferrocyanide, would be an organometal, although most chemists would regard it as inorganic. Despite the ambiguities, which will be present whatever the definition, the restrictive definition is usually followed.

12.110 Coordination Compounds

Compounds consisting of a central metal atom bonded to a specific number of ions or atoms, or neutral molecules. The metal and associated species act as a unit and tend to keep their identity in solution, although partial dissociation may occur.

12.120 Complexes

In inorganic and organometal chemistry, essentially synonomous with coordination compounds, except that this term is also applied to ions as well as compounds.

Examples. $Fe(CO)_5$; $Ca_2[Fe(CN)_6]$; $[Cr(H_2O)_6]Cl_3$. The ion $[Fe(CN)_6]^{4-}$ is best called a complex ion rather than a complex but the word ion is frequently omitted.

12.130 Ligands

The various ions, atoms, or neutral molecules that surround and are directly bonded to a central metal atom.

Examples. In the examples given in Sect. 12.120 the ligands are, respectively, carbon monoxide, cyanide ion, and water.

12.140 Chelating Ligands

Ligands that are coordinated to a central metal atom by two or more sigma bonds. Thus a ligand with two atoms bonded to the metal is called a **bidentate ligand**; with three such bonding sites, a **tridentate**; with four, a **tetradentate ligand**; etc.

Examples. Acetylacetonato anion, Fig. 12.140*a*, functions as a bidentate ligand, Fig. 12.140*b*; ethylenediaminetetraacetic acid (EDTA) could function as a hexadentate ligand but usually behaves as a pentadentate ligand with a free carboxy group Fig. 12.140*c*. The porphin nucleus, a flat 16-membered ring found in the important heme proteins (hemoglobin, myoglobin, cytochromes), consists of four linked pyrrole rings and thus is a tetradendate ligand. A substituted porphin, the iron porphyrin complex called heme, is shown in Fig. 12.140*d*.

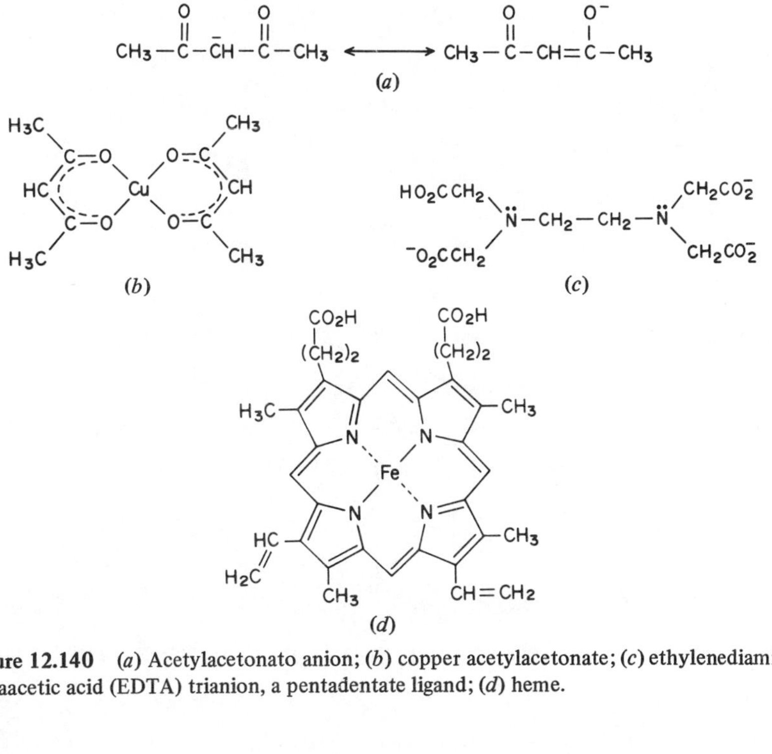

Figure 12.140 (*a*) Acetylacetonato anion; (*b*) copper acetylacetonate; (*c*) ethylenediaminetetraacetic acid (EDTA) trianion, a pentadentate ligand; (*d*) heme.

12.150 Bridging Ligands

Ligands in which an atom (or molecular species capable of independent existence) is simultaneously bonded to at least two metal atoms.

Examples. $Fe_2(CO)_9$, Fig. 12.150*a*, has three bridging carbonyl groups. The structure consists of two octahedra with a shared face, since three atoms are common to the two metals. The bridging group ligands are referred to as **face-sharing ligands** in such cases. Zeise's dimer, Fig. 12.150*b* has two bridging chlorine atoms. The structure consists of two square planes with a common edge and in general two such bridging ligands are called **edge-sharing ligands.** One bridging atom leads to a shared corner, $[(RuCl_5)_2O]^{4-}$, Fig. 12.150*c*. Molecular nitrogen can be a bridging ligand (see Section 12.320).

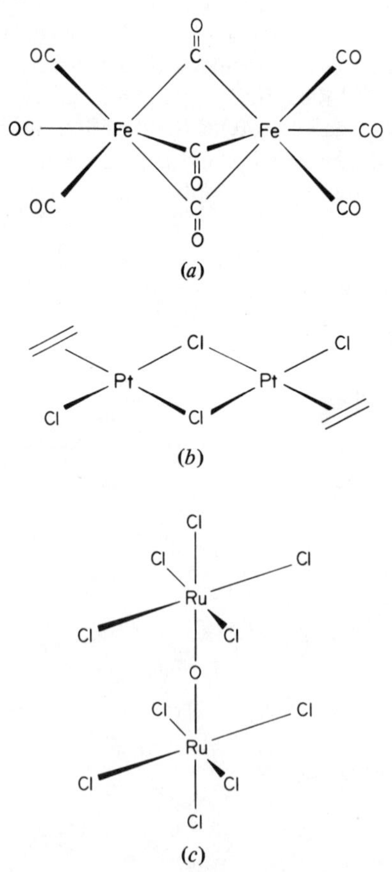

Figure 12.150 (*a*) $Fe_2(CO)_9$; (*b*) $[PtCl_2(C_2H_4)]_2$; (*c*) $[(RuCl_5)_2O]^{4-}$.

12.160 Organometal π Complexes

Organometal complexes having at least one organic ligand containing a π electron system that overlaps an orbital of the central metal atom (see also Sect. 12.170).

Example. The prototype is Zeise's salt $K[PtCl_3(C_2H_4)]$ (W. C. Zeise 1789–1847), the first π complex to be prepared (1828). The bonding in such complexes is assumed to consist of two component parts: a σ bond formed by overlap of a filled π orbital on the olefin with an empty metal orbital, Fig. 12.160*a*, and a second component part in which an appropriate filled *d* orbital of the metal overlaps an antibonding orbital of the ligand, Fig. 12.160*b*.

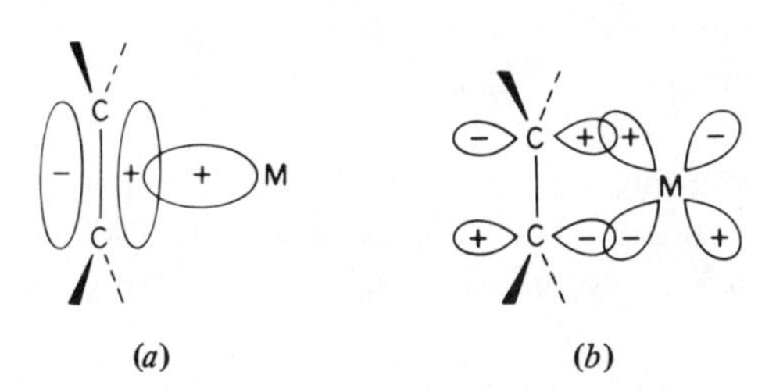

Figure 12.160 (*a*) σ Bonding in a π complex; (*b*) d-π^* back-bonding in a π complex.

12.170 Metal to Ligand Back-Bonding

The contribution to the overall bonding in π complexes which involves the overlap of the filled metal orbitals and vacant antibonding orbitals on the ligand (see also Sect. 2.770).

Example. The second component of the bonding in Zeise's salt, Fig. 12.160*b*.

12.180 Hapto (*Gk.* to Fasten)

A generic prefix attached to the name of a π bonded ligand specifying the number of carbon atoms in the π system of the ligand which are bonded to the central metal atom. If two atoms are bonded (e.g., Zeise's salt), the prefix is dihapto; if an allyl group is π bonded, it is trihapto; etc. If the ligand has a π system that is not involved in π bonding, it is a monohapto ligand. Most authors use the Greek letter eta, η, the lowercase form of the Greek letter for capital H, as an abbreviation for hapto, and thus di- and trihapto are η^2 and η^3, respectively. A few authors use the English *h* as the prefix.

Example. Ferrocene, bis(pentahaptocyclopentadienyl)iron(II), Fig. 12.180*a*; bis(pentahaptocyclopentadienyl)bis(monohaptocyclopentadienyl)titanium(IV), $[Ti(\eta^5\text{-}C_5H_5)_2(\eta^1\text{-}C_5H_5)_2]$, Fig. 12.180*b*.

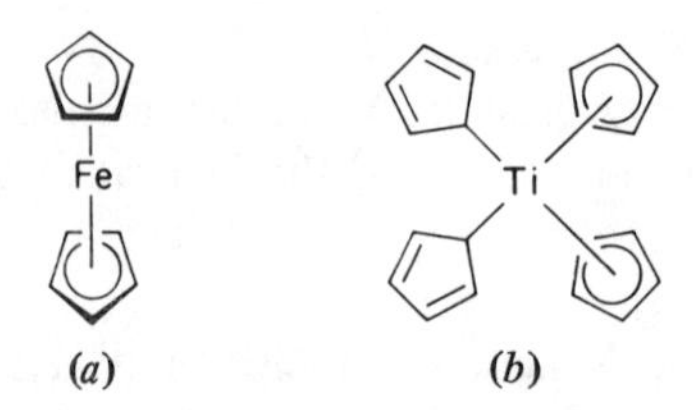

Figure 12.180 (*a*) Ferrocene; (*b*) $[Ti(\eta^5\text{-}C_5H_5)_2(\eta^1\text{-}C_5H_5)_2]$.

12.190 Coordination Number, CN

The number of ligand atoms bonded to the central metal atom of a coordination compound. The contribution to the CN of π bonded ligands, if present, may be considered to be one per pair of electrons involved.

Examples. The coordination number of Fe in $Fe(CO)_5$ is 5, in $[Fe(CN)_6]^{4-}$ it is 6, and in bis(acetylacetonato)copper(II), Fig. 12.140*b*, it is 4. Complexed ethylene contributes one to the coordination number of the metal to which it is attached.

12.200 Oxidation Number or Oxidation State (of Central Metal in Coordination Compounds)

The charge left on the central metal atom after all the ligands have been removed (theoretically) in their closed shell electronic configuration.

Examples. In dealing with complexes possessing ligands that have a separate existence as a neutral molecule or with ligands that form well characterized anions, there is generally very little trouble or ambiguity in assigning an oxidation number to the central metal. Thus in $[Cr(\eta^6\text{-}C_6H_6)_2]$ chromium is zero. In $K[PtCl_3(C_2H_4)]$, ethylene is removed with its π electrons and the three chlorines as anions. Inasmuch as the complex ion $[PtCl_3(C_2H_4)]$ has a 1− charge (evident from the presence of the counterion K^+), the net oxidation number of Pt is 2+. In $Fe_4[Fe(CN)_6]_3$, where there are three $[Fe(CN)_6]^{4-}$ groups, each Fe in the complex anion is +6 + (−4) = +2. In dealing with the organic ligands of organometal complexes that are σ bonded to the metal, the organic moiety is removed with the pair of bonding electrons, i.e., as R^-. Thus Pt in $[Pt(CH_3)_4]$ is 4+. According to this formalism, hydrogen attached to metals is removed as H^- regardless of whether such complexes possess hydridic character. Thus in $[CoH(CO)_4]$, Co is 1+. Even though this compound has some properties of a hydride, in aqueous solution it is a strong acid, giving the anion $[Co(CO)_4]^-$, where Co is 1−. An even more arbitrary situation arises with the organic ligands that are bonded to the metal by a delocalized π system such as in the cyclopentadienyl complexes. The closed shell configuration of such ligands involves 6 $p\pi$ electrons. Hence in $[Fe(\eta^5\text{-}C_5H_5)_2]$ Fe is regarded as 2+.

12.210 Effective Atomic Number, EAN

In a coordination complex, the number of electrons surrounding the central metal atom. It is equal to the atomic number of the metal minus its oxidation number plus the formal number of bonding electrons donated to the metal by the ligands.

Examples

Complex	Metal atomic no.	Metal oxid. no.	Ligand donation	EAN
$[Cr(\eta^6\text{-}C_6H_6)_2]$	24	0	12	36
$[PtCl_3(C_2H_4)]^{1-}$	78	+2	8	84
$[Fe(CN)_6]^{4-}$	26	+2	12	36
$[Co(H)(CO)_4]$	27	+1	10	36
$[Fe(\eta^5\text{-}(C_5H_5)_2]$	26	+2	12	36

12.220 Rule of 18

Transition metal complexes with EANs of 36, 54, and 86 isoelectronic with the rare gases Kr, Xe, and Rn, respectively, often exhibit unusual kinetic and thermodynamic stability. Chemical behavior involves the outer metal *dsp* orbitals and these are completely filled when 18 valence electrons are present ($d^{10}s^2p^6$). The rule that a transition metal in a complex tends to acquire this number of electrons is known as the rule of 18.

12.230 d^n Configuration

In a transition metal complex, the number of electrons in the valence orbitals of the metal, based on the assignment of its oxidation number. All the electrons beyond the closed shell configurations are assumed to be *d* electrons.

Examples. Pt(0) is d^{10}; in $[PtCl_3(C_2H_4)]^-$, Pt is Pt(II), hence has a d^8 configuration. Other examples:

No.	Complex	Oxidation state of M	d^n	$(d^n)^a$	CN^b
1	$[IrCl(CO)(PPh_3)_2]$	+1	d^8	(d^9)	4
2	$[CoH_3(PPh_3)_3]$	+3	d^6	(d^9)	6
3	$[Ni(CO)_4]$	0	d^{10}	(d^{10})	4
4	$[Co(CN)_5]^{3-}$	+2	d^7	(d^9)	5

[a] Configuration of zero valent metal.

[b] Coordination number.

12.240 Coordinative Unsaturation

When the central metal atom in a transition metal complex has an effective atomic number short of the rare gas structure, it is considered to be coordinatively unsaturated.

Examples. Of the complexes listed in Sect. 12.230, complexes 1 and 4 are coordinatively unsaturated. This is readily ascertained by simply multiplying the CN by 2 and adding the product to the number n of the d^n configuration; if the resulting number is less than 18 (the rule of 18), the metal is coordinatively unsaturated. Thus complex 2 in Sect. 12.230, $(6 \times 2) + 6 = 18$, is therefore coordinatively saturated. For complex 4 $(5 \times 2) + 7 = 17$. Since this complex is short one electron, it is not surprising that it reacts with many substrates by homolytic cleavage leading to radical intermediates. The value of the coordinatively unsaturated concept resides in predicting (albeit poorly) when a complex might function as a catalyst. Catalytic activity (in solution) requires that a substrate be coordinated to the metal site, and this is possible only if the metal is coordinatively unsaturated to begin with or becomes so under conditions of the reaction.

12.250 Oxidative Addition

The cleavage of a reactant molecule accompanied by addition of one or both its parts to the central metal atom of a complex. In the process σ bonds are formed between the metal and the entering ligand(s), which requires formal electron donation from the metal, resulting in an increase in both the coordination number of the metal as well as in its oxidation number.

Examples. A square planar Rh(I) d^8 complex may be converted to an octahedral d^6 complex by addition of RX: $[Rh^{I}Cl(CO)(PPh_3)_2] + CH_3Br \rightarrow [Rh^{III}Cl(Br)(CH_3)(CO)(PPh_3)_2]$. A d^7 five-coordinate complex may be converted to an octahedral d^6 complex: $2[Co^{II}(CN)_5]^{3-} + H_2 \rightarrow 2[Co^{III}(CN)_5H]^{3-}$

12.260 Reductive Elimination

In the usual case, the loss of two σ bonded ligands A and B from the central metal atom of a complex to form the molecule A—B. Because two σ bonds are broken and only one is formed in the process, the reaction constitutes a formal two electron reduction of the metal and a decrease in its coordination number. Reductive elimination is the reverse of oxidative addition.

Example. $[Rh^{III}Cl(C_2H_5)(H)(PPh_3)_3] \rightarrow [Rh^{I}Cl(PPh_3)_3] + C_2H_6$.

12.270 Insertion Reaction (Migratory Insertion)

The cleavage of a metal-ligand σ bond, M—L, in a complex A—M—L and the apparent insertion of a coordinated ligand A between M and L to give M—A—L.

The site vacated by the ligand A can be filled simultaneously with, or shortly after the insertion, by an incoming ligand.

Examples. Olefin insertion into an M—H bond, Fig. 12.270*a* and *b*; carbon monoxide insertion, Fig. 12.270*c*. In the case of $[Mn(CH_3)(CO)_4(PPh_3)] + CO \rightarrow [Mn(COCH_3)(CO)_4(PPh_3)]$, Fig. 12.270*c*, it has been demonstrated unequivocally that the reaction actually consists of a migration of CH_3 (with retention of stereochemistry if this group is chiral) onto a coordinated CO that does not become detached during the process. Hence this reaction may be called a migratory insertion to emphasize that the apparent insertion of CO between Mn and CH_3 is actually a migration of $^{-}\!:CH_3$, the new acyl group is attached to the metal at a site different from that occupied originally by methyl.

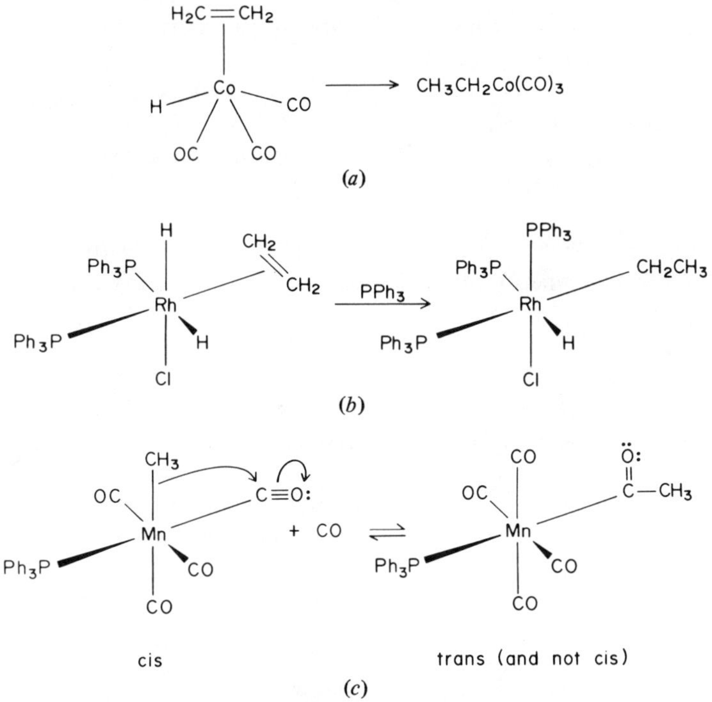

$$H(CO)_3Co(H_2C{=}CH_2) \longrightarrow CH_3CH_2Co(CO)_3$$

(*a*)

(*b*)

cis — trans (and not cis)

(*c*)

Figure 12.270 (*a*, *b*) Olefin insertion; (*c*) carbon monoxide insertion.

12.280 Catalytic Activation

A term used loosely to describe the conversion of a relatively inactive form of a chemical species to a more reactive form, usually by adsorption on a metal surface or by actual coordination of the metal.

Example. In hydrogenation reactions catalyzed by soluble transition metal complexes, hydrogen is assumed to be coordinated by overlap of its σ bond with an empty orbital on the metal. Back-donation (Sect. 12.170) of metal *d* electrons into the antibonding σ orbital is presumed to further weaken the H—H bond and thus activate the hydrogen prior to oxidative addition to generate a dihydrido species. Analogous reactions are assumed to occur when hydrogen is activated by heterogeneous catalysts.

12.290 Metallocenes

An unsystematic generic name for the class of organometal compounds having the structure $[M(\eta^5\text{-}C_5H_5)_2]$. Ferrocene, in which M = Fe, was the first such compound discovered, and it was given that common name because the *ene* suffix emphasized the unusual stability or ar*ene* character of the complex. The term metallocene has been extended by some chemists to include complexes in which the transition metal is "sandwiched" between other aromatic π electron systems, e.g., uranocene, $[U(\eta^8\text{-}C_8H_8)_2]$.

12.300 Ambident Ligands

Ligands possessing two or more different donor atoms, each having the potential of complexing with a metal (see Sect. 10.170).

Examples. The SCN^- ligand may be bonded as M—SCN (thiocyanato) or M—NCS (isothiocyanato); the NO_2^- group as M—NO_2 (nitro) or M—O—N=O (nitrito).

12.310 Dioxygen Complexes

Complexes in which molecular oxygen is coordinated to a metal as a ligating group. In such complexes the oxygen-oxygen bond is retained.

Example. Most dioxygen complexes involve symmetrical bonding of the oxygen atoms to the metal in a triangular arrangement, Fig. 12.310.

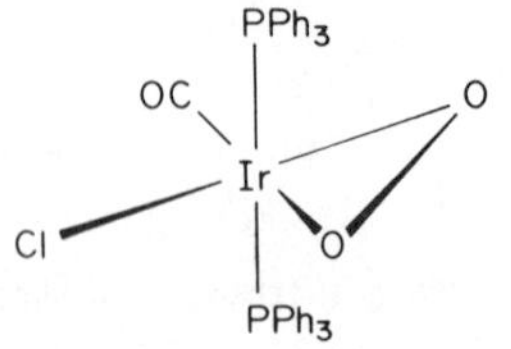

Figure 12.310 A dioxygen complex.

12.320 Dinitrogen Complexes

Complexes in which molecular nitrogen is coordinated to a metal as a ligating group. In such complexes the nitrogen molecule retains the nitrogen-nitrogen triple bond linkage.

Examples. In contrast to dioxygen complexes, the dinitrogen complexes involve end-on bonding with only one nitrogen atom bonded to a metal. Dinitrogen complexes in which N_2 is a bridging ligand, e.g., $[Ru(NH_3)_5{-}N{\equiv}N{-}Ru(NH_3)_5]^{4+}$, Fig. 12.320 are also known. Because CO and N_2 are isoelectronic, it was perhaps not unexpected that the latter should be able to function as a ligand, but it is generally accepted that N_2 is both a poorer donor and a poorer electron acceptor than CO.

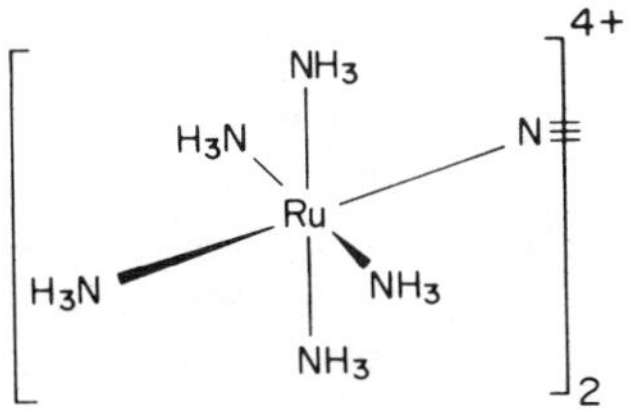

Figure 12.320 A dinitrogen complex.

12.330 Nomenclature of Metal Complexes

Preferred systematic nomenclature is embodied in the approved IUPAC rules for nomenclature of inorganic chemistry, and tentative rules for organometallic chemistry in IUPAC Information Bulletin #31. The rules deal with the naming of the ligands (uncharged or ionic) and the specific atom of attachment (if not obvious); the oxidation number of the metal; the charged or uncharged nature of the complex; and the stereochemistry, if relevant.

(a) The name of the central metal atom is placed last and, to avoid ambiguity, its oxidation number is given immediately thereafter in Roman numerals placed in parentheses.

(b) Ligands. The order of citation of ligands is: anionic; neutral; cationic. In each category alphabetical ordering is desirable. The names of all anionic ligands end in *o*, and usually all neutral and cationic ligands are used without change, except that coordinated water and ammonia are called aquo and ammine, respectively. The neutral ligands NO and CO are also exceptions for historical reasons; they are called nitrosyl and carbonyl, respectively. Alkyl and aryl groups that are σ bonded to the metal are given their conventional radicofunctional names (methyl, allyl, phenyl, etc.). The usual multiplying prefixes, di-, tri-, tetra-, penta-, etc., are used for simple ligands such as chloro, and bis-, tris-, tetrakis-, pentakis-, etc., for multiword ligands or ligands that already contain a multiplying prefix such as triphenylphosphine.

(c) Anionic complex ions are given the suffix *ate* appended to the English name (or a shortened version) of the metal atom. In a few cases (iron, copper, lead, silver, gold, tin), Latin stems are used (ferrate, cuprate, plumbate, argentate, aurate, stannate).

Examples. Although practice varies considerably, a systematic procedure for writing a line formula is the following: The symbol for the central atom is placed first, and the ligand structures are given in the same order (anionic, neutral, cationic and alphabetical with each category) as in naming. The formula of the whole complex ion or molecule is placed in brackets.

$Ca_2[Fe(CN)_6]$	Calcium hexacyanoferrate(II)
$K[PtCl_3(C_2H_4)]$	Potassium trichloro(ethylene)platinate(II)
$[Cr(C_6H_5NC)_6]$	Hexakis(phenylisocyanide)chromium(O)
$[Mn_2(CO)_{10}]$	Decacarbonyldimanganese(O)
$K_3[Co(CN)_5]$	Potassium pentacyanocobaltate(II)
$[Cu(CH_3COCHCOCH_3)_2]$	Bis(acetylacetonato)copper(II)
$[Fe(CO)_3(C_8H_8)]$	Tricarbonyl(cyclooctatetraene)iron(O)
$[Co(NH_3)_6]Cl_3$	Hexamminecobalt(III) chloride
$[Mn(COCH_3)(CO)_5]$	(Acetyl)pentacarbonylmanganese(I)
$[Re(C_5H_5)_2H]$	Bis(η^5-cyclopentadienyl)hydrido-rhenium(III)

12.340 Names of Bridging Groups

A bridging group is indicated by the Greek letter μ immediately before its name. Multiple bridging groups of the same kind are indicated by the usual multiplying prefix.

Examples. $[Ru(NH_3)_5{-}N{\equiv}N{-}Ru(NH_3)_5]Cl_4$ is μ-dinitrogenbis[pentammineruthenium(II)] tetrachloride; $Fe_2(CO)_9$ is tri-μ-carbonylbis[tricarbonyliron-(O)], and its common name is diiron ennacarbonyl, Fig. 12.150*a*; di-μ-dichloro-1,3-dichloro-2,4-bis(triphenylphosphine)diplatinum(II), Fig. 12.340.

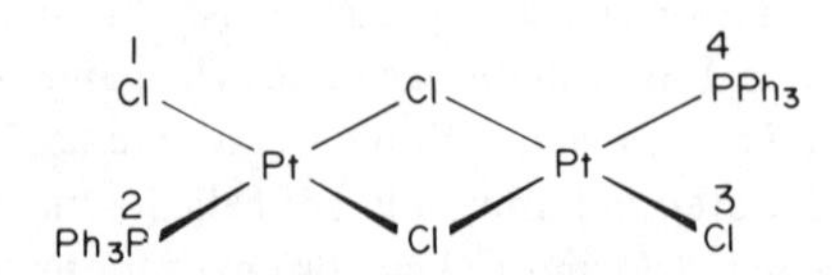

Figure 12.340 A di-μ-dichloro complex.

12.350 Metal Cluster Compounds

A limited (rather than infinite, as in bulk metal) group of metal atoms that are held together mainly by direct metal-metal bonding; each metal atom is bonded to at least two other metal atoms. The metal atoms frequently define a complete or nearly complete triangulated polyhedron. Some nonmetal atoms may be associated with the cluster.

Examples. [$Co_4(CO)_{12}$] (four metal atoms at the vertices of a tetrahedron), Fig. 12.350*a*; and [$Fe_5(CO)_{15}C$] (five metal atoms at the vertices of a square pyramid), Fig. 12.350*b*, an unusual metal carbonyl cluster in which a bare carbon atom is embedded.

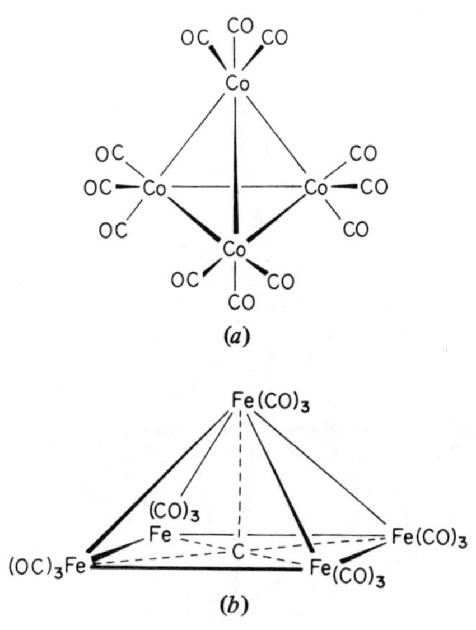

Figure 12.350 Metal cluster compounds: (*a*) tetrahedral; (*b*) square pyramidal.

12.360 Edge-Bridging Ligands

A ligand that bridges one edge of the polyhedron of a metal cluster.

Example. The octahedral metal cluster carbonyl, Fig. 12.360, which possesses one edge-bridging carbonyl.

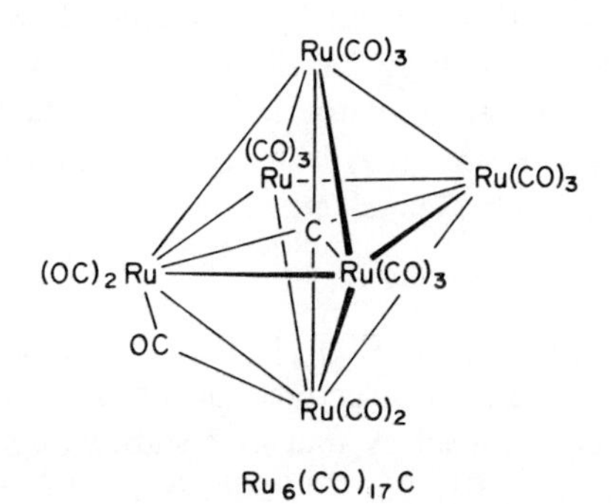

Figure 12.360 Metal cluster with one edge-bridging carbonyl.

12.370 Face-Bridging Ligands

A ligand that bridges one triangular face of a metal cluster polyhedron.

Example. The metal cluster $[Rh_6(CO)_{16}]$ has four face-bridging carbonyl groups staggered on the eight faces of an octahedron, Fig. 12.370 (top faces 1, 2, 5, and 3, 4, 5; bottom faces 1, 4, 6, and 2, 3, 6).

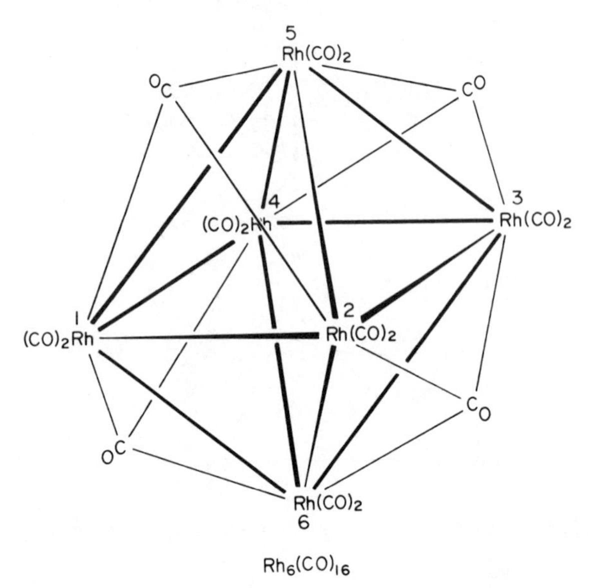

Figure 12.370 Metal cluster with four face-bridging carbonyls.

12.380 Closo-, Nido-, and Arachno-Clusters

Many cluster compounds including boron hydrides and carboranes (see Sect. 4.620) form complete or nearly complete triangular-faced polyhedra. These clusters form three structural types of polyhedra: closo- (closed); nido- (nest-liked); and arachno- (cobweb) structures. The closo-structures have a metal (or boron or carbon) atom at each vertex of the polyhedron. The nido-structures have incomplete polyhedra structures; they lack a skeletal atom at one of the vertices. The arachno-structures lack skeletal atoms at two vertices of an appropriate polyhedron.

Examples. The complexes $[Rh_6(CO)_{16}]$, Fig. 12.370, and $[Co_4(EtC{\equiv}CEt)(CO)_{10}]$, Fig. 12.380*a*, are closo-structures. In the latter the six vertices of the skeletal octahedron are occupied by four Co's and two C's. The metal carbonyl carbide cluster, $[Fe_5(CO)_{15}C]$, Fig. 12.350*b*, is a nido-structure. The five Fe atoms are at the vertices of an incomplete octahedron and thus there is absent one skeletal metal atom at one vertex. B_5H_9, Fig. 12.380*b*, is also a nido-structure, whose planar representation is shown in Fig. 12.380*c*. For purposes of the present classification, the platinum π-complex $[Pt(C_2H_4)(PPh_3)_3]$ may even be thought of as an arachno-structure with the missing two vertices of a trigonal bipyramid shown as circles in Fig. 12.380*d*.

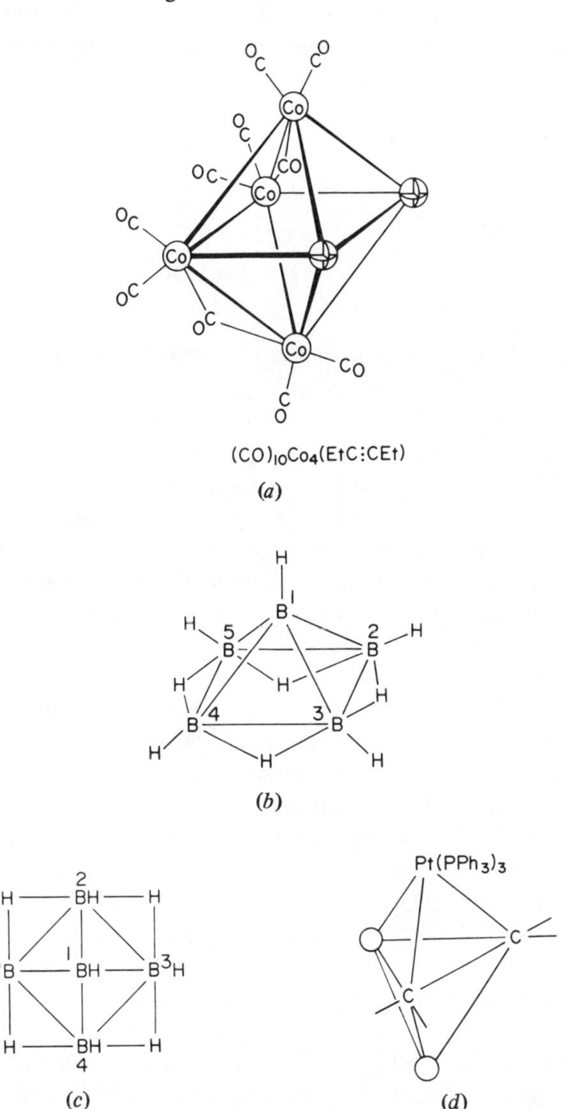

Figure 12.380 (*a*) A closo-structure; (*b*) a nido-structure, pentaborane (9); (*c*) planar projection of (*b*); (*d*) an arachno-structure.

12.390 Metallocycles

Compounds that consist of a cyclic array of atoms one of which is a metal atom. The most frequently encountered metallocycles are those consisting of three or four carbon atoms and a transition metal as part of the cycle.

Examples. Platinacyclobutane, Fig. 12.390*a*, and 3,4-dimethyltungstacyclopentane (usually the metal carries other ligands as well), Fig. 12.390*b*.

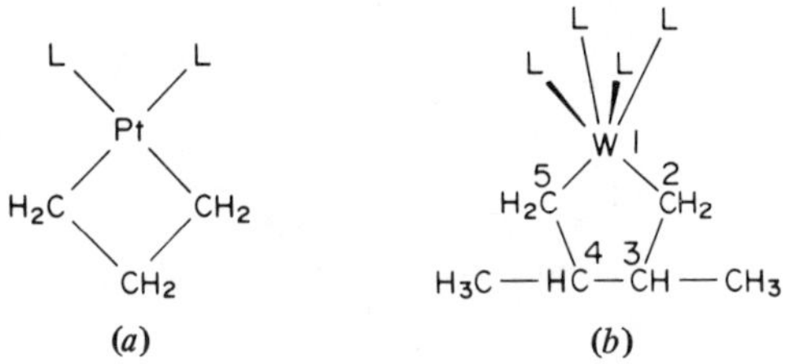

Figure 12.390 Metallocycles: (*a*) with platinum; (*b*) with tungsten.

12.400 Crystal Field Theory, CFT

The theory of bonding in metal complexes that views the interaction of the ligands and the metal as a strictly ionic or ion-dipole interaction resulting from electrostatic attractions between the central metal and the ligands. The ligands are regarded simply as point negative (or partially negative) charges surrounding a central metal atom; covalent bonding is completely neglected.

12.410 Crystal Field Splitting (of *d* Orbitals)

The splitting or separation of energy levels of the five degenerate *d* orbitals in a transition metal when the metal is surrounded by ligands arranged in a particular geometry with respect to the metal center. If the ligands (or charges they represent) were arranged spherically around the metal, all the *d* orbitals would be raised in energy relative to their energy in the isolated gaseous metal ion, but they all would remain degenerate. However, when the ligands assume particular geometries, e.g., square planar, tetrahedral, octahedral, the energies of the *d* orbitals are affected differently, resulting in *d* orbital splitting.

Examples. The octahedral complex $[CoF_6]^{3-}$ is shown in Fig. 12.410*a*, and Fig. 12.410*b* shows how the *d* orbitals are split in the octahedral field into two sets of orbitals: a lower energy, triply degenerate set given the designation t_{2g} and a higher energy, doubly degenerate set given the designation e_g (symmetry species in the O_h point group that characterizes an octahedron). The difference in energy Δ_0 (where 0 stands for octahedral) is the crystal field splitting energy.

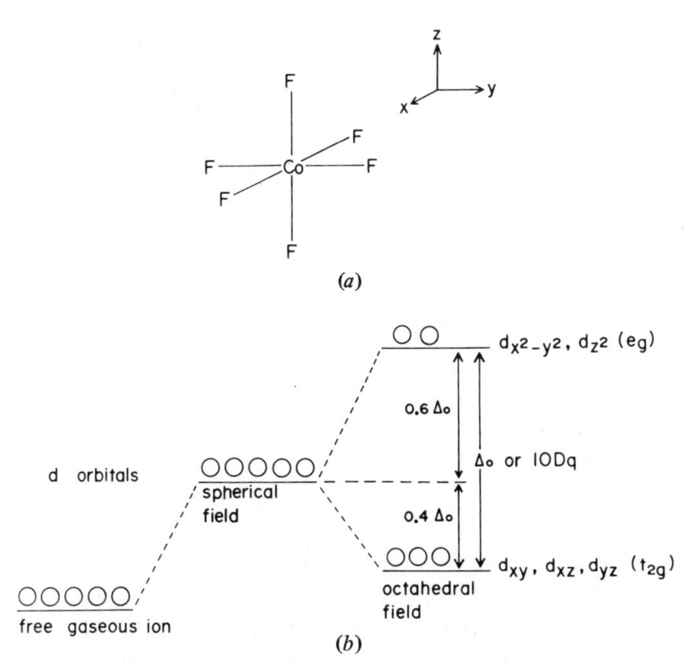

Figure 12.410 (*a*) The octahedral complex $[CoF_6]^{3-}$; (*b*) the relative energies of the *d* orbitals in a gaseous ion in a spherical field and in an octahedral field.

It is also designated $10D_q$ (the 10 is a multiplier of convenience). This energy difference can be determined experimentally (usually spectroscopically); its magnitude depends on the polarizability D of the metal (related to the charge and size of the metal ion) and particularly on the magnitude of the electronic and steric properties of the ligands, q. Ligands are frequently classified on the basis of the value of q, which is known as the ligand field splitting strength of the ligand. The decreasing order of ligand field strength of some common ligands is: $CN^- > NH_3 > H_2O > OH^- > F^- > Cl^- > Br^- > I^-$. In an octahedral field, for example, the $d_{x^2-y^2}$ and d_{z^2} set (e_g) of orbitals are repelled most because they point directly at the ligands whereas the d_{xz}, d_{yz}, d_{xy} set (t_{2g}) of d orbitals are repelled least because they point between the axes on which the ligands are located. In other geometries the splitting of the d orbitals is different. In a tetrahedral field the d splitting is exactly opposite to that of the octahedral field with the d_{xz}, d_{yz}, d_{xy} set being repelled more than the $d_{x^2-y^2}, d_{z^2}$ set.

12.420 Crystal Field Stabilization Energy, CFSE

The stabilization of a complex which results from electron occupation of the d orbitals split by the crystal field.

Examples. The crystal field splitting, which results from an octahedral environment, is displayed in Fig. 12.410*b*. If the metal ion were Co^{3+}, the electronic configuration is d^6. Each electron that goes into a t_{2g} orbital is assigned a value of $0.4\Delta_0$ and each electron that goes into an e_g orbital is assigned a value of $-0.6\Delta_0$. The six electrons go into the lower energy t_{2g} set, hence the crystal field stabilization energy for t_{2g}^6 is

$$6(0.4\Delta_0) + 0(-0.6\Delta_0) = 2.4\Delta_0$$

If the five *d* orbitals had remained degenerate, there would be no CFSE, and if the metal had a d^{10} configuration, the CFSE would be zero:

$$6(0.4\Delta_0) + 4(-0.6\Delta_0) = 0\Delta_0$$

12.430 Pairing Energy

The electron-electron repulsion energy that must be overcome to enable two electrons of opposite spin to occupy the same orbital.

Examples. The *d* electronic configuration of $[Co(NH_3)_6]^{3+}$ is t_{2g}^6 rather than a configuration that involves occupation of an e_g orbital, e.g., $t_{2g}^4e_g^2$. This arises because Δ_0 in this case is greater than the pairing energy. Rough estimates are that Δ_0 is about 66 kcal mol^{-1} (276 kJ mol^{-1}) and that the pairing energy is about 37 kcal mol^{-1} (155 kJ mol^{-1}) in this case.

12.440 Low Spin and High Spin Complexes

In any complex with a particular d^n configuration and a particular geometry, if the *n* electrons are distributed so that they occupy the lowest possible energy levels, the complex is a low spin complex. If some of the higher energy *d* orbitals are occupied before all the lower energy ones are completely filled, then the complex is a high spin complex.

Examples. In $[Co(NH_3)_6]^{3+}$ the electronic configuration is t_{2g}^6, as explained in Sect. 12.430, hence this is a low spin complex. In the complex $[CoF_6]^{3-}$ Co is again 3+ and again has a d^6 configuration. However, the crystal field splitting of the small, spherical F^- ligands is not nearly as great as that of the NH_3 ligands; in fact, the splitting (Δ_0) is somewhat less than the pairing energy. Accordingly, in building up the electronic configuration of Co^{3+} in this complex, one electron is placed in each of the three degenerate t_{2g} orbitals, then one electron is placed in each of the two degenerate e_g orbitals, since the pairing energy required to place two electrons in a t_{2g} orbital is greater than Δ_0. The sixth electron necessarily goes into a t_{2g} orbital, giving the high spin complex with an electronic configuration of $t_{2g}^4e_g^2$. Such a configuration involves four unpaired electrons and indeed $[CoF_6]^{3-}$ displays a paramagnetism corresponding to this configuration. Figure 12.440 summarizes this discussion.

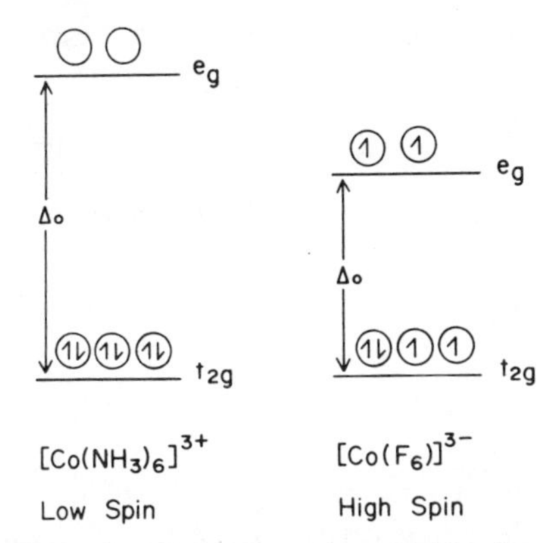

Figure 12.440 Octahedral low spin and high spin complexes.

12.450 Ligand Field Theory, LFT

A slightly modified crystal field theory (CFT) that takes into account certain parameters as variables rather than taking them as equal to the values found for free ions. One of the most important parameters is the spin-orbit coupling (Sect. 2.260) constant, which affects the magnetic properties of many ions in their complexes. For the reasons given above, LFT has also been called **"adjusted crystal field theory."**

12.460 Organometal Carbene

A dicovalent carbon compound (carbenes, Sect. 8.140) bonded to a metal atom.

Examples. Figure 12.460*a* shows a chromium carbene complex. Such a carbene compound has also been called a metal-stabilized carbenium ion, as the resonance structure 12.460*b* implies. The carbene may be regarded as a two electron donor similar to carbon monoxide, hence Cr in Fig. 12.460*a* has an EAN of 36, isoelectronic with $[Cr(CO)_6]$. Organometal carbenes in which a hetero atom is not bonded to the carbene carbon, e.g., the diphenylcarbene complex, Fig. 12.460*c*, have properties that suggest the carbene carbon is more electrophilic than in the heteroatom counterpart.

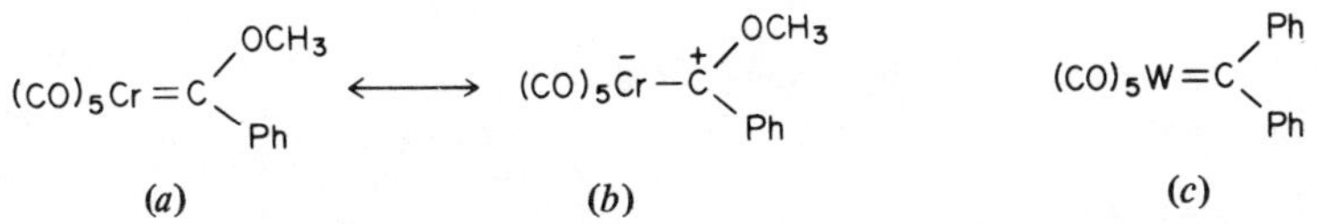

$(CO)_5Cr{=}C(OCH_3)(Ph)$ ⟷ $(CO)_5\overset{-}{Cr}{-}\overset{+}{C}(OCH_3)(Ph)$ $(CO)_5W{=}C(Ph)(Ph)$

(*a*) (*b*) (*c*)

Figure 12.460 (*a*) A chromium carbene complex; (*b*) a resonance structure of (*a*); (*c*) a tungsten diphenylcarbene complex.

12.470 Olefin Metathesis (*trans* Alkylidenation or Dismutation Reaction)

A reaction involving the exchange of alkylidene groups between two olefins:

$$R{-}CH{=}CH{-}R + R'{-}CH{=}CH{-}R' \rightleftharpoons 2R{-}CH{=}CH{-}R'$$

(R and/or R′ may also be H)

The reaction is catalyzed by transition metal complexes and presumably involves metal carbene intermediates.

Example. It is reported that the metathesis reaction

$$2CH_3CH{=}CH_2 \rightleftharpoons CH_3CH{=}CHCH_3 + CH_2{=}CH_2$$

is used commercially. Reactions carried out with stable carbenes suggest the intermediate formation of a metallocycle (Sect. 12.390), Fig. 12.470; such an intermediate rationalizes both the major and minor products of the reaction.

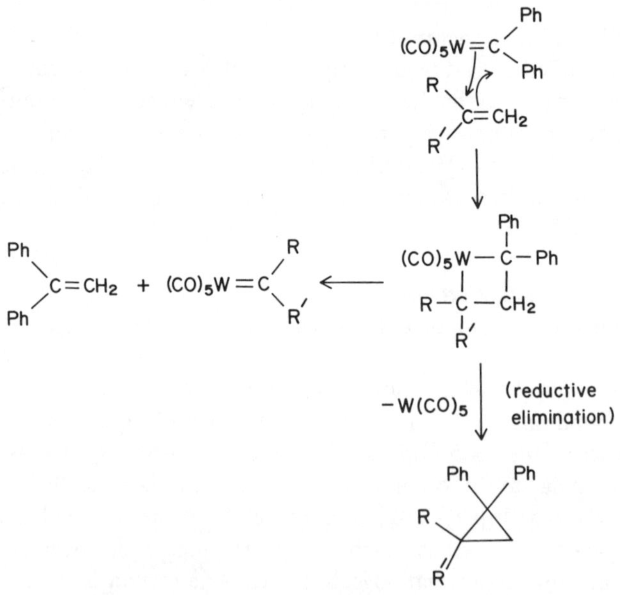

Figure 12.470 A mechanism of the olefin metathesis (or *trans* alkylidene or dismutation) reaction.

12.480 Organometal Carbyne, M≡C—R

A monocovalent carbon compound bonded to a metal atom.

Examples. $Br(CO)_4W{\equiv}C{-}Ph$. In a typical preparation a precursor carbene, $(CO)_5W{=}C(OC_2H_5)Ph$, is treated with BBr_3 in pentane at 228K, eliminating

CO and Br_2BOEt. Several other procedures are known but a precursor carbene is usually the starting compound.

12.490 Methyl Group Transfer (*trans* Methylation)

A reaction in which a methyl group attached to a transition metal is transferred to another acceptor molecule. The reaction is particularly important in vitamin B_{12} chemistry, involving enzymatic transfer of a methyl group bonded to cobalt in a reductive elimination reaction (Sect. 13.110).

Example. Methionine has been found to be formed from an enzyme bound B_{12}—CH_3 (E—Co—CH_3) and homocysteine:

$$E{-}Co^{n+}{-}CH_3 + HSCH_2CH_2CH(NH_2)CO_2H \longrightarrow$$
$$E{-}Co^{(n-1)+} + CH_2SCH_2CH_2CH(NH_2)CO_2H + H^+$$

12.500 Stereochemical Nonrigid Molecules

In the broadest sense, molecules that undergo rapid intramolecular rearrangement. In transition metal complex chemistry such rearrangements involve ground and transition states (or intermediates) viewed as idealized polygons and polyhedrons.

Examples. Although not a transition metal complex, the behavior of $(CH_3)_2$-NPF_4, Fig. 12.500, illustrates the concept. The original trigonal bipyramidal molecule goes through a square pyramidal intermediate in the rearrangement, as shown. The transformation can be demonstrated by ^{19}F nuclear magnetic resonance (NMR). At 25°C all the fluorine atoms are equivalent; the ^{19}F spectrum consists of one doublet due to P-F spin-spin coupling. At low temperatures, however, the ^{19}F spectrum shows two signals, each of which is a doublet of triplets (two sets of nonequivalent F's, one apical and one equatorial, each set split by the P atom and each set split by the two F's of the other set).

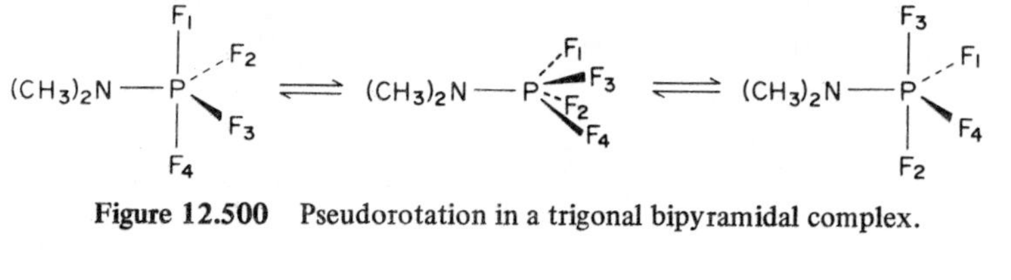

Figure 12.500 Pseudorotation in a trigonal bipyramidal complex.

12.510 Pseudorotation

The permutation of nuclear positions, as in the trigonal bipyramidal complexes shown in Sect. 12.500. This process is also known as the **Berry mechanism of rearrangement** (R. S. Berry).

12.520 *trans* Effect

In square planar complexes, the relative ability of a ligand to labilize the group *trans* to it in preference to the group *cis* to it in a substitution reaction.

Examples. The synthesis of *cis*- and *trans*-$[PtCl_2(NO_2)(NH_3)]^-$, Fig. 12.520, illustrates that the *trans* directing ability of NO_2^- is greater than that of Cl^- is greater than that of NH_3. The *trans* effect of some common ligands in decreasing order of strength is: $CN^- \simeq C_2H_4 \simeq CO > PR_3 > NO_2^- > I^- > Br^- > Cl^- > C_5H_5^- > RNH_2 > NH_3 > HO^- > H_2O$. The labilizing effect is not necessarily related to the weakening of the metal-ligand bond opposite the *trans* directing ligand, a thermodynamic effect, but is more likely a consequence of the reaction mechanism of displacement. The labilization accordingly should be more accurately called the **kinetic *trans* effect.** The exact nature of the *trans* effect is still being debated.

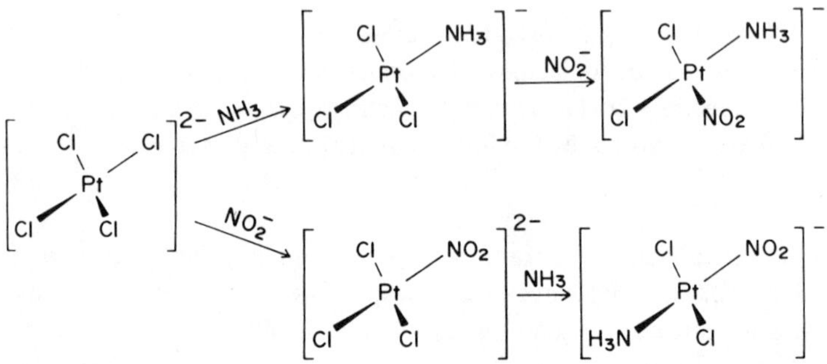

Figure 12.520 Preparation of *cis* and *trans* isomers using the trans effect principle.

13 Natural Products and Biosynthesis

CONTENTS

In the table of contents in this chapter there are many terms that appear in brackets after a main term. These bracketed terms are examples of or are related to, but not synonymous with, the corresponding main term; terms that are synonymous with the main term appear in parenthesis as in other chapters.

13.010 Natural Products

In the broadest sense any substance of biological origin. However, organic chemists tend to exclude those biological substances that are universally distributed in nature, substances that serve as building blocks for macromolecules via reversible condensation reactions, and the macromolecules derived from such condensations.

Example. Substances such as lipids, which are found in almost all living matter, are not usually considered natural products, but in many cases the component fatty acids do fall into the realm of natural products chemistry. Amino acids, proteins, nucleotides, nucleic acids, and to a lesser extent, sugars and starches are generally outside the rather ill-defined bounds of natural products chemistry.

13.020 Biosynthesis (Biogenesis)

The chemical pathway by which an organism synthesizes a natural substance.

13.030 Biomimetic Synthesis

A laboratory synthesis of a natural substance designed to mimic the biosynthesis, but conducted with the usual reagents of organic chemistry and without the aid of enzymes.

13.040 Prebiotic Chemistry

Chemistry that possibly may have led to the formation of the molecular building blocks of life under conditions that existed on earth before the evolution of the first living organisms.

13.050 Coenzyme (Prosthetic Group, Cofactor)

A relatively simple substance that acts in concert with a variety of different enzymes to effect some general chemical transformation such as oxidation, reduction, or decarboxylation.

Example. In the conversion of lactic acid to pyruvic acid, Fig. 13.050, the enzyme lactate dehydrogenase catalyzes the transformation with the aid of the coenzyme NAD (Sect. 13.070). NAD is involved in many other biological oxidations; however, a different enzyme is required for each of these transformations.

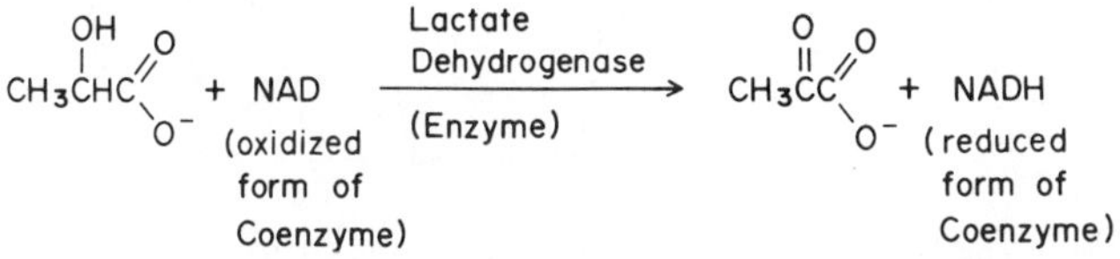

$$CH_3CH(OH)CO_2^- + \text{NAD (oxidized form of Coenzyme)} \xrightarrow[\text{(Enzyme)}]{\text{Lactate Dehydrogenase}} CH_3C(=O)CO_2^- + \text{NADH (reduced form of Coenzyme)}$$

Figure 13.050 The relationship between enzymes and coenzymes.

13.060 Vitamin

An essential dietary substance that cannot be synthesized by a particular organism and therefore must be added to the food supply of that organism.

Example. Humans cannot synthesize vitamin C (Sect. 13.180), so it must be obtained by eating foods that contain it, e.g., vegetables. On the other hand, carnivores such as cats can synthesize vitamin C and can survive with a diet low in vegetables. A substance that constitutes a vitamin for one organism need not be for another.

13.070 Nicotinamide Adenine Dinucleotide, NAD

The parent compound in the nicotinamide family of hydride-transfer coenzymes, Fig. 13.070.

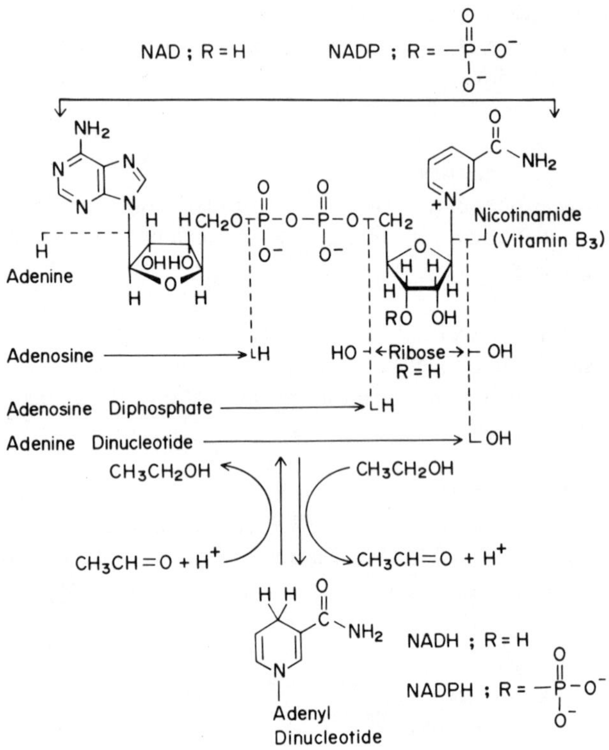

Figure 13.070 The constituents of the NAD family and their role in biological oxidation-reduction reactions.

Example. The reduced forms of the coenzymes, NADH and NADPH, serve as a source of hydride in biological reductions. The oxidized forms of the coenzymes, NAD and NADP, serve as a sink for hydride in biological oxidations. The requirements of the specific enzyme determine whether the NAD or NADP coenzyme is used (see also Sect. 13.090).

13.080 Riboflavin Adenosine Diphosphate, FAD [Riboflavin (Vitamin B_2)]

The principal member of the flavin family of hydride and electron-transfer coenzymes, Fig. 13.080.

Example. The interconversion between FAD and its reduced form $FADH_2$ is often coupled with the oxidation of thiols and the reduction of disulfides (see also Sect. 13.090).

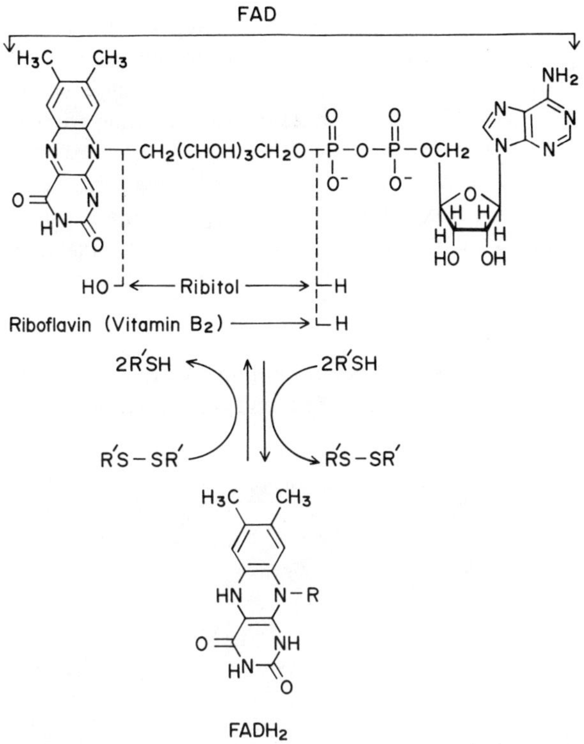

FAD

HO ← Ribitol → H

Riboflavin (Vitamin B_2) → H

2R′SH

R′S−SR′

$FADH_2$

Figure 13.080 The constituents of the FAD family and their role in biological oxidation-reduction reactions.

13.090 Adenosine Triphosphate, ATP

The universal phosphorylating coenzyme in biological systems, Fig. 13.090.

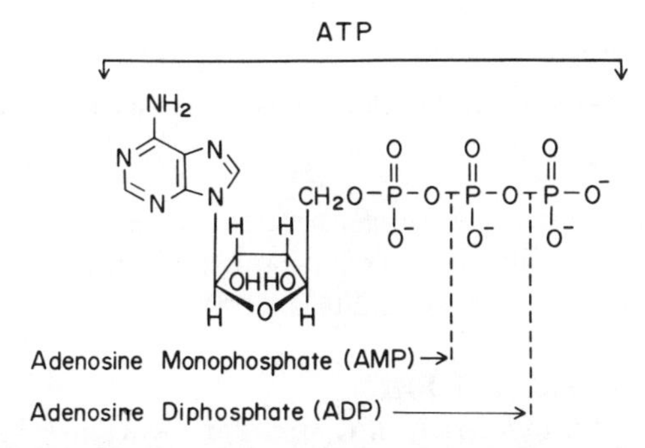

Figure 13.090 The constituents of the ATP family.

Example. All biological systems derive energy from one of two sources: photosynthesis (plants) or oxidative metabolism (plants and animals). This energy is stored in one of two forms: (1) a chemical reducing potential through the formation of NADH or $FADH_2$; (2) a chemical phosphorylating potential through the formation of ATP. One of the primary functions of this phosphorylating process is to provide reactive leaving groups in the form of phosphate esters. Thus phosphorylation of the hydroxy functional group provides a leaving group suitable for displacement by hydride in reductions or by other organic nucleophiles in the formation of carbon-carbon bonds (see Fig. 13.440).

13.100 *S*-Adenosylmethionine

This sulfonium salt is one of three important methyl-transfer coenzymes, Fig. 13.100. It functions in the transfer of methyl groups (see also Sect. 13.110).

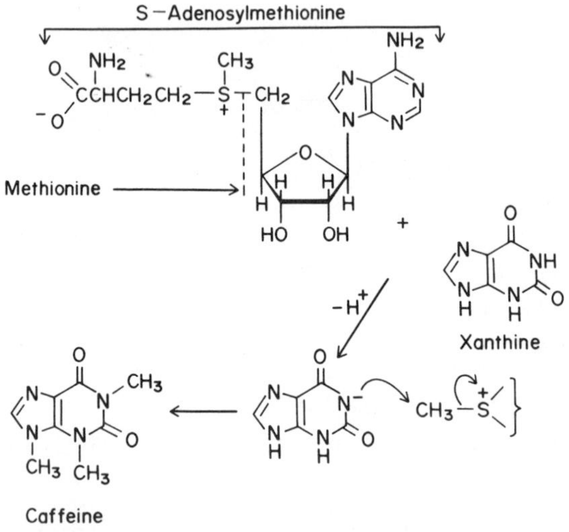

Figure 13.100 *S*-Adenosylmethionine and its role as a methyl-transfer coenzyme.

Example. A wide variety of oxygen, nitrogen, and carbon nucleophiles are methylated by *S*-adenosylmethionine in what is apparently an S_N2 reaction. The CH_3-transfer to xanthine is shown in Fig. 13.100.

13.110 Tetrahydrofolic Acid, FH_4

A well-known member of a family of C_1-transfer coenzymes, Fig. 13.110*a*. Various members of this family serve to transfer formate or formaldehyde equivalents and methyl groups.

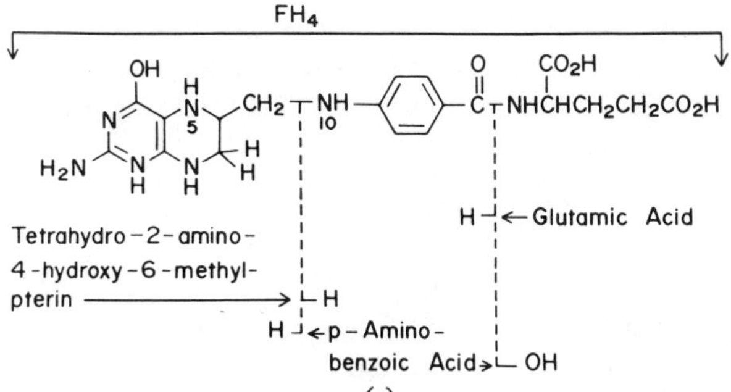

(*a*)

Example. The formaldehyde-transfer coenzyme is N^5, N^{10}-methylenetetrahydrofolic acid. This coenzyme may be either oxidized to the formate-transfer coenzyme N^{10}-formyltetrahydrofolic acid, or reduced to the methyl-transfer coenzyme N^5-methyltetrahydrofolic acid. This methyl-transfer coenzyme is the source of the *S*-methyl group in *S*-adenosylmethionine (Sect. 13.100), Fig. 13.110*b*.

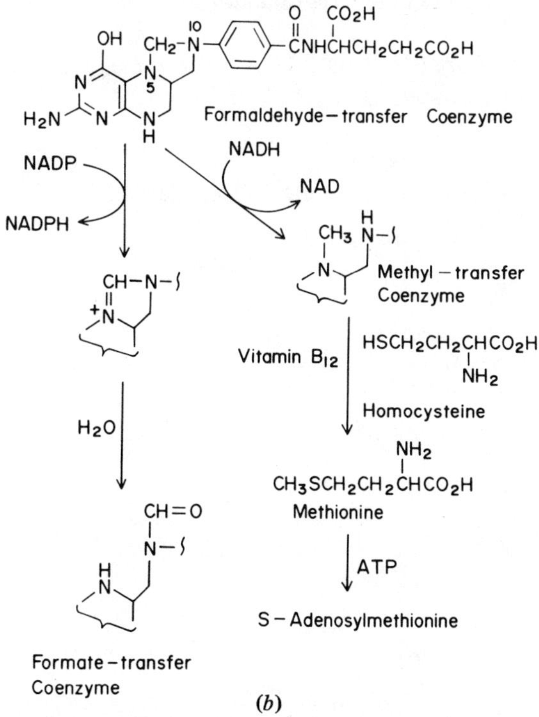

(*b*)

Figure 13.110 (*a*) The constituents of the tetrahydrofolic acid (FH_4) family; (*b*) the role of FH_4 as a C_1-transfer coenzyme.

13.120 Coenzyme A, CoASH

The coenzyme, Fig. 13.120, that activates carboxylic acids through the formation of a large family of acyl-transfer agents (RC(O)SCoA). The "A" stands for acylation.

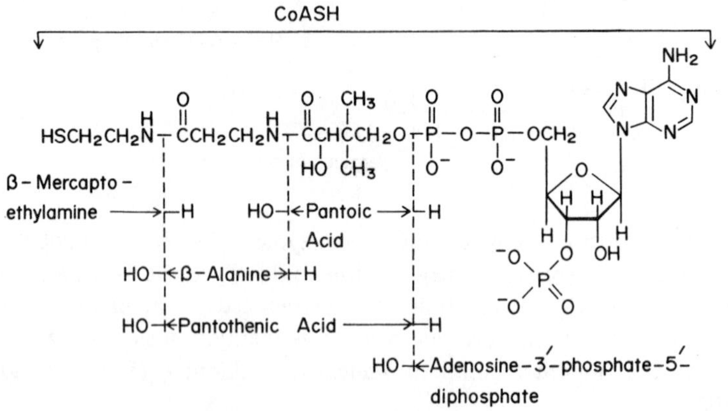

CoASH

HSCH2CH2NH–C(O)CH2CH2NH–C(O)CH(OH)C(CH3)2CH2O–P(O)(O⁻)–O–P(O)(O⁻)–OCH2 (adenosine, NH2)

β-Mercapto-ethylamine →H

HO←Pantoic Acid →H

HO←β-Alanine→H

HO←Pantothenic Acid →H

HO←Adenosine-3′-phosphate-5′-diphosphate

Figure 13.120 The constituents of coenzyme A.

Example. Acetyl CoA ($CH_3C(O)SCoA$) is the primary source of C_2-units in many biosynthetic pathways (Sect. 13.220).

13.130 Thiamine Pyrophosphate

A coenzyme that functions as an "active" aldehyde-transfer agent, Fig. 13.130.

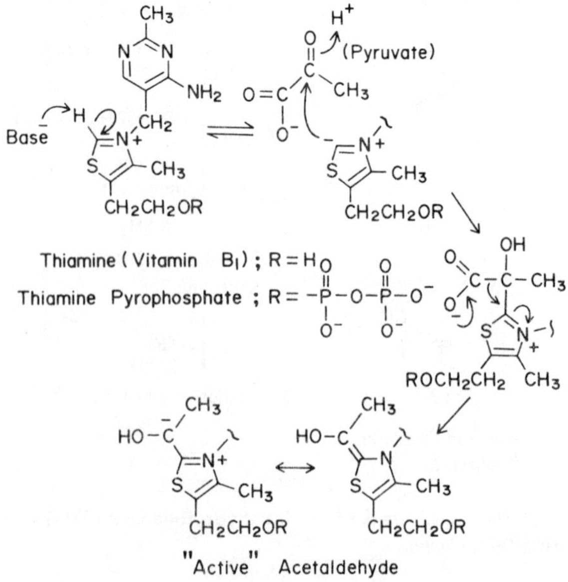

Base, (Pyruvate), CH2CH2OR, CH3, NH2

Thiamine (Vitamin B1); R = H

Thiamine Pyrophosphate; R = –P(O)(O⁻)–O–P(O)(O⁻)–O⁻

ROCH2CH2 CH3

"Active" Acetaldehyde

Figure 13.130 The formation of "active" acetaldehyde from thiamine.

Example. The absence in the diet of **vitamin B_1**, the thiazolium salt portion of this coenzyme, leads to the disease beriberi. In the first step of the biosynthesis of acetyl CoA, this coenzyme is required to generate "active" acetaldehyde, Fig. 13.130 (see also Fig. 13.150).

13.140 Umpolung (*G.*, Pole Reversal)

The reversal of polarity of an atom in a functional group through derivatization in such a way that the functional group may be regenerated following chemical modification.

Example. "Active" aldehyde derivatives (Sect. 13.130) are examples of biological umpolung. The carbon atom of the aldehyde functional group behaves as an electrophile in the classical chemistry of aldehydes. This carbon atom becomes a nucleophile upon formation of an "active" aldehyde derivative, Fig. 13.140. The term umpolung is more commonly used in organic synthesis.

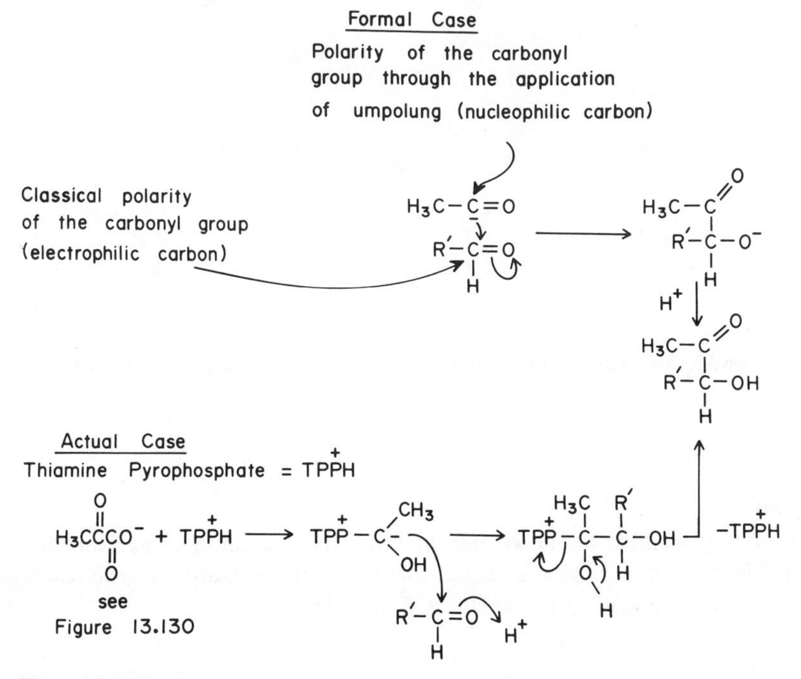

Figure 13.140 The reaction of "active" acetaldehyde as an example of umpolung.

13.150 Lipoic (or Thioctic) Acid

A reactive five-membered cyclic disulfide, Fig. 13.150, that serves as both a hydrogen-transfer and acyl-transfer coenzyme.

Example. An important role of lipoic acid is in the transfer of an acetyl group from thiamine to coenzyme A, Fig. 13.150.

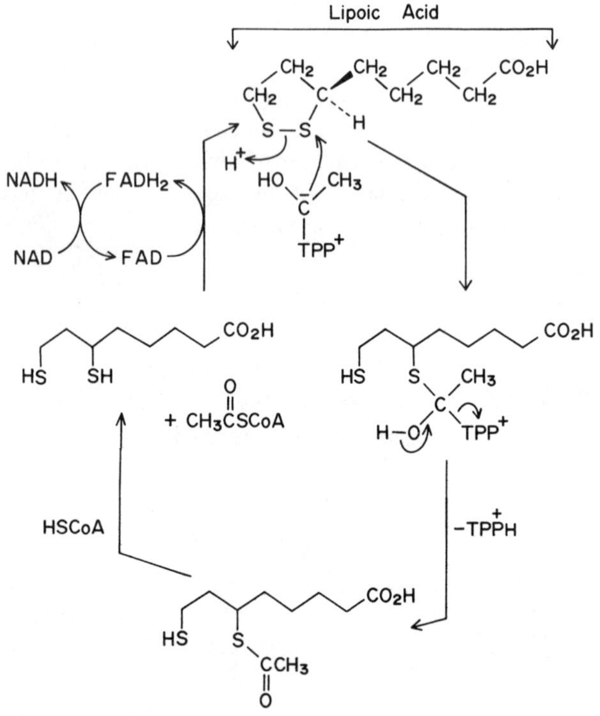

Figure 13.150 The role of lipoic (thioctic) acid in the transfer of acetyl groups.

13.160 Pyridoxal Phosphate

The primary member of the **vitamin B_6** or pyridoxine family of pyridine derivatives. These substances play a central role in such important biological reactions as transamination and decarboxylation, Fig. 13.160.

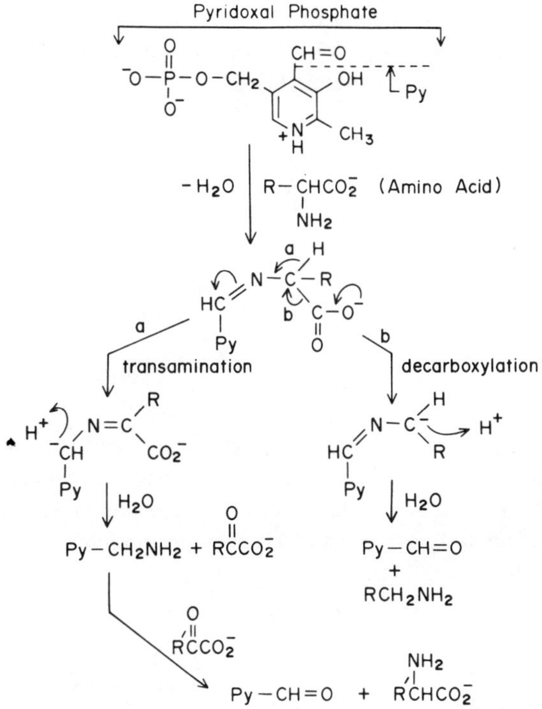

Pyridoxal Phosphate

$^{-}O-\overset{O}{\overset{\|}{P}}-O-CH_2$ ... CH=O ... OH ... Py ... $^{+}\underset{H}{N}$ ... CH_3

$-H_2O$ $R-\underset{NH_2}{CHCO_2^-}$ (Amino Acid)

a H

HC=N–C–R

b C–O⁻

Py O

a transamination b decarboxylation

H^+ $-N=C(R)CO_2^-$ ^{-}CH Py

$HC=N-C(H)R$ $\rightarrow H^+$ Py

H_2O H_2O

$Py-CH_2NH_2 + RCCO_2^-$ (with C=O) $Py-CH=O + RCH_2NH_2$

$R'CCO_2^-$ (with C=O)

$Py-CH=O + R'CHCO_2^-$ (with NH_2)

Figure 13.160 The role of pyridoxal phosphate in transamination and decarboxylation reactions.

13.170 Biotin

A cyclic urea that plays a central role in the transfer of "active" carbon dioxide in carboxylation-decarboxylation reactions, Fig. 13.170.

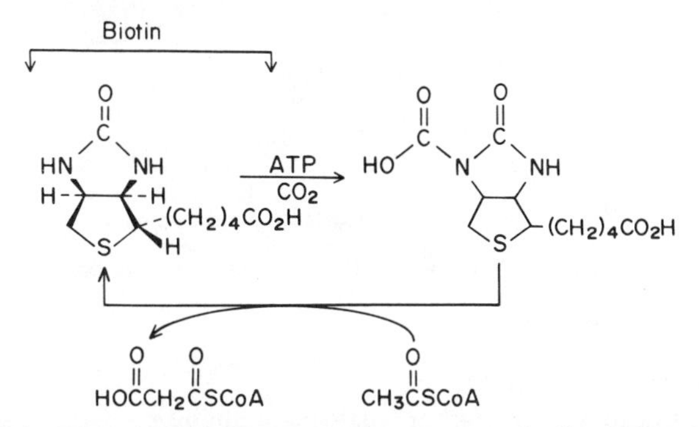

Figure 13.170 The role of biotin as a transfer coenzyme for "active" carbon dioxide.

13.180 L-Ascorbic Acid (Vitamin C)

A carbohydrate metabolite, Fig. 13.180, whose biochemical role is uncertain. Prolonged dietary deficiency in vitamin C leads to scurvy.

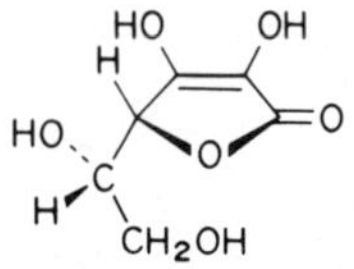

Figure 13.180 L-Ascorbic acid.

13.190 Fatty Acids

Long chain carboxylic acids produced by both plants and animals. The carbon chains in these acids are usually unbranched and usually contain an even number of carbon atoms. In solid animal fats these acids are present as triesters of glycerol and esters of cholesterol; in plant oils they are present as triglycerides and esters of sitosterols, Fig. 13.190.

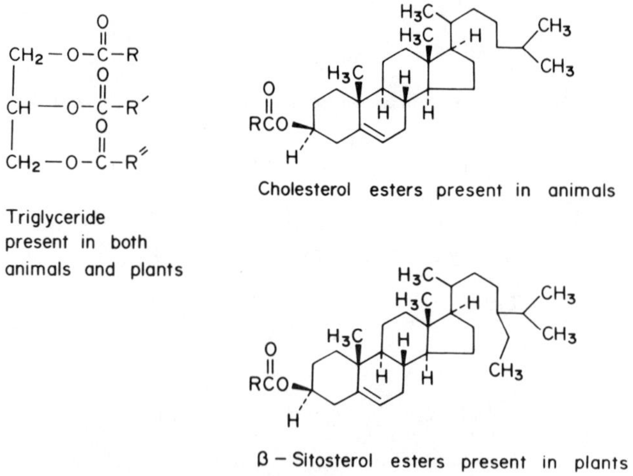

Triglyceride present in both animals and plants

Cholesterol esters present in animals

β – Sitosterol esters present in plants

Figure 13.190 Examples of animal and plant fats.

13.200 Saturated Fatty Acids

Fatty acids with long hydrocarbon chains containing no carbon-carbon double bonds. These acids are present in both animals and plants, Fig. 13.200.

$CH_3(CH_2CH_2)_nCO_2H$

n	number of carbons	name
1	4	Butyric (butanoic) acid
2	6	Caproic (hexanoic) acid
3	8	Caprylic (octanoic) acid
4	10	Capric (decanoic) acid
5	12	Lauric (dodecanoic) acid
6	14	Myristic (tetradecanoic) acid
7	16	Palmitic (hexadecanoic) acid
8	18	Stearic (octadecanoic) acid
9	20	Arachidic (eicosanoic) acid

Figure 13.200 Names of saturated fatty acids.

13.210 Unsaturated Fatty Acids

Fatty acids containing one or more carbon-carbon multiple bonds, Fig. 13.210. The stereochemistry of these olefinic linkages is usually *Z* (*cis*). Fatty acids containing more than one olefinic linkage, **polyunsaturated fatty acids,** are not synthesized by animals. Those polyunsaturated fatty acids related to linoleic acid are **essential fatty acids** in the diets of animals (see Sects. 13.240 and 13.250).

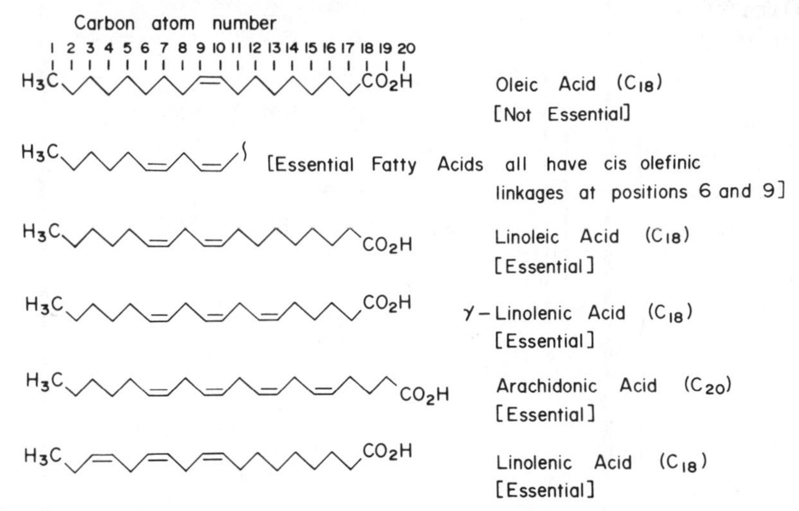

Figure 13.210 Structures and common names of essential unsaturated fatty acids (direction of numbering is opposite to that used in IUPAC nomenclature).

13.220 Fatty Acid Biosynthesis

Fatty acids are usually synthesized entirely from the acetyl groups (C_2-unit) of acetyl CoA. These C_2-units are added sequentially via: (1) Claisen condensation; (2) ketone reduction; (3) alcohol dehydration; (4) olefin reduction, Fig. 13.220.

Hence the resulting carbon chains are unbranched and contain an even number of carbon atoms. Unsaturated fatty acids may be synthesized by a variety of pathways depending on the organism: (1) deviation from the saturated fatty acid pathway at the dehydration step; (2) regiospecific and stereospecific oxidation of saturated fatty acids; (3) chain extension of an unsaturated fatty acid **starter unit**, Fig. 13.220.

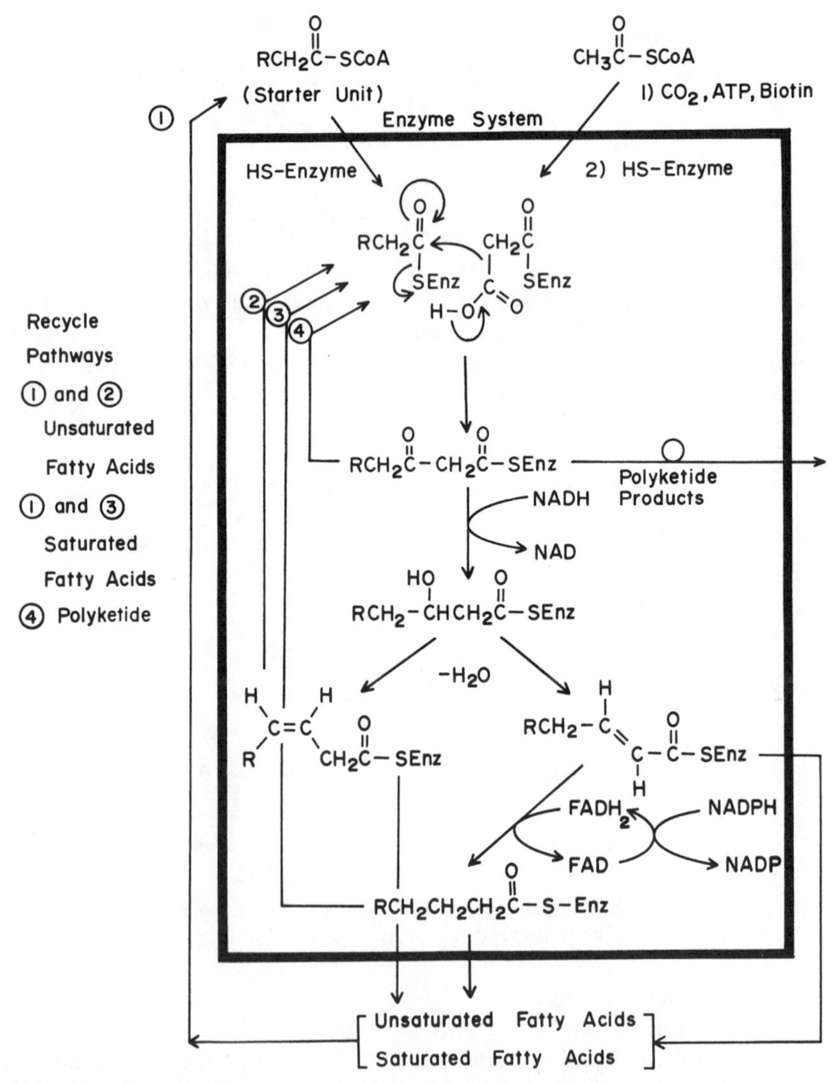

Figure 13.220 The biosynthetic mechanisms for the formation of fatty acids, polyketides, and related substances.

13.230 Lipids

A diverse class of biological substances that are soluble in nonpolar organic media.

Examples. The fatty acid esters shown in Fig. 13.190 constitute a major class of lipids. Related diesters of fatty acids and glycerol which also contain a phosphate ester chain are known as **phospholipids (phosphatides)**, Fig. 13.230*a*. Lipids of this type which have one polar chain and two nonpolar **(lipophilic)** chains tend to aggregate in aqueous medium to form **bilayers**, Fig. 13.230*b*. These bilayers are thought to provide the molecular foundation for many cell membranes.

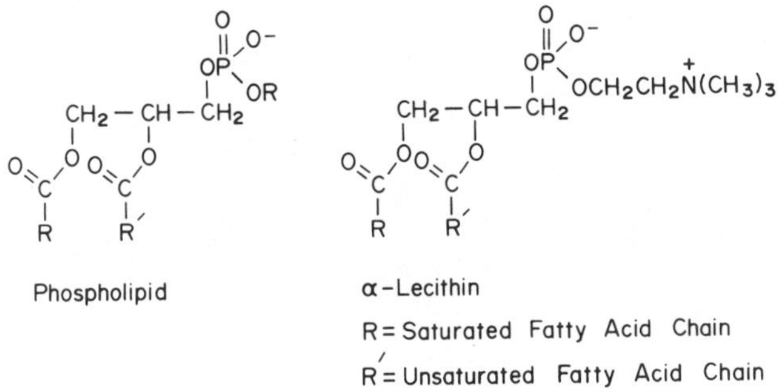

Phospholipid

α-Lecithin

R = Saturated Fatty Acid Chain

R′ = Unsaturated Fatty Acid Chain

(*a*)

Aqueous Medium (Polar Region)

Cell Membrane (Lipophilic Region)

Aqueous Medium (Polar Region)

(*b*)

Figure 13.230 (*a*) Lipids; (*b*) their aggregation into bilayers and cell membranes.

13.240 Prostaglandins

A family of complex fatty acid derivatives that are characterized by a carbon skeleton containing a cyclopentane ring. This ring is formed by the coupling of C-8 and C-12 of a straight chain, polyunsaturated C_{20} fatty acid, Fig. 13.240.

Example. The major classes of prostaglandins (PG) are distinguished by the functionalization of the cyclopentane ring, Fig. 13.240: **PGA**, cyclopentenone; **PGE**, β-hydroxycyclopentanone; **PGF**, cyclopenta-1,3-diol; **PGI**, the enol ether related to the PGF series (also referred to as **prostacyclin**). This classification is extended to indicate the number of double bonds in the side chains (the subscript following the class designation), PGA_2 and PGF_2, and to indicate the stereochemistry of the cyclopentane hydroxy groups (α if *cis*, and β if *trans* to the carboxy side chain), $PGF_{2\alpha}$. Many prostaglandins exhibit extreme biological activity which is thought to arise from the role of prostaglandins in regulating the passage (transport) of a variety of substances through cell membranes (see Sect. 13.610).

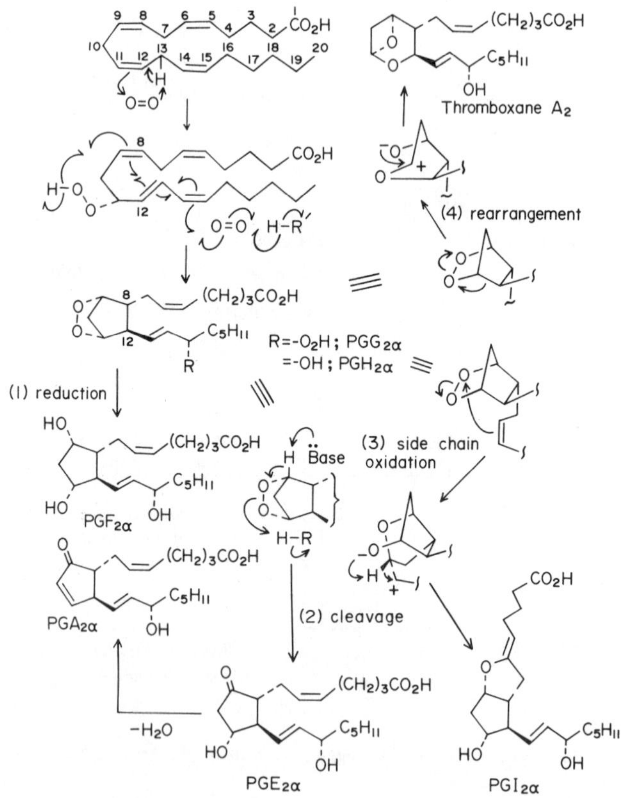

Figure 13.240 The biosynthetic mechanism for the formation of prostaglandins and related substances.

13.250 Prostaglandin Biosynthesis

The prostaglandins arise through the oxidative cyclization of C_{20} unsaturated fatty acids containing at least three double bonds (see essential fatty acids, Sect.

13.210, and Fig. 13.240). The prostaglandin endoperoxides that result, $\mathbf{PGG_{2\alpha}}$ and $\mathbf{PGH_{2\alpha}}$, undergo a number of further transformations: (1) reduction to the PGF series; (2) cleavage to the PGE series followed by dehydration to the PGA series; (3) intramolecular oxidation of the carboxylic acid side chain to form the PGI series; (4) rearrangement to the highly active **thromboxane $\mathbf{A_2}$** molecule.

13.260 The Acetate Hypothesis

Many naturally occurring aromatic (usually phenolic) compounds can be viewed as having been constructed biosynthetically through the linear combination of acetate (C_2) units (Sect. 13.270).

13.270 Polyketides

Poly-β-keto compounds that are never isolated as natural products themselves, but which, before being released by the enzyme systems that synthesize them, undergo facile conversions to aromatic and macrocyclic compounds, Fig. 13.270*a* and *b* (Sects. 13.280 and 13.290).

Example. Polyketides are synthesized by a variation of the fatty acid biosynthesis shown in Fig. 13.220. In the polyketide pathway the initially formed β-ketoester immediately reacts with another molecule of malonic ester rather than first undergoing reduction of the keto carbonyl group. A wide variety of different starter units may be incorporated into polyketide chains. In simple polyketides, acetate serves as the starter unit (Fig. 13.270*a*). However, other common starter units include fatty acids (Fig. 13.270*b*), compounds derived from shikimic acid (Fig. 13.380), and amino acids (Fig. 13.290*b*). In the vast majority of cases polyketide-derived natural products are made from a single polyketide chain rather than from several chains.

Simple Polyketide
$CH_3\dot{C}OSCoA$ (Starter Unit) 1-^{14}C-Acetic Acid
$CH_3\dot{C}OSCoA$
1) reduction
2) condensation
1) $-2H_2O$
2) enolization
6-Methylsalicyclic Acid

(*a*)

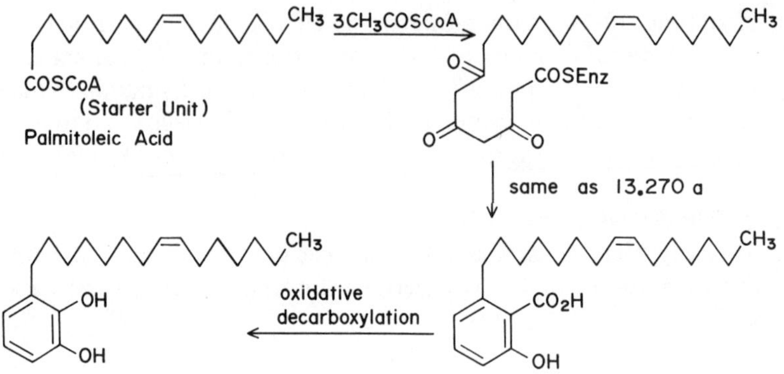

CH_3 $3CH_3COSCoA$ CH_3
COSCoA
(Starter Unit)
Palmitoleic Acid
COSEnz
same as 13.270 a
CH_3 CH_3
OH
oxidative decarboxylation
CO_2H
OH
OH
Urushiol II
(irritant oil of poison ivy, poison oak, and asiatic lacquer tree)
Anacardic Acid
(cashew nut oil)

(*b*)

Figure 13.270 The role of polyketides in the biosynthesis of aromatic natural products: (*a*) utilizing an acetate starter unit; (*b*) utilizing a fatty acid starter unit.

13.280 Tetracyclines

A family of potent antibacterial agents that are isolated from mold cultures. They have in common a highly functionalized 2-naphthacenecarboxamide carbon skeleton and are of polyketide origin, Fig. 13.280.

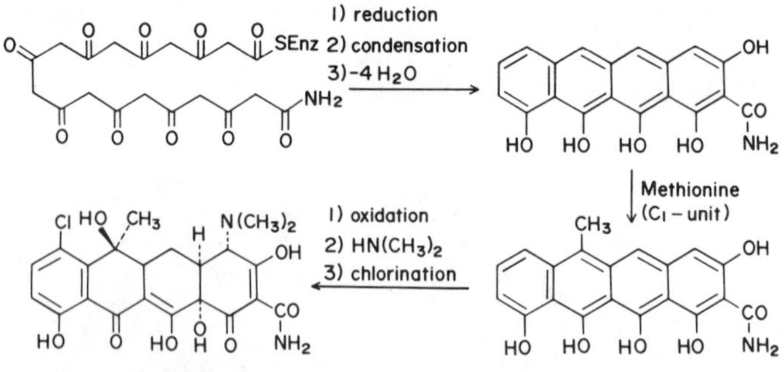

SEnz
NH_2
1) reduction
2) condensation
3) $-4\ H_2O$
OH
CO
NH_2
HO HO HO HO
Methionine
(C_1 – unit)
CH_3
Cl HO CH_3 H $N(CH_3)_2$
OH
1) oxidation
2) $HN(CH_3)_2$
3) chlorination
CO
NH_2
HO O HO H O
Chlorotetracycline

Figure 13.280 The biosynthesis of a tetracycline carbon skeleton.

13.290 Macrolides

A diverse group of highly functionalized, large ring molecules. Originally, this term was applied to macrocyclic lactones such as **erythromycin**, Fig. 13.290*a*.

More recently it has been extended to molecules in which the lactone is not part of the macrocyclic carbon skeleton such as lankacidin C, Fig. 13.290*b*, and to macrocyclic lactams and enol ethers such as rifamycin SV, a bridged dihydronaphthoquinone, or **ansa macrolide**, Fig. 13.290*c*.

Example. Lankacidin C is derived from a polyketide chain with a glycine starter unit. The methyl groups are supplied by methionine, Fig. 13.290*b*. This is in contrast to the biosynthesis of erythromycin, in which the methyl groups are derived from propionate via a polyketide chain composed of propionate units, Fig. 13.290*a*. Still other macrolides arise from heterogeneous polyketide chains constructed from acetate and propionate units.

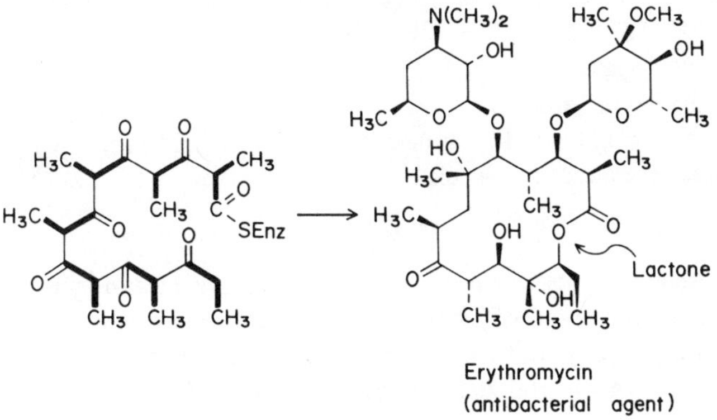

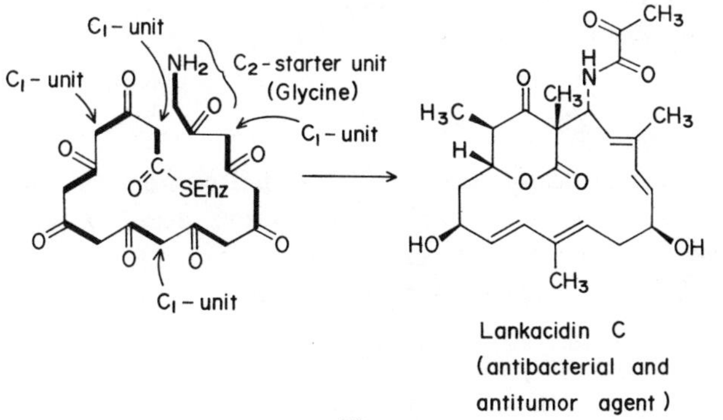

Figure 13.290 The biosynthesis of macrolides: (*a*) propionate-derived macrocyclic lactone; (*b*) an acetate-derived macrocyclic lactone.

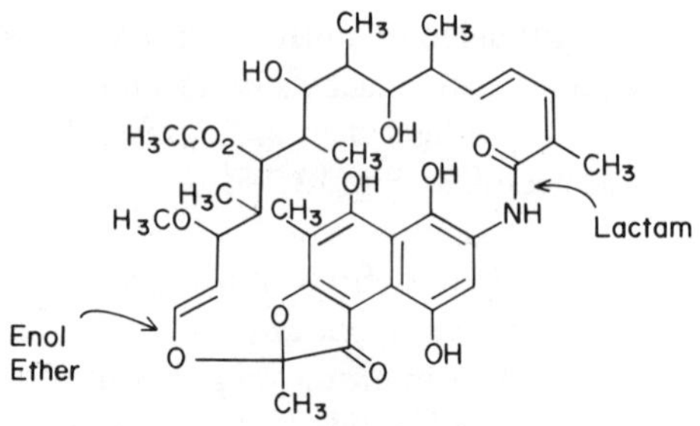

Rifamycin SV
(antibacterial agent)
(c)

Figure 13.290 (*continued*) The biosynthesis of macrolides: (*c*) a macrocyclic lactam enol-ether derived from both acetate and propionate (an ansa macrolide).

13.300 In Vivo (*L.* in Life)

A phrase used to indicate that a particular biochemical transformation has been achieved with the aid of living cells.

13.310 In Vitro (*L.* in Glass)

A phrase used to indicate that a particular biochemical transformation has been achieved without the aid of living cells. The transformation may have been accomplished either with cellfree extracts containing the appropriate enzymes, or with the usual reagents of organic chemistry without the aid of enzymes.

13.320 Shikimic Acid

A trihydroxycyclohexene carboxylic acid that is a strategic biosynthetic intermediate in the formation of phenyl and *p*-hydroxyphenyl groups (see Sect. 13.330).

13.330 Prephenic Acid

A 1,4-cyclohexadienedicarboxylic acid that is derived from shikimic acid (Sect. 13.320), and which undergoes either dehydrative decarboxylation to form phenylpyruvic acid, (pathway *a*) or oxidative decarboxylation to form *p*-hydroxyphenylpyruvic acid, (pathway *b*), Fig. 13.330.

Example. Shikimic acid pathways provide a major natural source of phenyl and *p*-hydroxyphenyl groups incorporated into a wide variety of biosynthetic building blocks, Fig. 13.330: benzoic acid derivatives (C_6-C_1 units); acetophenones and 2-phenylethyl derivatives (C_6-C_2 units); and **phenylpropanoids** (C_6-C_3 units).

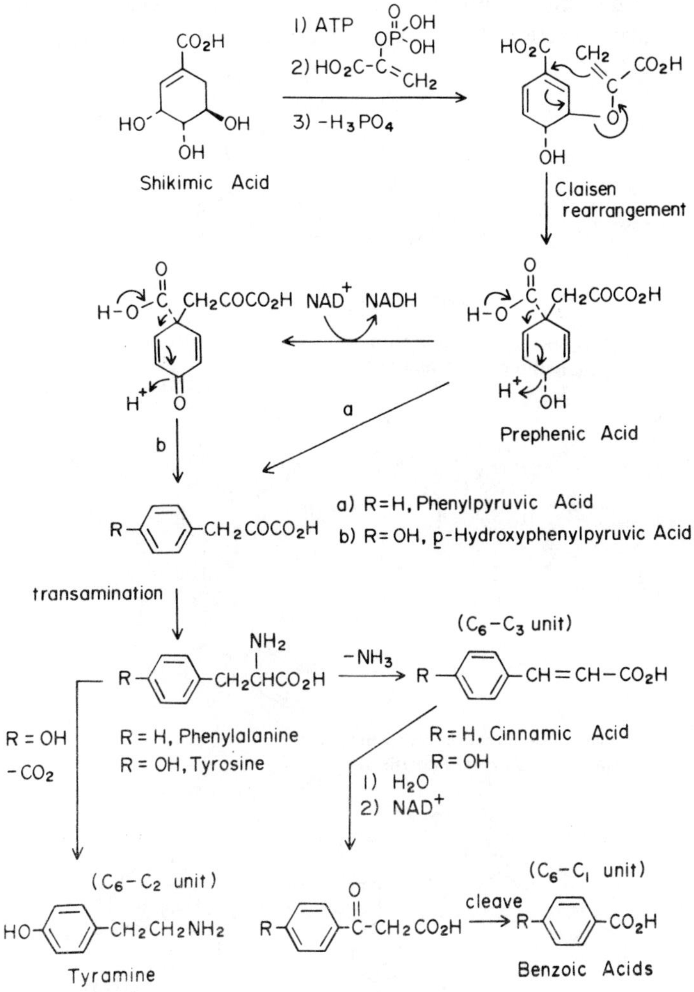

CO_2H

HO OH

OH

Shikimic Acid

1) ATP

O
‖
OP–OH
OH

2) HO_2C–C=CH_2

3) $-H_3PO_4$

HO_2C CH_2 CO_2H

C

O

OH

Claisen rearrangement

O
‖
H–O–C CH_2COCO_2H

H⁺ O

NAD^+ NADH

O
‖
H–O–C CH_2COCO_2H

H⁺ OH

Prephenic Acid

a

b

R–⟨⟩–CH_2COCO_2H

a) R=H, Phenylpyruvic Acid

b) R=OH, p-Hydroxyphenylpyruvic Acid

transamination

NH_2

R–⟨⟩–CH_2CHCO_2H

$-NH_3$

(C_6–C_3 unit)

R–⟨⟩–CH=CH–CO_2H

R = OH

$-CO_2$

R = H, Phenylalanine

R = OH, Tyrosine

R = H, Cinnamic Acid

R = OH

1) H_2O

2) NAD^+

(C_6–C_2 unit)

HO–⟨⟩–$CH_2CH_2NH_2$

Tyramine

O
‖
R–⟨⟩–C–CH_2CO_2H

cleave

(C_6–C_1 unit)

R–⟨⟩–CO_2H

Benzoic Acids

Figure 13.330 The biosynthetic mechanism for the formation of C_6-C_1, C_6-C_2, and C_6-C_3 units from shikimic acid.

13.340 Phenolic Coupling

The oxidative dimerization of phenolic compounds. Such coupling may be viewed as taking place via either a one electron oxidation followed by dimerization of two phenoxyl radicals, or a two electron oxidation followed by reaction of the resulting phenoxyl cation with a molecule of phenol or phenoxide ion,

Fig. 13.340. All possible coupling modes except oxygen-oxygen bond formation occur. Secondary reactions between the coupled moieties frequently take place (see Fig. 13.350).

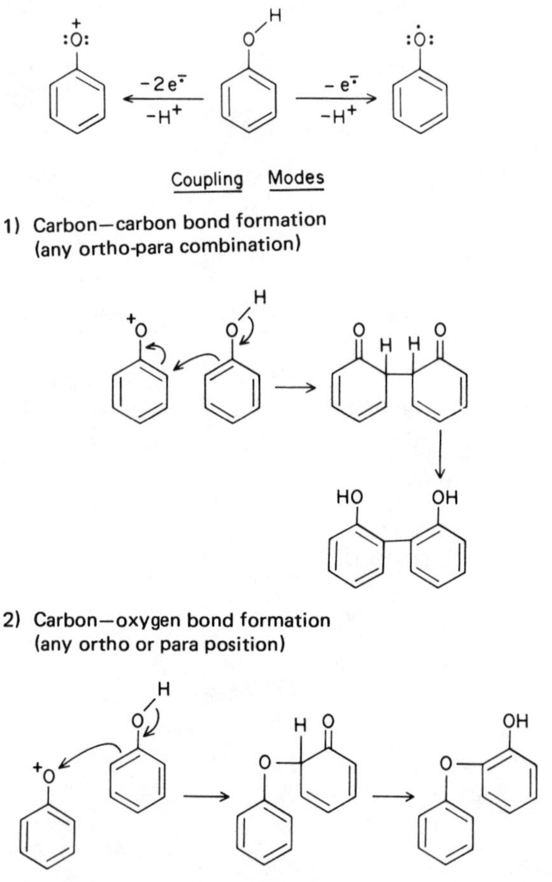

Coupling Modes

1) Carbon—carbon bond formation (any ortho-para combination)

2) Carbon—oxygen bond formation (any ortho or para position)

Figure 13.340 Possible modes of phenolic coupling.

Example. Phenolic coupling can be achieved both in vivo and in vitro. Its recent use in organic synthesis has been greatly stimulated by the realization that it is a very widespread reaction in plant biosynthesis. It is involved in the formation of lignans (Sect. 13.350), lignins (the woody substance of plants, Sect. 13.360), the constituents of several major families of alkaloids (Sects. 13.760 and 13.770), and a variety of other important natural products.

13.350 Lignans

A family of dimeric substances arising from phenolic coupling of C_6-C_3 units related to **coniferyl alcohol,** Fig. 13.350.

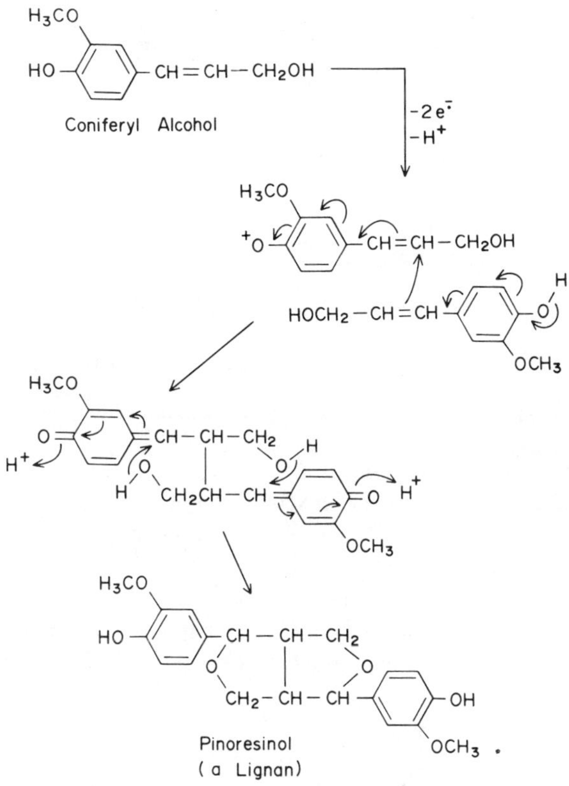

Figure 13.350 The biosynthesis of a lignan.

13.360 Lignins

Polymeric materials arising from the further phenolic coupling of the lignans (Sect. 13.350). These polymers help provide the structural strength of wood.

13.370 Flavonoids

A rather homogeneous family of plant metabolites and pigments that are developed around an aryl-substituted benzopyran carbon skeleton. The principal constituents of the flavonoids are shown in Fig. 13.370.

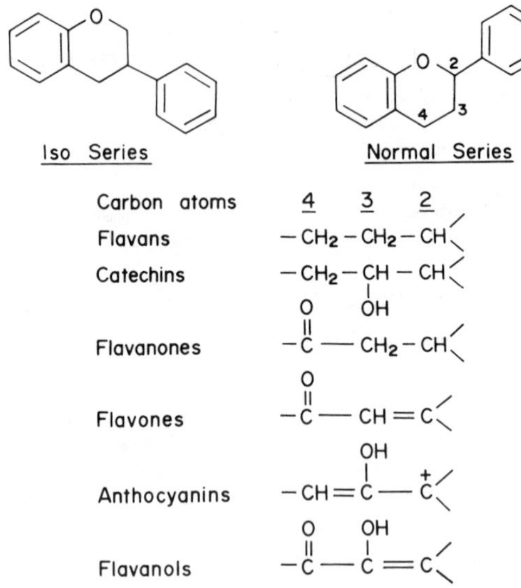
O
2
4
3
Iso Series
Normal Series
Carbon atoms
4 3 2
Flavans
$-CH_2-CH_2-CH<$
Catechins
$-CH_2-CH(OH)-CH<$
Flavanones
$-C(=O)-CH_2-CH<$
Flavones
$-C(=O)-CH=C<$
Anthocyanins
$-CH=C(OH)-C^{+}<$
Flavanols
$-C(=O)-C(OH)=C<$

Figure 13.370 The nomenclature of flavonoid substances.

13.380 Flavonoid Biosynthesis

The flavonoid carbon skeleton is derived from a heterogeneous polyketide (Sect. 13.270) in which a C_6-C_3 unit of shikimic acid origin (Sect. 13.320) serves as a starter unit, Fig. 13.380. This is indicated by the hydroxylation pattern; a *p*-hydroxyphenyl group is often associated with the shikimic acid pathway, and the *m*-hydroxy groups of the flavanone nucleus are associated with the polyketide pathway. **Isoflavonoid compounds** (Fig. 13.370) arise from flavonoids via a 1,2-phenyl migration.

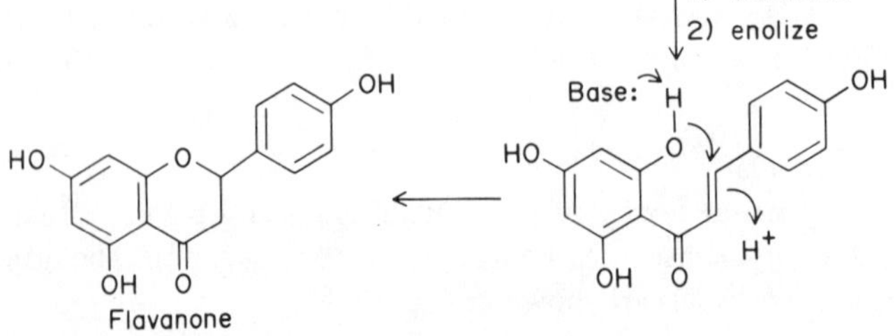
SCoA
HO
Starter Unit
$3CH_3CSCoA$
OH
SCoA
H
:Base
1) condense
2) enolize
Base:
H
O
HO
OH
H^+
OH O
HO
O
OH
OH O
Flavanone

Figure 13.380 The biosynthetic mechanism for the formation of flavanones.

13.390 Terpenes (Turpentine, *Gk.* Terpentin)

A structurally diverse family of compounds with carbon skeletons composed exclusively of isopentyl (isoprene) C_5-units. Terpenes are subdivided into the following classes:

Number of isopentyl C_5-units	Class
1 (C_5)	Hemiterpenes
2 (C_{10})	Monoterpenes (often referred to as terpenes, see Sect. 13.460)
3 (C_{15})	Sesquiterpenes (Sect. 13.480)
4 (C_{20})	Diterpenes (Sect. 13.520)
5 (C_{25})	Sesterterpenes (Sect. 13.530)
6 (C_{30})	Triterpenes (Sect. 13.540)

For further elaboration see Sect. 13.410.

13.400 Terpenoids

A large family of substances, encompassing not only terpenes, but also compounds of terpene origin that do not have carbon skeletons composed exclusively of isopentyl C_5-units. In this latter class the original terpene skeleton may have been altered through rearrangements, degradative loss of carbon atoms, or introduction of additional carbon atoms of nonterpene origin (see Sect. 13.410).

13.410 The Isoprene Rule

As originally expressed, the **regular isoprene rule** states that all terpenes have carbon skeletons formed by the head-to-tail linkage of isopentyl (isoprene) C_5-units, Fig. 13.410. Although these C_5-units do have the carbon skeleton of isoprene, Fig. 13.410, isoprene is not a natural product, nor is it in anyway involved in terpene biosynthesis (see Sect. 13.440). As modified by such modern considerations, the **biogenetic isoprene rule** states that all terpenoid compounds are derived through a rational sequence of mechanistic events (usually involving carbocation chemistry) starting from a few key acyclic terpenes (geraniol, Sect. 13.440; farnesol, Sect. 13.440; geranylgeraniol, Sect. 13.520; and squalene, Sect. 13.550). The acyclic terpenes are usually composed of head-to-tail linked isoprene units, but head-to-head linkages are encountered sometimes, as in the case of squalene. (For examples of the application of this modified rule of terpene biosynthesis see Sects. 13.470, 13.540, and 13.600).

Example. In Fig. 13.410 the structures of several terpenes have been dissected into their constituent isoprene units. The arrows indicate the head-to-tail linkages between the isoprene units. The carbon skeleton of the acyclic terpene precursors referred to in the biogenetic isoprene rule becomes apparent by follow-

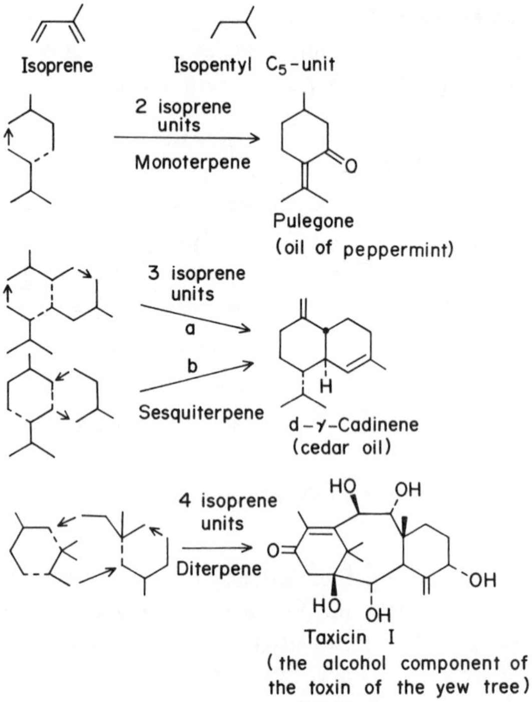

Isoprene

Isopentyl C_5-unit

2 isoprene units

Monoterpene

Pulegone (oil of peppermint)

3 isoprene units

a

b

Sesquiterpene

d-γ-Cadinene (cedar oil)

4 isoprene units

Diterpene

HO OH O OH HO OH

Taxicin I (the alcohol component of the toxin of the yew tree)

Figure 13.410 Terpenes with carbon skeletons described by the regular isoprene rule.

ing the course of these arrows. The dotted lines represent secondary linkages formed during the cyclization of the acyclic precursors. In some cases it is possible to find more than one chain configuration that might lead to the same carbon skeleton, as indicated for the sesquiterpene in Fig. 13.410. Based on the simple considerations evolved here, it is not possible to decide which configuration represents the true biogenetic route. However, the fact that at least one such dissection is possible is a requirement for a substance to be classified a terpene.

13.420 Mevalonic Acid

3-Methyl-3,5-dihydroxypentanoic acid, a strategic biosynthetic intermediate in the formation of all terpenoids (see Fig. 13.440).

13.430 Mevalonic Acid Biosynthesis

The carbon skeleton of mevalonic acid is formed via an aldol-type condensation of acetyl CoA with acetoacetyl CoA (which might be regarded as a diketide, Sect. 13.270). (*R*)-Mevalonic acid results from the reduction of this acetyl CoA carboxyl function to an alcohol (Fig. 13.440).

13.440 Geraniol

3,7-Dimethyl-2(*E*), 6-octadien-1-ol, an alcohol, and its *Z* isomer, **nerol**, are the simplest acyclic terpene precursors referred to in the biogenetic isoprene rule (Sect. 13.410). From these substances all monoterpenes are formed directly, and the larger acyclic terpene precursors are derived through the addition of further isopentyl C_5-units.

Example. The head-to-tail coupling of isopentyl C_5-units occurs via the reaction of **dimethylallyl pyrophosphate** with **isopentenyl pyrophosphate**, as shown in Fig. 13.440. Both of these substances are derived from mevalonic acid. The coupling product, **geranyl pyrophosphate**, may undergo chain extension to **farnesyl pyrophosphate** through reaction with a further molecule of isopentenyl pyrophosphate. This chain extension process can be repeated to afford even larger acyclic terpene precursors.

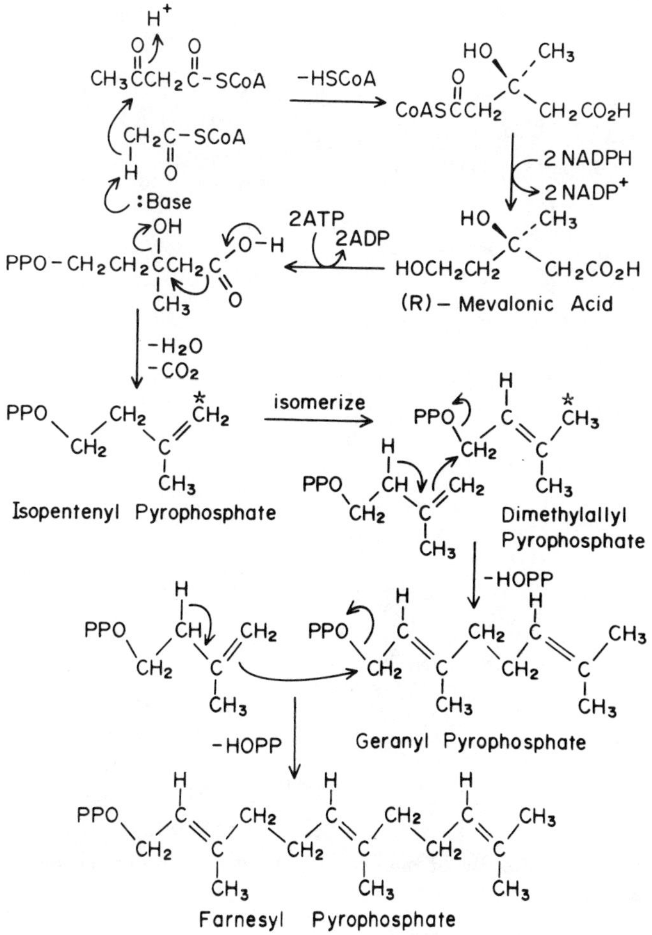

Figure 13.440 The biosynthetic mechanism for the formation of farnesyl pyrophosphate via mevalonic acid.

13.450 Essential Oils

The odoriferous oils that can be distilled from various plant components. These oils are usually complex mixtures that are rich in terpenes and serve as the raw material for the isolation of many terpenoid compounds.

13.460 Monoterpenes

Substances derived from geranyl pyrophosphate (two isopentyl C_5-units) in accordance with the isoprene rule (Sect. 13.410).

Examples. Monoterpenes are pungent oils or volatile solids that are usually isolated from plants. They are widely used as flavorings and fragrances. For examples, see Figs. 13.410 and 13.470.

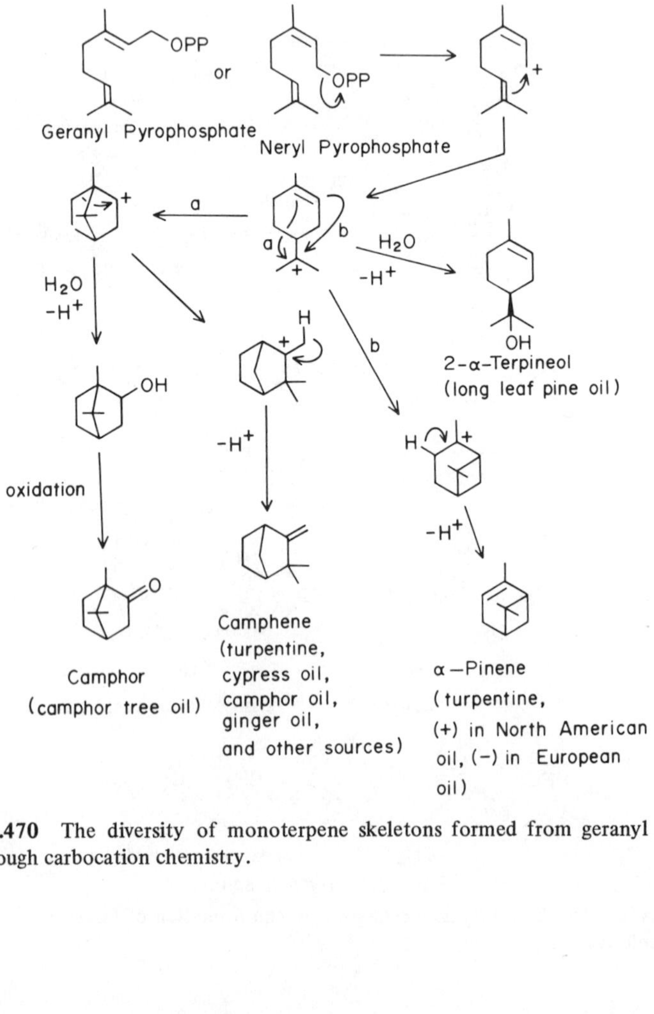

Figure 13.470 The diversity of monoterpene skeletons formed from geranyl pyrophosphate through carbocation chemistry.

13.470 Monoterpene Biosynthesis

Monoterpenes are formed from either geranyl or neryl pyrophosphate, Fig. 13.470. The biosynthesis is usually initiated with the ionization of the pyrophosphate leaving group. The resulting carbocation then undergoes any one of a variety of cyclizations with available olefin sites. The subsequent carbocation chemistry is that usually observed in vitro, e.g., alkyl and hydride shifts, and further cyclizations with the remaining double bonds, as indicated in Fig. 13.470. A sequence is terminated by the quenching of the carbocation either by reaction with water to form an alcohol or by the loss of a proton to form a new olefin. Frequently these transformations scramble the carbon skeletons to the extent that it is no longer possible to dissect a complete set of isoprene units from the skeleton of the product. Processes very similar to those outlined in Fig. 13.470 also are involved in the biosynthesis of the larger terpenes.

13.480 Sesquiterpenes

Substances derived from farnesyl pyrophosphate (Fig. 13.440; three isopentyl C_5-units) in accordance with the isoprene rule.

Example. These substances are pungent oils and solids that are usually isolated from plants and are widely used as flavorings and fragrances (Figs. 13.410 and 13.500).

13.490 Pheromones (*Gr.* Pherin, To Bear; Hormôn, Stimulation)

Compounds that provide a basis of chemical communication. Such compounds are released into the environment by one organism, where they are detected by another organism and elicit a specific response from the recipient.

Examples. Insect pheromones have been most thoroughly investigated, although other animals apparently produce similar agents. Pheromones serve a variety of functions. **Sex attractants** and **aggregation pheromones** are responsible for bringing individuals of the opposite sex together for mating purposes, **alarm pheromones** warn of danger and incite to battle, and **trail pheromones** mark the path to sources of food. In a given species the sex attractants tend to be mixtures of related substances, but the sex attractant of one species may differ drastically from those of another. For instance, the attractant of the pink cotton bollworm moth, a destructive cotton pest, is a 1:1 mixture of hexadecadienyl acetates, Fig. 13.490*a*. These substances are clearly of fatty acid origin. On the other hand, the attractant of the boll weevil, another cotton pest, is a mixture of four monoterpenes. The most interesting of these is the cyclobutane, grandisol, which apparently is derived via the novel biosynthesis shown in Fig. 13.490*b*.

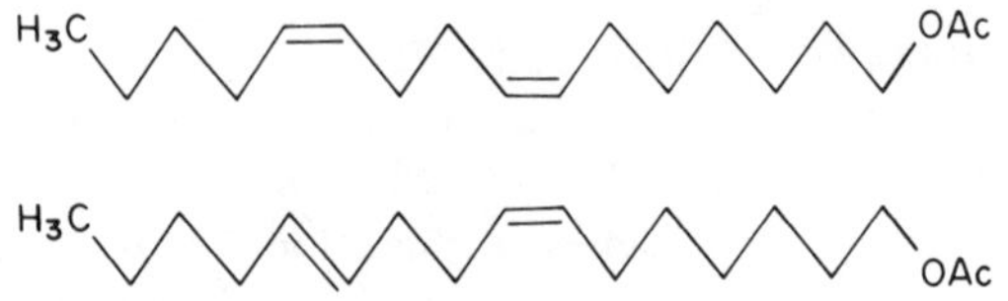

1:1 mixture (Gossyplure)

(*a*)

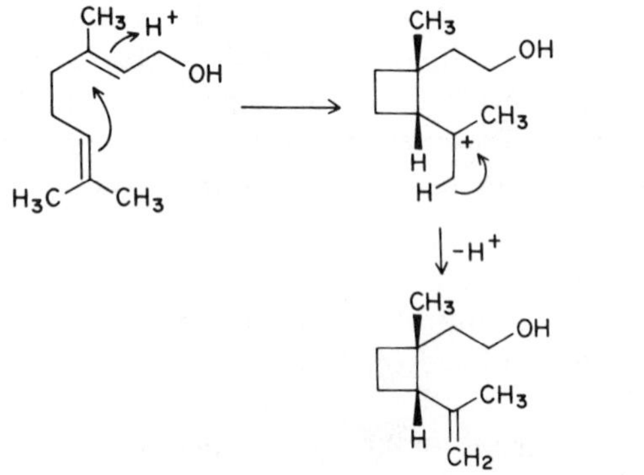

Grandisol
(isolated from 50 kg of excrement from 5 million boll weevils)

(*b*)

Figure 13.490 Insect sex pheromones: (*a*) of fatty acid origin; (*b*) of terpene origin.

13.500 Juvenile Hormones of Insects

Substances that are produced by insects during the larval stages of their life cycle, and which serve to maintain the insect in their larval form. Reduction in the level of juvenile hormone production is accompanied by metamorphosis into the adult form.

Examples. A juvenile hormone of the cecropia moth, Fig. 13.500*a*, is apparently of sesquiterpenoid origin. The additional carbon atom in the terminal ethyl group arises from the replacement of one acetate unit by a propionate unit in the mevalonate-farnesol biosynthetic pathway. The juvenile hormone of a Czechoslovakian beetle, (+)-juvabione, Fig. 13.500*b*, is also of sesquiterpene origin. Both American and Czechoslovakian balsam fir trees also synthesize this material, apparently as a defense mechanism against these pests. Exposure of this beetle to balsam fir wood or wood products such as paper prevents metaphorphosis into the adult form before death.

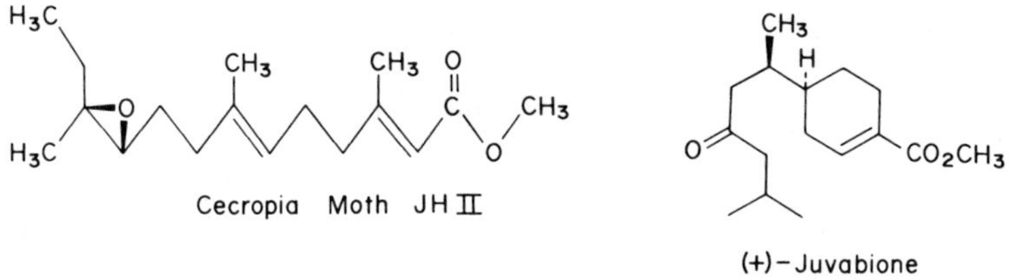

Cecropia Moth JH II

(+)-Juvabione

(*a*) (*b*)

Figure 13.500 Insect juvenile hormones: (*a*) of the cecropia moth; (*b*) of a Czechoslovakian pine beetle.

13.510 Defense Substances of Arthropods

Any of a wide variety of substances that are ejected by arthropods when they are molested. These chemical warfare agents have distinctly unpleasant properties and are delivered by being blown, sprayed, squirted, foamed, or spit at the molester.

Example. All manner of noxious agents are employed in this capacity including hydrogen cyanide from the lateral glands of the millipede, and hot benzoquinone from a variety of arthropods. Among the most structurally interesting are the terpenoids (Fig. 13.510): dolichodial, a lachrymatory oil from ants; a sesquiterpenoid allene, an ant repellant found in grasshopper spit; and cantharidin (Spanish fly), a blistering agent and notorious aphrodisiac which has been fatal in doses of about 10 mg, isolated from blister beetles.

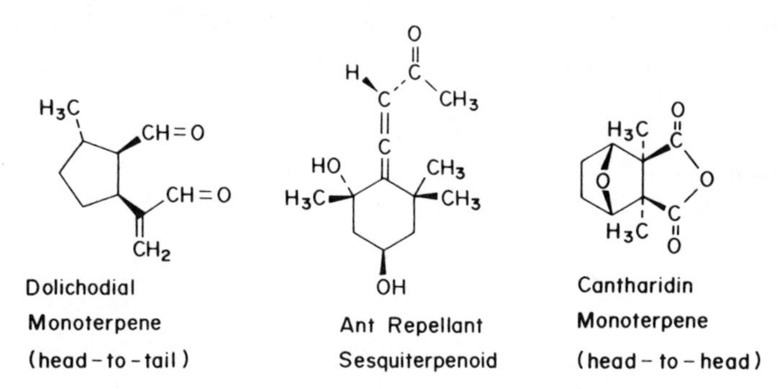

Figure 13.510 Arthropod defense substances.

13.520 Diterpenes

Substances derived from geranylgeranyl pyrophosphate, the C_{20} analog of geranyl pyrophosphate (four isopentyl C_5-units), in accordance with the isoprene rule (Fig. 13.410).

13.530 Sesterterpenes

Substances derived from geranylfarnesyl pyrophosphate, the C_{25} analog of geranyl pyrophosphate (five isopentyl C_5-units), in accordance with the isoprene rule.

Example. Sesterterpenes are a surprisingly rare class of terpenes. Only two cyclic carbon skeletons are presently known. The majority of representatives have the ophiobolane skeleton, ceroplasteric acid, Fig. 13.530.

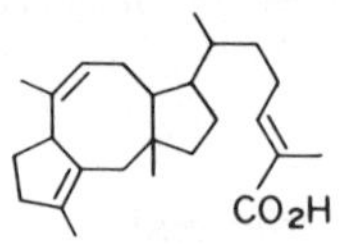

CO2H

Ceroplasteric Acid
(scale insect wax)

Figure 13.530 A sesterterpene.

13.540 Triterpenes

Substances derived from farnesylfarnesyl pyrophosphate, the C_{30} analog of geranyl pyrophosphate (six isopentyl C_5-units), in accordance with the isoprene rule.

Example. These substances are high melting solids, usually isolated from plant waxes and the heartwood of trees. The biosyntheses of these substances are characterized by the most extensive carbocation rearrangements reported. Such a rearrangement is that of β-amyrin to friedelin, Fig. 13.540. No less than seven synchronous 1,2-alkyl and 1,2-hydride shifts are involved in this transformation. All of these shifts take place stereospecifically *trans* along the triterpene backbone. Stereospecific migrations of this type are frequently encountered in terpene chemistry. It is an amazing fact that this transformation not only occurs in the biosynthesis of friedelin, but that it can also be realized in vitro as well by simply treating β-amyrin with acid.

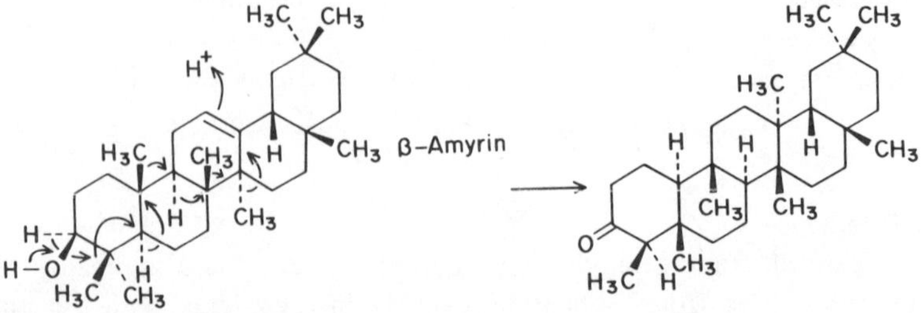

H3C CH3
H+
H3C CH3 H CH3 β-Amyrin
H CH3
H
H–O H
H3C CH3

H3C CH3
H3C
H H H CH3 Friedelin
CH3 CH3
O CH3
H3C H

Figure 13.540 An extensive triterpene carbocation rearrangement in which each 1,2-migration proceeds with *trans* stereochemistry.

13.550 Squalene

A symmetrical C_{30} hexaene formed through the head-to-head coupling of two farnesyl units. This substance is the key biosynthetic intermediate in the formation of the steroids (see Figs. 13.560 and 13.600).

13.560 Squalene Biosynthesis

The unusual head-to-head coupling of two farnesyl units occurs via the formation of the cyclopropyl intermediate **presqualene pyrophosphate**, Fig. 13.560. This novel intermediate then rearranges to squalene through a cyclopropylcarbinyl carbocation that is quenched in a highly stereospecific reduction step.

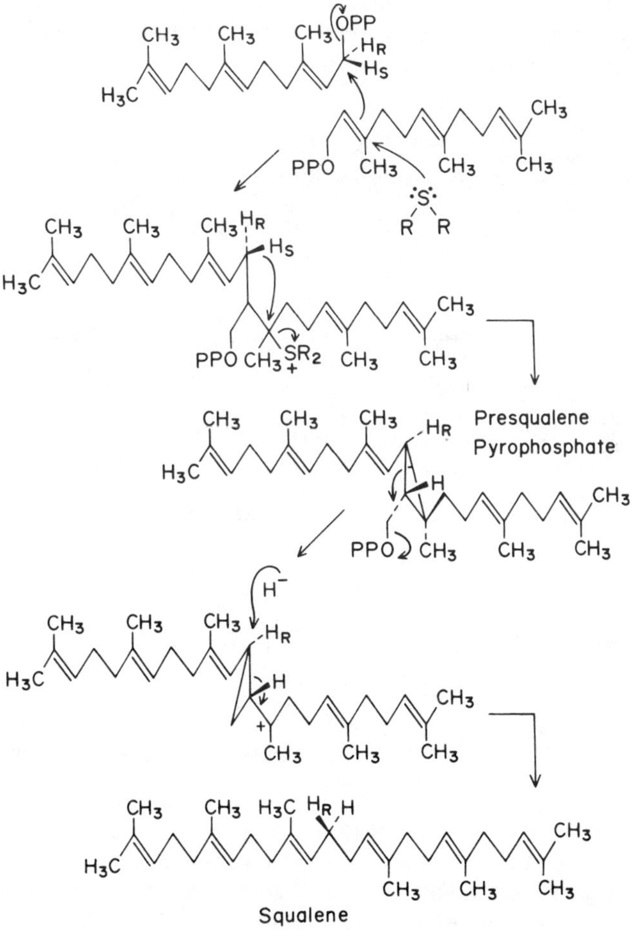

Figure 13.560 The biosynthetic mechanism for the formation of squalene via presqualene; the mechanism of head-to-head coupling of terpene units.

13.570 Carotenoids

A family of symmetrical C_{40} polyenes formed through the head-to-head coupling of two geranylgeranyl pyrophosphate units.

Example. β-Carotene is the best known member of the carotenoids. It is a ubiquitous natural product found in both animals and plants, and is the most important precursor of the A vitamins (see Fig. 13.580).

13.580 Retinol (Vitamin A_1)

A C_{20} polyene alcohol derived from β-carotene via a symmetrical oxidative cleavage at the central double bond, Fig. 13.580. The aldehyde derivative, **retinal**, plays a central role in the chemistry of vision. The 11-*cis* isomer of retinal combines with the protein **opsin** via Schiff base formation. The resulting pigment **rhodopsin** is the primary light sensor in the eye.

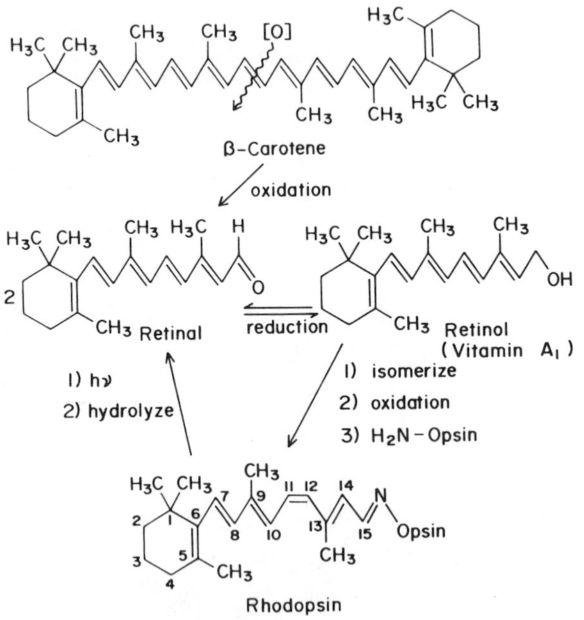

Figure 13.580 The biosynthesis of visual pigment from β-carotene.

13.590 Steroids (*Gk.* Stereos, Solid)

A family of C_{27}-C_{18} secondary alcohols involving a hydrogenated 1,2-cyclopentanophenanthrene carbon skeleton, Fig. 13.590. A C_8 terpenoid side chain is attached to the 17 position in the D-ring in animal steroids. In plant steroids **(phytosterols)** this same side chain has been modified by the addition of one or two extra carbon atoms at the 24 position. Both plant and animal steroids exist in which this side chain has been partially or completely degraded.

Examples. **Cholesterol**, Fig. 13.590*a*, is the most abundant steroid found in man; a 180 lb man contains about 240 g of cholesterol. The phytosterol, **β-sitosterol** (see Fig. 13.190), is one of the most common plant steroids. The adrenal cortical hormone, cortisol, Fig. 13.590*b*, is a steroid in which the side chain in the 17 position has been degraded to two carbon atoms (see also Sect. 13.620).

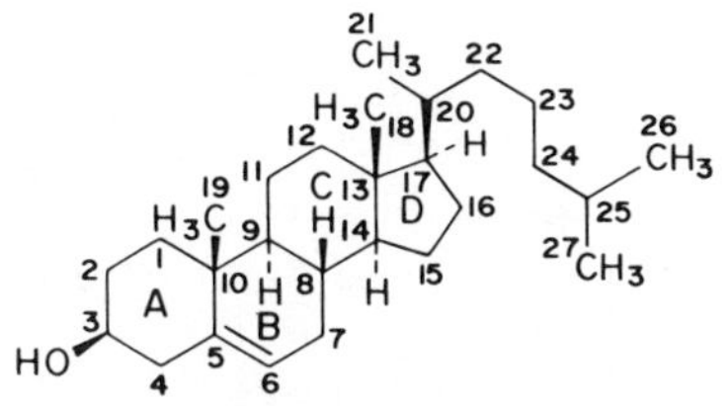

(β–group projecting up ; α–group projecting down)

(*a*)

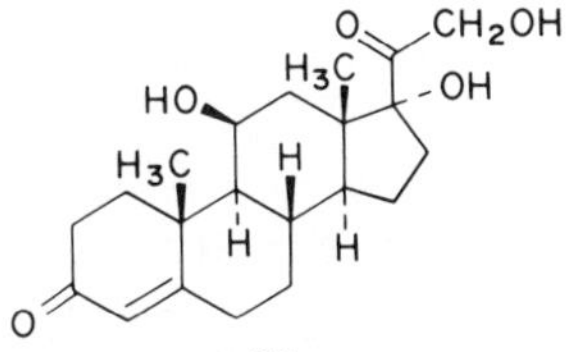

(*b*)

Figure 13.590 Steroids: (*a*) ring and position designations for a steroid with the C_8 side chain in the 17 position intact; (*b*) with this side chain degraded to two carbon atoms.

13.600 Steroid Biosynthesis

The entire steroid ring system is formed in a single spectacular series of carbocation cyclizations followed immediately by a series of 1,2-hydride and methyl shifts, Fig. 13.600. This process begins with 2,3-epoxysqualene and terminates with the primitive steroid, **lanosterol**; no compounds of intermediate structure have ever been detected. Lanosterol then is modified to the typical steroid carbon skeleton through the reduction of the double bonds in the 8 and 24 positions and the oxidative removal of the methyl groups in the 4 and 14 positions. Although steroids have traditionally been placed in a class of their own, the elucidation of their biosynthesis now clearly establishes them as triterpenoid compounds.

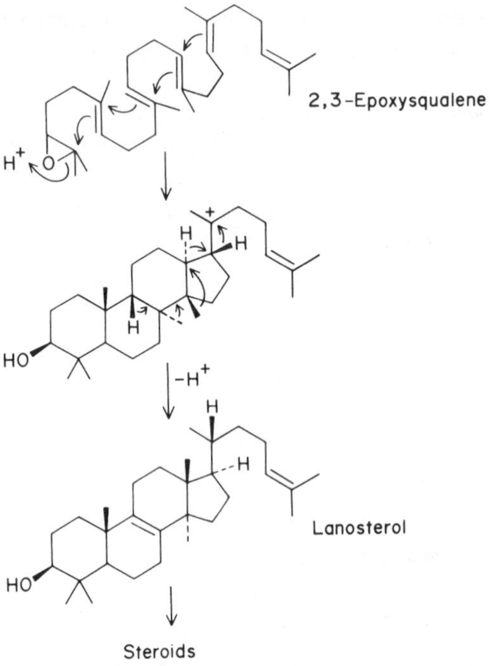

Figure 13.600 The biosynthetic mechanism for the formation of the steroid skeleton from 2,3-epoxysqualene.

13.610 Hormone (*Gr.* Hormôn, To Stimulate)

Any substance that is produced by specialized glands or tissues and is released into the blood to be later absorbed by the cells of the organism. Upon entering the cell, the hormone serves to regulate the cell chemistry.

Examples. There are several biochemical roles that hormones may play. They may alter the ease of passage of substances through cell membranes (see Sect. 13.240) or they may activate specific genes and trigger the synthesis of certain proteins. There are several classes of hormones, one of which, the steroid hormones, is described in Sect. 13.620.

13.620 Steroid Hormones

A diverse family of steroids that function as animal hormones.

Examples. **Adrenal cortical hormones** such as **cortisol** (Fig. 13.590*b*) regulate the mineral balance and sodium retention in the cells of higher animals. **Male sex hormones (androgenic hormones)** such as **testosterone**, Fig. 13.620*a*, regulate the development of secondary sex characteristics (beard growth, deep voice, etc.) and maturation of sperm cells. **Female sex hormones** such as **estrone**, Fig.

13.620*b*, regulate egg development and the menstrual cycle, and maintain pregnancy. Steroid hormones also play a strategic role in the regulation of the life processes of lower animals. For instance, insect **molting hormones** such as **ecdysone**, Fig. 13.620*c*, regulate the shedding of the exoskeleton (skin) and the metamorphosis of the larva into the adult (see also Sect. 13.500).

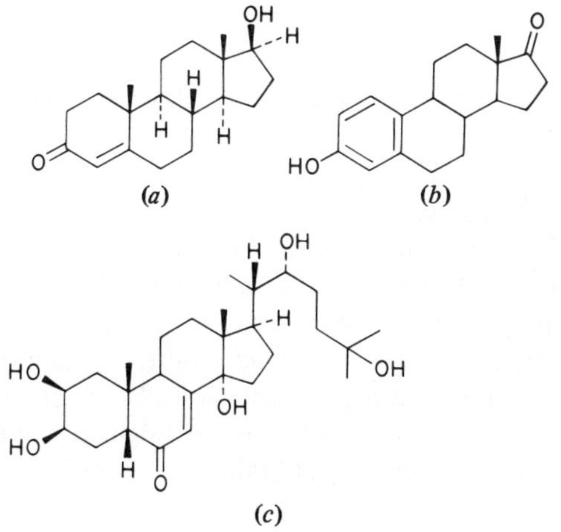

Figure 13.620 Steroid hormones: (*a*) testosterone; (*b*) estrone; (*c*) ecdysone.

13.630 Homo-

A prefix commonly used in chemical nomenclature to indicate the addition of a skeletal atom to a well-known structure.

Example. D-Homotestosterone, Fig. 13.630, has one more carbon atom inserted into the D-ring of testosterone, Fig. 13.620*a*.

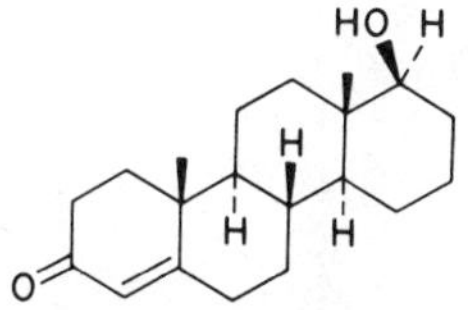

Figure 13.630 A D-homosteroid.

13.640 Nor-

A prefix commonly used in chemical nomenclature to indicate the removal of a skeletal atom from a well-known structure.

Example. 19-Nortestosterone, Fig. 13.640, does not have the methyl group that is designated C-19 in testosterone, Fig. 13.620*a*.

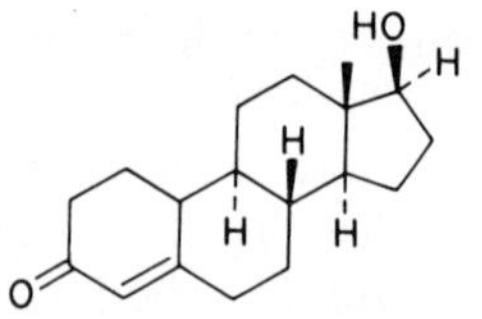

Figure 13.640 A 19-norsteroid.

13.650 Seco-

A prefix commonly used to indicate the ring cleavage product of a well-known structure (see Sect. 13.660).

13.660 Calciferol (Vitamin D_2)

One of a series of triene alcohols that are involved in calcium transport and bone development. Lack of this vitamin during early childhood, either through diet deficiencies or through inadequate exposure to sunlight, leads to the syndrome rickets, which is characterized by bone deformities. Vitamin D_2 is derived from the plant sterol **ergosterol** via a photochemical cleavage of the B-ring to form 9,10-secoerogosterol, followed by a thermal 1,7-sigmatropic shift, Fig. 13.660. In higher animals these processes occur in the skin or feathers.

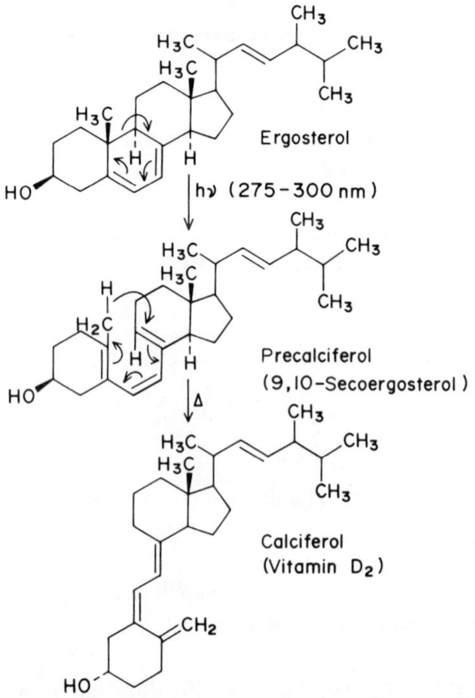

Figure 13.660 The biosynthetic mechanism for the formation of vitamin D_2.

13.670 α-Tocopherol (Vitamin E) (*Gk.* Tokos, Childbirth; Pherein, To Bear)

A cyclic ether derived from a hydroquinone containing a saturated side chain of terpene origin, Fig. 13.670. Its exact biochemical function is uncertain, but it is necessary for the maintainance of muscle tissue. Muscular dystrophy may be associated with a deficiency in this vitamin.

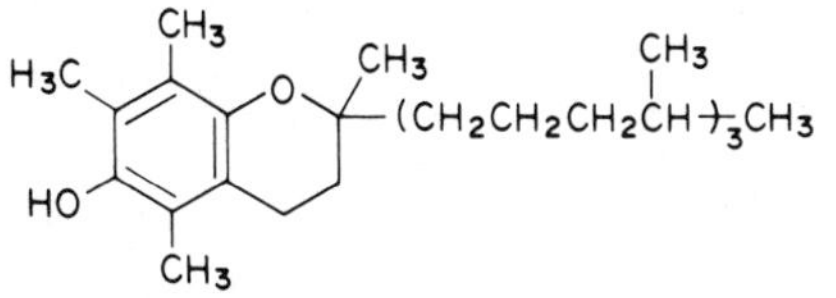

H_3C, CH_3, CH_3, CH_3, $(CH_2CH_2CH_2CH)_3CH_3$, HO, CH_3

Figure 13.670 Vitamin E.

13.680 Plastoquinones (PQ-*n*)

A family of dimethylbenzoquinones that differ only in the number (*n*) of isoprene units in their terpene side chains, Fig. 13.680. These quinones are found mainly in the chloroplasts of plants, where they are required in the oxidation-reduction processes **(electron transport)** associated with photosynthesis.

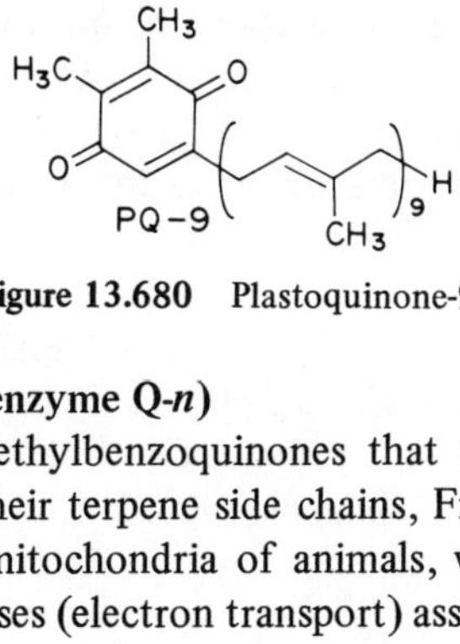

CH_3, H_3C, O, O, H, 9, PQ-9, CH_3

Figure 13.680 Plastoquinone-9.

13.690 Ubiquinones (Coenzyme Q-*n*)

A family of dimethoxymethylbenzoquinones that differ only in the number (*n*) of isoprene units in their terpene side chains, Fig. 13.690. These quinones are found mainly in the mitochondria of animals, where they function in the oxidation-reduction processes (electron transport) associated with respiration. In photosynthesizing bacteria they replace the plastoquinones of higher plants (Sect. 13.680).

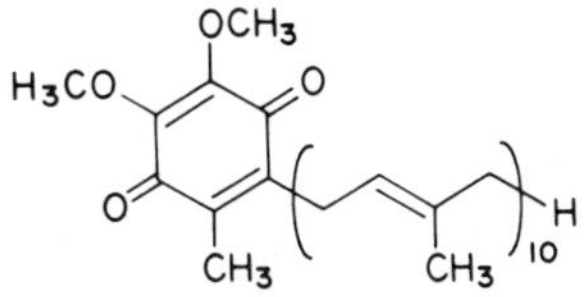

OCH_3, H_3CO, O, O, H, 10, CH_3, CH_3, Q-10

Figure 13.690 Ubiquinone-10.

13.700 Menaquinones (Vitamin K_2) (MK-*n*)

A family of methylnaphthoquinones that differ only in the number (*n*) of isoprene units in their terpene side chains, Fig. 13.700. The menaquinones are derived from **phylloquinone**, a plant quinone, through side chain replacement, Fig. 13.700. In higher animals menaquinones function to maintain blood clotting factors. Phylloquinones possibly play an important role in both **photo-** and **oxidative phosphorylation**, the former in the synthesis of ATP (Sect. 13.090) in photosynthesis and the latter in respiration.

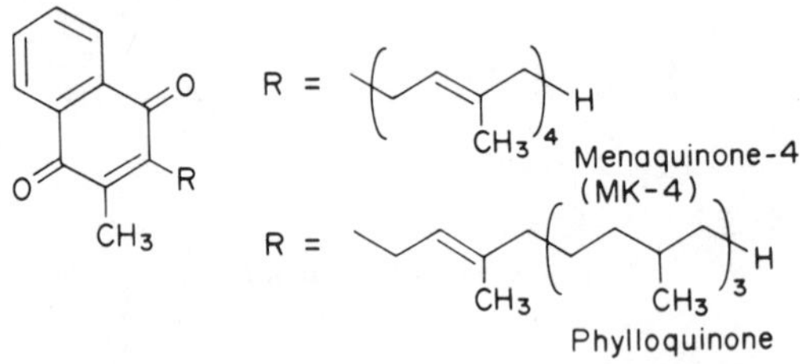

Figure 13.700 The characteristic side chains of menaquinones and phylloquinones.

13.710 Alkaloids (*Gr.* eidos, Similar, Thus Alkali Similar)

An immense group of biogenetically unrelated families of organic, nitrogenous bases. Most alkaloids are derived from amino acids, although a substantial number are formed through the incorporation of a nitrogen atom into a carbon skeleton of polyketide, shikimate, or terpenoid origin. Alkaloids are found mainly in plants, but a few have been isolated from animals and microorganisms. Because of the extensive variety of alkaloid structure types, only a few of the major classes are defined in the following sections.

13.720 Catecholamines

A family of simple alkaloids with a β-phenylethylamine structure. This family is derived from phenylalanine or tyrosine and includes such important hormones as **epinephrine (adrenalin)** and **dopamine**, Fig. 13.720.

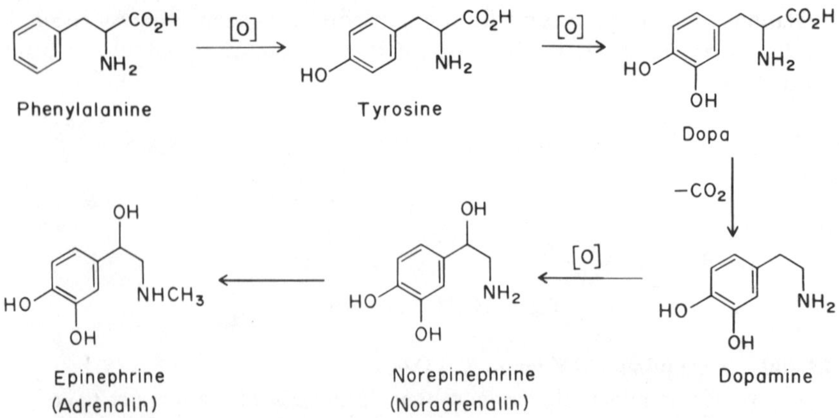

Figure 13.720 The biosynthesis of the major catecholamines from phenylalanine.

13.730 Pyridine Alkaloids

A biogenetically diverse class of alkaloids containing a pyridine ring. The best known member of this family is **nicotine**, Fig. 13.730, which is produced in the growing root tips of the tobacco plant.

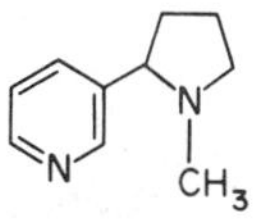

N
N
CH_3

Figure 13.730 Nicotine.

13.740 Piperidine Alkaloids

A biogenetically diverse class of alkaloids containing a piperidine ring.

Examples. The well-known alkaloid from the hemlock, **coniine**, Fig. 13.740*a*, with which Socrates was put to death, is a member of this class. Its carbon skeleton is derived exclusively from acetate units. The superficially similar alkaloid **pelletierine** arises through a more conventional biosynthetic pathway from the amino acid **lysine**, Fig. 13.740*b*. The **lupin alkaloids** are elaborated in a similar fashion through a series of Mannich reactions from more than one lysine unit, Fig. 13.740*c*.

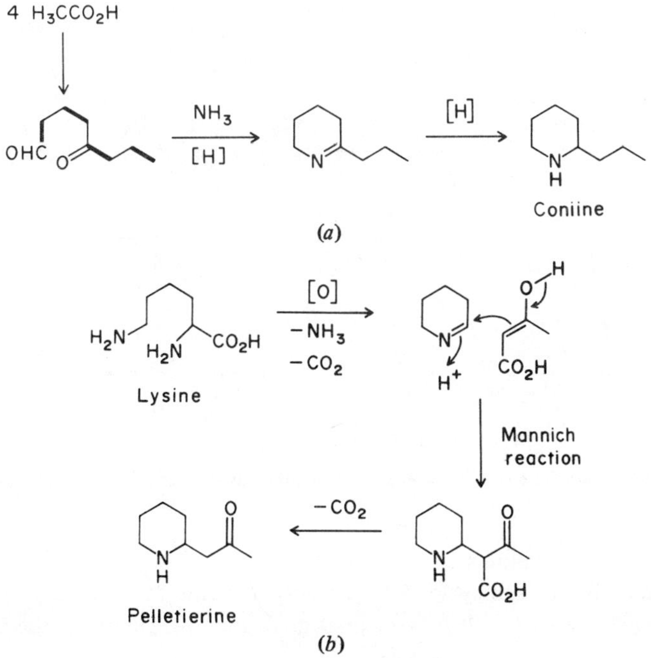

4 H_3CCO_2H
OHC O
NH_3
[H]
N
[H]
N
H
Coniine

(*a*)

H_2N H_2N CO_2H
Lysine
[O]
$-NH_3$
$-CO_2$
N
H^+
O H
CO_2H
Mannich reaction
N H O CO_2H
$-CO_2$
N H O
Pelletierine

(*b*)

Figure 13.740 The biosynthesis of piperidine alkaloids: (*a*) from a polyketide; (*b*) from lysine and a polyketide.

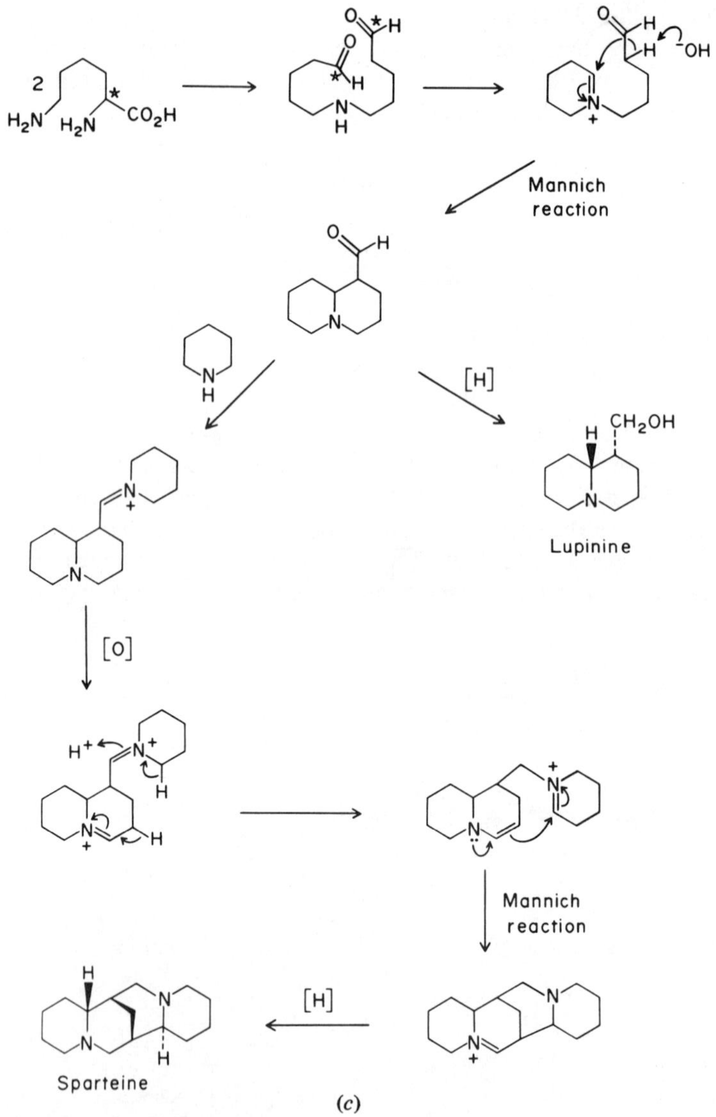

(c)

Figure 13.740 (*continued*) The biosynthesis of piperidine alkaloids: (*c*) from lysine units alone.

13.750 Tropane Alkaloids

A large family of 8-azabicyclo[3.2.1]octane alkaloids derived from the amino acid **ornithine**, Fig. 13.750, in a manner analogous to the biosynthesis of the piperidine alkaloids from lysine (Sect. 13.740). The best known member of this family is the alkaloid **cocaine**.

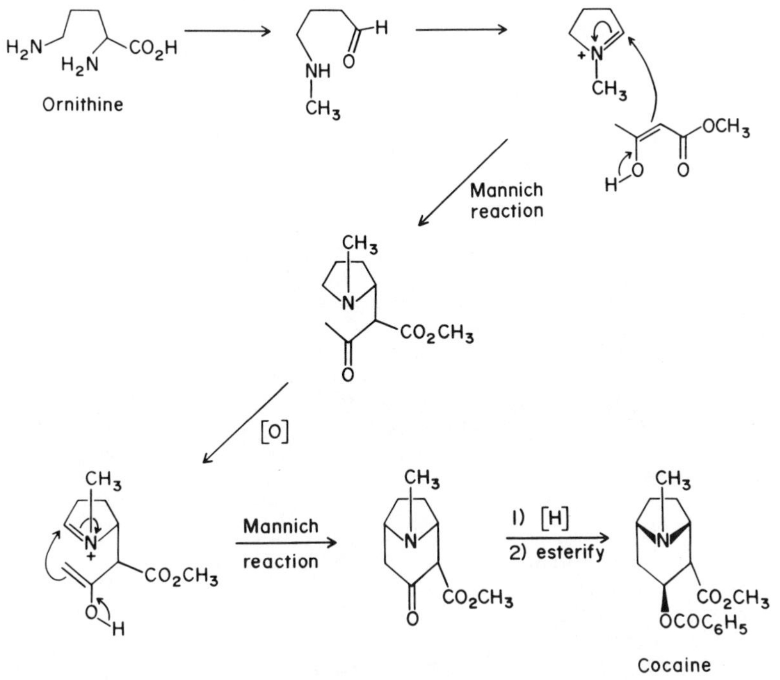

Figure 13.750 The biosynthesis of the tropane alkaloids from ornithine and a polyketide (acetoacetic ester).

13.760 Amaryllidaceae (or Daffodil) Alkaloids

A family of alkaloids derived from phenylalanine and tyrosine via phenolic coupling (Sect. 13.340) of the C_6-C_1-N-C_2-C_6 precursor **norbelladine**, Fig. 13.760.

Example. This family provides one of the best examples of the power of the phenolic coupling concept in the correlation of the structures of natural products. The three subfamilies of alkaloids in this class become differentiated at the phenolic coupling step by bond formation (dashed arrows in Fig. 13.760) between different positions in the phenolic rings of norbelladine (see also Sect. 13.770).

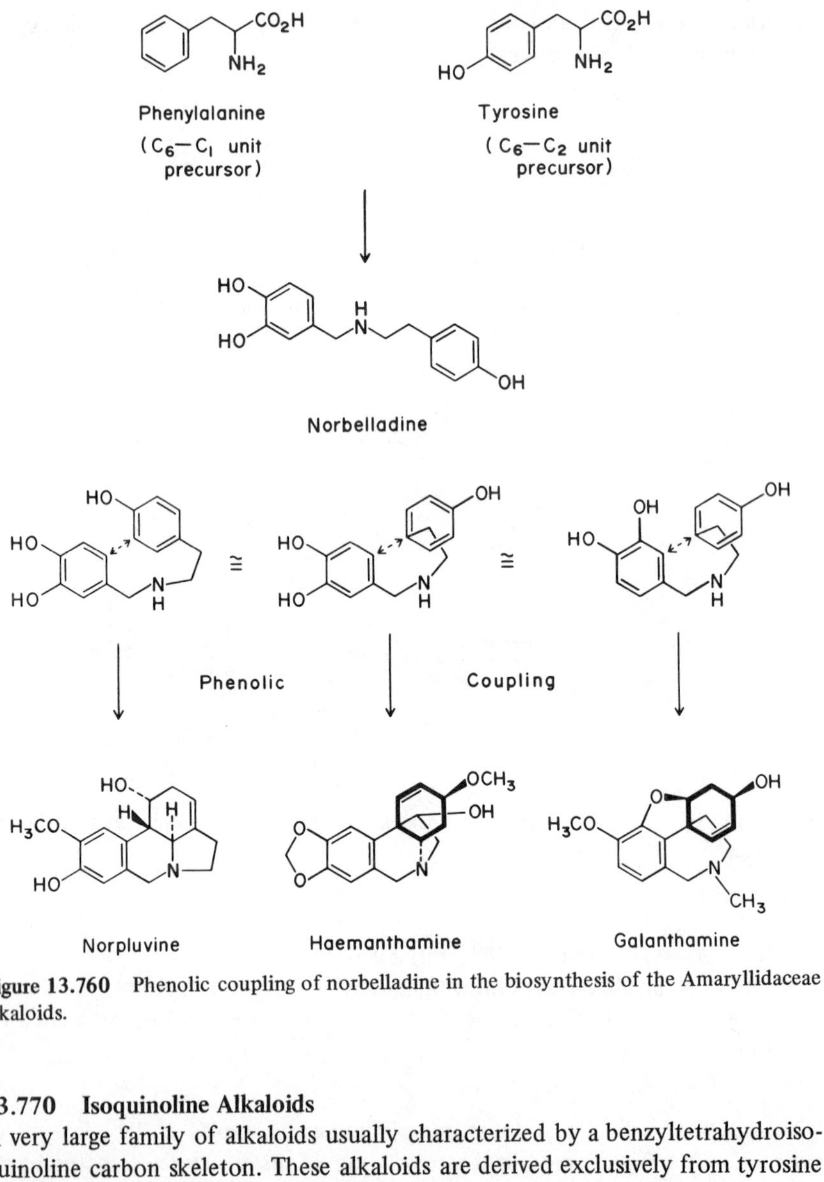

Phenylalanine (C_6—C_1 unit precursor)

Tyrosine (C_6—C_2 unit precursor)

Norbelladine

Phenolic Coupling

Norpluvine Haemanthamine Galanthamine

Figure 13.760 Phenolic coupling of norbelladine in the biosynthesis of the Amaryllidaceae alkaloids.

13.770 Isoquinoline Alkaloids

A very large family of alkaloids usually characterized by a benzyltetrahydroisoquinoline carbon skeleton. These alkaloids are derived exclusively from tyrosine via phenolic coupling (Sect. 13.340) of the C_6-C_2-N-C_2-C_6 precursor **norlaudanosoline**, Fig. 13.770.

Examples. The well-known alkaloids of the opium poppy **(thebaine, codeine,** and **morphine)** belong to this family, Fig. 13.770.

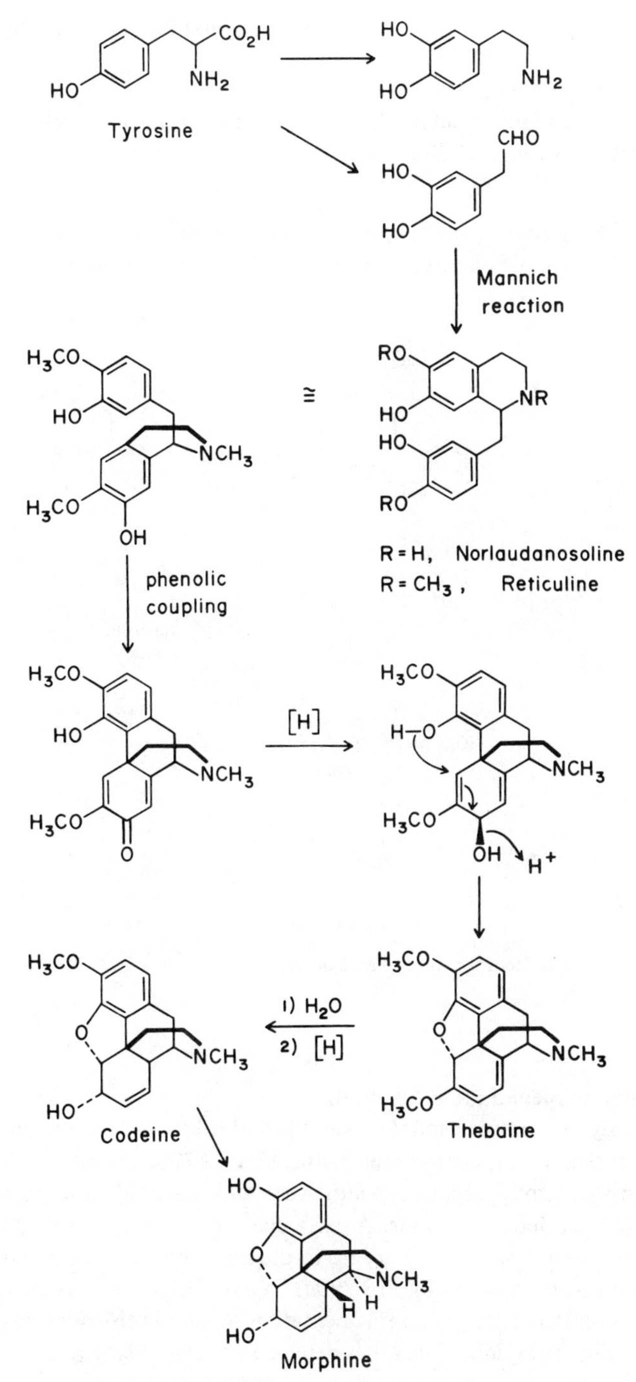

Figure 13.770 The biosynthetic mechanism for the formation of morphine.

13.780 Ergot Alkaloids

A small family of indole alkaloids derived from the amino acid tryptophan and a single isoprene unit, Fig. 13.780.

Example. The potent hallucinogenic agent **lysergic acid** is the best known member of this alkaloid class which is produced by ergot fungus.

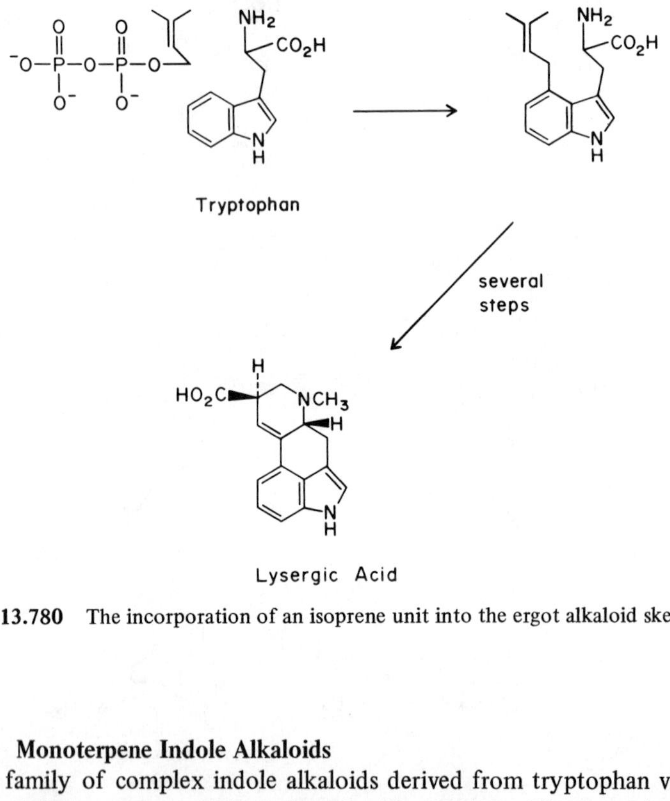

Tryptophan

several steps

Lysergic Acid

Figure 13.780 The incorporation of an isoprene unit into the ergot alkaloid skeleton.

13.790 Monoterpene Indole Alkaloids

A large family of complex indole alkaloids derived from tryptophan via tryptamine and the monoterpenoid **secologanin**, Fig. 13.790*a*. There are three major branches to this family: the **Corynanthe-strychnos alkaloids**, the **Aspidosperma alkaloids**, and the **Iboga alkaloids.** A particular indole alkaloid is assigned to one of the three family branches according to the arrangement of its nontryptamine carbon skeleton, as shown in Fig. 13.790*a*. Two of three branches are thought to become differentiated through a fragmentation to, and Diels-Alder reaction of, a vinylindole dihydropyridine known as **dehydrosecodine**, Fig. 13.790*a*.

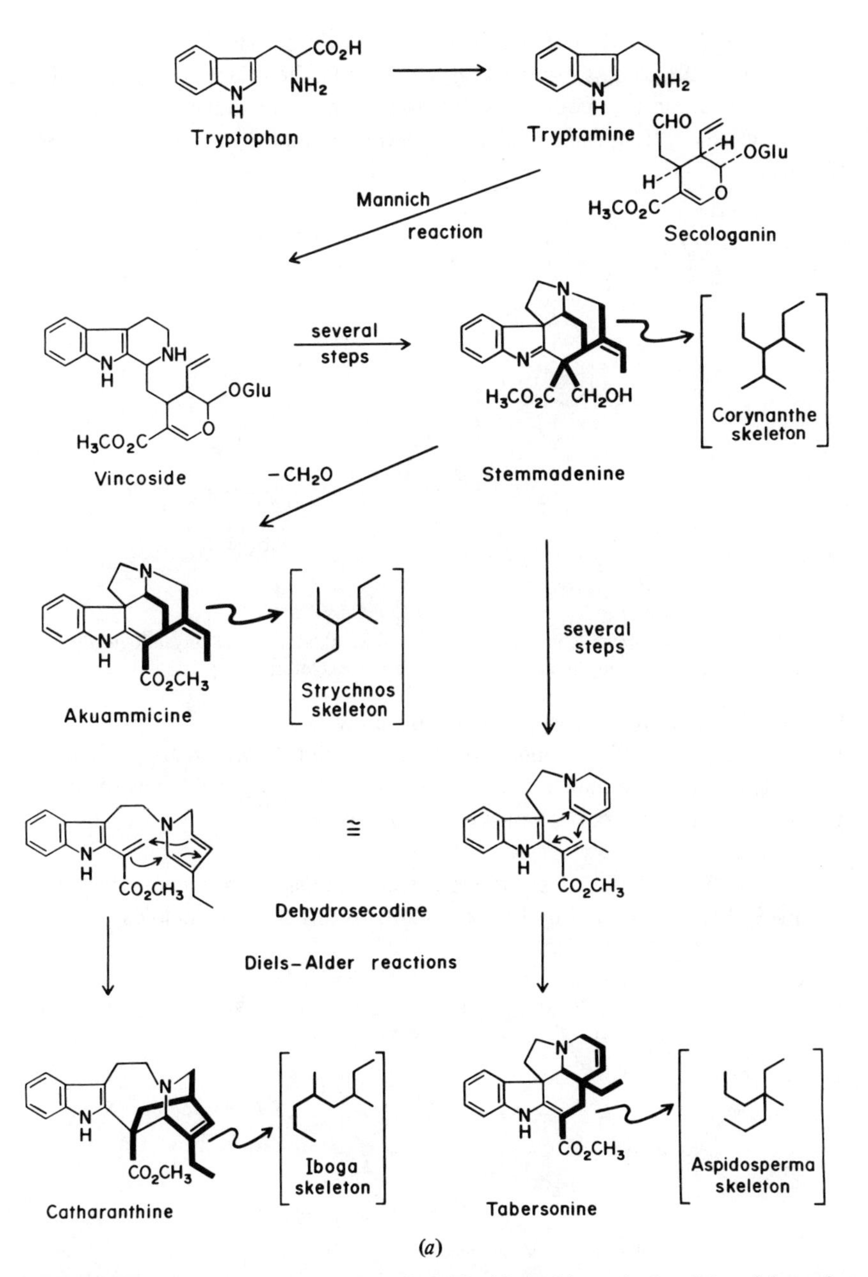

Figure 13.790 The monoterpene indole alkaloids: (*a*) the biosynthetic origins of the major families.

Examples. This family contains alkaloids that exhibit extreme physiological activity such as the toxic substances **strychnine** and the calabash-curare alkaloids used as dart and arrow poisons by the South American Indians. The first tranquilizer was the East Indian alkaloid **reserpine,** Fig. 13.790*b* (see also Sect. 13.800).

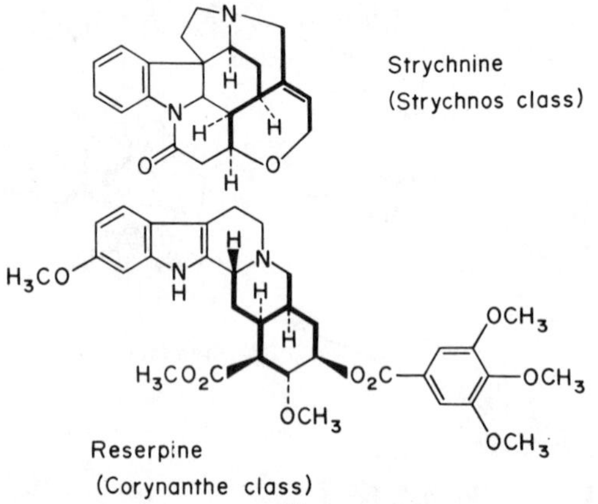

Strychnine
(Strychnos class)

Reserpine
(Corynanthe class)

(*b*)

Figure 13.790 (*continued*) The monoterpene indole alkaloids: (*b*) physiologically active examples, with emphasis on the nontryptamine carbon skeleton.

13.800 Dimeric Monoterpene Indole Alkaloids

These alkaloids are of extremely complex structure. However, they often arise via straightforward Mannich reactions between the terpenoid portion of one monomer and the aryl portion of the other monomer.

Examples. Two of the most effective antileukemia drugs known, vincaleukoblastine and leurocristine, Fig. 13.800, belong to this class of alkaloids.

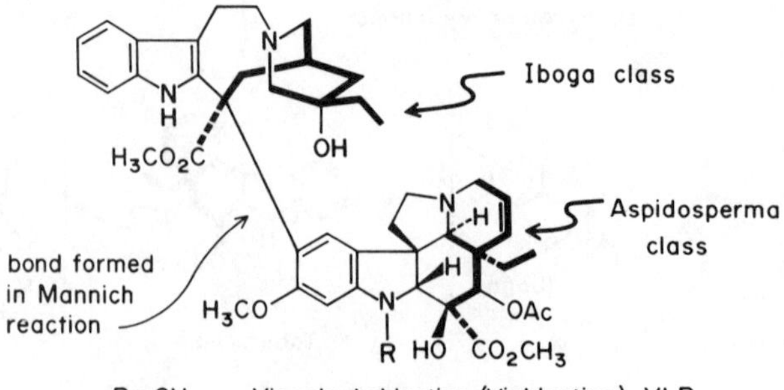

Iboga class

Aspidosperma class

bond formed in Mannich reaction

R = CH_3 Vincaleukoblastine (Vinblastine) VLB

R = CHO Leurocristine (Vincristine) VCR

Figure 13.800 Typical dimeric monoterpene indole alkaloids, with emphasis on the nontryptamine carbon skeletons.

13.810 Quinoline Alkaloids of Indole Alkaloid Origin

A group of alkaloids that have in common a quinoline ring substituted with a terpenoid moiety of the same skeletal arrangement found in the indole alkaloids (Sect. 13.790).

Examples. The quinoline rings in these alkaloids are formed by oxidation and rearrangement of an indole moiety in an indole alkaloid precursor, Fig. 13.810. The two prominent alkaloids of this type are **quinine**, which was for many years the only effective drug for treatment of malaria, and **camptothecin**, which has shown anticancer activity.

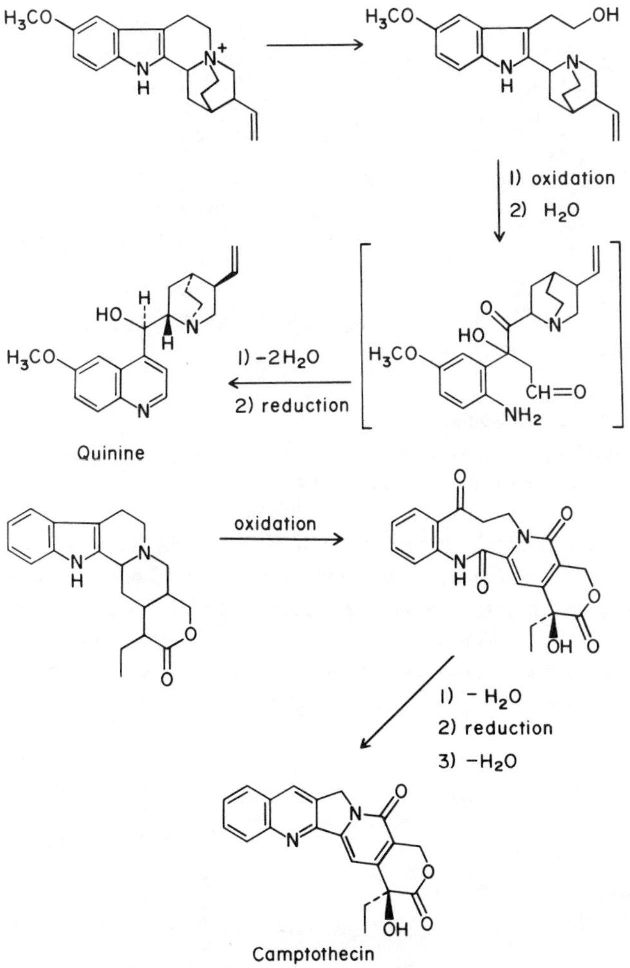

H_3CO N   H N^+ → H_3CO OH   N   H   N

1) oxidation
2) H_2O

HO   H   N   H H_3CO N

Quinine

1) $-2H_2O$
2) reduction

H_3CO O   N   HO   CH=O NH_2

N   H   N   O   O

oxidation

O   N   O   N   H   O   O   OH   O

1) $-H_2O$
2) reduction
3) $-H_2O$

N   O   N   O   OH   O

Camptothecin

Figure 13.810 Quinoline alkaloids of indole alkaloid origin.

13.820 Terpene Alkaloids

A diverse class of alkaloids that are derived from the incorporation of ammonia or a simple amine into a preformed terpene skeleton.

Examples. Ethanolamine has apparently been incorporated into a diterpene skeleton to form the alkaloid **veatchine**, Fig. 13.820*a*. This same diterpene skeleton has undergone rearrangement and extensive hydroxylation, but is still discernable in the highly toxic Indian arrow poison **aconitine** (2–5 mg is fatal to man), Fig. 13.820*b*.

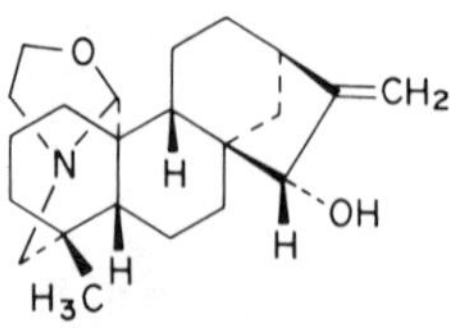

Veatchine

(*a*)

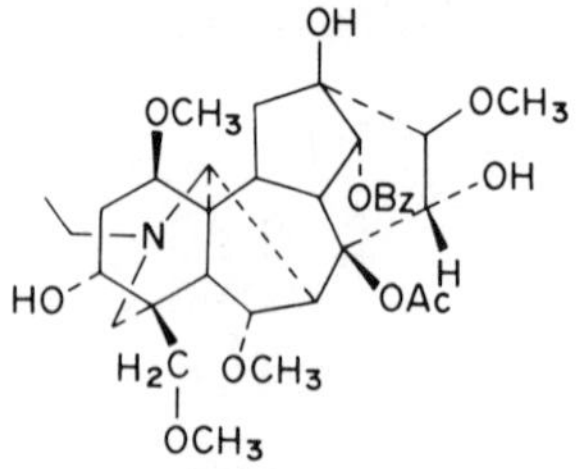

Aconitine
(the most toxic substance of plant origin)

(*b*)

Figure 13.820 Terpene alkaloids: (*a*) veatchine; (*b*) aconitine, the most toxic known non-peptide substance of plant origin.

13.830 Steroid Alkaloids

A class of alkaloids that are derived from the incorporation of ammonia or a simple amine into a preformed steroid skeleton.

Examples. These alkaloids are found in both plants and animals. The steroid skeleton is clearly evident in both **samandarine**, Fig. 13.830*a*, isolated from salamander skin, and **batrachotoxin**, Fig. 13.830*b*, isolated from the skin of South American tree frogs. The latter substance is one of the most toxic nonprotein toxins known (2 μg is fatal to a mouse) and is used by the Colombian Indians as an arrow poison.

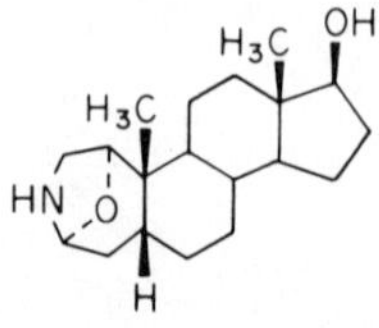

Samandarine

(*a*)

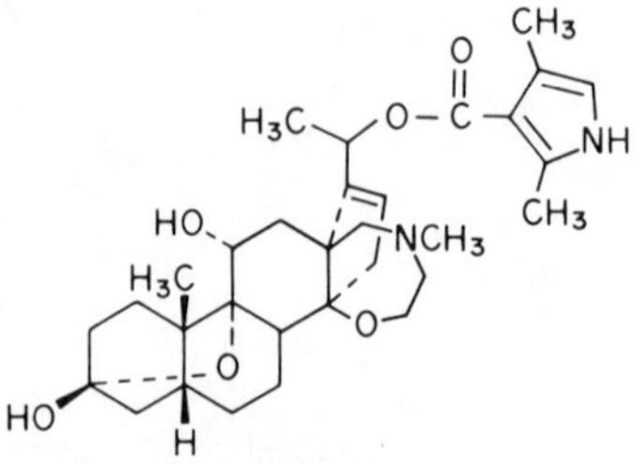

Batrachotoxin

(*b*)

Figure 13.830 Steroidal alkaloids: (*a*) samandarine; (*b*) batrachotoxin.

13.840 Penicillins

A family of important antibiotics containing a thiazolidine ring fused to a labile β-lactam, Fig. 13.840. The various penicillins all have the same ring system and differ only in the nature of the acyl side chain. The details of the biosynthetic formation of the ring system are not known. However, a tripeptide serves as the primary precursor. Thus penicillin-N arises from L-α-aminoadipic acid, L-cysteine, and L-valine, as shown in Fig. 13.840.

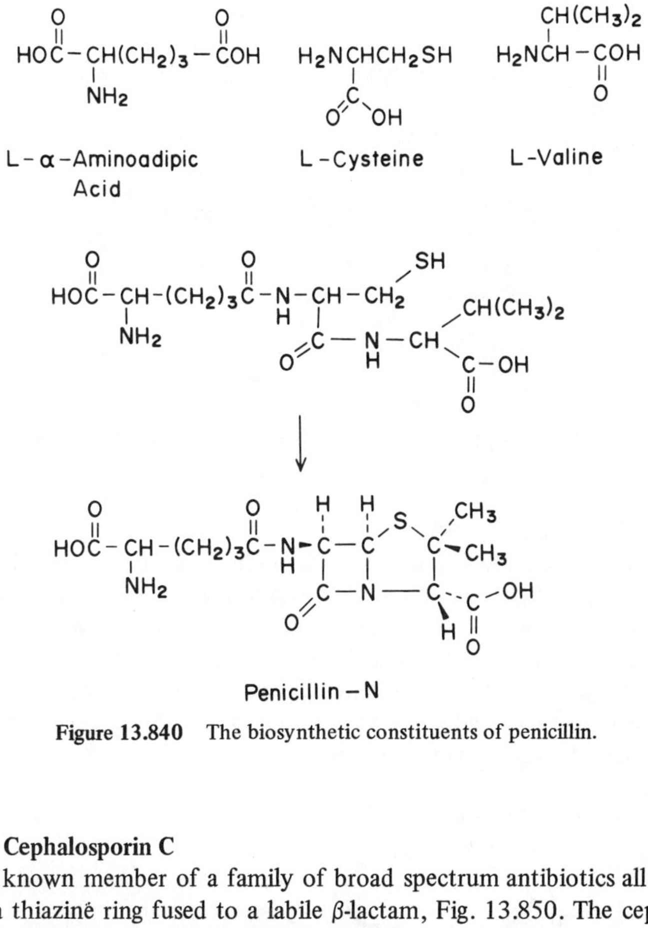

L-α-Aminoadipic Acid   L-Cysteine   L-Valine

Penicillin-N

Figure 13.840 The biosynthetic constituents of penicillin.

13.850 Cephalosporin C

The best known member of a family of broad spectrum antibiotics all of which contain a thiazinė ring fused to a labile β-lactam, Fig. 13.850. The cephalosporins differ only in the structure of the acyl side chain. These substances probably are derived from the corresponding penicillins through the rearrangement of the penicillin sulfoxide shown in Fig. 13.850.

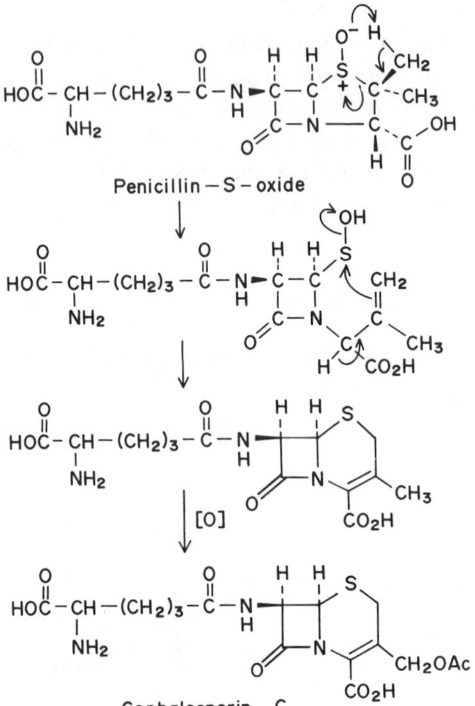

Penicillin – S – oxide

Cephalosporin C

Figure 13.850 The biosynthetic conversion of a penicillin into the corresponding cephalosporin skeleton.

13.860 Porphyrins

A family of biologically important metal binding macrocycles, Fig. 13.860. The porphyrin macrocycle contains the 17-membered aromatic array of 18 π electrons outlined in Fig. 13.860 plus three additional olefinic double bond shunts. These substances or closely related analogs play a central role in respiration (heme, Sect. 13.880, and cytochrome C, Sect. 13.880), photosynthesis (chlorophylls, Sect. 13.910), and crucial biosynthetic transformations such as methyl transfer (Sect. 13.110) (vitamin B_{12}, Sect. 13.950).

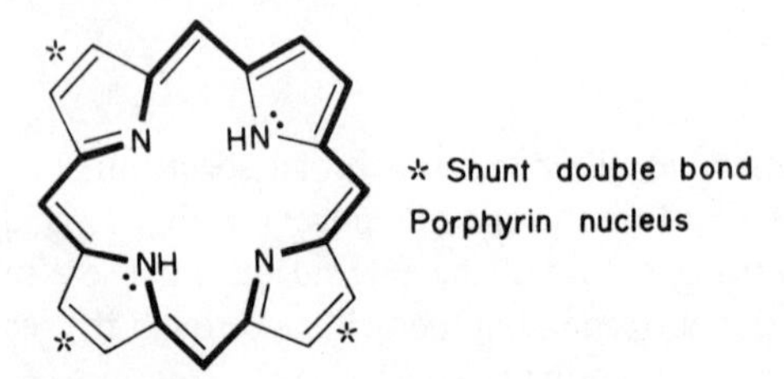

Figure 13.860 The aromatic 18 π electron system incorporated into the porphyrin nucleus.

13.870 Porphyrin Biosynthesis

The formation of the porphyrin nucleus stems from the condensation of two molecules of δ-**aminolevulinic acid** to yield the substituted pyrrole **porphobilinogen**, Fig. 13.870*a*. This pyrrole then undergoes an unusual head-to-head coupling followed by migration of the aminomethyl "head" group to afford a rearranged dipyrrylmethane, Fig. 13.870*b*. The incorporation of two more prophobilinogen units and cyclization via conventional Mannich reactions lead to the porphyrin skeleton. The substituents and oxidation state of the nucleus then are modified in the following sequence. **Uroporphyrinogen III** undergoes decarboxylation of the acetic acid side chains to produce the methyl groups of **coproporphyrinogen III**, Fig. 13.870*c*. This substance in turn undergoes oxidative decarboxylation of the two nonadjacent propionic acid side chains to produce the vinyl groups of **protoporphyrinogen IX** which is finally oxidized to **protoporphyrin IX**. This compound is the parent porphyrin from which other porphyrins are derived either through metal atom incorporation or side chain modification.

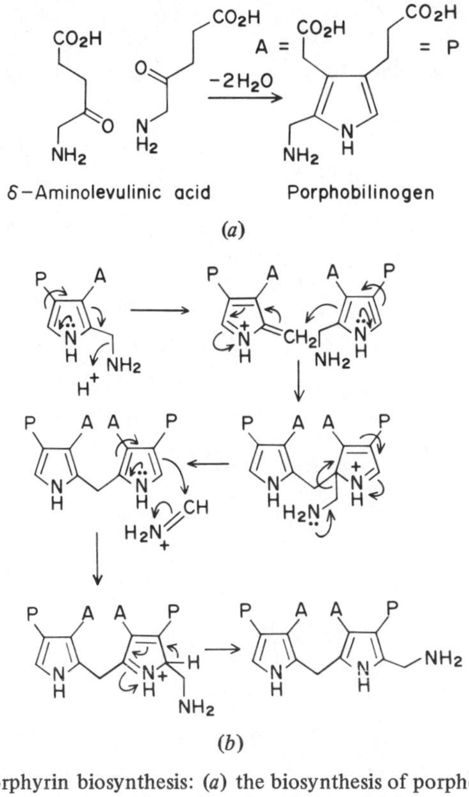

CO_2H CO_2H A = CO_2H CO_2H = P

$-2H_2O$

δ-Aminolevulinic acid Porphobilinogen

(*a*)

(*b*)

Figure 13.870 Porphyrin biosynthesis: (*a*) the biosynthesis of porphobilinogen; (*b*) the rearrangement involved in the formation of the first dipyrrylmethane intermediate.

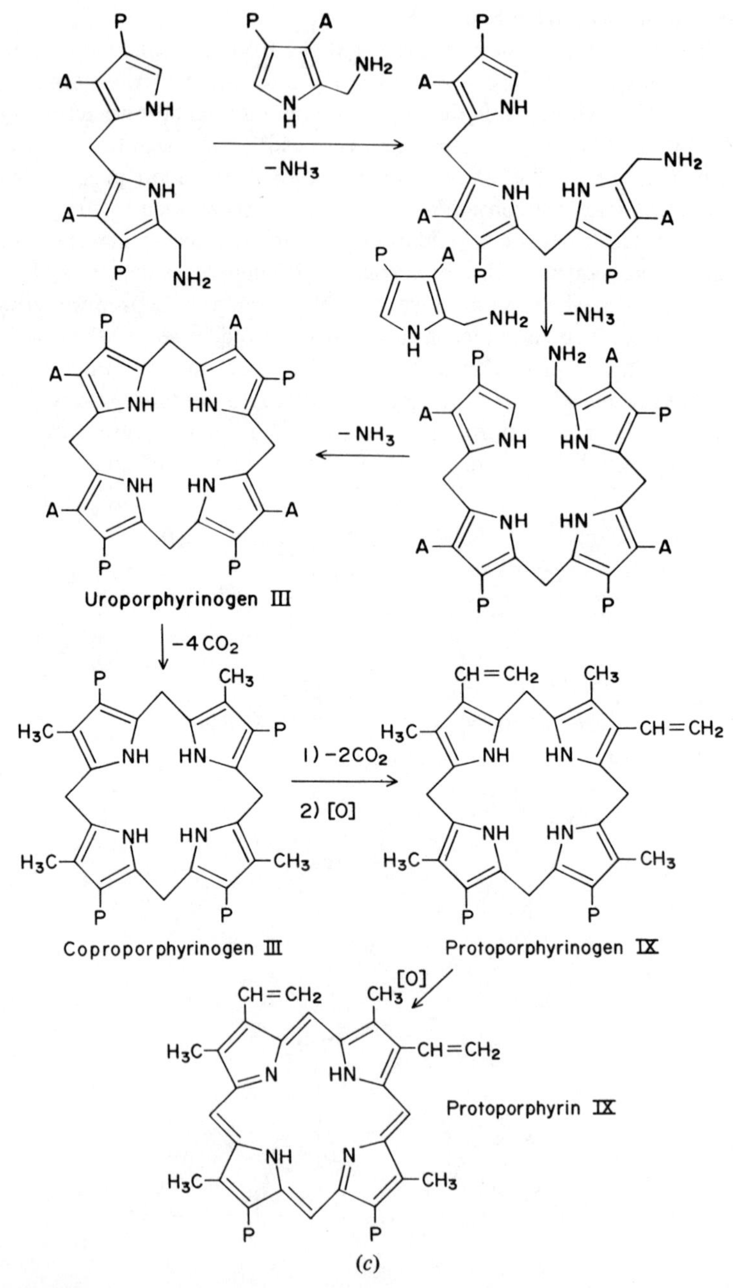

(c)

Figure 13.870 *(continued)* Porphyrin biosynthesis: (*c*) the cyclization and cyclized intermediates involved in porphyrin biosynthesis.

13.880 Heme

The ferrous iron containing protoporphyrin IX (Sect. 13.870), Fig. 13.880 (see Sect. 12.140).

Example. This porphyrin is an important cofactor for a large number of oxygen-transport proteins and oxidation enzymes. In the oxygen-transport agent myoglobin it is loosely bonded to the protein through hydrogen bonds involving the propionic acid side chains. In the well-known electron-transport enzyme **cytochrome C**, the heme is covalently bonded to the protein through the addition of protein thiol groups to the vinyl side chains of the heme, Fig. 13.880.

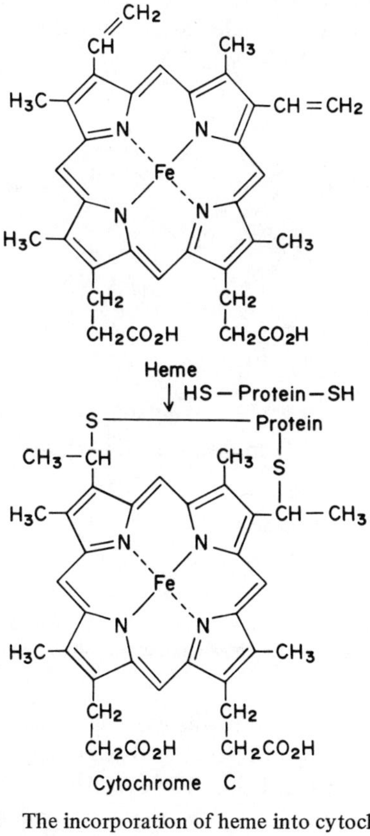

Figure 13.880 The incorporation of heme into cytochrome C.

13.890 Oxophlorins

Porphyrin derivatives in which one of the **meso carbon atoms** has been oxidized to a carbonyl group (see Fig. 13.910*a* for meso carbon atom assignments, α, β, γ, δ).

Example. These substances are thought to be intermediates in the oxidative degradation of the porphyrins to the bile pigments (see Sect. 13.960).

13.900 Chlorins

Porphyrin derivatives in which one of the shunt double bonds has been reduced.

Example. The best known case of this type of modified porphyrin nucleus is found in the chlorophylls (see Sect. 13.910).

13.910 Chlorophylls

A family of magnesium containing chlorins in which one of the propionic acid side chains has undergone a condensation with the chlorin nucleus to form a cyclopentanone ring, Figs. 13.910*a* and *b*.

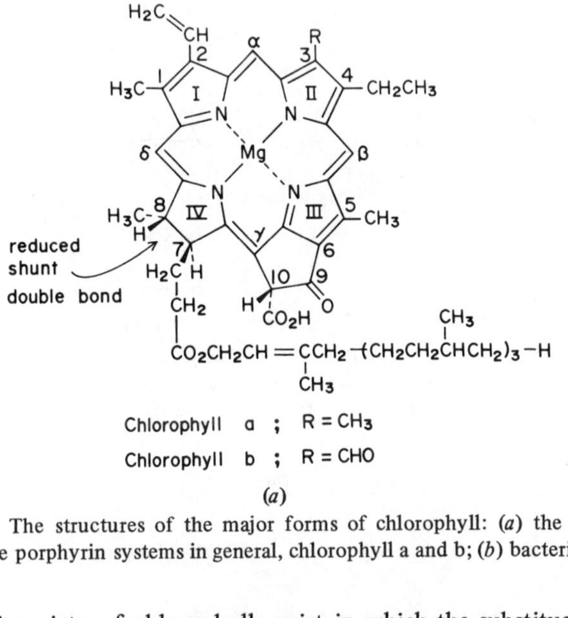

Chlorophyll a ; $R = CH_3$

Chlorophyll b ; $R = CHO$

(*a*)

Figure 13.910 The structures of the major forms of chlorophyll: (*a*) the nomenclature used to describe porphyrin systems in general, chlorophyll a and b; (*b*) bacterio-chlorophyll a.

Examples. A variety of chlorophylls exist in which the substituents at the 2 and 3 positions differ, Fig. 13.910*a*. In **bacteriochlorophylls**, a second shunt double bond has been reduced, Fig. 13.910*b*. Upon loss of the magnesium ion, these chlorophylls are transformed to the corresponding **pheophytins.** Chlorophylls are found in the chloroplasts of plants, where they serve as the primary chromophore for the gathering of light energy to be used in photosynthesis. Chlorophyll molecules play two roles in this process. Most of the chlorophyll molecules serve simply to gather the light energy and are called **antenna chlorophyll.** This energy is transferred to a few specialized chlorophyll molecules which in higher plants are called **P-700** (pigment absorbing light at 700 nm).

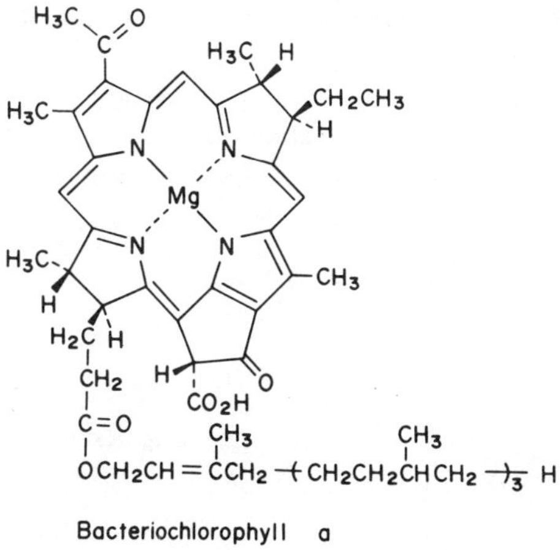

Bacteriochlorophyll a

(*b*)

P-700 is thought to be a chlorophyll dimer formed by the association of two chlorophyll molecules with a water molecule, and is the species actively involved in the oxidation of water to oxygen (see Sect. 13.920).

13.920 Photosynthesis

The highly complex process by which light energy is used to oxidize H_2X (H_2O in plants, H_2S, or other oxidizable substrates in bacteria) to X_n (O_2 or S_8, respectively) and reduce carbon dioxide to carbohydrates. The photosynthetic process produces NADH (bacteria) or NADPH (plants) and ATP as the primary products. These substances in turn function as reagents for the fixation of carbon dioxide.

13.930 Corroles

A modified porphyrin nucleus in which one of the meso carbon atoms has been extruded, Fig. 13.930.

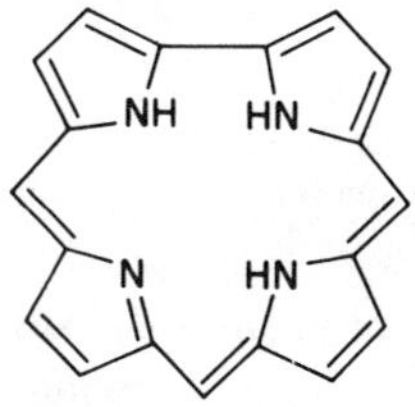

Figure 13.930 The corrole nucleus.

13.940 Corrins

A highly reduced corrole that constitutes the basic ring system of vitamin B_{12} (Sect. 13.950).

13.950 Vitamin B_{12} (Cyanocobalamin or α-(5,6-Dimethylbenzimidazolyl) Cyanocobamide)

A cobalt containing corrin in which the cobalt(III) atom is coordinated to a 5,6-dimethylbenzimidazole and a cyanide ion, Fig. 13.950*a*. In the coenzyme form the cyanide ion is replaced by a 5′-deoxyadenosine moiety, Fig. 13.950*a*.

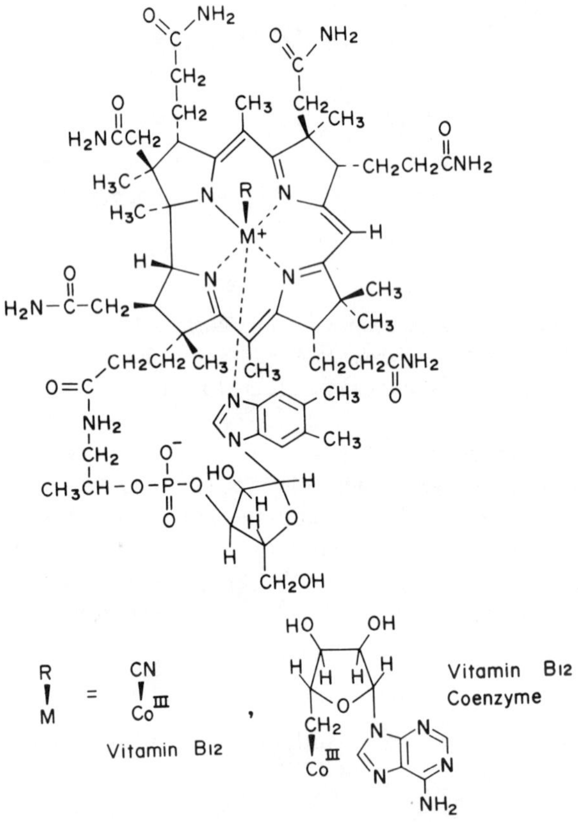

(*a*)

Figure 13.950 Vitamin B_{12}: (*a*) its structure; (*b*) important rearrangements that it catalyzes; (*c*) a possible mechanistic outline for the methylmalonate-succinate interconversion.

Example. The richest sources of vitamin B_{12} are treated sewage sludge, manure, and dried river mud. The reduced form of vitamin B_{12} (B_{12s}) containing cobalt(I) displays extraordinary nucleophilicity and is involved in the formation of the coenzyme from vitamin B_{12}. This coenzyme is the only known compound found in nature to date which contains a covalent carbon-metal bond, and it is required for a number of apparently unrelated biological conversions. Among the most interesting of these are the carbon skeleton rearrangements shown in Fig. 13.950*b*. These transformations apparently all involve the initial formation of a carbon-cobalt bond in the substrate followed by the transposition of an alkyl group and the cobalt atom, Fig. 13.950*c*. The biosynthesis of the corrin nucleus is not yet fully understood. It apparently diverges from the porphyrin biosynthetic pathway at the uroporphyrinogen III stage, Fig. 13.870*c*.

(*b*)

(*c*)

13.960 Bile Pigments

A family of tetrapyrrylmethane derivatives that are formed in the oxidative degradation of the porphyrin nucleus, Fig. 13.960.

Examples. The best known members of this class of animal pigments are those derived from the protoporphyrin IX nucleus of heme (Sect. 13.880). The degradation is initiated by the oxidation of the heme to an oxophlorin (Sect. 13.890). This is followed by the formation of a labile peroxide, Fig. 13.960, which undergoes fragmentation to carbon monoxide and **biliverdin IX α**. The designation IX α indicates that protoporphyrin IX was cleaved at the α meso carbon atom (see Fig. 13.910*a* for meso carbon atom assignments, $\alpha, \beta, \gamma, \delta$). Biliverdin IX α subsequently is reduced to **bilirubin IX α**.

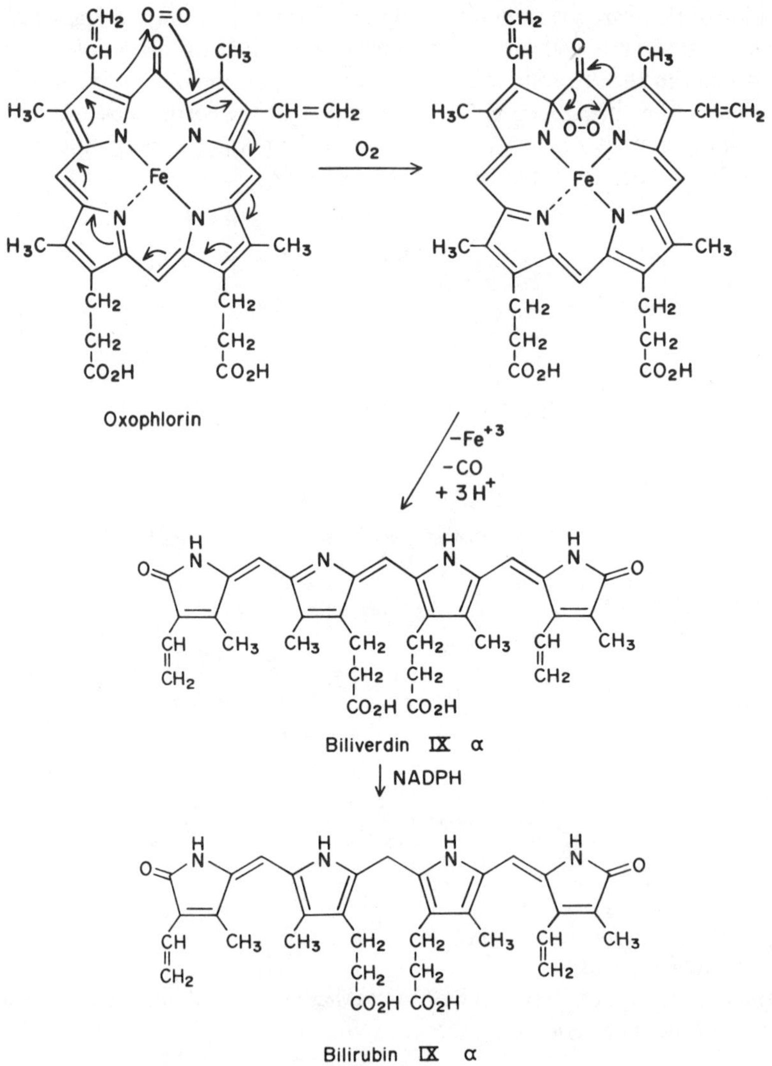

Oxophlorin

O_2

$-Fe^{+3}$
$-CO$
$+3H^+$

Biliverdin IX α

NADPH

Bilirubin IX α

Figure 13.960 The biosynthesis of the bile pigments.

13.970 Phytochromes

A family of substances of plant origins that is analogous to the bile pigments of animals (Sect. 13.960).

Example. These substances serve a variety of important functions in plants. **Phycocyanobilin,** Fig. 13.970, is an auxiliary light-harvesting pigment for photosynthesis. They also play a role in **photoperiodism**, the coordination of plant metabolism with the light cycle. Germination of seeds appears to involve these substances. Thus a seed may lay dormant for hundreds of years and germinate only upon exposure to red light.

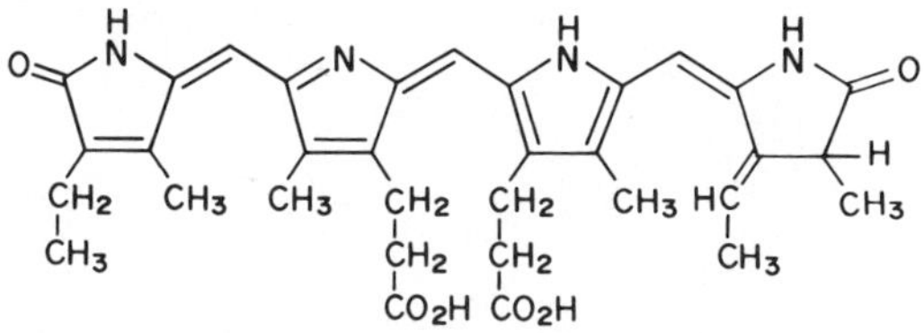

Phycocyanobilin

Figure 13.970 Phytochromes, the plant analogs of the bile pigments.

14 Polymers

CONTENTS

14.010 Polymer

A relatively high molecular weight substance composed by repetitively linking together small molecules, called monomeric units, in sufficient number such that the addition or removal of one or several units does not change the properties of the substance. An **oligomer** resembles a polymer except that the number of linked units is considerably smaller and removal of one or a few units does affect its properties. Thus there is no specific number of units that must be linked to qualify the resulting substance as a polymer. Polymers composed of as few as 30 and as many as 100,000 units are not uncommon, whereas oligomers may range from 4 to about 15 linked monomers.

14.020 Constitutional Repeating Unit

The smallest repeating structural unit of which the polymer is a multiple. The unit is (for single strand polymers) a bivalent group.

Examples. Consider the polymer chain shown in Fig. 14.020*a*. It is possible to select either unit A or unit B as the constitutional repeating unit. According to IUPAC rules [*Pure Appl. Chem.*, **40**, No. 3, 477–91 (1974)], unit B is chosen and the polymer is written as shown in Fig. 14.020*b*. This choice gives the substituent the lowest locant number in a numbering system constructed to increase in the conventional left to right direction. If R is phenyl, the systematic IUPAC name is poly(1-phenylethylene). In the IUPAC system the polymer is considered as an essentially endless chain and the manner of its construction is of no importance. However, in considering the chemistry of polymerization, it is clear that the initiation and propagation steps (Sect. 10.630) both proceed via the most stable radical (or ion); and this is the most, rather than the least, substituted radical (or ion). Hence, if importance is attached to the mechanism of the growth of the polymer, there is justification for regarding the constitutional repeating unit as unit A in Fig. 14.020*a* rather than unit B and, accordingly, the polymer is written as in Fig. 14.020*c*. We use unit A, $-\!(\!CH_2{-}CH(R)\!)_n\!-$, hereafter for writing the structure of vinyl polymers because it is more satisfying in rationalizing the chemistry of their formation.

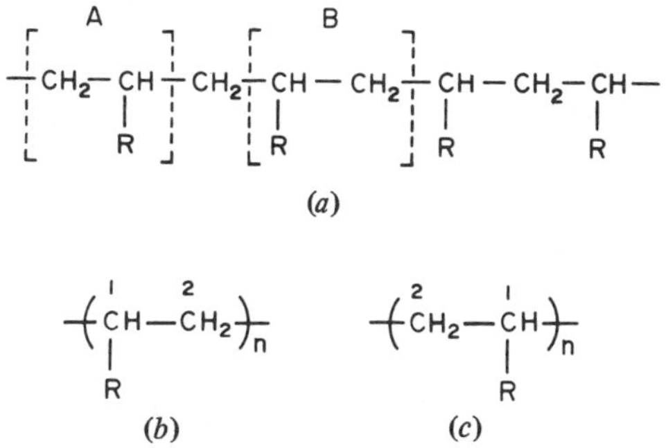

A B

$-CH_2-\underset{R}{CH}-CH_2-\underset{R}{CH}-CH_2-\underset{R}{CH}-CH_2-\underset{R}{CH}-$

(*a*)

$\left(\underset{R}{\overset{1}{C}H}-\overset{2}{C}H_2\right)_n$ $\left(\overset{2}{C}H_2-\underset{R}{\overset{1}{C}H}\right)_n$

(*b*) (*c*)

Figure 14.020 Constitutional repeating unit of a polymer: (*a*) possible units A and B; (*b*) the IUPAC unit; (*c*) the preferred unit.

14.030 Names of Polymers

Linear polymers of unspecified length are named by prefixing *poly* to the name (placed in parentheses except for several simple cases) of the smallest repeating monomeric structural unit.

Examples. Semisystematic or trivial names are given to many common polymers such as polystyrene and polyacrylonitrile. The IUPAC names for these polymers are poly(1-phenylethylene) and poly(1-cyanoethylene), respectively. Even though we choose, contrary to IUPAC selection, the structural unit for vinyl polymers shown in Fig. 14.020*c*, we can still retain the IUPAC nomenclature for such polymers by numbering the carbons from right to left as shown in that figure. If the common name is an obvious source-based name, the polymer is written as derived from that source, e.g., $-(CH_2-CH_2)_n-$ is polyethylene although the systematic name and structure is poly(methylene), $-(CH_2)_n-$. The structure, common name, and commercial name for frequently encountered vinyl and several other type polymers are given in Table 14.030.

14.040 Monomeric Unit

The group of atoms, derived from a monomer, comprising the unit that is repeated in the polymer; synonymous with constitutional repeating unit.

Examples. In Table 14.030 the structures of the monomeric units are the group of atoms shown inside the parentheses. An obvious ambiguity arises in the case of $-(CH_2)_n-$. If the monomer is diazomethane, CH_2N_2, the monomeric bivalent unit is $-CH_2-$ and the polymer is polymethylene (source name), but if the monomer is ethylene, $CH_2{=}CH_2$, then the monomeric unit is $-CH_2-CH_2-$ and the polymer is polyethylene (source name).

Table 14.030 Some common polymers

Structure	Trivial or semisystematic name	Commercial name	Abbreviation
$-(CH_2-CH_2)_n-$	Polyethylene		PE
$-(CH_2-CH(CH_3))_n-$	Polypropylene		PP
$-(CH_2-C(CH_3)_2)_n-$	Polyisobutylene		PIB
$-(CH_2-CH{=}CH-CH_2)_n-$	Polybutadiene		
$-(CH_2-C(CH_3){=}CH-CH_2)_n-$	Polyisoprene		
$-(CH_2-CH(C_6H_5))_n-$	Polystyrene		PS
$-(CH_2-CH(CN))_n-$	Polyacrylonitrile		PAN
$-(CH_2-CH(OH))_n-$	Poly (vinyl alcohol)		PVAL
$-(CH_2-CH(OCOCH_3))_n-$	Poly (vinyl acetate)		PVAC
$-(CH_2-CH(Cl))_n-$	Poly (vinyl chloride)		PVC

$-\!\!(CH_2-CF_2)_n-$ (structure: CH_2–C with two F substituents)	Poly(vinylidene fluoride)		PVDF
$-\!\!(CF_2-CF_2)_n-$	Poly(tetrafluoroethylene)	Teflon	
$-\!\!(CH_2-CH(CO_2CH_3))_n-$	Poly(methyl acrylate)		PMA
$-\!\!(CH_2-C(CH_3)(CO_2CH_3))_n-$	Poly(methyl methacrylate)	Lucite, Plexiglas, Perspex	PMMA
$-\!\!(O-CH_2)_n-$	Polyformaldehyde (n = ~30–100)		
$CH_3CO-\!\!(O-CH_2)_n-OCOCH_3$		Delrin	
$-\!\!(OCH_2-CH_2)_n-$	Poly(ethylene oxide)		
$-\!\!(CF_2-CClF)_n-$	Poly(1-chloro-1, 2, 2-trifluoroethylene)	KEL-F	
$CH_2{=}CCl_2 + CH_2{=}CHCl$ (random copolymer)		Saran	
$-\!\!(CH_2-CH(F))_n-$	Poly(vinyl fluoride)	Tedlar	
$CF_2{=}CF-CF_3 + CH_2{=}CF_2$ (random copolymer)		Viton	
$-\!\!(O-C_6H_4-C(CH_3)_2-C_6H_4-O-C({=}O))_n-$	Polycarbonate		
$-\!\!(CH_2-CH(OH))_n- + n-C_3H_7CHO$	Poly(vinyl butyral)		

14.050 Polymerization

The process that converts monomers to polymers.

14.060 Free Radical Polymerization

The polymerization of monomers initiated by a free radical. Such polymerization involves a chain reaction (Sect. 10.630).

Example. The polymerization of vinyl chloride is initiated by thermal decomposition of an initiator such as AIBN (Sect. 10.670), Fig. 14.060. The addition of the radical initiator R· to vinyl chloride can theoretically result in the formation of $R{-}CHCl{-}CH_2\cdot$; however rather than this primary radical, the more stable secondary radical is formed and each succeeding monomer is added regiospecifically as shown.

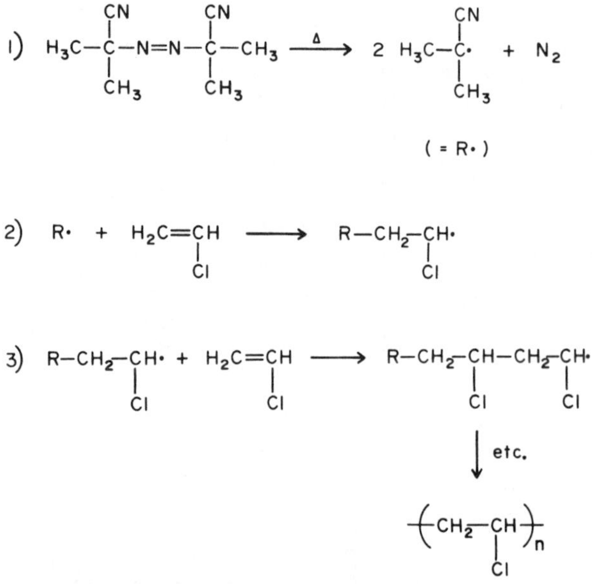

$$1)\quad H_3C{-}C(CN)(CH_3){-}N{=}N{-}C(CN)(CH_3){-}CH_3 \xrightarrow{\Delta} 2\ H_3C{-}\dot{C}(CN)(CH_3) + N_2 \quad (= R\cdot)$$

$$2)\quad R\cdot + H_2C{=}CHCl \longrightarrow R{-}CH_2{-}\dot{C}HCl$$

$$3)\quad R{-}CH_2{-}\dot{C}HCl + H_2C{=}CHCl \longrightarrow R{-}CH_2{-}CHCl{-}CH_2{-}\dot{C}HCl \xrightarrow{\text{etc.}} {+}CH_2{-}CHCl{+}_n$$

Figure 14.060 Free radical polymerization of vinyl chloride.

14.070 Head-to-Tail Orientation (of Vinyl Polymers)

The polymer structure that involves the bonding of the most substituted carbon atom (the head of the monomeric unit) to the least substituted carbon atom (the tail of the adjacent unit). In such orientation the substituent appears on alternate carbon atoms.

Example. The free radical polymerization of vinyl chloride, Fig. 14.060. Head-to-tail polymerization is the preferred orientation of all vinyl polymers.

14.080 Head-to-Head, Tail-to-Tail Orientation (of Vinyl Polymers)

The polymer structure that is generated by addition of the most substituted terminal carbon of the growing chain to the most substituted carbon of the next monomer unit (head-to-head) immediately followed by the addition of the least substituted terminal carbon of the growing chain to the least substituted carbon of the succeeding monomer (tail-to-tail). It is possible that a growing polymer can have an occasional orientation of this nature, but the possibility of obtaining a regular H → H—T → T—H → H · · ·, etc., chain is remote. Head-to-head orientation results in the substituent appearing on adjacent carbon atoms.

Example. Figure 14.080, an unlikely polymer structure resulting from head-to-head, tail-to-tail, head-to-head, etc., polymerization.

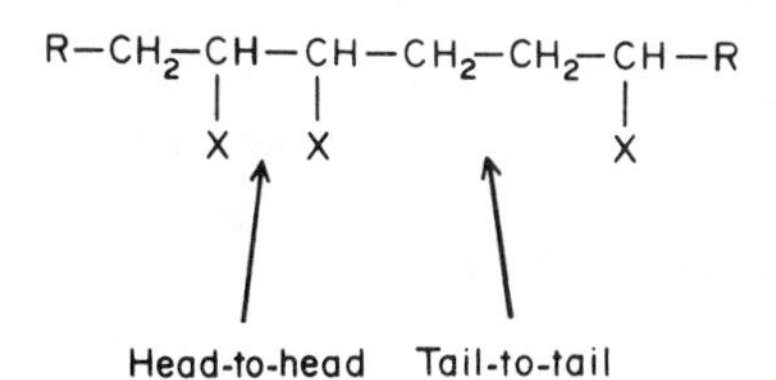

Figure 14.080 Head-to-head, tail-to-tail orientation in a vinyl polymer.

14.090 Addition Polymer

Polymers can be classed into two broad types, addition polymers and condensation polymers (*vide infra*, Sect. 14.380). An addition polymer results from a polymerization process involving the stepwise consecutive addition of monomers (either identical or different) to a chain with no loss of atoms accompanying the addition.

Examples. The free radical polymerization of vinyl chloride, Fig. 14.060; poly(ethylene oxide), Fig. 14.090*a*; a polyurethane polymer, Fig. 14.090*b*.

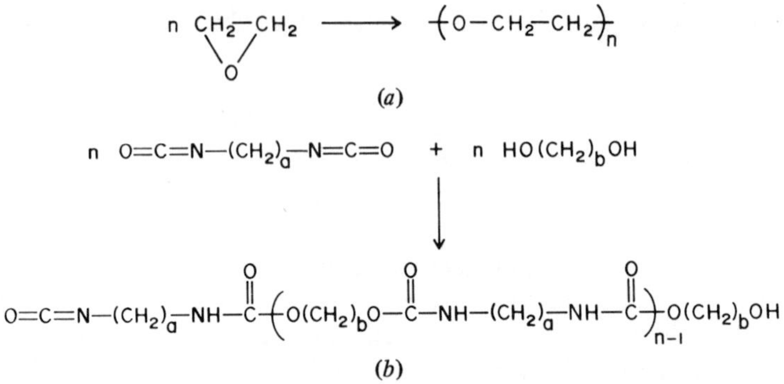

$$n\ \underset{\diagdown O \diagup}{CH_2-CH_2} \longrightarrow -\!\!\left(O-CH_2-CH_2\right)\!\!-_n$$

(*a*)

$$n\ O{=}C{=}N-(CH_2)_a-N{=}C{=}O \quad + \quad n\ HO(CH_2)_bOH$$

$$\downarrow$$

$$O{=}C{=}N-(CH_2)_a-NH-\overset{O}{\overset{\|}{C}}\left(O(CH_2)_bO-\overset{O}{\overset{\|}{C}}-NH-(CH_2)_a-NH-\overset{O}{\overset{\|}{C}}\right)_{n-1}O(CH_2)_bOH$$

(*b*)

Figure 14.090 Addition polymers: (*a*) poly(ethylene oxide); (*b*) a polyurethane polymer.

14.100 Linear (or Single Strand) Polymer

A polymer in which the chain backbone is formed by consecutive additions whereby the units become arranged in a linear chain. The smallest repeating unit in the single strand polymer is a bivalent group.

Examples. Polyethylene, poly(vinyl chloride), and all other polymers discussed to this point. The bivalent repeating unit in most vinyl polymers is $-\!\!\left(CH_2-CH(R)\right)\!\!-_n$ and R is H, CH_3, Ph, CO_2CH_3, Cl, $OCOCH_3$, OH, etc.

14.110 Homopolymer

A polymer made from a single species of monomer.

Examples. Poly(vinyl chloride) and polyethylene.

14.120 Copolymer

A polymer made from a mixture of two or more species of monomers. The two or more monomeric units may be distributed in random sequence **(random copolymer)** or, alternatively, the monomeric units may be distributed in alternating sequence **(alternating copolymers).**

Examples. One of the most common elastomers (Sect. 14.610) is made from a mixture of 75% butadiene (B) and 25% styrene (S). A portion of the resulting random copolymer may be represented by · · · BBSBBBBBSBSB · · · . (This differs from a block polymer, Sect. 14.150.) On the other hand, styrene and methyl methacrylate form an alternating copolymer and this occurs for the following reasons. In the growing radical chain, if the end group were the styrene (benzylic) moiety, the electron releasing properties of the phenyl group would tend to give nucleophilic character to the radical. Accordingly, such a radical would prefer to attack the relatively electron deficient methylene group of the methyl methacrylate (in preference to a competing styrene monomer) owing to the electron withdrawing properties of the carbomethoxy group. On the other hand, if the growing radical chain had an ester grouping at the end, the tertiary carbon radical would be electrophilic and would preferentially attack a styrene monomer (rather than another acrylate monomer) whose polarization tends to place a partial negative charge on the methylene group of styrene. Accordingly, a polymer is built up of alternating units, Fig. 14.120.

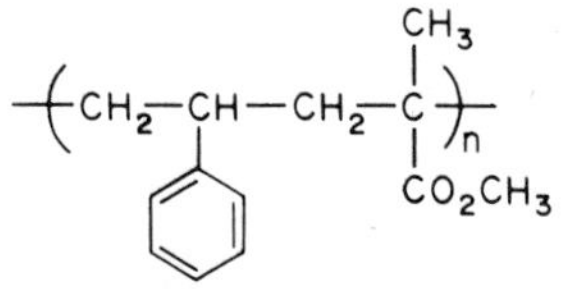

$$-\!\!\left(CH_2-\underset{C_6H_5}{CH}-CH_2-\overset{CH_3}{\underset{CO_2CH_3}{C}}\right)_{\!n}\!\!-$$

Figure 14.120 An alternating copolymer of styrene and methyl methacrylate.

14.130 Chain Transfer

The termination of a polymer chain with the simultaneous initiation of a new chain. The technique is frequently used to generate polymers of shorter chain length than would otherwise be obtained in the absence of a chain transfer agent.

Example. When styrene is polymerized in the presence of carbon tetrachloride, the polystyrene has a relatively low molecular weight owing to the reaction sequence shown in Fig. 14.130.

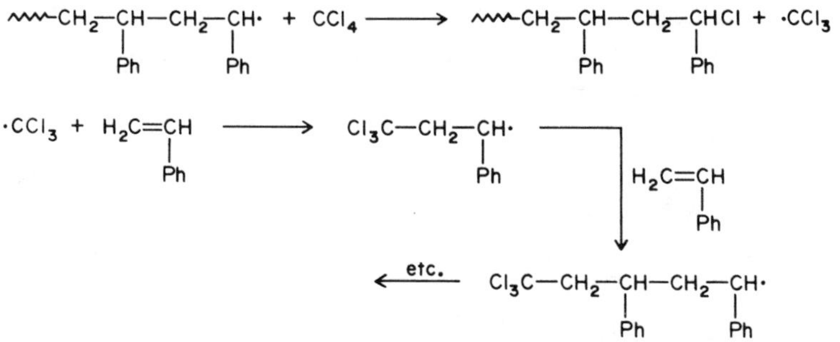

$$\sim CH_2-\underset{Ph}{CH}-CH_2-\underset{Ph}{CH}\cdot + CCl_4 \longrightarrow \sim CH_2-\underset{Ph}{CH}-CH_2-\underset{Ph}{CHCl} + \cdot CCl_3$$

$$\cdot CCl_3 + H_2C{=}\underset{Ph}{CH} \longrightarrow Cl_3C-CH_2-\underset{Ph}{CH}\cdot \xrightarrow{H_2C=\underset{Ph}{CH}} Cl_3C-CH_2-\underset{Ph}{CH}-CH_2-\underset{Ph}{CH}\cdot \xrightarrow{etc.}$$

Figure 14.130 Chain transfer.

14.140 Branched Polymer

A homopolymer in which occasional chains are attached to the backbone; i.e., the polymer occasionally grows branches instead of growing as a continuous chain.

Examples. Branching may occur as a result of chain transfer between a growing but rather short polymer chain with another and longer polymer chain, as shown in Fig. 14.140*a*. Branching may also occur if the radical end of a growing chain abstracts a hydrogen from a carbon atom four or five carbons removed from the end, as shown in Fig. 14.140*b*. This phenomenon is called **back-biting.** In the special case of polyethylene, methyl branching occurs presumably because of hydrogen migration, as shown in Fig. 14.140*c*.

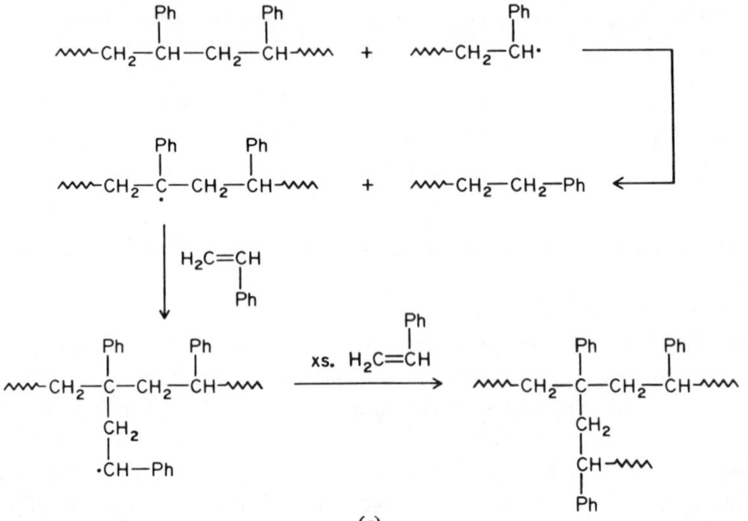

$$\sim CH_2-CH(Ph)-CH_2-CH(Ph)\sim \; + \; \sim CH_2-\dot{C}H(Ph) \longrightarrow \sim CH_2-\dot{C}(Ph)-CH_2-CH(Ph)\sim \; + \; \sim CH_2-CH_2-Ph$$

$$\xrightarrow{H_2C=CH(Ph)} \sim CH_2-C(Ph)(CH_2-\dot{C}H-Ph)-CH_2-CH(Ph)\sim \xrightarrow{\text{xs. } H_2C=CH(Ph)} \sim CH_2-C(Ph)(CH_2-CH(Ph)\sim)-CH_2-CH(Ph)\sim$$

(*a*)

$$Ph-\dot{C}H-CH_2-CH(Ph)-CH_2-CH(Ph)-CH_2\sim \longrightarrow \sim CH_2-\dot{C}(Ph)-CH_2-CH(Ph)-CH_2-CH_2-Ph$$

$$\xrightarrow{H_2C=CH(Ph)} \sim CH_2-C(Ph)(CH_2-\dot{C}H-Ph)-CH_2-CH(Ph)-CH_2-CH_2-Ph$$

$$\xrightarrow{\text{xs. } H_2C=CH(Ph)} \sim CH_2-C(Ph)(CH_2-CH(Ph)\sim)-CH_2-CH(Ph)-CH_2-CH_2-Ph$$

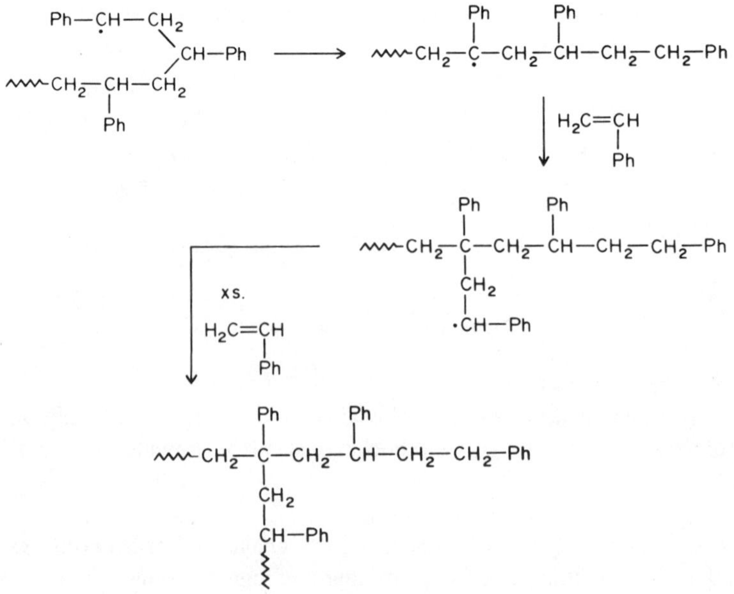

(*b*)

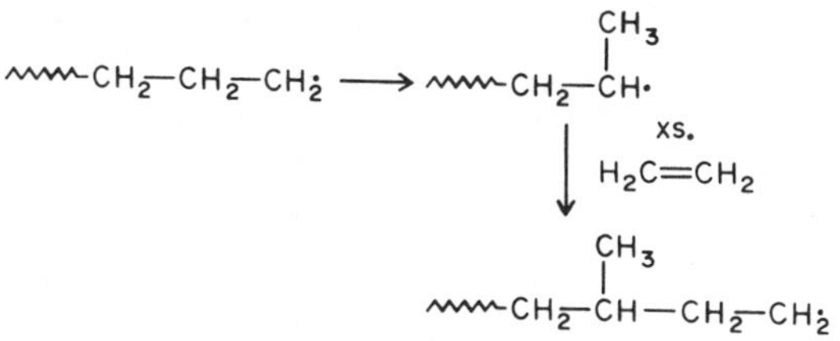

$$\sim\sim CH_2{-}CH_2{-}CH_2\cdot \longrightarrow \sim\sim CH_2{-}CH(CH_3)\cdot$$

$$\downarrow \text{xs. } H_2C{=}CH_2$$

$$\sim\sim CH_2{-}CH(CH_3){-}CH_2{-}CH_2\cdot$$

(*c*)

Figure 14.140 Branched chain polymer formation: (*a*) By intermolecular chain transfer; (*b*) by intramolecular back-biting; (*c*) methyl branching in polyethylene.

14.150 Block Polymer

A copolymer whose structure consists of a homopolymer attached to chains of another homopolymer.

Examples. If two different monomer units are designated A and B, the block polymer can have the structure AAAABBBBBAAA · · · . Block polymers frequently form multiphase systems, giving rise to unusual mechanical properties such as impact resistance.

14.160 Graft Polymer

A copolymer in which the chain backbone is made up of one kind of monomer and the branches are made up of another kind of monomer.

Examples. Polystyrene may be grafted onto a poly(butadiene) polymer by treating the latter with styrene in the presence of a radical initiator. The reactions are shown in Fig. 14.160*a*. The schematic formula for a graft polymer made from a single strand polymer A and a grafting monomer B is shown in Fig. 14.160*b*.

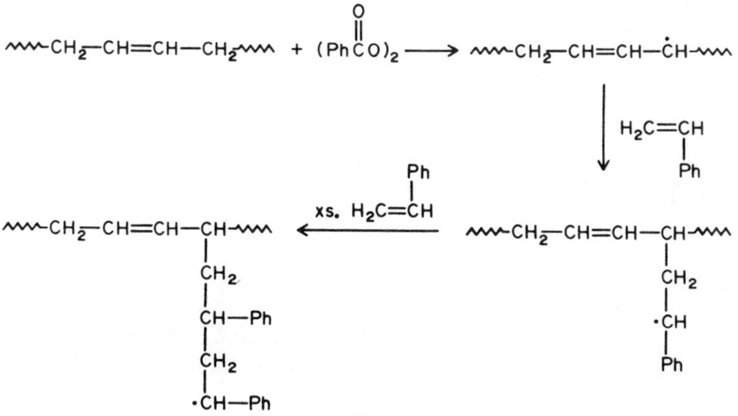

$$\sim\sim CH_2{-}CH{=}CH{-}CH_2\sim\sim + (Ph\overset{O}{\overset{\|}{C}}O)_2 \longrightarrow \sim\sim CH_2{-}CH{=}CH{-}\dot{C}H\sim\sim$$

$$\downarrow H_2C{=}CH{-}Ph$$

$$\sim\sim CH_2{-}CH{=}CH{-}CH(CH_2{-}\cdot CH{-}Ph)\sim\sim$$

$$\downarrow \text{xs. } H_2C{=}CH{-}Ph$$

$$\sim\sim CH_2{-}CH{=}CH{-}CH(CH_2{-}CH(Ph){-}CH_2{-}\cdot CH{-}Ph)\sim\sim$$

(*a*)

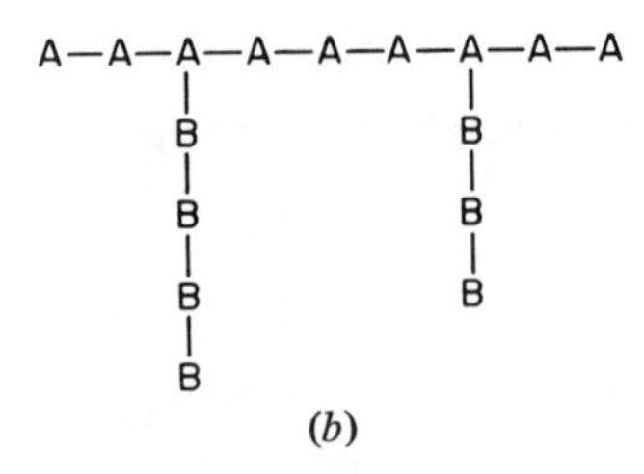

(*b*)

Figure 14.160 Graft polymers: (*a*) reactions leading to a graft polymer; (*b*) schematic representation of a graft polymer having a main chain of monomer A units and branches of B monomer units.

14.170 Cross-Linking

The linking of two independent polymer chains by a grouping that spans or links two chains.

Examples. Cross-linked polystyrene. In practice cross-linking of linear polystyrene is achieved by adding some divinylbenzene to the styrene. When the two vinyl groups on the same ring each participate in the chain propagation, a cross-link is formed by that molecule of divinylbenzene, Fig. 14.170*a*. A randomly spaced cross-linked network is shown schematically in Fig. 14.170*b*.

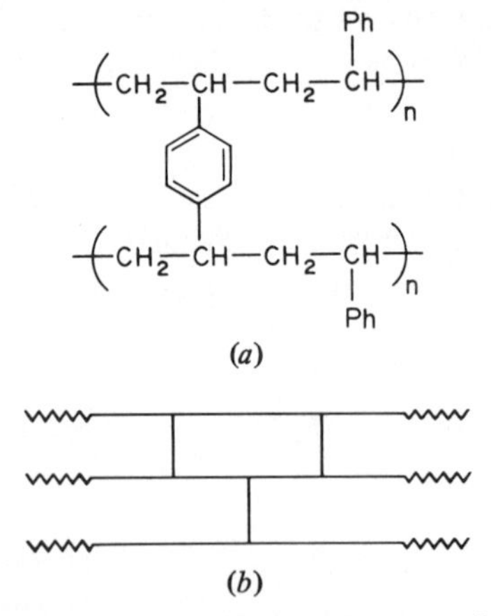

Figure 14.170 (*a*) A cross-link in polystyrene; (*b*) a cross-linked polymer network.

14.180 Ladder Polymer

The polymers described thus far possess bivalent constitutional repeating units. When the repeating units have a tetravalent structure, a polymer consisting of two backbone chains regularly cross-linked at short intervals (a double stranded polymer) results; this is called a ladder polymer.

Examples. The simplest possible constitutional repeating unit of a ladder polymer is shown in Fig. 14.180*a*. A ladder polymer can be constructed so that a single strand polymer is first formed, followed by the construction of the second strand to give the double strand structure of the ladder polymer. In one such case, vinyl isocyanate, Fig. 14.180*b*, is treated at -55° with NaCN to give the linear polymer, Fig. 14.180*c*, which on subsequent treatment with a radical initiator gives the ladder polymer shown in Fig. 14.180*d*. The tetravalent constitutional repeating unit is shown in Fig. 14.180*e*, and the general schematic structure of a ladder polymer is shown in Fig. 14.180*f*.

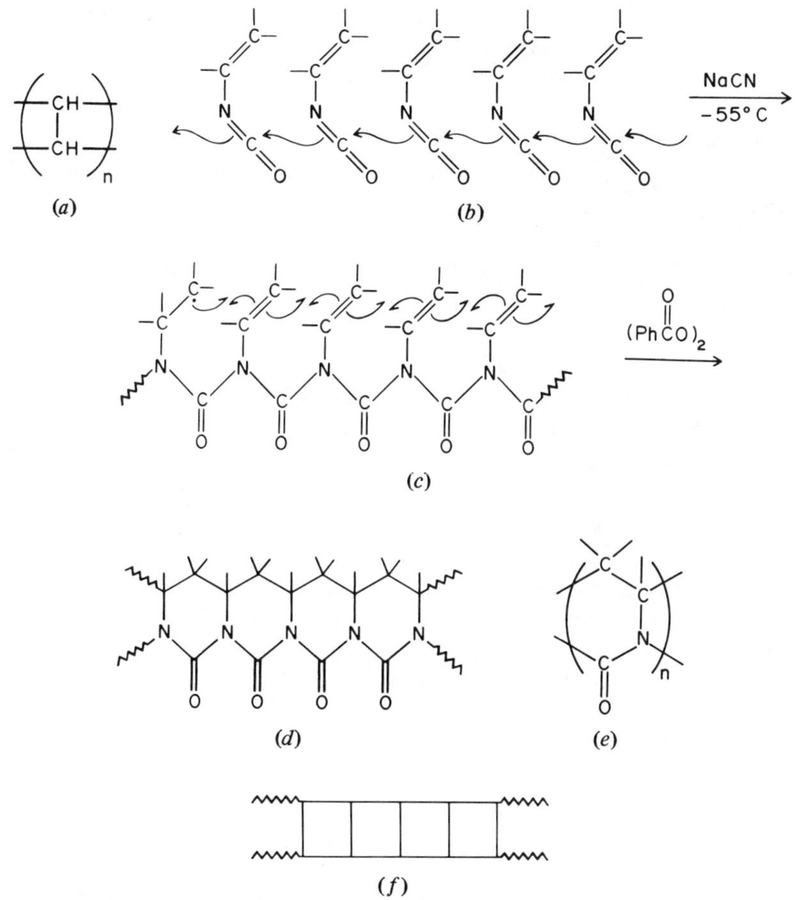

Figure 14.180 Ladder polymers: (*a*) the simplest repeating unit; (*b*) vinyl isocyanate polymerization; (*c*) the single strand polymer polymerizing further to give (*d*) the ladder polymer; (*e*) the repeating tetravalent constitutional unit; (*f*) schematic representation of a ladder polymer.

14.190 Emulsion Polymerization

A polymerization process (usually free radical) in which the water-insoluble monomers are allowed to react with each other in an aqueous medium containing an emulsifying agent. The polymeric material separates as a coagulated latex and is processed further.

14.200 Cationic Polymerization

The polymerization of monomers initiated by proton donors.

Examples. The common mineral acids, particularly H_3PO_4, are commonly used, but the polymers produced generally possess quite low molecular weights. The reaction between Lewis acids and water is also used to provide a proton source.

14.210 Anionic Polymerization

The polymerization of monomers initiated by anions, usually carbanions, or by the amide ion, $^-NH_2$.

Examples. The carbanions may be generated from sodium and a reactive hydrocarbon, e.g.,

$$PhCH_3 + 2\,Na \longrightarrow Ph\bar{C}H_2Na^+ + NaH$$

or from sodium or lithium alkyls, R^-M^+, prepared from alkyl halides. Anionic polymerization is generally a slow reaction and is effective when the substrate monomer is either a diene such as butadiene or vinyl compound with a strong electron withdrawing group attached to the double bond, Fig. 14.210*a*. Anionic polymerization can also be initiated by an active metal reacting with a polynuclear aromatic compound to form a radical-anion. The radical-anion can then transfer the electron to the vinyl monomer to form a new radical-anion that can add monomers from both ends or, more commonly, can dimerize to form a dianion that then undergoes anionic polymerization at both ends. Such a sodium-naphthalene initiated polymerization of styrene is shown in Fig. 14.210*b*.

$$\underline{n}\text{-BuLi} + H_2C{=}C(CH_3)(CO_2CH_3) \longrightarrow \underline{n}\text{-Bu—}CH_2\text{—}\ddot{C}^-(CH_3)(CO_2CH_3) \longrightarrow \left(CH_2\text{—}C(CH_3)(CO_2CH_3)\right)_n$$

(*a*)

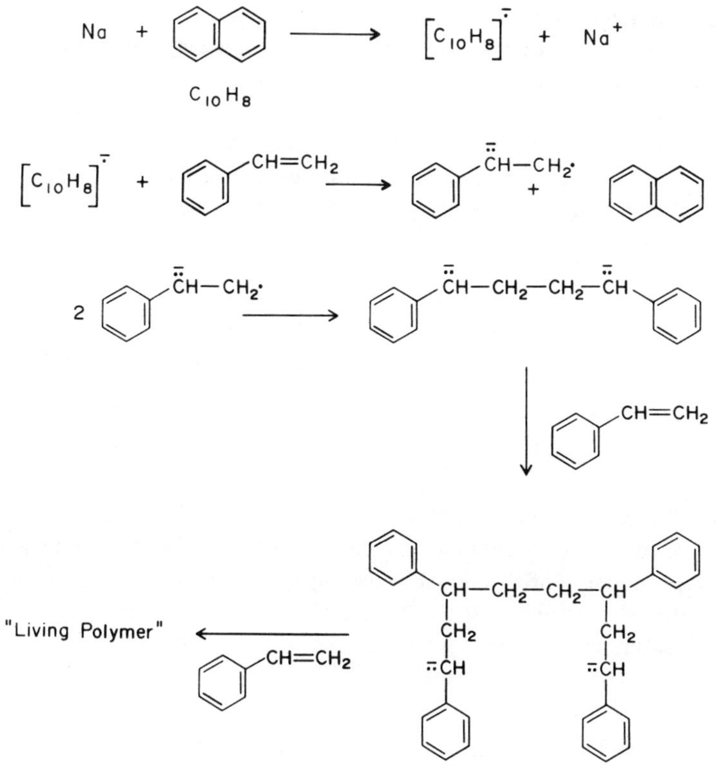

$$Na + C_{10}H_8 \longrightarrow [C_{10}H_8]^{\overline{\cdot}} + Na^+$$

$$[C_{10}H_8]^{\overline{\cdot}} + C_6H_5CH{=}CH_2 \longrightarrow C_6H_5\ddot{C}H{-}CH_2\cdot + C_{10}H_8$$

$$2\,C_6H_5\ddot{C}H{-}CH_2\cdot \longrightarrow C_6H_5\ddot{C}H{-}CH_2{-}CH_2{-}\ddot{C}HC_6H_5$$

$$\downarrow C_6H_5CH{=}CH_2$$

$$C_6H_5\ddot{C}HCH_2{-}CH(C_6H_5){-}CH_2{-}CH_2{-}CH(C_6H_5){-}CH_2\ddot{C}HC_6H_5$$

"Living Polymer" ← $C_6H_5CH{=}CH_2$

(*b*)

Figure 14.210 Anionic polymerization: (*a*) *n*-butyllithium initiated polymerization of methyl methacrylate; (*b*) sodium-naphthalene initiated polymerization of styrene to give a living polymer.

14.220 Living Polymer

A polymer formed under conditions such that there are no termination steps possible, resulting in a chain that continues to grow in response to each added increment of monomer.

Examples. Living polymers (Michael Szwarc, 1909-) are usually formed under carefully controlled anionic polymerization conditions where no impurities (e.g., water) are present to terminate the chain. The formation of a living polymer of styrene is shown in Fig. 14.210*b*. When all the styrene has polymerized, a new monomer may be added with the result that the new monomer, e.g., butadiene, adds to both ends of the polystyrene to form a block polymer of styrene and butadiene. The living polymer is "killed" by the addition of an impurity, such as water or other protic compounds that react with carbanions.

14.230 Configurational Base Unit

The constitutional repeating unit in a polymer possessing at least one site whose stereoisomerism is specified.

Example. In polypropylene the constitutional repeating unit is $—CH_2—CH(CH_3)—$ and the two possible configurational base units are shown in Fig. 14.230. These configurational base units are enantiomeric.

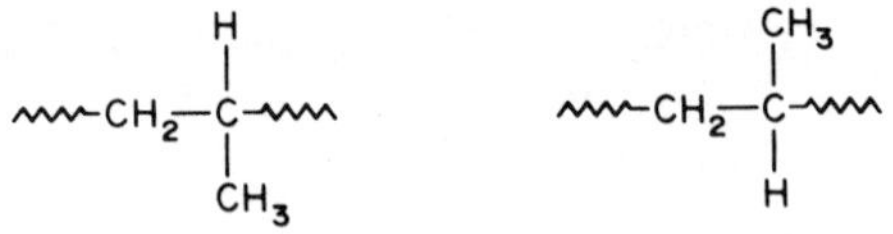

H | ~~~CH_2—C~~~ | CH_3 CH_3 | ~~~CH_2—C~~~ | H

Figure 14.230 Plane projection representation of enantiomeric configurational base units of polypropylene.

14.240 Tactic (*Gk.* Tactikos, Order) Polymer

A polymer in which the configuration is specified around one or more stereoisomeric sites in the **configurational repeating unit** (the smallest set of successive configurational base units). If there are multiple chiral centers in the configurational repeating unit, at least one of them must be configurationally defined for the polymer to be a tactic polymer.

14.250 Isotactic Polymer

A tactic polymer having only one configurational base unit in the repeating unit. In a vinyl polymer, $-\!\!\left[CH_2\text{-}CH(R) \right]_n-$, this requires that all the similar substituents on the chiral atoms in the polymer chain appear on the same side of the chain. In isotactic polymers the configurational repeating unit is identical with the configurational base unit.

Examples. A portion of the polymer chain of isotactic polypropylene is shown in Fig. 14.250*a*. If one wished to make an assignment of absolute configuration at every chiral center, then one would need to consider the end groups. If the polymer chain were terminated at the left end of the chain by an atom or group

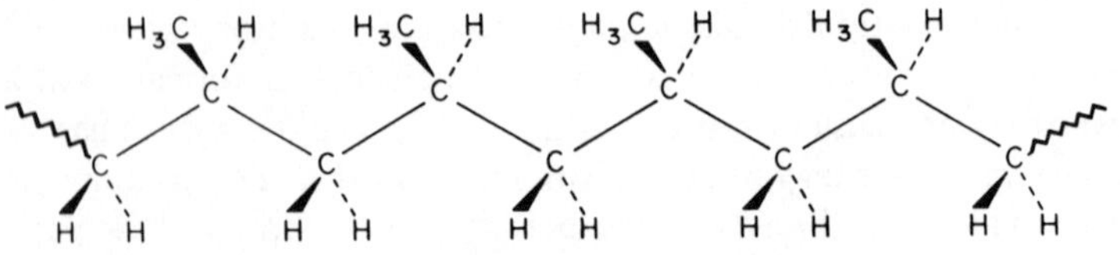

H_3C H H_3C H H_3C H H_3C H
C C C C
C C C C C
H H H H H H H H H H

(*a*)

with a higher priority (Sect. 5.290) than the terminal group on the right end, then the configuration at each chiral center of this isotactic polypropylene would be (*S*) and this polymer chain would theoretically be optically active. However, the end groups of the polymer chain represent such a small fraction of the molecule that their presence can be neglected. If the end groups are neglected (or even if they are equivalent), then the chain, whatever its length, has a plane of symmetry and is optically inactive. The polymer has multiple chiral centers but, because it has a plane of symmetry, it may be considered to be a meso compound (Sect. 5.690). If each chiral center on the left side of the middle of the chain has an (*R*) configuration, then every chiral center of the right side of the middle of the chain has the (*S*) configuration! And because the chain has a plane of symmetry, the mirror image of the chain shown in Fig. 14.250*a* (which would have all the methyl groups on the far side of the chain) is identical and not enantiomeric with the chain of Fig. 14.250*a*. The same type of argument with respect to optical activity can be made with respect to syndiotactic (Sect. 14.260) vinyl polymers, and in fact no optically active vinyl polymers having chiral centers in the backbone chain are known. However, if an olefin possesses a chiral center in the substituent group, optically active isotactic polymers can be prepared (using Ziegler-Natta catalysis, Sect. 14.290). When a racemic mixture of the monomer is used resolution can lead to an optically active polymer, Fig. 14.250*b*.

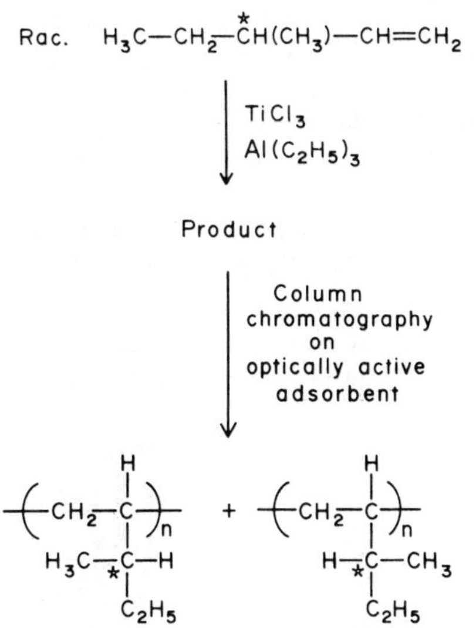

Rac. $H_3C-CH_2-\overset{*}{C}H(CH_3)-CH=CH_2$

$\downarrow$ $TiCl_3$, $Al(C_2H_5)_3$

Product

$\downarrow$ Column chromatography on optically active adsorbent

$\left(CH_2-\underset{H_3C-\overset{*}{C}(H)-C_2H_5}{\overset{H}{C}}\right)_n + \left(CH_2-\underset{H-\overset{*}{C}(CH_3)-C_2H_5}{\overset{H}{C}}\right)_n$

(*b*)

Figure 14.250 Isotactic polymers: (*a*) polypropylene; (*b*) an optically active polymer.

14.260 Syndiotactic (*Gk.* Syndyo, Two Together) Polymer

A tactic polymer whose configurational repeating unit contains successive configurational base units that are enantiomeric.

Examples. Syndiotactic polypropylene, Fig. 14.260*a*; the chiral centers have opposite configurations. In the polymer chain successive methyl groups would appear alternately on one side and then on the other side of the chain. If, in addition to the enantiomeric configurational sites, the repeating unit contains another, but undefined chiral center, it is still considered to be syndiotactic, Fig. 14.260*b*, but such a polymer is not stereoregular.

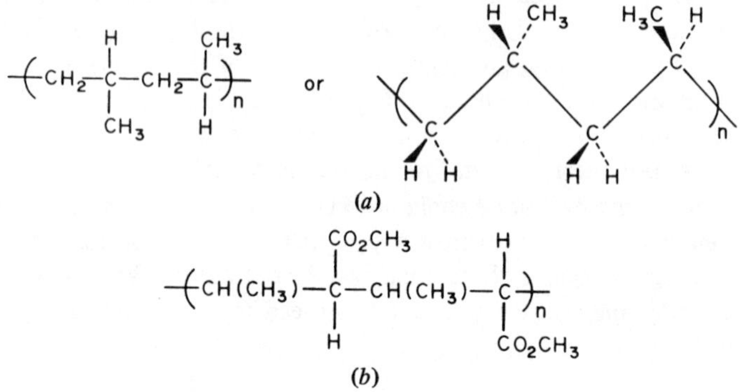

$$\left(CH_2-C(H)(CH_3)-CH_2-C(CH_3)(H)\right)_n$$

or

(zigzag chain: C with H, H; C with H, CH_3; C with H, H; C with H_3C, H; $)_n$)

(*a*)

$$\left(CH(CH_3)-C(CO_2CH_3)(H)-CH(CH_3)-C(H)(CO_2CH_3)\right)_n$$

(*b*)

Figure 14.260 Syndiotactic polymers: (*a*) syndiotactic polypropylene; (*b*) a nonstereoregular syndiotactic polymer.

14.270 Stereoregular Polymer

A polymer in which *every* stereoisomeric center has a defined configuration.

Examples. Isotactic and syndiotactic polypropylene. According to IUPAC rules, in a polymer such as $-\!\!\left[CH(CO_2CH_3)CH(CH_3)\right]_n\!\!-$, if only one of the main chain stereoisomeric sites of each constitutional repeating unit is defined, Fig. 14.270*a*, then the polymer, Fig. 14.270*b*, is isotactic but not stereoregular. A stereoregular polymer is always tactic but a tactic polymer need not be stereoregular.

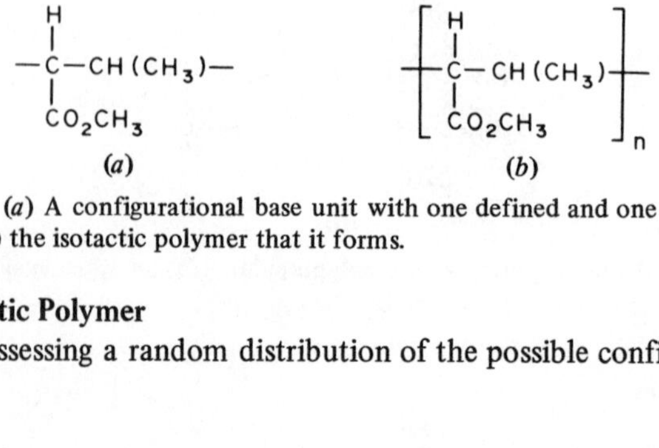

$$-C(H)(CO_2CH_3)-CH(CH_3)- \qquad \left[C(H)(CO_2CH_3)-CH(CH_3)\right]_n$$

(*a*) (*b*)

Figure 14.270 (*a*) A configurational base unit with one defined and one undefined stereoisomeric site; (*b*) the isotactic polymer that it forms.

14.280 Atactic Polymer

A polymer possessing a random distribution of the possible configurational base units.

Examples. Atactic polypropylene, Fig. 14.280*a*. According to the IUPAC definition, an atactic polymer is one having a random distribution of *equal numbers* of the possible configurational base units. Most polymer chemists call a regular polymer that does not possess a configurationally specified repeating unit an atactic polymer even though an equal number of the configurational base units may not be present. It should be pointed out that a polymer conceivably can be tactic without being either iso- or syndiotactic, e.g., Fig. 14.280*b* (this polymer contains three configurational base units in the configurational repeating unit).

```
         CH3      CH3      H        CH3      H        H
         |        |        |        |        |        |
~~~~CH2--C--CH2---C--CH2---C--CH2---C--CH2---C--CH2---C~~~~
         |        |        |        |        |        |
         H        H        CH3      H        CH3      CH3
```

(*a*)

```
        CH3       CH3       H
        |         |         |
 ( CH2--C--CH2----C--CH2----C )
        |         |         |   n
        H         H         CH3
```

(*b*)

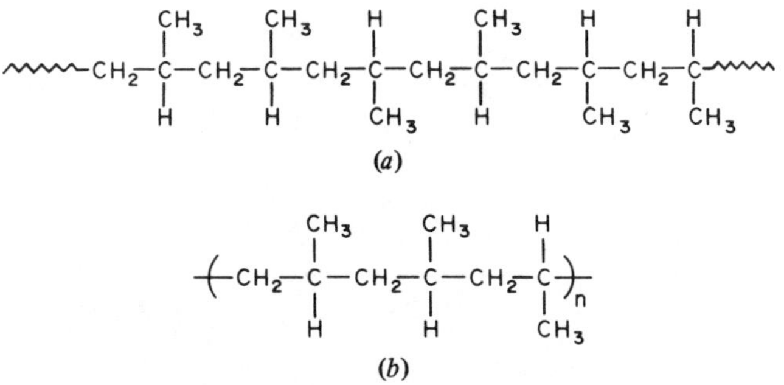

Figure 14.280 (*a*) Atactic polypropylene; (*b*) tactic polypropylene showing a configurational repeating unit that is neither isotactic nor syndiotactic.

14.290 Ziegler-Natta Catalysis (Coordination Polymerization)

The catalytic system discovered by K. Ziegler (1898-1973) and developed by G. Natta (1903-1979) (both Nobel Laureates, 1963), consisting of a bimetallic coordination complex, e.g., $Et_3Al:TiCl_4$, which leads to stereoregular isotactic polymers of terminal olefins of high density and high crystallinity.

Examples. The structure of the active catalytic bimetallic coordination complex is not known. Presumably the $TiCl_4$ is reduced by the $AlEt_3$. It is very likely that the olefin to be polymerized is coordinated to the titanium ion, although aluminum alkyls by themselves are known to be oligomerization catalysts. The active catalyst is probably a reduced, insoluble $TiCl_3$ with the $AlEt_3$ chemisorbed on the $TiCl_3$. The chain growth mechanism has, at least until very recently, been assumed to be that shown in Fig. 14.290*a*. The key step in the mechanism is assumed to be olefin insertion into a metal-alkyl bond. Recently it has been suggested that Ziegler-Natta catalysis may involve metal-carbene formation, followed by carbene addition to coordinated olefin to give a metallocyclobutane, the same type of reaction involved in the metathesis of olefins (Sect. 12.470), Fig. 14.290*b*. In order for the metal-carbene intermediate to form in the polymerization reaction, a 1,2-hydrogen shift from the α carbon to the metal is required, followed by metallocyclobutane formation and then the reverse 1,2-hydrogen shift, Fig. 14.290*c*.

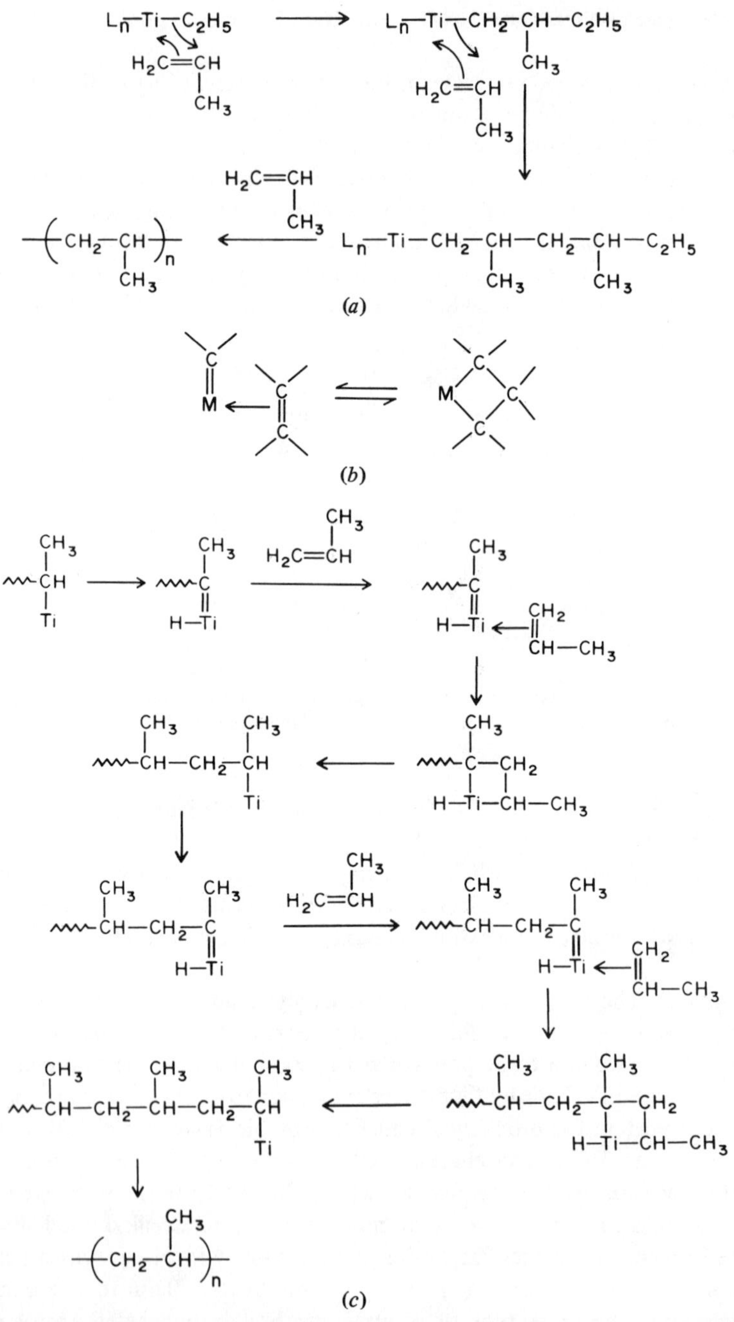

(a)

(b)

(c)

Figure 14.290 Zeigler-Natta catalysis: (*a*) olefin insertion mechanism; (*b*) metallocyclobutane intermediate in metathesis; (*c*) metallocyclobutane intermediate in polypropylene polymerization.

14.300 Ring Opening Cycloalkene Polymerization

The polymerization of certain cycloalkenes to give acyclic polyolefin polymers.

Examples. The ring opening polymerization is assumed to proceed via carbene intermediates (see olefin metathesis, Sect. 12.470) and is catalyzed by, e.g., an appropriate combination of tungsten hexachloride, diethylaluminum chloride and ethanol. The polymerization of cyclopentene to give an elastomer (Sect. 14.610) known commercially as polypentamer is shown in Fig. 14.300*a*. 1-Methylcyclobutene is polymerized by a preformed carbene complex, $(C_6H_5)_2C{=}W(CO)_5$, to give a polymer having essentially the same structure as *cis*-1,4-poly(isoprene), an isotactic polymer, Fig. 14.300*b*.

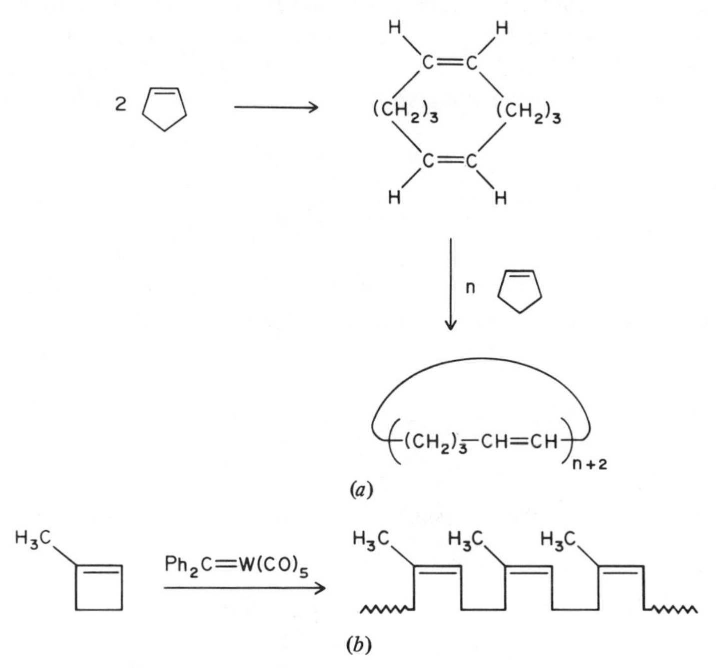

Figure 14.300 Ring opening polymerization: (*a*) polypentamer from cyclopentene; (*b*) *cis*-1,4-poly (isoprene) from 1-methylcyclobutene.

14.310 Degree of Polymerization

The number of monomeric repeating units in a polymer, e.g., the subscript n in polyethylene $-\!\!(CH_2-CH_2)_n\!\!-$.

14.320 Regiospecific Polymerization

The polymerization process in which chain growth proceeds by successive additions at one preferred site of the monomer.

Example. Head-to-tail orientation of substituted vinyl polymers (Sect. 14.070) is a result of regiospecific polymerization.

14.330 Molecular Weights of Polymers

Any finite sample of polymer contains a mixture of chains of different molecular weights. The experimental measurement of molecular weight thus gives only an average molecular weight; there will be many chains with molecular weights greater and many chains with molecular weights smaller than this average. However, several different kinds of averages are possible.

14.340 Number-Average Molecular Weight, $\overline{M}_n$

Experimental procedures that reflect the colligative properties of a sample, i.e., the number of molecules in a known weight of sample, lead to the number-average molecular weight, $\overline{M}_n$. The total number of moles in a polymer sample is the sum over all molecular species of the number of moles N_i of each species present:

$$\sum_{i=1}^{\infty} N_i$$

The total weight w of the sample is the sum of the weights of each particular molecular species, which for each species i is equal to N_i times its molecular weight M_i:

$$w = \sum_{i=1}^{\infty} w_i = \sum_{i=1}^{\infty} N_i M_i$$

The number-average molecular weight is the weight of sample per mole:

$$\overline{M}_n = \frac{w}{\sum_{i=1}^{\infty} N_i} = \frac{\sum_{i=1}^{\infty} N_i M_i}{\sum_{i=1}^{\infty} N_i}$$

Examples. Suppose a sample of polyethylene contains only two molecular species in the following amounts: four moles of a pentamer (mol. wt. = 5 × 28 = 140) and nine moles of a decamer (mol. wt. = 10 × 28 = 280):

$$\overline{M}_n = \frac{4(5 \times 28) + 9(10 \times 28)}{13} = 237$$

$\overline{M}_n$ is very sensitive to changes in the percentage of low molecular weight species in a polymer and relatively insensitive to changes in the percent of high molecular weight species. Common molecular weight measurements that depend on colligative properties are vapor pressure lowering, boiling point elevation, and osmotic pressure. Most molecular weights of polymer solutions are determined by various kinds of osmometry.

14.350 Weight-Average Molecular Weight, $\overline{M}_w$

This measure of molecular weight is obtained experimentally by light scattering techniques. The measurement depends on the fact that the intensity of scattering is proportional to the square of the particle mass. $\overline{M}_w$ is calculated as follows:

$$\overline{M}_w = \frac{\sum_{i=1}^{\infty} N_i M_i^2}{\sum_{i=1}^{\infty} N_i M_i}$$

where the quantities are defined as in the preceding section.

Examples. Assuming the same sample as in the preceding section:

$$\overline{M}_w = \frac{4[(5 \times 28)^2] + 9[(10 \times 28)^2]}{4(5 \times 28) + 9(10 \times 28)} = 255$$

Note that $\overline{M}_w$ is greater than $\overline{M}_n$ and this is almost always true. $\overline{M}_w$ is very sensitive to changes in the fraction of high molecular weight species. The difference between $\overline{M}_n$ and $\overline{M}_w$ can also be illustrated by comparing their values when, on one hand, equal concentrations of two species are present in a polymer and, on the other hand, when equal weights of the same two species are present, Table 14.350. As can be seen from the table, the number-average molecular weight of an equal molar mixture is equal to the weight-average molecular weight of an equal weight mixture. The statistical distribution of molecular weights in a typical polymer is shown in Fig. 14.350, where the percentage of polymer per unit interval of molecular weight is plotted against the molecular weight.

Table 14.350 Combinations of species A (mol. wt. = 40,000) and species B (mol. wt. = 100,000)

Mol. wt.	Equal molar	Equal weight
$\overline{M}_n$	70,000	57,143
$\overline{M}_w$	82,857	70,000

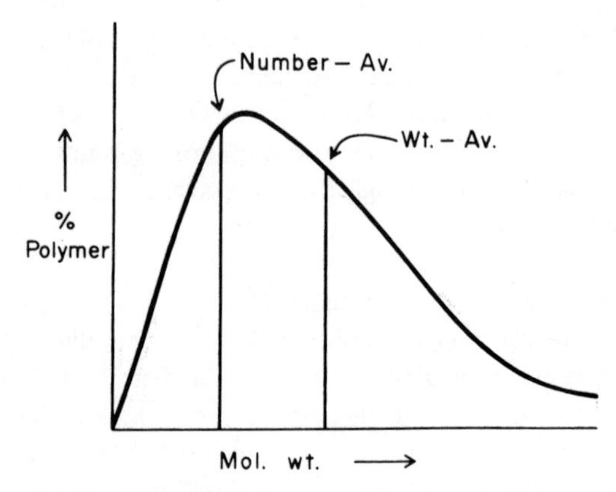

Figure 14.350 The distribution of molecular weights in a polymer.

14.360 Low-Density Polyethylene

A branched polyethylene polymer, partially crystalline, that melts at about 115°C and has a density of 0.91-0.94 g cm^{-3}. The branches can be as long as the main chain or as short as one to four carbon atoms, but most of the short branches consist of ethyl and *n*-butyl groups. Low-density polyethylene is prepared at 1000-3000 atm with traces of oxygen to initiate the polymerization.

14.370 High-Density Polyethylene

Essentially a linear polymer; it is highly crystalline, melts at about 135°C, and has a density of 0.95-0.97 g cm^{-3}. Linear polyethylene has greater tensile strength and hardness than the low-density variety. It is frequently prepared via Ziegler-Natta catalysis (Sect. 14.290).

14.380 Condensation Polymer

The polymer that results when bifunctional monomers react with each other through the intermolecular loss of small molecules such as water or alcohols.

Examples. Polyesters, Fig. 14.380*a*, and polyamides, Fig. 14.380*b*.

$$n\ HO{-}\overset{\overset{O}{\|}}{C}{-}(CH_2)_a{-}\overset{\overset{O}{\|}}{C}{-}OH \quad + \quad n\ HO(CH_2)_bOH$$

$$\downarrow$$

$$HO{-}\overset{\overset{O}{\|}}{C}{-}(CH_2)_a{-}\overset{\overset{O}{\|}}{C}{\left(O(CH_2)_bO{-}\overset{\overset{O}{\|}}{C}{-}(CH_2)_a{-}\overset{\overset{O}{\|}}{C}\right)_{n-1}}{-}O(CH_2)_bOH \quad + \quad (2n-1)H_2O$$

(*a*)

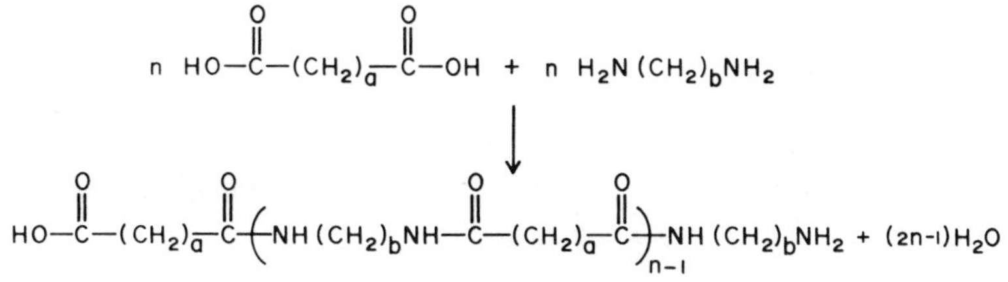

$$n\ HO-\overset{O}{\overset{\|}{C}}-(CH_2)_a-\overset{O}{\overset{\|}{C}}-OH + n\ H_2N(CH_2)_bNH_2$$

$$\downarrow$$

$$HO-\overset{O}{\overset{\|}{C}}-(CH_2)_a-\overset{O}{\overset{\|}{C}}\left(NH(CH_2)_bNH-\overset{O}{\overset{\|}{C}}-(CH_2)_a-\overset{O}{\overset{\|}{C}}\right)_{n-1}NH(CH_2)_bNH_2 + (2n-1)H_2O$$

(*b*)

Figure 14.380 Condensation polymers: (*a*) polyester ($a = 4, b = 2$), poly(ethylene adipate); (*b*) polyamide ($a = 4, b = 6$), poly(hexamethyleneadipamide, nylon 66).

14.390 Nylon

A generic term for any long chain synthetic polyamide polymer. Practically all nylons are capable of being formed into a filament in which the structural chains are oriented in the direction of the axis of the filament.

Examples. The polyamide shown in Fig. 14.380*b* is a nylon. Such nylons are usually designated by a two digit number that indicates the number of carbon atoms in the diamine component followed by the number of carbon atoms in the diacid component. Thus a nylon prepared from hexamethylenediamine and adipic acid, Fig. 14.380*b* ($a = 4, b = 6$), is a six-six (66) nylon. Polyamides can also be prepared from ω-amino acids, in which case they are given a single digit designation. The self-condensation nylon prepared from ϵ-aminocaproic acid, or its lactam, is designated nylon 6, Fig. 14.390. The lactam polymerization requires a small quantity of initiator, such as water. Nylons 3, 4, 6 (Perlon), 7 (Enant), 8, 11 (Rilsan), and 12 have commercial value and are prepared either from a lactam or from an ω-amino acid.

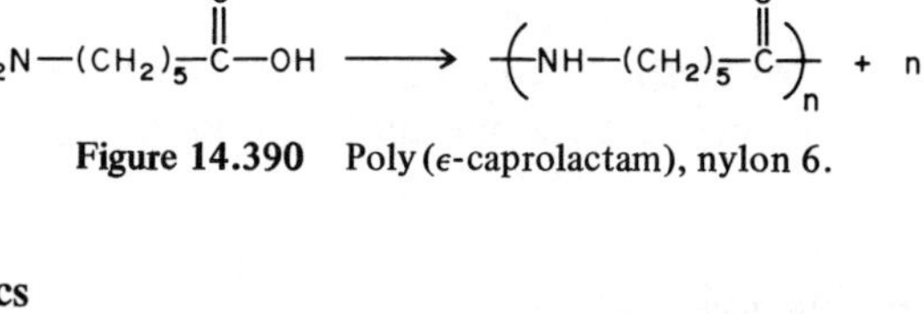

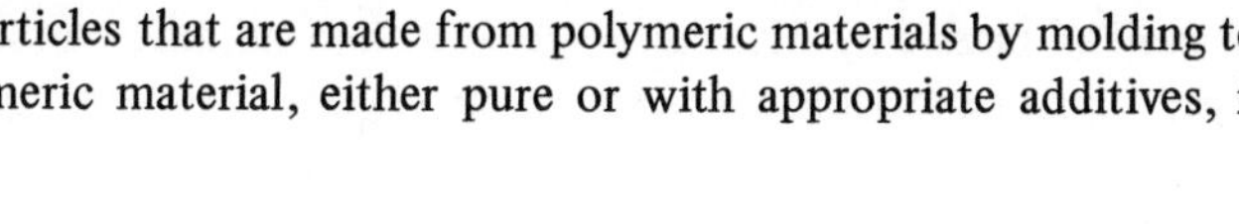

$$\text{(cyclic lactam: } HN-C(=O)-CH_2-CH_2-CH_2-CH_2-CH_2 \text{ ring)} \xrightarrow{\Delta} \left(NH-(CH_2)_5-\overset{O}{\overset{\|}{C}}\right)_n$$

$$n\ H_2N-(CH_2)_5-\overset{O}{\overset{\|}{C}}-OH \longrightarrow \left(NH-(CH_2)_5-\overset{O}{\overset{\|}{C}}\right)_n + n\ H_2O$$

Figure 14.390 Poly(ϵ-caprolactam), nylon 6.

14.400 Plastics

Finished articles that are made from polymeric materials by molding techniques. The polymeric material, either pure or with appropriate additives, is called a resin, and the molding to give a plastic is usually achieved by subjecting the resin to heat and/or pressure.

14.410 Plasticizer

A relatively nonvolatile inert compound mechanically incorporated into a polymer to increase its flexibility and workability.

14.420 Thermoplastics

Plastics that reversibly soften and harden on heating and cooling. The polymer in such plastics usually is a single strand or linear polymer with few if any cross-links and such polymers either swell appreciably, or are soluble, in organic solvents.

14.430 Thermosetting Plastics

Plastics that are permanently hardened and do not reversibly soften on heating. Such plastics contain a cross-linked polymer network that extends through the plastic article, making it stable to heat and insoluble in organic solvents.

Example. Phenol-formaldehyde plastics were the first thermosetting materials invented. Such plastics are called Bakelite after their discoverer L. Baekeland (1863–1944). The structure is shown in Fig. 14.430.

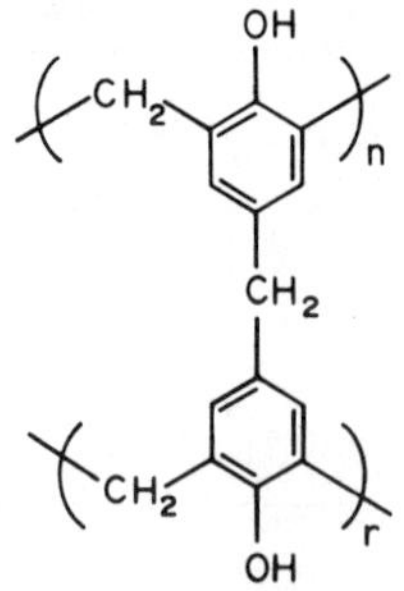

OH
CH_2
n
CH_2
CH_2
r
OH

Figure 14.430 The structure of Bakelite.

14.440 Molding

Forcing a finely divided plastic to conform to the shape of a cavity or form or mold, usually by applying heat and pressure to make the plastic flow into the form.

14.450 Compression Molding

Forming a molded plastic by placing the polymer in a stationary molding cavity and forcing the matching mold into the polymer while it is being heated. Both thermoplastic and thermosetting polymers may be compression molded, but if the former is used, the mold must be cooled before the pressure is released.

14.460 Injection Molding

Forming a molded plastic by heating the polymer, usually in a cylinder carrying a plunger, until it can flow and then forcing it, hydraulically, through a nozzle into a relatively cold, closed mold cavity. Most thermoplastic materials are molded by this technique. The most common configuration of an injection molding machine is a reciprocating screw machine. The polymer feed is first softened by heating and delivered to the mold by a screw arrangement rotating in a cylinder.

14.470 Epoxy Resins

Relatively low molecular weight polymers formed by condensing an excess of epichlorhydrin with a dihydroxy compound, most frequently 2,2,-bis(*p*-hydroxyphenyl)propane, commonly called bisphenol A.

Example. The condensation shown in Fig. 14.470 leaves epoxy end groups that are then reacted in a separate step (cured) with nucleophilic compounds (alcohols, acids, or amines). For use as a household adhesive, the epoxy resin and the curing resin, usually a polyamine, are packaged separately and mixed together immediately before use.

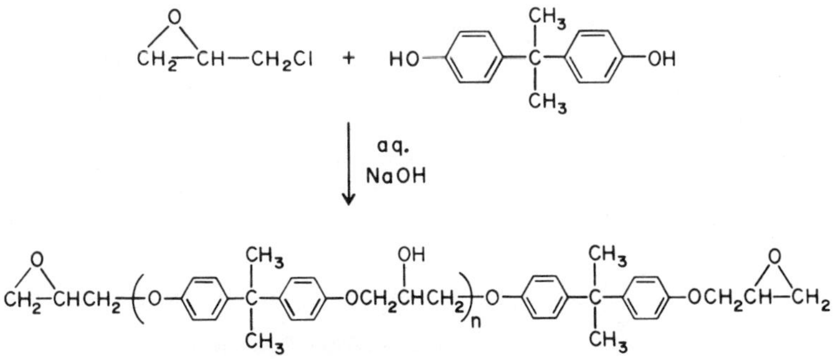

Figure 14.470 The formation of an epoxy resin.

14.480 Foamed Polyurethanes

The formation of a polyurethane polymer, Fig. 14.090*b*, accompanied by the liberation of carbon dioxide formed according to the typical reaction:

$$R{-}N{=}C{=}O + H_2O \longrightarrow RNHCO_2H \longrightarrow RNH_2 + CO_2$$

The gas liberated during the polymerization causes the polymer to foam or froth and thereby generates a finished product that is cellular in structure.

Examples. In the most simple model a prepolymer prepared from a polymeric diol and excess diisocyanate is treated with water (or carboxylic acids) in the presence of a tertiary amine catalyst to give the reactions shown in Fig. 14.480. The formation of the amino group occurs while free isocyanate groups are still present, and the amine-isocyanate reaction leads to disubstituted urea formation and chain lengthening. The protons on nitrogen atoms of the urethane and urea linkages can also react with free isocyanate groups to generate cross-linkages. Urethane foams can be made in either flexible or rigid form, the latter being highly cross-linked, very rigid and very strong.

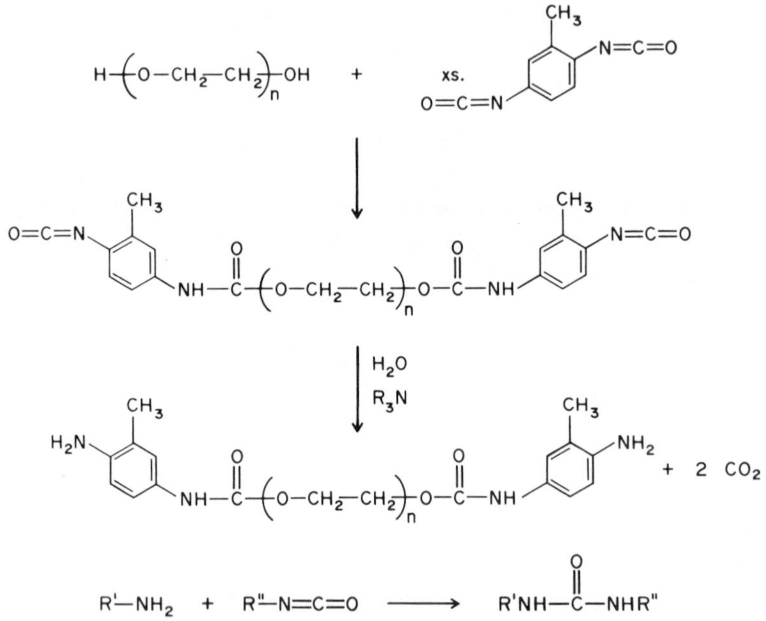

$$H{-}(O{-}CH_2{-}CH_2)_n{-}OH \quad + \quad \text{xs. } O{=}C{=}N{-}C_6H_3(CH_3){-}N{=}C{=}O$$

$$\downarrow$$

$$O{=}C{=}N{-}C_6H_3(CH_3){-}NH{-}\overset{O}{\overset{\|}{C}}{-}(O{-}CH_2{-}CH_2)_n{-}O{-}\overset{O}{\overset{\|}{C}}{-}NH{-}C_6H_3(CH_3){-}N{=}C{=}O$$

$$\downarrow H_2O,\ R_3N$$

$$H_2N{-}C_6H_3(CH_3){-}NH{-}\overset{O}{\overset{\|}{C}}{-}(O{-}CH_2{-}CH_2)_n{-}O{-}\overset{O}{\overset{\|}{C}}{-}NH{-}C_6H_3(CH_3){-}NH_2 \quad + \quad 2\ CO_2$$

$$R'{-}NH_2 \quad + \quad R''{-}N{=}C{=}O \longrightarrow R'NH{-}\overset{O}{\overset{\|}{C}}{-}NHR''$$

Figure 14.480 Reactions leading to foamed polyurethanes.

14.490 Silicones

Low molecular weight polymers containing Si—O—Si linkages that are produced by intermolecular condensation of silanols. The silanols in turn are generated by hydrolysis of halosilanes.

Examples. Silicone oils, typically, are made by reactions represented in Fig. 14.490. The silicones are not wet by water, hence are used for waterproofing.

They have excellent low temperature properties (i.e., they don't become viscous or crystalline) and remarkable stability to high temperatures.

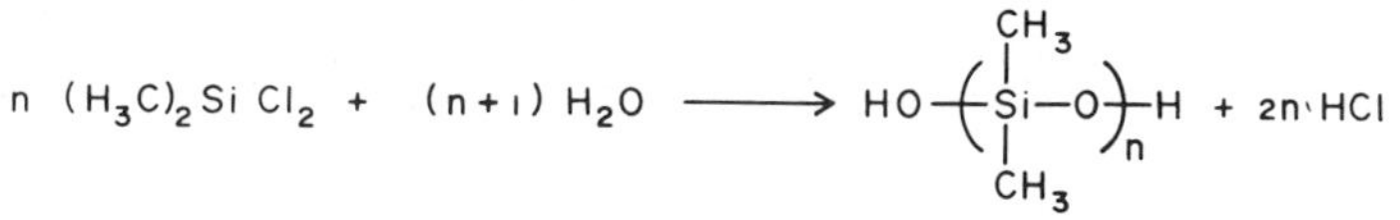

$$n\ (H_3C)_2SiCl_2 + (n+1)\ H_2O \longrightarrow HO\!\left(\begin{matrix} CH_3 \\ | \\ Si-O \\ | \\ CH_3 \end{matrix}\right)_n\!H + 2n\ HCl$$

Figure 14.490 Silicone formation.

14.500 Fibers

Slender, threadlike materials, natural or synthetic, that can be woven into fabrics. They are characterized by great tensile strength in the longitudinal direction of the fiber. The classification of a substance as a fiber is rather arbitrary; usually a fiber is a material that has a length at least 100 times its diameter. Of course, artificial or synthetic fibers can be made into any desired ratio of length to diameter.

Examples. Cotton, wool, silk, and flax have lengths 1000-3000 times their diameter; coarser fibers such as jute and hemp have lengths 100-1000 times their diameter. The polymer molecules of a fiber must be at least about 100 nm long (when extended), hence have a molecular weight of at least 10,000.

14.510 Animal Fibers

Fibers made up of protein molecules manufactured by animals or insects; principally wool and silk.

14.520 Wool

The hair of certain animals, notably sheep. The hair consists of proteins made from about 20 different amino acids. The hair strands are considered to be made up principally of interacting protein chains, highly cross-linked through hydrogen bonding, sulfur bridges, or saltlike ($RCO_2^- \cdots {}^+NH_3R'$) ionic interactions or combinations of these.

14.530 Silk

Fibrous proteins produced by spiders, butterflies and moths. The cocoon spun by the larva of the moth *Bombyx mori* provides most of the silk used in fabrics. The threads of *B. mori* silk consist of a highly oriented fibrous protein, called **fibroin.** Fibroin has a high content (86%) of the small amino acids glycine, alanine, and serine.

14.540 Leather

Most of the skin, bone, and muscle of animals consists of the complex protein called **collagen.** Collagen has a high content of the amino acids proline and hydroxyproline and resembles a block copolymer. The skin of animals is converted into leather by the introduction of stable cross-links between the collagen strands in a process called **tanning.** Formaldehyde is one of the most simple tanning agents.

14.550 Vegetable Fibers

Vegetable fibers are obtained from the stems, leaves, or seed hairs of plants, the chief fibrous constitutent of which is the polymer cellulose (Sect. 6.500).

14.560 Cotton

A pure form of cellulose obtained from the seed hairs of the plant *Gossypium spp.* Its structure consists of a partially crystalline polymer containing four cellobiose units in the unit cell. Hydrolysis of cotton with hot, dilute, mineral acid ultimately yields D-glucose.

14.570 Spinning

The conversion of bulk polymer to fiber form. Usually the polymer is dissolved and then metered through orifices into a vertical, heated chamber where the volatile solvent is removed in a stream of air and the remaining threadlike polymer is pulled out of the chamber as a yarn. This is called **dry spinning.** In the **wet spinning** process the polymer is coagulated or precipitated from solution and the polymer removed as yarn. In the **melt spinning** process polymer chips are melted and the melt is pumped through an extrusion nozzle and into a cooling chamber. The resolidified polymer is removed as yarn. The extrusion nozzle is called a spinneret and may consist of a 3 in. steel disk about $\frac{1}{4}$ in. thick into which is drilled 50–60 holes 0.010 in. or less in diameter. Spinning develops molecular orientation due to the intermolecular interaction between polar groups on neighboring chains. As the filaments become solid, they are brought together and wound on spools.

14.580 Denier

A measure of the size of spun yarn equal to the weight in grams of 9000 m of the fiber. The filaments that emerge from a spinneret may be 2 or 3 deniers and the size of the thread that is made by bringing such filaments together may vary from 15 to 500 deniers. The smaller the denier, the finer is the thread.

14.590 Rayon

The fiber that results from solubilizing cellulose and then regenerating it by some form of precipitation. Rayon is thus a **regenerated cellulose. Viscose rayon** is obtained by reacting the hydroxy groups of cellulose with carbon disulfide in the presence of alkali to give xanthates. When this solution is spun into an acid medium, the reaction is reversed and the cellulose is coagulated.

Examples. Cellulose such as cotton linters (the fibers adhering to the cotton seeds) is treated with carbon disulfide (approximately one mole per C_6 unit), and the mixture is dissolved in 3% NaOH to give a viscous solution called **viscose** in which the average chain length is 400-500 C_6 units. After a certain period of aging the solution is put through a spinneret and the filaments passed through a bath of $NaHSO_4$ and additives to give a regenerated cellulose. The reactions involved are shown in Fig. 14.590. Solubilization of the cellulose can also be achieved by treatment with cupric ammonium hydroxide, prepared from copper sulfate and ammonia. **Cuprammonium rayon** is obtained by spinning the solution into an acid coagulating bath.

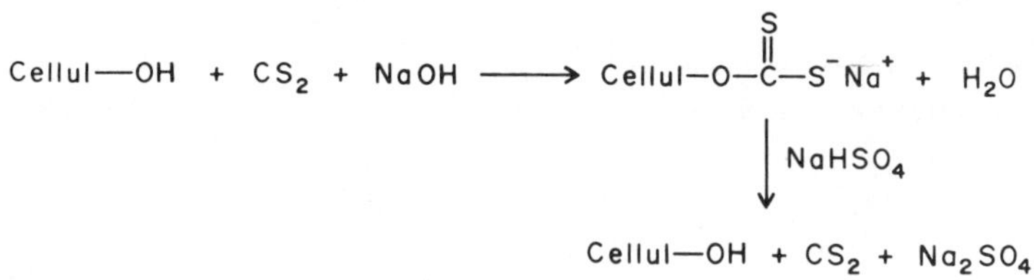

$$\text{Cellul—OH} + CS_2 + NaOH \longrightarrow \text{Cellul—O—}\overset{\overset{S}{\|}}{C}\text{—}S^-\,Na^+ + H_2O$$

$$\downarrow NaHSO_4$$

$$\text{Cellul—OH} + CS_2 + Na_2SO_4$$

Figure 14.590 Reactions for preparing viscose rayon.

14.600 Synthetic Fibers

Crystalline polyamides, polyesters, poly(acrylonitrile), polyolefins, polyurethanes, and practically all polyvinyl type polymers can be spun into fibers. These are all synthetic fibers. Graft polymers can also be spun.

Examples. The structures and names of some commercial synthetic fibers are listed in Table 14.600. Synthetic fiber-forming polymers are characterized by high tensile strength and high modulus of stiffness. Some fibers possess these properties as well as resistance to heat to an unusually high degree, and their use enables the production of light-weight reinforced structures and components. This is particularly true of some of the aromatic polyamides that have been given the generic name **aramids.** One of these, Kevlar, poly(*p*-benzamide), Table 14.600 can be fabricated into cables and ropes that possess a breaking stress higher than steel wire of identical diameter but are only approximately one-sixth the weight of the steel wire. Most **carbon fibers** are produced by the controlled thermal treatment of acrylic fibers. The acrylic is heated in air at 200–300°C while held under tension to give orientation. The oxidized fibers are then carbonized in an inert atmosphere by gradual heating to 1500°C. Finally, the carbon is graphitized by heating to 2500–3000°C.

Table 14.600 Some commercial synthetic fibers

Fiber	Name
1. Polyamides	
(a) $\text{—(NH(CH}_2)_5\text{CO)}_n\text{—}$	Nylon 6
(b) $\text{—(NH(CH}_2)_6\text{NHCO(CH}_2)_4\text{CO)}_n\text{—}$	Nylon 66
(c) 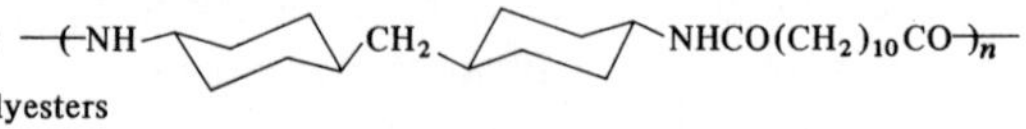 $\text{—(NH—C}_6\text{H}_{10}\text{—CH}_2\text{—C}_6\text{H}_{10}\text{—NHCO(CH}_2)_{10}\text{CO)}_n\text{—}$	Qiana
2. Polyesters	
(a) 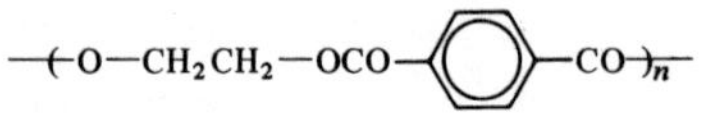 $\text{—(O—CH}_2\text{CH}_2\text{—OCO—C}_6\text{H}_4\text{—CO)}_n\text{—}$	Terylene; Dacron; Mylar
(b) 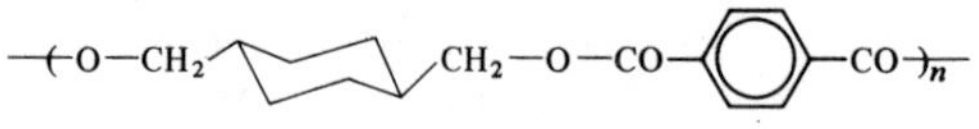 $\text{—(O—CH}_2\text{—C}_6\text{H}_{10}\text{—CH}_2\text{—O—CO—C}_6\text{H}_4\text{—CO)}_n\text{—}$	Kodel
(c) 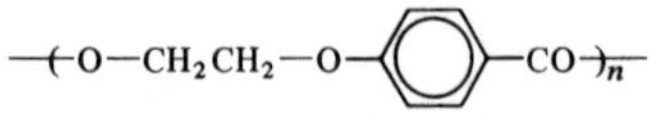 $\text{—(O—CH}_2\text{CH}_2\text{—O—C}_6\text{H}_4\text{—CO)}_n\text{—}$	A-Tel
3. Acrylic Fibers and Mod(ified)acrylic Fibers	
(a) $\text{CH}_2\text{=CH(CN)}$ (85%) + other vinyl monomers (15%)	Orlon, Acrilan, Creslan
(b) $\text{CH}_2\text{=CH(CN)}$ (~55%) + $\text{CH}_2\text{=CH(Cl)}$ (~45%)	Dynel
4. Elastomeric Fibers	
(a) Polyurethanes: 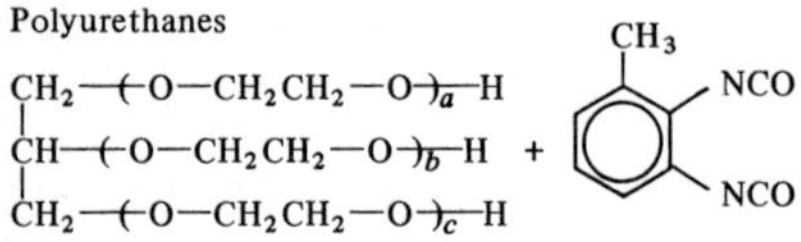 $\text{CH}_2\text{—(O—CH}_2\text{CH}_2\text{—O)}_a\text{—H}$; $\text{CH—(O—CH}_2\text{CH}_2\text{—O)}_b\text{—H}$; $\text{CH}_2\text{—(O—CH}_2\text{CH}_2\text{—O)}_c\text{—H}$ + $\text{CH}_3\text{C}_6\text{H}_3\text{(NCO)}_2$	Spandex
5. Aramids	
(a) 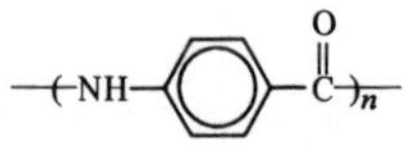 $\text{—(NH—C}_6\text{H}_4\text{—C(=O))}_n\text{—}$	Kevlar

14.610 Elastomers.

A generic name for polymers that exhibit rubberlike elasticity; i.e., they can be stretched several hundred percent under tension and, when the stretching force is removed, they retract rapidly and recover their original dimensions. The retractive force is entropy based, as the polymer returns to its most probable conformation after being stretched.

14.620 Rubber

The name given to the solid polymeric material isolated from the white fluid or latex found in a variety of plants and trees. It was called rubber by Joseph Priestley (1733-1804), who used it to erase or rub out pencil marks. It now has a generic meaning and applies to the class of substances, regardless of origin, that have elastomeric properties.

14.630 Natural Rubber

The rubber isolated from natural sources such as trees and plants. The principal commercial source is from the rubber tree, *Hevea brasiliensis.* It is a polymer of isoprene, $-\!\!(CH_2-CH(CH_3)=CH-CH_2)_n\!\!-$, and the configuration around each double bond is *cis* (or *Z*), Fig. 14.630.

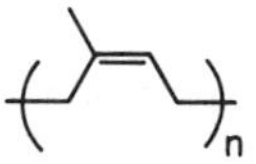

Figure 14.630 Natural rubber polymer.

14.640 Gutta-percha

A natural rubber obtained from the latex cells of certain specific plants. It is a polymer of isoprene, but its configuration around each double bond is *trans* (or *E*), Fig. 14.640. In contrast to the all-*cis* diastereoisomer, natural rubber (Sect. 14.630), it is hard and horny.

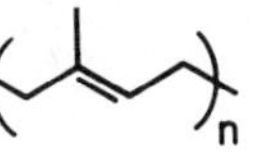

Figure 14.640 Gutta-percha polymer.

14.650 Vulcanization

Originally the reaction of rubber hydrocarbons with elemental sulfur at about 150-200°C to produce sulfur cross-links between the chains of the rubber polymer. This process (Charles Goodyear, 1800-1860) makes the rubber flexible, insoluble, and elastic. In general any agent that causes cross-linking of the linear rubber polymer, e.g., certain peroxides, is a vulcanization agent. When finely divided fillers, called reinforcing agents, such as carbon black are added to give the vulcanized rubber abrasion and tear resistance, the mixture can be used in tire manufacture.

14.660 Synthetic Rubbers

The polymers with rubberlike characteristics prepared synthetically from dienes or olefins. Other polymers such as polyurethanes, fluorinated hydrocarbons, and polyacrylates can be synthesized to generate rubbers with special properties.

Examples. Styrene-butadiene rubber (SBR) is prepared from the free radical copolymerization of one part by weight of styrene and three parts of butadiene.

The butadiene is incorporated by both 1,4-(80%) and 1,2-addition (20%), and the configuration around the double bond of the 1,4-adduct is about 80% *trans*. It is a random copolymer and a typical portion of the polymer is shown in Fig. 14.660; several types of commercial synthetic rubbers are listed in Table 14.660.

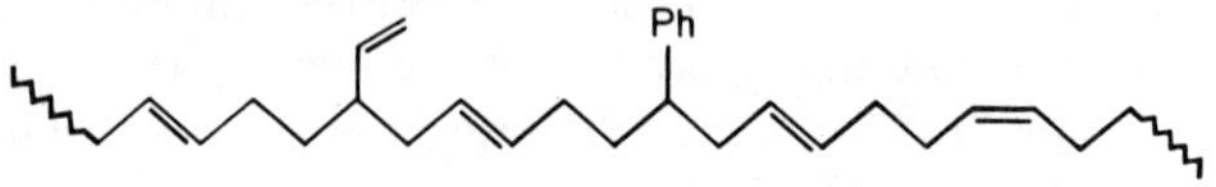

Ph

Figure 14.660 A portion of the butadiene-styrene copolymer chain.

Table 14.660 Some commercial synthetic rubbers

Name	Monomers
GRS, Buna S, SBR	1,3-Butadiene and styrene
Neoprene	Chloroprene (2-chloro-1,3-butadiene)
cis-Polybutadiene, BR	1,3-Butadiene ⟶ (mostly) 1,4-addition; *Z* config.
cis-Polyisoprene, IR	Isoprene ⟶ 1,4-addition (only); *Z* config.
Butyl rubber, IIR, GRI	Isobutylene (acid catalyzed)
Nitrile Rubber, NBR, GRN, Buna N	1,3-Butadiene (2 parts) + acrylonitrile (1 part)
Ethylene-propylene rubber, EPR	Ethylene (60 parts) + propylene (40 parts)
Polysulfide Rubbers, Thiokol ST	1,2-Dichloroethane + Na_2S_4 ⟶ $HS(CH_2CH_2SSSS)_nCH_2CH_2SH$
Silicone Rubbers	$(CH_3)_2SiCl_2 \xrightarrow{\text{several steps}} -\!\!(Si(CH_3)_2-O)_n\!-$
Tygon	Poly(vinyl chloride) + special plasticizers

14.670 Foamed Rubber

Rubber that has been mixed with a compound so that on heating the softened mixture liberates a gas such as nitrogen, forming a low-density, cellular rubber full of gas bubbles. The most commonly used nitrogen generator is the thermally unstable *N,N'*-dinitrosopentamethylenetetramine, Fig. 14.670. When the nitrogen is liberated, the resulting product acts as a cross-linking agent.

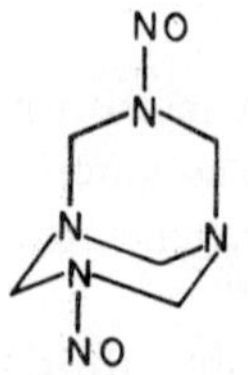

NO
N
N N
N
NO

Figure 14.670 Nitrogen precursor (*N,N'*-dinitrosopentamethylenetetramine or 3,7-dinitroso-1,3,5,7-tetraazobicyclo[3.3.1]nonane) used for foaming rubber.

14.680 Elastomeric Fibers

Fibers that possess elastic properties. The most common such fibers are polyurethanes (Sect. 14.090), which are prepared from polymeric polyols (e.g., Fig. 14.680), and diisocyanates.

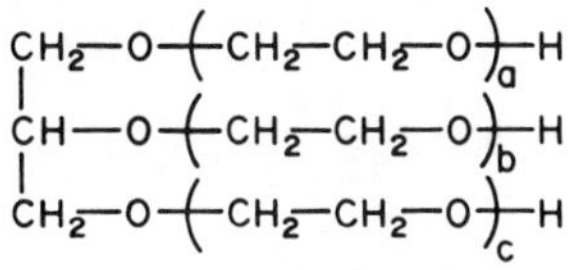

$$
\begin{array}{l}
CH_2-O-(CH_2-CH_2-O)_a-H \\
| \\
CH-O-(CH_2-CH_2-O)_b-H \\
| \\
CH_2-O-(CH_2-CH_2-O)_c-H
\end{array}
$$

Figure 14.680 Polymeric polyol used to prepare polyurethane elastomeric fibers.

14.690 Poly(phosphazenes), $-(N{=}PX_2)_n-$

Although the early polymers of this structure were completely inorganic (X = Cl), the halogens are replaceable by organic groups and a host of such polymers containing organic groupings is now known (e.g., X = CH_3, OR, OCH_2CF_3, $NHCH_2CO_2C_2H_5$).

Examples. The poly(phosphazenes) can be prepared from the reaction between NH_4Cl and PCl_5, Fig. 14.690. All the chlorines can be replaced by alkoxy or aryloxy groups to give poly[bis(alkoxy or aryloxy)] phorphazenes. The polymers have unusual properties, especially high flexibility.

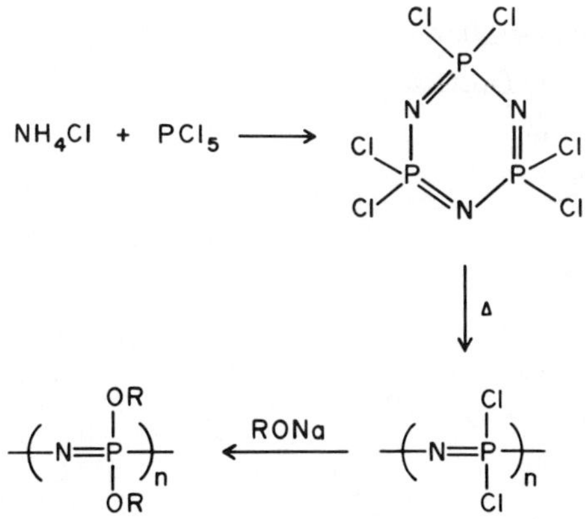

$$NH_4Cl + PCl_5 \longrightarrow (NPCl_2)_3 \text{ (cyclic trimer)} \xrightarrow{\Delta} -(N{=}PCl_2)_n- \xrightarrow{RONa} -(N{=}P(OR)_2)_n-$$

Figure 14.690 The preparation of poly[bis(alkoxy or aryloxy)] phosphazenes.

15 Fossil Fuels, Synthesis Gas and Industrial Processes

CONTENTS

15.010 Petroleum

An oily, colored (from amber to black, depending on origin) complex liquid mixture of many organic compounds, principally hydrocarbons, found beneath the earth's surface and believed to have been developed from decaying organic material in a marine environment.

15.020 Crude Oil

Untreated petroleum as it is extracted from the earth. Quantities are usually expressed in barrels, defined in the petroleum industry as containing 42 gal. Crude oil is frequently fractionally distilled without prior treatment to give the various boiling fractions shown in Table 15.020.

15.030 Sour Crude

Crude oil containing relatively large amounts of sulfur compounds, hence requiring special refining techniques. A typical Indonesian oil may contain less than 0.1% sulfur and is classed as a **sweet crude**, but the sour crude from Saudi Arabia may contain more than 1.5% sulfur.

Table 15.020 Composition of crude oil

Boiling point range, °C	Name of fraction	Specific gravity
30–180	**Gasoline**[a]	0.75
150–260	**Kerosine**[b]	0.78
180–400	**Gas Oil**[c]	0.84
	Fuel Oil[d]	0.96

[a]The gasoline fraction obtained directly from crude oil before any processing is called **straight-run gasoline** and/or **naphtha.**

[b]This fraction is frequently used as a heating oil in space heaters.

[c]This fraction is used as a **diesel fuel.**

[d]This fraction is usually further fractionally distilled in vacuum to give distillate fractions used for central heating of homes (heating oils, numbers 1, 2, 3, and 4). The residual material, often called **bunker oils** (numbers 5 and 6), are heavy viscous oils that are used as boiler fuels for power production.

15.040 API Gravity

An arbitrary scale adopted by the American Petroleum Institute (API) for comparing the density (hence, in part, the quality) of petroleum oils:

$$°\text{API} = \frac{141.5}{\text{specific gravity } (60°\text{F}/60°\text{F})} - 131.5$$

15.050 Thermal Cracking

The reduction in molecular weight of various fractions of crude oil through pyrolysis. Thermal cracking is mainly used to produce a mixture rich in ethylene and propylene. Typically a naphtha fed stock (b.p. 30–180°C) is passed through a coiled tube heated by a furnace. Process conditions are: temperature of 750–900°C; short contact or residence time at temperature (approximately 0.1–0.5 s); low feed partial pressure achieved by mixing with steam; and rapid quenching.

Example. The mechanistic scheme that best explains the course of the conversion is the free radical chain mechanism consisting of the steps initiation; propagation, and termination. This sequence is shown in Fig. 15.050. Of course, not all the possible reactions are shown. Because steam is used, this process is also called **steam cracking.** The steam diluent favors unimolecular processes, hence reduces the rate of termination. When steam cracking units are operated to produce acetylene as the desired product, the temperature is raised to above 1200°C because unlike other hydrocarbons the free energy of formation of acetylene decreases with increasing temperature.

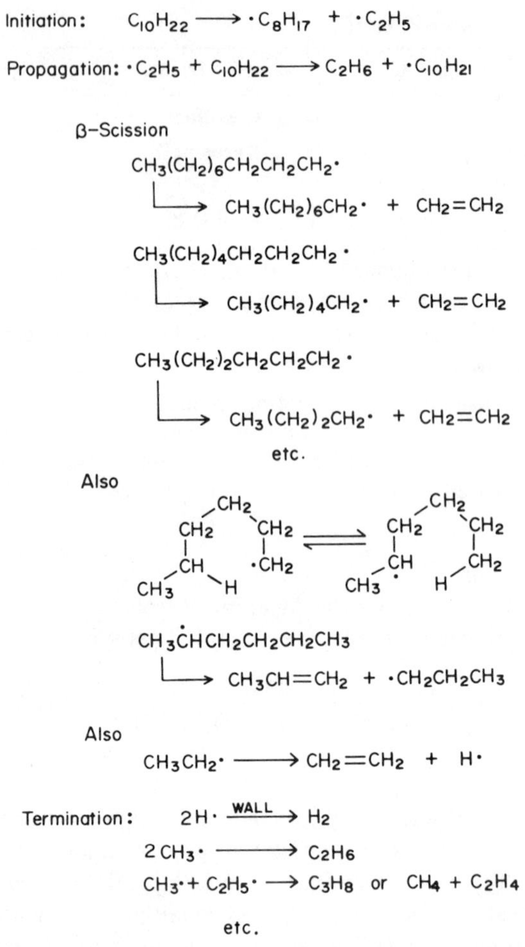

Initiation: $C_{10}H_{22} \longrightarrow \cdot C_8H_{17} + \cdot C_2H_5$

Propagation: $\cdot C_2H_5 + C_{10}H_{22} \longrightarrow C_2H_6 + \cdot C_{10}H_{21}$

β-Scission

$$CH_3(CH_2)_6CH_2CH_2CH_2\cdot \longrightarrow CH_3(CH_2)_6CH_2\cdot + CH_2{=}CH_2$$

$$CH_3(CH_2)_4CH_2CH_2CH_2\cdot \longrightarrow CH_3(CH_2)_4CH_2\cdot + CH_2{=}CH_2$$

$$CH_3(CH_2)_2CH_2CH_2CH_2\cdot \longrightarrow CH_3(CH_2)_2CH_2\cdot + CH_2{=}CH_2$$

etc.

Also

$$CH_3CH(H)CH_2CH_2CH_2CH_2\cdot \rightleftharpoons CH_3\dot{C}HCH_2CH_2CH_2CH_3$$

$$CH_3\dot{C}HCH_2CH_2CH_2CH_3 \longrightarrow CH_3CH{=}CH_2 + \cdot CH_2CH_2CH_3$$

Also

$$CH_3CH_2\cdot \longrightarrow CH_2{=}CH_2 + H\cdot$$

Termination:

$$2H\cdot \xrightarrow{\text{WALL}} H_2$$

$$2\,CH_3\cdot \longrightarrow C_2H_6$$

$$CH_3\cdot + C_2H_5\cdot \longrightarrow C_3H_8 \text{ or } CH_4 + C_2H_4$$

etc.

Figure 15.050 A typical free radical pyrolysis in thermal cracking.

15.060 Catalytic Cracking

The reduction in molecular weight of various fractions of crude oil, achieved by passing them over an active catalyst, usually at about 450–600°C. The catalysts are usually crystalline zeolites (Sect. 15.070) which function as strong acids. The molecular weight reduction involves carbocation intermediates.

Example. The generation of the initiating carbocation can proceed either by hydride abstraction from an alkane by the catalyst or by protonation of an olefin by surface protons on the catalyst. The small amount of olefin required can be formed by a thermal reaction. The carbocation resulting from either of these initiating processes will undergo β-scission to generate a primary carbocation. The primary carbocation will undergo 1,2-hydride migration to form a more stable secondary carbocation, and degradation proceeds by β-cleavage, Fig. 15.060. Such a mechanism accounts for the preponderance of propylene in catalytic cracking. Thermal cracking involves radical reactions and because radicals are not prone to undergo analogous 1,2-shifts, such cracking leads to a preponderance of ethylene, Fig. 15.050, even though β-scissions are also involved. Other reactions are of course possible.

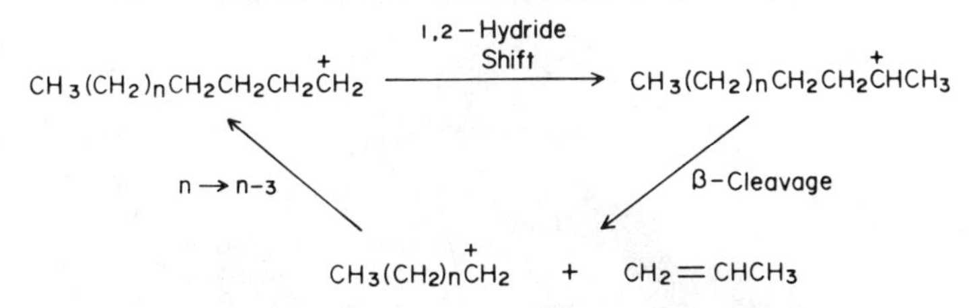

Figure 15.060 The carbocation mechanism for catalytic cracking.

15.070 Zeolites

A family of naturally occurring crystalline aluminosilicate minerals. The structure consists of a three-dimensional network of tetrahedra linked by shared oxygen atoms. The molecular formula may be expressed in terms of the oxides of aluminium and silicon by the molecular formula $M_{x/n}[(AlO_2)_x \cdot (SiO_2)_y] \cdot mH_2O$ where M is the cation of valence n, m is the number of water molecules, and $x + y$ is the total number of tetrahedra in the crystallographic unit cell. The framework of tetrahedra forms a network of uniform channels such as that shown in Fig. 15.070*a* for the synthetic zeolite **mordenite** ($y/x = 10$, 12 tetrahedra in a ring). These channels have room to accommodate neutral molecules of many kinds whose cross section does not exceed certain diameters (for mordenite this is 8.0 Å). Product selectivity occurs because only molecules of a certain size can diffuse in and out of the channels. The cations are located close to the apertures and changes in the cation result in substantial changes in pore size and hence adsorbent characteristics. The zeolites may be used as ion exchange resins, inclusion agents, drying agents (some zeolites may include as much as 265 ml of H_2O per g, which can be driven out by heating to 350°C without disturbing the structure) and catalysts. The naturally occurring **faujasite** zeolites having the formula $(Na_2Ca)_{32}[(AlO_2)_{64} \cdot (SiO_2)_{128}] \cdot 216\ H_2O$, Fig. 15.070*b*, have been studied extensively because of their catalytic properties. These are large pore zeolites; **erionite** and **chabazite** are naturally occurring small pore size zeolites. Most zeolites in use today are prepared synthetically.

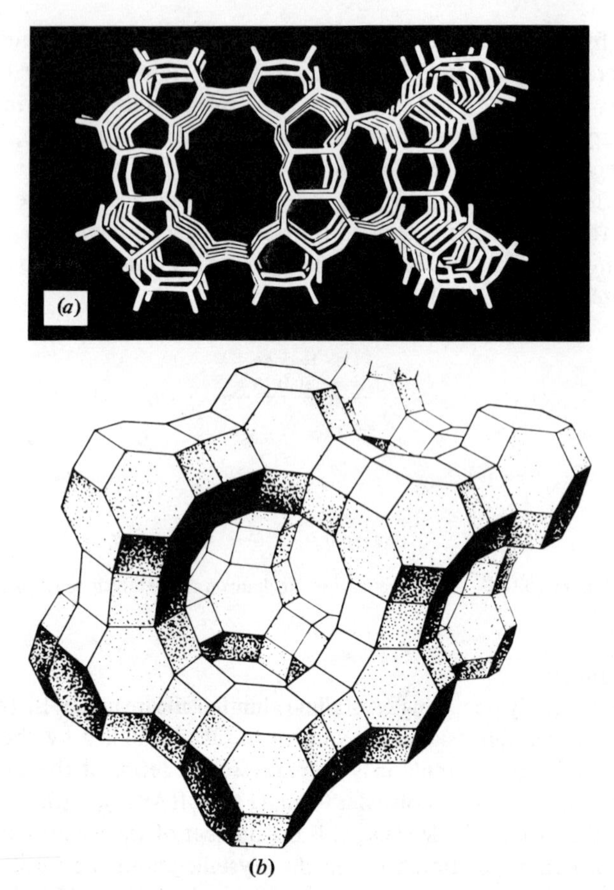

Figure 15.070 (*a*) The spatial arrangement of tetrahedra in the zeolite, mordenite (courtesy Norton Chemical Process Products); (*b*) the zeolite, faujasite (courtesy H. Heineman).

15.080 Molecular Sieves

Synonymous with zeolites.

15.090 Heterogeneous Catalysis

A catalytic system in which the catalyst constitutes a separate phase. The catalyst is usually a solid and the reactants and products are either gases or liquids. The catalytic reaction occurs at the surface of the catalyst, hence the rate of the reaction is sensitive to the surface area (Sect. 7.700) of the catalyst.

15.100 Homogeneous Catalysis

A catalytic system in which the catalyst and all the reactants are molecularly dispersed in one phase.

15.110 Dual Function (or Bifunctional) Catalyst

A catalyst that possesses two different catalytic sites, hence can catalyze two different kinds of reactions.

Examples. In the reforming reactions Sect. 15.150) of hydrocarbons a zeolite can be prepared that also has platinum on the surface, and therefore reactions that are acid catalyzed as well as noble metal catalyzed may proceed.

15.120 Fixed Bed Operation

A catalytic processing operation in which the catalyst is in the form of solid particles held stationary in a reactor. The reacting gases or liquids are pumped over the catalyst.

15.130 Fluidized Bed Operation

A catalytic processing operation in which the catalyst in the form of small solid particles is maintained in suspension by the upward flow of a gas at an appropriate velocity. This mode of operation ensures very intimate contact between the catalyst and reactants and allows for very rapid heat transfer.

15.140 Hydrocracking

Catalytic cracking processes carried out in the presence of 150-200 atm of hydrogen at about 450°C.

15.150 Reforming

A combination of dehydrogenation, isomerization, and hydrocracking reactions performed on distillates from crude oil by passing them over dual functional catalysts. The object is to prepare gasoline of high quality (high octane number, Sect. 15.170). Operating conditions are chosen to maximize aromatic content. Platinum metal (0.3-1.0 wt %) supported on pure, high surface area alumina (Al_2O_3) was for many years the preferred catalyst; the platinum acts as a catalytic site for hydrogenation-dehydrogenation reactions, and the alumina as a Lewis acid site that catalyzes isomerization reactions. Presumably both sites are capable of catalyzing hydrocracking reactions. Typical operation conditions are temperatures of 500-525°C and pressures of 10-40 atm. Feedstocks may have an octane number of 60, whereas the product (called **reformate**) can have an octane number of 100. In the past decade, a bimetallic catalyst consisting of platinum-rhenium has gradually replaced the single metal system. About half the gasoline used in the United States is produced by reforming reactions. In the older descriptions of fuel compositions, it was common to characterize fuels in terms of their percentages of **paraffins** (alkanes), **naphthenes** (cycloalkanes), and aromatics.

Examples. Typical reforming reactions for C-6 hydrocarbons are shown in Table 15.150 along with equilibrium constants and heats of reaction at 500°C for each of the reactions.

Table 15.150 Reforming reactions of C-6 hydrocarbons at 500°C

Reaction	K^a	ΔH mol^{-1}	
		kJ	kcal
cyclohexane ⇌ benzene + $3H_2$	6×10^5	221	52.8
CH_3–cyclopentane ⇌ cyclohexane	8.6×10^{-2}	-15.9	-3.8
$CH_3(CH_2)_4CH_3$ ⇌ benzene + $4H_2$	7.8×10^4	266	63.6
$CH_3(CH_2)_4CH_3$ ⇌ $CH_3CH_2CH_2CH(CH_3)CH_3$	1.1	-5.9	-1.4
$CH_3(CH_2)_4CH_3$ ⇌ $CH_3CH_2CH(CH_3)CH_2CH_3$	7.6×10^{-1}	-4.6	-1.1
$CH_3(CH_2)_4CH_3$ ⇌ $CH_3CH_2CH_2CH_2CH{=}CH_2 + H_2$	3.7×10^{-2}	130	31.1

[a]Partial pressures in atmospheres.

15.160 Alkylate

In the petroleum industry this refers to the product produced by the acid-catalyzed reaction between an olefin and an alkane **(alkylation)**. The reaction proceeds via carbocation intermediates and leads to a highly branched, low molecular weight product (alkylate) with a very high octane number.

Example. 2,2,4-Trimethylpentane (incorrectly called isooctane) is prepared by treating a mixture of isobutane (in excess) and isobutylene with a proton source derived from either H_2SO_4 or HF plus $AlCl_3$, according to the mechanism shown in Fig. 15.160.

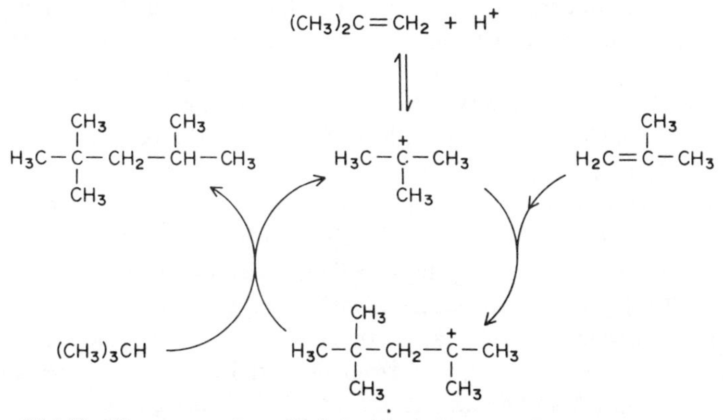

Figure 15.160 The preparation of 2,2,4-trimethylpentane (isooctane) by alkylation.

15.170 Octane Number

A number usually ranging from 60 to 100 assigned to a gasoline sample which reflects its relative performance as a fuel in the internal combustion engine. It is the volume percent of 2,2,4-trimethylpentane (100 octane) in a blend with heptane (0 octane) that gives the same intensity of engine **knock** (the sound produced by premature ignition of the fuels) under the same operating conditions as does the fuel sample being tested. A single cylinder engine with variable compression ratio is used for the evaluation. Octane numbers obtained in this test engine operating under mild engine severity and low speed (600 rpm, 52°C inlet temperature) are called **research octane numbers.** The research octane number (RON) for a typical leaded gasoline of regular grade is about 93 (1979). Octane numbers obtained with an engine operating under high engine severity and high speed (900 rpm, 150°C) are somewhat less than RONs and correlate well with actual road tests. Such octane numbers are called **motor octane numbers** (MON) and a gasoline with an RON of 93 might show a MON of 85. The octane number that is required to be posted at the gasoline pump is the average of the two octane numbers, (RON + MON)/2 (the **antiknock index**), which in the above example would be (85 + 93)/2 = 89. The **blending octane number** is the number assigned to an additive to gasoline that reflects the effect such an additive has on the octane number of the gasoline. The addition of 10 volume percent of ethanol to a gasoline having a RON of 87 results in a mixture (see Gasohol, Sect. 15.690) having a RON of 91.8. The blending research octane number of ethanol can be calculated to be 135:

$$(87 \times 0.9) + (y \times 0.1) = 91.8$$

$$\text{Blending RON}, y = 135$$

15.180 Leaded Gasoline

Gasoline to which has been added small quantities of liquid **tetraethyllead (TEL).** The TEL used commercially is a mixture of tetraethyllead, $Pb(C_2H_5)_4$ (63%); 1,2-dibromoethane, $BrCH_2CH_2Br$ (26%); and 1,2-dichloroethane, $ClCH_2CH_2Cl$ (9%); plus a red dye (2%). The purpose of the bromo compound (and to a lesser extent, the chloro compound) is to "scavenge" the lead oxide formed in the automobile cylinder by reacting with it to form volatile lead compounds. The addition of TEL (1-3 ml gal^{-1}) increases the research octane number by about seven numbers (points). Motor octane numbers over 100 can be secured by using 2,2,4-trimethylpentane to which some TEL has been added. An alternative anti-knock agent is η^5-methylcyclopentadienyltricarbonylmanganese, $[Mn(\eta^5\text{-}C_5H_4CH_3)(CO)_3]$ (in the trade called **MMT). Methyl *tert*-butyl ether (MTBE)** is becoming increasingly popular as an octane booster to replace metal organic compounds. It is readily prepared from methanol and isobutylene and is reported to have an RON of about 125 and an MON of about 100.

15.190 Cetane Number

A number given to a diesel fuel sample which is a measure of its performance in a compression engine. It is the volume percent of hexadecane (cetane) in a blend with 1-methylnaphthalene that gives the same performance in a standard compression engine as the fuel sample under the same operating conditions. Hexadecane is given an arbitrary cetane number of 100 and 1-methylnaphthalene a number of 0. The diesel engine charges air into a chamber and compresses it during intake. Liquid fuel is then injected into the compressed air, whereupon combustion occurs, generating the energy required for locomotion. Fuels with cetane numbers greater than 45 are required for good performance in a compression engine.

15.200 Diesel Fuel

A mixture of hydrocarbons boiling in the range 180–240°C, hence considerably less volatile than gasoline. In contrast to gasoline good diesel fuels are low in aromatic content and high in straight chain alkane content.

15.210 Space Velocity

The volume of feed per unit volume of catalyst per unit of time at operating conditions; the usual unit is hr^{-1}.

15.220 Contact Time

The time the feed is in contact with the catalyst; it is calculated (in seconds) by dividing the space velocity (hr^{-1}) into 3600s hr^{-1}.

15.230 Percent Conversion

The percent conversion is $[(F_0 - F_t)/F_0] \times 100$ where F_0 is the molar equivalents of feed before reaction, and F_t is the molar equivalents of feed at time t.

15.240 Selectivity

The ratio of the quantity of desired product over the quantity of feed converted. The percent selectivity is $[P_t/(F_0 - F_t)] \times 100$ where P_t is the molar equivalents of desired product at time t and F_0 and F_t are as defined above.

15.250 Yield

The fraction of feed converted at time t, multiplied by the selectivity; or stated equivalently, the quantity of desired product over the quantity of feed. The percent yield is $[(F_0 - F_t)/F_0] \times P_t/(F_0 - F_t) \times 100 = (P_t/F_0) \times 100$ where all the terms are the same as defined above.

15.260 Space Time Yield

The number of moles of product P per liter of catalyst per hour. In heterogeneous catalytic systems the bulk volume of the catalyst is measured, and in homogeneous catalytic systems the total volume of the feed solution containing dissolved catalyst is measured.

15.270 Turnover Number

In heterogeneous catalysis the number of moles of feed or substrate converted per number of catalytic sites. In homogeneous catalysis the number of moles of feed or substrate converted per mole of catalyst.

15.280 Turnover Rate

The turnover number per unit of time (usually the second, but the unit must be specified).

15.290 Lubricating Oils

The oils used to reduce friction between solid moving parts, usually obtained by vacuum distillation of a high boiling fraction (~375-475°C) of crude oil. Synthetic lubricating oils are becoming more common; typically these are esters of sebacic (1,10-decadioic) acid.

15.300 Greases

High boiling petroleum oils thickened with metallic soaps, such as lithium stearate, which are the salts of fatty acids.

15.310 Coal

Buried plant remains that have been chemically altered and consolidated through geologic time by heat and pressure. Coals exhibit a wide range of properties because of the different kinds and quantities of plant material and intermixed mineral matter from which they were formed and the extent of physical and chemical alteration before, during, and after burial.

15.320 Rank of Coal

A classification of coal based on the relative degree of alteration or metamorphosis of the plant material. The precursor to coal from the metamorphosis of wood and other plant material is peat, which is not considered coal. The least altered peat is the lowest rank (lignite) coal and the most altered is the highest rank (anthracite) coal, hence the rank of the coal corresponds roughly to its relative age. The classification of coals is made on the basis of the "fixed carbon" for the high rank coals and the calorific value for the lower rank coals, Table 15.320. The geologic age is also reflected in the elemental composition of the coal; older, more altered coals show progressively lower oxygen content and lower hydrogen to carbon ratios than younger, less altered coals.

15.330 Proximate Analysis of Coal

The direct analytical determination of three quantities (**moisture**, **volatile matter**, and **ash**) and the calculation of a fourth quantity called the **fixed carbon**. Moisture is obtained from loss of weight of a sample heated at 104–110°C; volatile matter as the percent loss in weight, other than moisture, when a separate sample is heated to 950°C; and ash as the percent residue remaining after combustion of a sample in a high temperature (~1500°C) furnace. The fixed carbon is 100 minus the sum of the other three quantities. All the analyses are carried out under specified controlled conditions.

15.340 Heating Value of Coal

The amount of heat liberated per unit weight, obtained by combustion at constant volume in an oxygen bomb calorimeter. The usual units of measurement are BTU's per pound of coal. [**British thermal unit**; the amount of energy necessary to raise the temperature of 1 lb of water by 1°F at 39.2°F (4°C); one Btu is equivalent to 1055.6 J or 252.3 cal.]

15.350 Banded Coal

Coal that appears to the eye to consist of a series of distinguishable bands whose visual classification and characterization constitute the **petrographic** descriptions of coal. These bands or layers are of three kinds, vitrain, attrital coal, and fusain, distinguishable by their degree of luster or shininess.

15.360 Vitrain

Shiny, lustrous, black layers in banded coal.

15.370 Fusain

Dull, black layers resembling charcoal in which the original plant tissue structure is preserved. It is very brittle.

Table 15.320 Classification of coal by rank

Classification	Group	Fixed carbon (%)[a]	Volatile matter (%)[a]	Caloric value (Btu lb^{-1})[b]
Anthracite	Metaanthracite	>98	–	
	Anthracite	92–98	2–8	
	Semianthracite	86–92	8–14	
Bituminous	Low volatile bituminous	78–86	14–22	
	Medium volatile bituminous	69–78	22–31	
	High volatile A bituminous	<69	>31	14,000
	High volatile B bituminous			13,000–14,000
	High volatile C bituminous			11,500–13,000
Subbituminous	Subbituminous A[c]			10,500–11,500
	Subbituminous A			10,500–11,500
	Subbituminous B			9,500–10,500
	Subbituminous C			8,300–9,500
Lignite	Lignite A			6,300–8,300
	Lignite B			<6300

[a]Dry, mineral matter-free basis, see Sects. 15.330 and 15.470.

[b]Mineral matter-free basis.

[c]A Subbituminous A coal that has agglomerating properties.

15.380 Attrital Coal

The layers of coal that vary in luster and are usually intermediate in this respect as compared to the other constituents because it is the matrix in which vitrain and fusain bands are embedded. It has a striated or rough texture. It is visibly heterogeneous, and accordingly some petrographers further classify attrital coal into **clairain** (bright luster similar to vitrain) and **durain** (dull luster).

15.390 Nonbanded Coal

Fine, granular coal in which bands are absent. The principal kinds of nonbanded coal are **cannel coal** and **boghead coal**, which are differentiated on the basis of the constitution of the waxy component **(exinite)** that is present.

15.400 Petrographic Composition, Coal Macerals

The classification of coals based on microscopic examination (200 or more magnification) and the observation of the distinctive kinds of organic matter (macerals) that are present in a cross-section sample. The macerals are analogous to minerals in inorganic rocks but do not have the stoichiometric integrity of composition characteristic of minerals. The kind of macerals present can be determined from reflectance measurements, the magnitude of which are mainly a measure of the increasing proportion and number of rings in polynuclear aromatic structures.

15.410 Mineral Matter in Coal

The various minerals associated with coal. Low temperature ashing using an oxygen plasma (radio frequency activated) at low pressure permits standard x-ray analysis of the original minerals associated with the coal sample.

15.420 Ash

The residue remaining from a coal sample after heating it in a muffle furnace (~1500°C). This treatment converts most of the mineral matter to metal oxides.

15.430 Fly Ash

The small solid particles (particulates) of noncombustible mineral residue carried out of a bed of coal by the gases formed during combustion.

15.440 Coke

The residue that remains when bituminous coal is heated rather rapidly at high temperature (~800°C) in the absence of air (or in the presence of a limited supply of air). The coal softens as it is heated and goes through a plastic stage in which the volatile gases are eliminated, causing the coke to have a porous structure.

15.450 Free Swelling Index, FSI

A measure of the caking characteristics of a coal. A standard sample is heated rapidly to 800°C and cooled. If the residue is a powder, it is given a value of 0; if it is coherent but does not support a 500 g weight, it is given a FSI of $\frac{1}{2}$. Coherent residues that do support the weight are graded from 1 to 9 in half units whose size is dependent on comparison with a standard chart.

15.460 Delayed Coking

The process of forming coke by relatively long heating at relatively low (~500°C) temperature.

15.470 MAF Coal

Moisture and ash-free coal. When the percentage of carbon, hydrogen, nitrogen, sulfur, etc., is determined directly on a sample of coal, the results are frequently normalized to the MAF basis to reflect the percentage of these elements in terms of the organic content of the coal. In proximate analysis reports, the equivalent term dmmf (dry, mineral matter-free) is often used.

15.480 Bergius Process

The high temperature (~400°C), high pressure (~200 atm) direct hydrogenation of coal (suspended in a recycle oil so that the slurry fed to the reaction consists of about 1.5 parts oil to 1 part coal) to convert it to a liquid product suitable for processing into motor fuels. The process was demonstrated to be feasible by F. Berguis and was used by Germany in World War II to provide more than half of the fuel supplies used in transportation and combat. Inexpensive catalysts such as iron oxides may be used, but better results are obtained with tin and zinc salts, principally the chlorides.

15.490 Pott-Broche Process

The dissolution or dispersion of coal by heating its suspension in an inert gas at 450°C in high boiling (350-450°C) aromatic solvents. The mixture can be filtered while hot to remove ash and inerts, and the solvent recovered by vacuum distillation. It is uncertain whether a true solution or a colloidal suspension is obtained by this treatment. The process is named after the two German investigators who were largely responsible for researching the process, A. Pott and H. Broche.

15.500 Solvent-Refined Coal, SRC

A process designed to convert high sulfur (>1.0%), high mineral matter (3.5-15%) coals to a product that is relatively low in sulfur and mineral matter and therefore environmentally acceptable as a boiler fuel. Pulverized coal is mixed with a coal-derived recycle solvent (~2:1 solvent/coal), and the slurry is treated with hydrogen at ~400°C and ~100 atm H_2 with no added catalyst. From the dissolver vessel the mixture passes to a separator, where gases are bled off and the liquid mixture filtered or centrifuged. The organic material in the solids is gasified for recycle hydrogen and the liquid product is distilled to recover recycle solvent. The residue from the distillation, solvent-refined coal (SRC), softens at ~150°C and has a heating value of about 16,000 Btu lb^{-1}. It can be pulverized and used like coal. In a modification (SRC II) the filtration is avoided by distilling off the recycle oil, recycling a portion of the residue and gasifying the remainder.

15.510 H-Coal Process

(Named after Hydrocarbon Research, Inc.) A catalytic hydroliquefaction process for converting high sulfur coal to boiler fuels and synthetic oil (**syn crude**). Dried, crushed coal (minus 60 **mesh**, i.e., sufficiently fine to pass through a screen having 60 holes per linear in.) is slurried with recycle oil, pressured to 200 atm, and then compressed hydrogen is added. The preheated mixture is charged continuously to the bottom of a modified fluid bed (Sect. 15.130) reactor containing a cobalt-molybdate catalyst. The upward passage of the reaction mixture maintains the catalyst in a fluidized state. The reactor is kept at ~360°C. After the gas and light liquids are removed, the mixture is fed to a flash separator to yield a liquid product as distillate and a residue that is centrifuged in a hydroclone (*vide infra*). The oil from this separator is recycled along with some of the lighter distillate fractions. The solid material can be treated in a variety of ways to obtain additional oil or char.

15.520 Hydroclone

A solid-liquid separating device in which the suspension is passed through a tight conical vortex and the solids are separated by the centrifugal force that results from the passage. It is an example of a cyclone extractor.

15.530 Exxon Donor Solvent Process

A hydroliquefaction process in which the principal conversion of coal to oil is achieved by hydrogen transfer from a donor solvent to the coal in the absence of added catalyst. After liquefaction, the converted coal and solvent are vacuum distilled and a portion of the overhead sent to a catalytic hydrogenation unit to restore the hydrogen donating ability of the recycle solvent. It is estimated that over 3 barrels of oil per ton of dry coal can be made by this process developed by the Exxon Corporation.

15.540 Natural Gas

A gaseous fossil fuel consisting of about 60-80% methane, 5-9% ethane, 3-18% propane, and 2-14% higher hydrocarbons. Nitrogen, carbon dioxide, and hydrogen sulfide are usually present in relatively small quantities. The energy content varies from about 900 to 1300 Btu SCF^{-1} (standard cubic foot).

15.550 Liquefied Natural Gas, LNG

Natural gas that has been cooled to about -160°C for shipment or for storage as a liquid.

15.560 Liquefied Petroleum Gas, LPG

A mixture of propane and butane (also called **bottled gas**) recovered either from natural gas or petroleum refineries and having an energy content of 2000-3500 Btu SCF^{-1}.

15.570 Coal Gasification

The chemical transformation of coal to gas, principally carbon monoxide and hydrogen, along with some methane and carbon dioxide and other gases depending on the particular gasification process.

15.580 Synthesis Gas

A mixture of carbon monoxide and hydrogen. Synthesis gas can be prepared, theoretically, from any carbon source by treatment with steam at about 950°C:

$$C(\text{amorphous}) + H_2O(g) \longrightarrow CO(g) + H_2(g) \qquad \Delta H = 162.8 \text{ kJ mol}^{-1} \; (38.9 \text{ kcal mol}^{-1})$$

Since the reaction is highly endothermic, oxygen is added to the steam to provide heat from the combustion of carbon to CO_2.

15.590 Substitute Natural Gas, SNG

A gas prepared from coal (or naphtha) which has approximately the same energy as natural gas, roughly 1000 Btu SCF^{-1} and can be fed into the gas pipeline system. Pipeline gas is principally methane, and since coal has atomic $H/C = \sim 0.8$ and methane has $H/C = 4$, the preparation of SNG from coal requires the addition of hydrogen. Processes for producing SNG involve the preparation of a synthesis gas that, after a clean-up, is sent to a water-gas shift converter to increase the proportion of hydrogen. After an intense purification, the hydrogen-rich mixture is sent to a methanation unit to provide a gas that is essentially the equivalent (~1000 Btu SCF^{-1}) of natural gas.

15.600 Water-Gas Shift

The catalytic reaction

$$CO + H_2O \rightleftharpoons H_2 + CO_2 \qquad \Delta H = -41.0 \text{ kJ mol}^{-1} \; (-9.8 \text{ kcal mol}^{-1})$$

The reaction is usually used in conjunction with synthesis gas production to shift the concentration of hydrogen in the gas mixture from approximately 1:1 to a higher concentration (3:1) required for conversion to methane. Reaction temperature is about 325°C.

15.610 Methanation

The conversion of synthesis gas to methane at about 375°C:

$$CO + 3H_2 \rightleftharpoons CH_4 + H_2O \qquad \Delta H = -206.3 \text{ kJ mol}^{-1} \; (-49.3 \text{ kcal mol}^{-1})$$

15.620 Lurgi Coal Gasification

A proven commercial process for coal gasification developed by the Lurgi Company (Germany). The gasifier unit operates at 28 atm (called low or medium pressure depending on who is describing the process) and 570°–700°C. Crushed coal is fed into a vertical reactor from the top, and steam and oxygen are fed into the bottom. In the first stage of the reaction the crushed coal is contacted by the rising synthesis gas and volatile matter is driven off in this stage. In the second stage the dry char reacts with the steam and oxygen to produce the synthesis gas. The remaining ash is removed through a rotating grate at the bottom of the reactor.

15.630 Lurgi Slagging Gasifier

A Lurgi gasifier operated at higher temperatures so that the ash runs off as a liquid slag. The disadvantages of the Lurgi process are that it must operate with noncaking coals and the size of the units must be relatively small. When oxygen (rather than air) is used, the synthesis gas produced has a heating value of about 300 Btu SCF^{-1} and the efficiency (Btu in the outgas per Btu in feed and fuel) is about 60%.

15.640 Koppers-Totzek Gasifier

A commercial gasifier that can utilize all types of coal. A mixture of pulverized, predried coal and oxygen is introduced through four coaxial burners into a horizontal reactor at 1650°F and atmospheric pressure.

15.650 CO_2-Acceptor Process

A gasification process that utilizes calcium oxide, along with the coal, to react with the CO_2 produced to form calcium carbonate in an exothermic reaction.

15.660 Producer Gas (Power Gas)

A low Btu gas (~150 Btu SCF^{-1}) produced in gasifiers that operate with air instead of oxygen. Because it is uneconomical to transport such low quality gas, it is used on-site for operating turbines in electrical generation.

15.670 Steam Reforming

The catalytic conversion of methane or other low molecular weight hydrocarbons to carbon monoxide, carbon dioxide, and hydrogen. The two principal reactions are

$$CH_4 + H_2O \rightleftharpoons CO + 3H_2 \qquad \Delta H = +206.3 \text{ kJ mol}^{-1} \; (49.3 \text{ kcal mol}^{-1})$$

$$CH_4 + 2H_2O \rightleftharpoons CO_2 + 4H_2 \qquad \Delta H = +163 \text{ kJ mol}^{-1} \; (39 \text{ kcal mol}^{-1})$$

Such reactions are frequently used in coal processing plants to supply hydrogen and/or synthesis gas if they are used in the processing.

15.680 Coal Gas (Town Gas)

The gas produced by the pyrolysis of coal in a limited supply of air. It contains hydrogen, carbon monoxide, carbon dioxide, nitrogen, some oxygen, and a range of low boiling hydrocarbons and has a Btu SCF^{-1} of about 450. It is largely of historical interest.

15.690 Gasohol

An acronym for a mixture of *gas*oline and alc*ohol*. It is usually a mixture of 10 vol% ethanol and gasoline. Pure ethanol has a research octane number of 105 and a blending research octane number of 135 (Sect. 15.170). The similar mixture with methanol is also called gasohol. However, water compatibility, corrosion, and toxicity problems associated with methanol make it a less popular choice.

15.700 Fischer-Tropsch Process

The process developed by F. Fischer (1877-1948) and H. Tropsch (1889-1935) in Germany about 1925 which involves the catalytic conversion of synthesis gas to hydrocarbons and low molecular weight oxygenated compounds. The overall reaction is highly exothermic:

$$n\text{CO} + 2n\text{H}_2 \longrightarrow (\text{CH}_2)_n + n\text{H}_2\text{O} \qquad \Delta H = -192 \text{ kJ mol}^{-1} \; (-46 \text{ kcal mol}^{-1})$$

Typical catalysts are supported cobalt-thoria or supported iron catalysts. The reaction temperature is about 250-300°C and pressures range from 1 atm to about 20 atm. A large commercial plant is in operation in South Africa.

15.710 Methanol Synthesis

The reaction of synthesis gas at 300-400°C and 200-300 atm over a zinc oxide-based catalyst promoted with chromium oxide to produce methanol in high selectivity. Presently a low pressure process (50-100 atm) using a mixture of copper, zinc, and aluminum oxides is preferred. The reaction is

$$CO + 2H_2 \rightleftharpoons CH_3OH \qquad \Delta H = -109 \text{ kJ mol}^{-1}$$
$$(-26 \text{ kcal mol}^{-1})$$

15.720 Ammonia Synthesis

Commercially, ammonia is usually prepared from nitrogen and hydrogen by the **Haber process.** (F. Haber, 1868-1934).

$$N_2 + 3H_2 \rightleftharpoons 2NH_3 \qquad \Delta H = -92 \text{ kJ mol}^{-1}$$
$$(-22 \text{ kcal mol}^{-1})$$

It is a high temperature (400-550°C), high pressure (100-1000 atm) reaction that is carried out with an iron-based catalyst (Fe_3O_4) containing small amounts of K_2O and Al_2O_3. The impetus to build coal gasification plants in many areas of the world arises from the desire to provide hydrogen (via synthesis gas followed by the water-gas shift, followed by CO_2 removal) for conversion to ammonia to be used as a fertilizer.

15.730 Ammoxidation of Propylene

The conversion of propylene to acrylonitrile by treatment with oxygen and NH_3:

$$C_3H_6 + NH_3 + \tfrac{3}{2}O_2 \longrightarrow CH_2{=}CH{-}CN + 3H_2O$$
$$\Delta H = -515 \text{ kJ mol}^{-1}$$
$$(-123 \text{ kcal mol}^{-1})$$

The reaction is carried out in a fluidized bed reactor at about 450°C at 10-20 sec contact time. Either a bismuth molybdate or an antimony oxide promoted with a uranium salt are used as catalysts. By-products are carbon dioxide (from combustion), acetonitrile, acrolein, acetone, and acetaldehyde.

15.740 Oxo Reaction (Hydroformylation)

The conversion of olefins to aldehydes containing one carbon more than the starting olefin, by reacting the olefin with synthesis gas in the presence of a cobalt (or rhodium) catalyst:

$$CH_3CH{=}CH_2 + CO + H_2 \xrightarrow{Co_2(CO)_8} CH_3CH_2CH_2CHO + CH_3CH(CH_3)CHO$$

Usual conditions consist of adding a soluble cobalt compound to the olefin feed and reacting the mixture with synthesis gas at 160-200°C and 200 atm. The overall stoichiometry corresponds to the addition of H and CHO across the olefinic double bond. The reaction is not only commercially important as a route to alcohols but is the first example of a major reaction that involves transition metal homogeneous catalysis. The active catalyst is hydridotetracarbonylcobalt(I), $HCo(CO)_4$. (This compound has also been called hydrogen tetracarbonylcobaltate(-I) because of its protic character, and some authors prefer to call it cobalt hydrocarbonyl to avoid a decision as to the formal oxidation number of cobalt.) Both straight chain and branched chain aldehydes are formed. Major steps in the catalytic cycle (isomer formation neglected) are shown in Fig. 15.740. The reaction is much more complex than shown, with a great many equilibria involved. The overall rate of the reaction is proportional to $[H_2]$ [Olefin] [Co] and inversely proportional to [CO]. Selectivity to straight chain product can be enhanced by adding triaryl or trialkylphosphines to the system; under such conditions $HCo(CO)_3(PR_3)$ is presumed to be the active catalyst. Rhodium compounds, e.g., $RhCl(PPh_3)_3$, give faster rates than cobalt salts and milder conditions can be used in such cases.

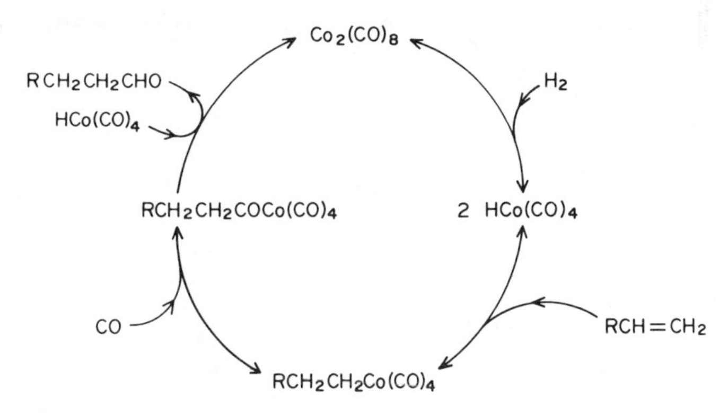

Figure 15.740 Simplified catalytic scheme for hydroformylation of olefins.

15.750 Wacher Process

The catalytic conversion of ethylene to acetaldehyde in aqueous solution using Pd^{2+} as an oxidant:

$$CH_2{=}CH_2 + H_2O + Pd^{2+} \longrightarrow CH_3CHO + Pd + 2H^+$$

The palladium is regenerated by a two-electron redox reaction with Cu^{2+}:

$$Pd + 2Cu^{2+} \longrightarrow 2Cu^+ + Pd^{2+}$$

The copper in turn is reoxidized by air:

$$2H^+ + \tfrac{1}{2}O_2 + 2Cu^+ \longrightarrow 2Cu^{2+} + H_2O$$

Summation of the three reactions gives

$$CH_2{=}CH_2 + \tfrac{1}{2}O_2 \longrightarrow CH_3CHO$$

The reaction is carried out continuously in an aqueous medium and has stimulated many studies with respect to its intimate mechanism. Most investigators agree that a hydroxyethylpalladium(II) species, $HOCH_2CH_2$—Pd—, is involved.

15.760 Acetic Acid Synthesis

The reaction of methanol with carbon monoxide in the presence of a soluble rhodium catalyst provides a homogeneous synthesis of acetic acid with extremely high selectivity:

$$CH_3OH + CO \xrightarrow{Rh} CH_3CO_2H$$

The reaction requires the presence of methyl iodide for best results. The proposed catalytic cycle for the conversion of the steady state concentration of methyl iodide to acetyl iodide is shown in Fig. 15.760. Acetyl iodide reacts with water to give acetic acid and hydrogen iodide and the latter converts methanol to the methyl iodide involved in the catalytic cycle.

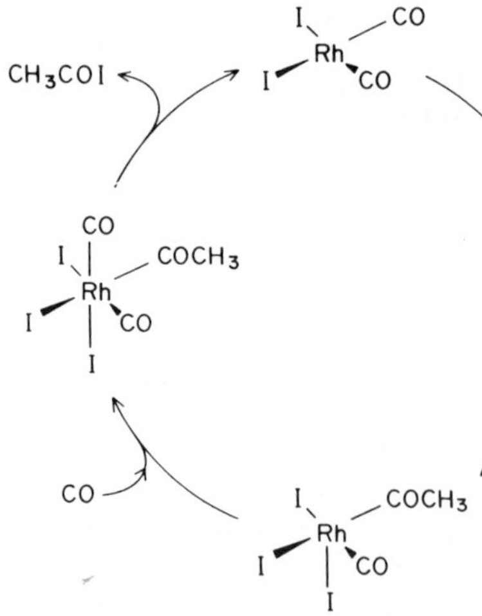

Figure 15.760 Catalytic cycle involved in the acetic acid synthesis.

15.770 Oil Shale

A laminated sedimentary rock that contains 5-20% organic material.

15.780 Shale Oil

The oil that can be obtained from oil shale by heating or retorting. Commercial processes have been developed for this purpose and yields of about 30 gal ton^{-1} of oil shale have been obtained.

15.790 Kerogen

The waxy organic material in shale which is the precursor of the oil obtained upon pyrolysis.

15.800 Fischer Assay

A standardized analytical retorting procedure used to evaluate the quantity of shale oil that can be produced from a particular oil shale.

15.810 Tar Sand

A naturally occurring mixture of tar, water, and sand. The tar probably results from petroleum quite near the surface of the earth which has been partially devolatilized. The large tar sand deposit called the Athabasca deposit in Alberta, Canada, is presently being used for commercial extraction. The mined tar sand is treated with hot water and steam to separate the tar as a liquid. A typical sample ton of tar sand contains about 80 lb of recoverable tar.

15.820 Fuel Cells

An electrochemical (voltaic) cell in which electricity is produced directly from the combustion of a fuel. Common fuels are hydrogen, carbon monoxide, and methane.

Examples. In a typical fuel cell hydrogen and oxygen are bubbled through porous carbon electrodes that are impregnated with active catalysts such as platinum metal. The electrodes are immersed in an aqueous solution of sodium hydroxide. The reactions are

$$\text{anode:} \quad 2H_2(g) + 4OH^- \longrightarrow 4H_2O(l) + 4e^-$$

$$\text{cathode:} \quad 4e^- + O_2(g) + 2H_2O(l) \longrightarrow 4OH^-$$

$$\text{cell reaction:} \quad 2H_2(g) + O_2(g) \longrightarrow 2H_2O(l) \quad \mathcal{E}^\circ = 1.229\text{V}$$

The gases are thus continuously consumed, and since the cell is maintained at an elevated temperature, the water evaporates as it is formed.

Index

Index